SO-CFO-766

Battle Against Extinction

Battle Against Extinction

Native Fish Management in the American West

Edited by W. L. Minckley
and James E. Deacon

The University of Arizona Press
Tucson & London

Published with the generous support of the Desert Fishes Council
and Arizona State University.

The University of Arizona Press
Copyright © 1991
The Arizona Board of Regents
All Rights Reserved

96 95 94 93 92 91   6 5 4 3 2 1

Library of Congress Cataloging-in-Publication Data

Battle against extinction : native fish management in the American
    West / edited by W. L. Minckley and James E. Deacon.
        p. cm.
    Includes bibliographical references and index.
    ISBN 0-8165-1221-3
    1. Rare fishes—West (U.S.)—Congresses. 2. Fishery conservation—
West (U.S.)—Congresses. I. Minckley, W. L. II. Deacon, James E.
QL617.73.U6B38 1991
333.95′616′0978—dc20   91-6977
                                                            CIP

British Library Cataloguing-in-Publication Data
A catalog record for this book is available from the British Library

CONTENTS

Foreword

The plight of aquatic life in a society dominated by Big Science and massive technology first came into my view in the early 1960s when I was secretary of the interior. The poisoning of fishes from above Flaming Gorge Dam on the Green River caught my attention in 1962, and fisheries experts in my department described for me the efforts then under way to wipe out invading sea lampreys in the Great Lakes that had decimated the native fishes of that vast waterway. The snail darter–Tellico Dam controversy in Tennessee also was emerging as a major environmental issue. I later watched with interest as questions pitting the need to maintain water levels in Devil's Hole against development in Ash Meadows, Nevada, were answered in the U.S. Supreme Court in favor of the habitat and its unique pupfish.

In my view, although California condors and big mammals may have received more attention, native fishes and their proponents were as important as any other group of organisms or scientists in the developing conservation movement. The Green River controversy was pivotal in solidifying the resolve of numerous scientists to fight for endangered species legislation. The Devil's Hole decision has become the precedent for important cases involving federal reserved water rights and questions of interrelations of ground and surface water in western water law. And exemption of the snail darter from protection under the Endangered Species Act of 1973 taught legislators a new tactic. Such an action, in one form or another, has been repeated for too many other organisms—the Concho water snake in Texas, the Mount Graham red squirrel in Arizona, and soon, perhaps, the spotted owl in the Pacific Northwest, if some people have their way.

It is unfortunate that we must deal at the level of individual species. This forces us to focus attention on single parts of ecosystems, while ecosystems themselves should be the subjects of our efforts. Endangered species are nonetheless the messengers of change, and we must heed their messages.

Water to irrigate agriculture has always been recognized as the key to human occupation and development of the arid American West. At first its use was on the local level, supported by muscle and sweat to carve out individual liveli-hoods. When the weather cooperated, settlers did well, but drought or flood too often paralyzed their most diligent efforts. There were dreams of averaging out the extremes by storing the floods and releasing them slowly.

Beginning in the years after Congress enacted the Reclamation Act of 1903, a series of major engineering projects were built that changed the region forever. Major federal water developments—giant dams and diversions, irrigation works, and power plants—allowed burgeoning development in the West after World War II. But spin-offs of these marvels demanded more water each year in a place where little water exists. The two-headed monster of avarice and greed replaced the intentions of our fathers, who were far more sensitive to the needs of the region than the entrepreneurs and growth-boomers of today.

The native fishes forming the subject of this book are only part of the spec-tacular natural system caught up in this maelstrom of change. I was involved in formulating many of the water projects that affected them, in an era when few probing, environmental questions were asked about big-dam plans. I am dis-turbed that these valuable resources are now used for other than their original intent.

Diverting precious water for filling recreational lakes or using hydroelectric power to drive fountains fifty meters or more into the sky to extend oppor-tunities for urban sprawl are, I fear, setting a stage for ecological catastrophes in the water-poor deserts surrounding Phoenix and Las Vegas. Unbridled de-velopment in Tucson, anticipating completion of the Central Arizona Project, placed unacceptable demands on the local ecosystem, as well as exceeding even the city's future water supplies. The use of Glen Canyon Dam to provide peak-ing power has endangered the ecosystem in Grand Canyon National Park. Tide-like fluctuations pass through the canyon, eroding beaches, endangering recre-ationists, and further jeopardizing the native fishes already reeling from the physical and biological impacts of reservoirs both up- and downstream.

Americans, and humans in general for that matter, are exploiters of the envi-ronment. This is especially true in newly colonized areas because there is at first so much land and so many natural resources that no one sees an end to the supply. Those days are gone. Colonization and exploitation must end, because the planet upon which all known life depends will soon reach the limit of its ability to support the only civilized species ever known to have evolved.

Few places are as fortunate in this regard as the American West. Here it is not too late to apply restraint and thus retain and perpetuate the natural features that make the West a special place. Large tracts of public lands enhance the possibility of this happening, and despite the federal emphasis on development over the past decade, the American public is demanding conservation of these resources more than ever before.

Federal and state laws have been passed in response to these demands, and it is up to administrators and politicians to expedite and sharpen the application of existing legislation and provide new initiatives to achieve this goal. Whether they know it or not, this will prove to be one of the steps toward developing an equity between humans and their environment that will be sorely needed in the decades ahead. In the process they will influence, through example, a wiser development of the balance between resource use and human existence. As citizens of the planet we must educate our children to do what we have not done well—to act as stewards of the Earth and all its inhabitants and to pass the planet on to their offspring in a better condition than they found it. The more we know, the better we can apply ourselves to such aims.

This book contains vital guidelines for this effort. It is far more than a plea for conservation of natural aquatic habitats and fishes. It deals as well with the philosophy and ethics of conservation, and the need to document both failures and successes of the past as guides to the future. Most importantly, it thrusts a little-understood group of animals before the public eye in an ecosystem context and presents positive recommendations for maintenance of biodiversity through their conservation.

Stewart L. Udall
Santa Fe, New Mexico

Preface

The papers included in this volume were presented at a symposium of the same name, organized and held to commemorate the twentieth anniversary of the Desert Fishes Council at its annual meeting held at Death Valley National Monument, Furnace Creek, California, on 17 and 18 November 1988. Participants in that event not only contributed to what constitutes the first compilation on the application of principles of the new science of conservation biology to fishes, but they also emphasized fishes in deserts, an apparent paradox.

Fishes are relatively rare under desert conditions, as should be the case since water is at a premium, yet about a third of the native fish fauna of North America lives in the arid western part of the continent. Such fishes are now in severe competition for places to live, since colonization and exploitation of desert lands by humans requires development, use, and consequent depletion of water. Western fishes and other aquatic organisms experienced violent and widespread disruption as human populations increased after World War II.

The early 1960s was a period of near despair for those of us dealing with the western fishes. Aquatic habitats and species were disappearing at ever-increasing rates, and little was being done about it. With accelerated effort and new legislation protecting habitats and imperiled species alike, the late 1960s was a time for rejoicing. In the decade following passage of the Endangered Species Act of 1973, researchers and managers operated on the black side of the ledger: native fish programs were developed, needed research was funded and carried out, and recovery plans were prepared. Species and habitats were saved, and positive programs of refuge establishment, propagation, reintroduction, and other recovery actions were initiated and solidified.

Each action produces a reaction, however, and the middle and late 1980s saw a resurrection of emphasis on development, often in direct conflict with conservation efforts. Fewer action programs were commenced, and some ground to a standstill; less independent research was funded, legislation was amended to accommodate greater political and bureaucratic control over recovery activities and research, regional endangered species offices were decentralized, and many of the most dedicated and experienced endangered species managers were transferred to other programs or otherwise rendered ineffective. Species have again begun to decline and habitats to deteriorate, and we and others are deeply concerned.

Therefore, our goals in organizing the symposium and editing this volume were threefold. First, it seemed clear that considerable progress has been made in understanding the basic biology of western fishes. There was a major void in information when we commenced working with this fauna about three decades ago. It was time to attempt synthesis, generalization, and identification of gaps that still exist in the knowledge of the biology of native fishes.

Second, action programs toward saving individual endangered species, and legislation geared toward those ends, are clearly evolving toward the broader perspective of perpetuating ecosystems. *Biodiversity* and *conservation biology* are becoming common words and concepts. It also seemed time to demonstrate in a single volume that community and ecosystem approaches to conservation had long ago been espoused by aquatic biologists in the West. Species cannot be conserved and perpetuated without appropriate habitats, and an aquatic habitat cannot persist without some ecosystem order in its terrestrial surroundings. To the eye, aquatic systems in arid lands have a sharper demarcation from the land around them than those in more mesic zones, yet intimate links exist that must not be broken if ecosystem integrity is to be maintained.

Third, successes for some animals and habitats, punctuated by failures for others, have become far more complex as new habitats, species, and human players enter the scene. As a result of change in administrative philosophies, endangered species and other conservation programs of the 1960s and 1970s are under attack. This is especially true in arid lands, where continuing and expanded development of water resources forms the basis for economic and population expansion. Aquatic resources are inexorably eroded when development proceeds without limits. The "Battle against Extinction" must be rejoined; the contributions in this volume form the basis for such action.

Considerable latitude was exercised by different authors in their interpretation of just what parts of the North American continent are classified as "West." Some restricted their coverage to west of the continental divide, others dealt with species occurring generally in arid lands west of the Mississippi River, which include large areas in New Mexico and Texas that drain into the Atlantic. The term *desert* is also used loosely, and, especially when linked with fishes, should best be construed as a region rather than a habitat type. Thus even the trout in cold streams on a heavily forested mountain are surrounded by regional desert and are accepted by some as "desert fishes." Although such may appear as transgressions that offend the purist, we hope they will be generally acceptable.

The most discouraging thing about editing and contributing to a book of this nature is the realization that one may be working mostly for an audience that is already convinced of the values of conservation and perpetuation of natural systems. Conservationists agree as a group, with minor permutations, that diversity must be maintained for the welfare of the biosphere, as well as for the welfare of humans. Those who embrace other philosophies are often just as firmly convinced of their alternative views, and only education based on tangible data and logical, documentable results of research and observation can be expected to change their minds. This volume provides such information, and we hope it is widely used as a reference to provide examples of what has been learned, and accomplished, in dealing with an obscure group of animals that depend on water in an improbable place.

W. L. Minckley
Arizona State University

James E. Deacon
University of Nevada, Las Vegas

Acknowledgments

We first thank Michelle Dotson and Stephen K. Johnson, who as student assistants provided much-needed editorial and other assistance without which this book might not have been completed. Their cheerful diligence at menial and tedious tasks was remarkable to behold and is gratefully acknowledged.

In addition to the contributors to this volume, most of whom were called upon to comment on their cocontributors' manuscripts, the following persons acted as reviewers for papers published here: F. W. Allendorf, R. J. Behnke, J. Bennett, W. Bennett, A. Brasher, D. E. Brown, D. J. Brown, L. Brown, D. G. Buth, C. A. Carlson, D. Castleberry, J. P. Collins, W. R. Courtenay, Jr., D. A. Crosby, M. D. Deacon, M. S. Deacon, B. D. DeMarais, T. E. Dowling, P. W. Eschmeyer, T. Hopkins, L. Kaeding, C. Karp, D. Livermore, E. J. Loudenslager, W. Miller, D. C. Morizot, M. Parker, A. Riggs, B. Sada, D. W. Sada, E. Strange, S. P. Vives, E. Wikramanayke, B. Williams, C. M. Williams, H. Williamson, and M. Zallen.

Death Valley National Monument provided the hall and hosted the symposium, as well as hosting many previous and important meetings of the Desert Fishes Council. Monument Superintendent Edward Rothfuss, earlier superintendents, and other monument personnel are offered our special thanks, and those of the Desert Fishes Council.

The Department of Zoology and Center for Environmental Studies, Arizona State University, and the Department of Biology, University of Nevada, Las Vegas, supported word-processing supplies, copying, postage, preparation of final illustrations, and other costs, which were not inconsequential. Funds donated by the members of the Desert Fishes Council (especially W. I. Follett) and Arizona State University helped defray publication costs and keep the final product price at a more reasonable level. We are deeply indebted for their assistance. Barbara Terkanian and Charles Kazilek drafted and lettered the figures, for which we are grateful. The advice and support of Barbara Beatty Williams and Marie Webner of the University of Arizona Press saw the book to its fruition.

Many agencies supported our research on western fishes and sent their personnel to contribute to council meetings; we acknowledge and thank them all. We especially appreciate the efforts, expertise, and friendship of E. P. Pister, who has provided years of hard labor for the cause of desert fish conservation in general and the Desert Fishes Council in particular. Without his continuous and infectious enthusiasm the resource and its proponents would be far more insecure than they are. Last, and sincerely, we thank our respective families for continuing understanding and support, over the years and during the production of this volume.

. . . to our respective physicians and surgeons.

Battle Against Extinction

SECTION   I

The Subjects and Their Plight

Chapter 1 deals with the chronology and geography of original descriptions of the freshwater fishes of western North America. The pattern of discovery paralleled exploration and early exploitation, first in the 1700s by sea from the north and northwest, spreading inland with the Hudson's Bay Company, then south along coastal California by the early 1800s to meet the Spaniards, who had already penetrated to California from the south. Scientific exploration of the vast inland deserts and mountains came later, during and following the middle of the nineteenth century, and was related to military expeditions, the gold rush of 1849, surveys for the United States–Mexican boundary, and searches for wagon and railroad routes through the rugged and inhospitable terrain. Some of the most prominent scientists of the day described the fishes, and their contributions are briefly reviewed.

The second part of the first chapter delineates the precipitous decline toward extinction of this unique and splendidly isolated fish fauna, largely as a result of the application of modern technology to water use. At first, individual species disappeared. Now, whole subfaunas are collapsing and are being replaced by non-native fishes tolerant of modified conditions. Predictions are that fishes of the better-watered and more faunally saturated eastern parts of the continent will soon respond similarly to impacts of development, if they are not already doing so.

The second chapter provides glimpses of the past extracted from field notes recording personal experiences in the American West between 1915 and 1950, the time immediately before major changes in aquatic systems were recognized. Scientific and anecdotal information on the region, its fauna, and some of its people is included. We find, perhaps unexpectedly, that there was an abundance of native fishes in deserts. Ample water was present in special places like springs rising from valley floors and streams isolated in canyons tributary to long-desiccated lakes, and major rivers flowed strongly from the mountains to the sea.

These two chapters introduce some of the problems that exist in the battle to conserve western American fishes. Hard data are mixed with nostalgia, failures are buffered by success in some arenas, and the scene is set for essays, historical documentations, details on methods and organizations needed for conservation, and case histories, which comprise most of the remainder of this book.

A

B

E

F

C

D

Some distinguished North American ichthyologists whose work dealt with fishes in the American West: (A) Spencer Fullerton Baird, 1823–1887 (photograph from C. L. Hubbs 1964); (B) Charles Girard, 1822–1895 (photograph from LaRivers 1962); (C) William O. Ayres, 1817–1891 (photograph from Myers 1964; original from the California Historical Society); (D) David Starr Jordan, 1851–1931 (photograph from Jordan 1905); (E) John Otterbein Snyder, 1867–1943 (photograph from files of the Smithsonian Institution); and (F) Carl Leavitt Hubbs, 1894–1979 (photograph by D. E. McAllister, 1970).

Chapter 1

Discovery and Extinction of Western Fishes: A Blink of the Eye in Geologic Time

W. L. Minckley and Michael E. Douglas

Introduction

The nineteenth century was the period when the vast arid region of North America west of the Rocky Mountains was mapped and settled, and when important historical written records were initiated and maintained. It was also the time when most western freshwater fishes were discovered. Trained scientists assigned as physicians and naturalists to military campaigns, government boundary surveys, and expeditions seeking transportation routes collected and cataloged the biota as part of their official duties.

These efforts in natural history were not the government's response to demands for environmental impact statements, for the time for those was in the distant future. The western fauna was relatively untouched at this point, although even then some were expressing concern for its conservation (Mitchill 1981). Rather, naturalists were assigned in accord with British tradition, because the Victorian government of England considered science an obligation of the state. Charles Darwin himself (Darwin 1968 [1859]) served as chronicler and scientist on H.M.S. *Beagle* during its expedition to the Galápagos Islands and South America. Thus, the United States merely followed its Anglo Saxon heritage as voucher specimens were garnered, preserved, and prepared by naturalists, to be crated and either shipped overland or by sea to museums for processing.

As valuable specimens accumulated, the laborious work of species descriptions, floral and faunal inventories, and curatorial maintenance occupied many well-known scientists for decades. Several such scientists made, or at least greatly increased, their reputations based on studies of the western American biota. The wealth of accumulated specimens to this day attracts researchers with avid interests in achievements of the period.

Environmental conditions in western North America have changed dramatically since the arrival of the Spaniards and later explorers. Desert rivers once bordered by galleries of cottonwoods and willows are now perennially dry, forced into canals, or incised in natural channels that pass between barren cut banks. Streams that raged in flood and vanished in drought now fluctuate only mildly in response to rates of human-determined water deliveries. Sloughs and backwaters, formerly vegetated by cattails and sedges, were drained as water tables dropped, except where they were artificially maintained by dams and irrigation works. As a result of these activities (R. R. Miller 1946a, 1961), most native western fishes have become threatened or endangered. The habitat destruction and alterations of large rivers accompanying development have negatively affected long-lived "big-river fishes,"

while our unending thirst for water has desiccated springs and lakes of desert basins,
thereby destroying or greatly reducing aquatic
habitats and short-lived species. Where surface waters persist, or where they are artificially increased through impoundment, we
find introduced, non-native fishes replacing
native forms. The chapters of this book represent not only an overview but indeed a chronicle of these activities and their impacts on
individual species of the unique and ancient
assemblage of western North American fishes.

In the pages of this introductory chapter we
focus first on the explorers and scientists who
discovered the western North American ichthyofauna. Our initial goal is to place discovery and description of the fish fauna into a
historical context that parallels the opening
of the American West. We thus construct a
chronological history in which description of
the fauna is related to exploration, settlement,
and emergence of ichthyology as a science in
North America. The second part of this chapter jumps to the present, then into the future,
as we summarize the ever-intensifying threats
to the North American ichthyofauna. We review estimates of absolute and relative numbers of fishes in jeopardy of extinction, and
briefly discuss the patterns of endangerment
that developed as a result of regional settlement and exploitation.

We thank J. P. Collins, B. D. DeMarais,
P. C. Marsh, and S. P. Vives for reading and
improving the manuscript (Vives, especially,
for assistance with literature), and J. E. Williams and R. R. Miller for providing unpublished manuscripts for our use.

Discovering the Fishes of Western North America

The Fauna

There are approximately 810 species of native
fishes breeding in fresh waters of North Amer-

ica north of (but including) the Río Grande
de Santiago and Río Pánuco basins of southern Mexico (see contributions in D. S. Lee et
al. 1980, 1983; and Hocutt and Wiley 1986).
Excluding transcontinental forms, about 170
species occur west of the Rocky Mountain
axis, compared with 600 in waters draining
east from that divide. Only about 40 species
(ca. 5% of the total fauna) occur both east
and west of the continental divide; 28 (70%)
of these live far to the north, attaining transcontinental distributions by passing through
estuaries or coastal seas. Evolution of the depauperate western fauna has been tied to a
long history of disruptive geologic and climatic events, all of which substantially reduced
the diversity, availability, and reliability of
aquatic habitats (G. R. Smith 1981b; Minckley et al. 1986).

The modern western ichthyofauna is further
characterized by many endemic subfaunas,
most of which also result from geologic and
climatic disruptions of aquatic habitats (R. R.
Miller 1959; G. R. Smith 1978). The smallest
of these are single endemic species restricted
to springs, streams, or individual lakes of endorheic intermontane basins. Larger, more
complex aquatic systems often have two or
more subfaunas represented, reflecting the
fact that modern river drainages commonly
comprise two or more original sub-basins
brought together by geologic events (McPhail
and Lindsey 1986; Minckley et al. 1986;
M. L. Smith and Miller 1986). For example,
the upper Colorado River watershed has a
subfauna distinct from that in its lower part
(Gila River basin), while distinctive "middle"
Colorado River fishes (R. R. Miller 1959;
R. R. Miller and Hubbs 1960) are associated
with another, formerly independent, system
separated prehistorically from both the upper
and lower parts.

At the largest scale, major drainage basins
have few fish species in common, and those
which do usually share species that: (1) can

travel through seawater, (2) occupy montane tributaries subject to interbasin stream piracy, or (3) are confined to areas of high latitude but low relief, where divides between basins are weakly developed. All these factors aided and abetted the splendid isolation of western fishes, not only from related species in other parts of the continent but just as frequently from sister taxa within the region.

Discovery and Description: Geography and Chronology

Naturalists working before 1800 described only 20 (13.2%) of the 151 western American fish species recognized by Lee et al. (1980) in their *Atlas of Freshwater Fishes of the United States and Canada*. Most were circumpolar in distribution and important for food or commerce, caught from the great coastal fisheries that were then (as now) exploited in subarctic seas. Most have type localities in Europe or the Soviet Union (Fig. 1-1). About 30% of them were named by Linnaeus (1758), the father of modern taxonomy, in *Systema Naturae*, and 40% by Linnaeus's colleague Johann Julius Walbaum (1792), who edited Peter Artedi's *Genera Piscium: Ichthyologiae Pars III* and added descriptive footnotes. A handful of other authors described the remainder.

From 1801 to 1850, sixteen more taxa (10.6%) were named. A few were from the northwestern United States, but most (75%) were again collected farther north in association with the British Hudson's Bay Company and expeditions to locate the fabled Northwest Passage (Dymond 1964), a fictitious waterway purportedly connecting the Atlantic and Pacific oceans and thus providing a prime trade route to the Orient. The search for this passage was fueled by dreams of historical fame, for the first navigation of such a route would achieve a reward of twenty thousand pounds sterling offered by Great Britain. John Richardson described nine fish species in his *Fauna Boreali-Americana* (1836) after serving as a naval surgeon and naturalist with Sir John Franklin on two separate searches for the passage. Fortunately, Richardson did not participate in Franklin's third expedition, which disappeared with all hands in 1843. Nearly fifty additional expeditions searched for them, but never a trace was found.

Explorations between 1851 and 1900 were accompanied by an increase in scientific collecting, and as a result, more than half (79 of 151) of the regional fishes were described in this period: 29 from the Pacific Northwest (both the United States and Canada), 21 from California, and most of the remainder from the Intermountain Great Basin and southern deserts (Fig. 1-1).

Many great names in ichthyology participated in this flurry of discovery. Charles Fredrick Girard, a Frenchman who came to the United States to study with Professor Louis Agassiz (C. L. Hubbs 1964), described twenty species alone, and another twelve in coauthorship with Spencer Fullerton Baird. Most were from specimens collected during the United States–Mexico boundary surveys of 1849–1855, and the Pacific Railroad surveys of 1851–1858. Girard received his medical degree from Georgetown College in 1856 and returned to France in 1865, where he practiced medicine for the rest of his life (LaRivers 1962).

Baird was highly competent as a scientist, promoter of scientific endeavor, and administrator. Beginning in 1850, he served as first assistant secretary of the new Smithsonian Institution, succeeding to secretary in 1878 when Joseph Henry, the first designate, died. In 1871 Baird initiated the United States Fish Commission, forerunner of the Bureau of Fisheries and the Fish and Wildlife Service, and acted as the first commissioner of fisheries. As commissioner, he fostered a program of transplanting fishes throughout the United States, often outside their native ranges, and including fish such as the com-

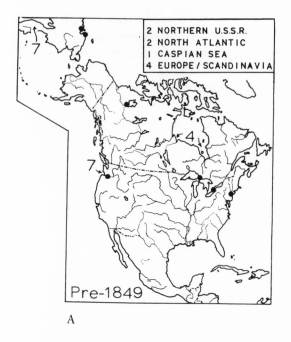

A

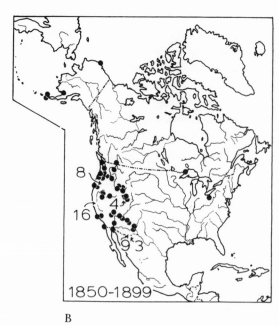

B

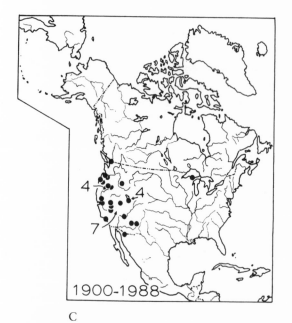

C

Fig. 1-1. Geographic and temporal distribution of type localities designated in original descriptions of fishes of western North America: (A) prior to 1800; (B) 1850–1899; and (C) 1900–1988. Clusters of species in B were described: (1) from the lower Columbia River, mostly by Richardson from the vicinity of Fort Vancouver; (2) from Arizona and New Mexico in the 1850s, mostly by Girard, Baird, and Cope from material collected during the U.S. and Mexican boundary survey and the later surveys for transportation routes; and (3) from Sacramento–San Joaquin basin (San Francisco area) by Ayres and Girard following the California gold rush. Gilbert, Jordan, and Carl and Rosa Eigenmann were most active in the later 1800s at scattered localities in California and Nevada.

mon carp (*Cyprinus carpio*). If he could only know of the adverse effects of such actions on indigenous species, some of which he described, we are convinced he would feel remorse. Baird was also instrumental in influencing Louis Agassiz's son Alexander, the copper-mining magnate and scientist, to purchase land at Woods Hole and establish the Marine Biological Laboratory. Baird died there in 1887 after a brief final tour of the facilities by wheelchair (R. V. Bruce 1987).

Charles Henry Gilbert named six species independently and six more with other authors, among whom were David Starr Jordan and Barton Warren Evermann. Gilbert was a critical and precise scientist, while Jordan was driven, knowledgeable, and deeply intuitive, with a prodigious memory. Both were longtime associates at Butler and Indiana universities. Jordan rose to be president at Indiana, and when he was appointed president of the new Leland Stanford University, he recruited Gilbert to chair its zoology department. Their collaboration resulted in the monumental *Synopsis of the Fishes of North America* (D. S. Jordan and Gilbert 1883). Jordan, trained by Louis Agassiz, who lured him to fishes from an early interest in microbes and marine algae, began publishing on fishes in 1874, producing 645 ichthyological papers and 1400 other general works before his death in 1931.

Agassiz had come to the United States from Switzerland in 1846 as a well-established and prestigious scientist, and his influence on Jordan is not surprising. Agassiz's vast accomplishments in ichthyology, and science in general, made lasting impressions on people and institutions in both Europe and America. Most of today's students of fishes can trace their academic lineage back to Jordan, and through him to the insight and brilliance of Louis Agassiz (C. L. Hubbs 1964). Among his many accomplishments in the United States, Agassiz founded the Museum of Comparative Zoology at Harvard and was deeply involved

in the formation of both the American Association for the Advancement of Science in 1848 and the National Academy of Sciences in 1863 (Lurie 1960; R. V. Bruce 1987).

Evermann was another of Jordan's longtime associates who was markedly effective in promoting research and in bibliographic compilation (the first of which is becoming, for some, an occupation in itself, and the latter an almost lost art). Major contributions of his collaboration with Jordan were *Fishes of North and Middle America* (D. S. Jordan and Evermann 1896b–1900) and two associated checklists (D. S. Jordan and Evermann 1896a; D. S. Jordan et al. 1930), all of which remain important ichthyological references. Evermann also served in key positions in the U.S. Fish Commission and the California Academy of Sciences.

Edward Drinker Cope, whose career broadly overlapped Jordan's, was an independent scientist who rarely collaborated. Although he was a prominent authority on fishes, Cope's expertise and contributions are perhaps more widely recognized in herpetology and paleontology, and in the latter especially with reference to his long and bitter rivalry with Othniel Charles Marsh over dinosaurs (Colbert 1984). He published approximately 1400 works, essentially without the benefit of assistants or coauthors, and in the process named eight of the fishes described from our region.

William O. Ayres, a Boston physician who followed the stampede of forty-niners westward in search of California gold, also worked alone to describe six species, mostly from specimens obtained in markets along the Pacific Coast. He was one of the first physicians in San Francisco, the first ichthyologist in California, and a founder of the California Academy of Sciences (G. S. Myers 1964; Briggs 1986).

Rosa Smith, Jordan's student and one of the earliest women contributors to American ichthyology, described one western fish before her marriage to Carl H. Eigenmann, then collaborated with her husband to describe seven more between 1851 and 1900. Rosa Smith Eigenmann later became the first woman president of the American Association for the Advancement of Science. Carl Eigenmann was another of Jordan's students who extensively researched the ichthyofauna of South America, using many specimens collected originally by Louis Agassiz (G. S. Myers 1964). Other naturalists such as Charles Conrad Abbott, Tarleton Hoffman Bean, Theodore Nicholas Gill, John Otterbein Snyder, Cloudsley M. Rutter, and Seth Eugene Meek described one or two western species while working mostly with other ichthyofaunas. All are nonetheless familiar names to those who work with western fishes. If the compilations of Lee et al. (1980, 1983) had included the fishes of Mexico, Meek's name would be prominent among describers of the western American fish fauna.

Between 1901 and 1950, twenty-two additional species were recognized. Major surveys of river basins in the western United States were performed by John O. Snyder from Gilbert and Jordan's group at Stanford University. Snyder described nine species still recognized today. Eigenmann, Evermann, Meek, Rutter, and Norman B. Scofield were also involved in survey activities, and Rutter described two species from the Pacific Northwest. In later years, Snyder went on to direct the Marine Biological Laboratory at Woods Hole and served as chief of California's Bureau of Fish Conservation.

Carl Leavitt Hubbs, Gilbert's student and one of Jordan's last major associates, was Snyder's young assistant on the 1915 survey of the Bonneville Basin. Hubbs was destined to become a major force in American (and world) ichthyology and to remain so for more than sixty years. Singly, and with his student, colleague, and (ultimately) son-in-law Robert Rush Miller, Hubbs named nine of the twenty other fishes recognized in this period. Hubbs, Miller, their families, and their associates collected throughout most of the American West (Miller et al., *this volume*, chap. 2), and Miller soon rose to become the leading authority on freshwater fishes of the western United States and Mexico. Eight of the last fifteen species we include in this analysis were named by Miller, singly or in coauthorship with Hubbs, during the period 1951–1980 (Fig. 1-1).

Finally, one of the remaining two species described during the period 1901–1950 was named by Leonard Peter Schultz, and the Devils Hole pupfish (*Cyprinodon diabolis*), which figured so strongly in later efforts to conserve native fishes in the region (Deacon and Williams, *this volume*, chap. 5), was named by the biometrician Joseph H. Wales. Four sculpins (*Cottus* spp.) were described by Reeve Maclaren Bailey and Carl Elder Bond in the last three decades (1951–1980) of our coverage; a chub of the genus *Gila* was described by Bond and his student Jack E. Williams; and John D. Hopkirk and Donald Evan McAllister each named a species, both from California.

Extermination of the Fishes of Western North America

From this treatment of species descriptions, tinged with the nostalgia and excitement of exploration and discovery, we pass on to the less enjoyable topic of species extinctions. A note of caution must be expressed before results of our analysis can be discussed, however, for direct comparisons of numbers of imperiled fishes are not as simple as the comparisons in our preceding exercise. Differences of opinion exist as to the taxonomic validity of various imperiled forms.

The U.S. Endangered Species Act of 1973 wisely encouraged and included listing and recognition not only of full species as imperiled, but of subspecies and undescribed populations as well. This presents an operational problem similar to the cliché of comparing apples and oranges. Full species described in the distant past and recognized today have survived the test of long-term scrutiny, and most are indeed accepted as valid taxa. Many subspecies and most undescribed forms are not so generally accepted, however, and their subspecific names or manuscript recognition are referred to even less. Most subspecies have distinguishing characteristics and geographic ranges less well defined than those of full species. Clines and intergradation between subspecies also cause problems: What does one do with an intermediate population? Undescribed taxa are even more difficult, since without thorough study and description their singular characteristics and distributions, as well as relationships to other taxa, may be unknown.

Some may argue that the imperiled subspecies and undescribed taxa of western North America are less important than "full" species. We reject such a premise, because in any geographic area and at any level of differentiation, each taxon has developed in response to unique situations and survives in a special place, thus being the most fit for a set of local conditions. Even, perhaps especially, in a fauna replete with relics of better-watered times, as in western North America (M. L. Smith 1981), each isolate represents an entity shaped by a unique set of environmental parameters. Each is irreplaceable. Every single population, be it an isolated deme, subspecies, or species, constitutes an evolutionary unit of some degree of importance, and thus is "worthy" of perpetuation (see Rolston, *this volume*, chap. 6).

Disregarding such problems for now, about half (105 [47%] of 224) of the species, subspecies, and undescribed forms now listed for the United States as threatened or endangered, or being considered for listing by the U.S. Fish and Wildlife Service (USFWS 1989c, d), occur west of the continental divide (Table

1-1). Four species (1 west and 3 east of the Rockies) were removed from the official list due to extinction, and 26 others, exterminated before or in spite of the Endangered Species Act (Table 1-2), were not included in the compilations. If these are added, 254 taxa have either disappeared or are considered in some danger of extinction, 122 from western North America and 132 from east of the continental divide. Mexican species and subspecies are not included in these statistics, except for the few that also occur in the United States.

Agency compilations do not necessarily reflect the true biological status of all the many taxa that may be imperiled (official listing, for example, may take place years after the taxon is proposed, thus our inclusion of candidates for listing). We examined other more general works dealing with species of special concern to rectify this imbalance and obtain a clearer perspective on the geography of species extinctions (Table 1-1).

Unlike the official listings, compilations dealing with the continental fauna (thus excluding J. E. Williams et al. [1985], who restricted their coverage to arid lands) indicate substantially more imperiled taxa east of the continental divide than west. This might be expected because of the larger overall fauna, but it also may be surprising to some. Losses of formerly widespread and abundant eastern species like the harelip sucker (*Lagochila lacera*), endemic whitefishes (*Coregonus* spp.) and blue pike (*Stizostedion vitreum glaucum*) from the Great Lakes (Table 1-2), and aurora trout (*Salvelinus fontinalis timagiensis*), which disappeared from its natural habitats because of acid rain (Parker and Brousseau 1988), attest to the broad and serious perturbations there. Many eastern North American species are in trouble in substantial proportions of their ranges (J. E. Williams et al. 1989), and only their extensive geographic distributions in a well-watered region prevent more taxa from being imperiled.

Table 1-1. Taxa (including species, subspecies, and undescribed forms) of threatened, endangered, and extinct North American freshwater fishes. Totals are similar but not equivalent to those originally published, since some species were excluded to adhere to our criterion of freshwater breeding.

Geographic coverage	Authority and areas	Numbers of imperiled taxa	Extinct taxa	Totals
United States	R. R. Miller 1972a[1]			
	western	65	—	65
	eastern	228	—	228
	east + west	7	—	7
Totals		300	—	300
Canada, United States	Deacon et al. 1979			
	western	86	—	86
	eastern	147	—	147
	east + west	2	—	2
Totals		235	—	235
Canada, Mexico, and United States	Ono et al. 1983			
	western	74	12	86
	eastern	91	9	100
	east + west	2	0	2
Totals		167	21	188
Arid lands: Mexico and Western United States	J. E. Williams et al. 1985			
	western	105	9	114
	eastern	60	6	66
	east + west	1	0	1
Totals		166	15	181
United States and Canada	J. E. Johnson 1987a[1]			
	western	135	—	135
	eastern	370	—	370
	east + west	7	—	7
Totals		512	—	512
Canada, northern Mexico, and United States	USFWS 1989c, d[2]			
	western	105	6	111
	eastern	119	10	129
	east + west	0	0	0
Totals		224	16[3]	240
Canada, Mexico, and United States	J. E. Williams et al. 1989			
	western	135	—	135
	eastern	218	—	218
	east + west	3	—	3
Totals		356	—	356

[1]Both R. R. Miller's (1972a) and J. E. Johnson's (1987a) listings included any taxon considered threatened, endangered, or of special concern by any of the fifty states and thus were more politically comprehensive than

A common pattern may be developing. When pervasive, long-term environmental stress was applied in the past to a small western river or spring, individual species disappeared. Under environmental stresses caused by continued development on a regional scale, not only species but whole faunas of major western river basins are now collapsing. As regionwide perturbations such as acid rain continue to mount in eastern North America, the more sensitive species first, then major blocks of species of that diversified fauna will predictably follow suit.

Based on percentages, the western fauna is clearly more endangered than the fauna east of the Rocky Mountains. Recall that only about 150 full species were included by Lee et al. (1980, 1983) in the fish fauna of the West, while eastern species numbered almost 600. Thus, even with a number of listed fishes being subspecies of the same forms, the total of 122 taxa considered to be in some sort of trouble is an impressive statistic! Regional aridity and ever-increasing demands for water by a burgeoning human population soon may annihilate most of this distinct but poorly understood fauna. The question is, can we take the chance of allowing such diversity to pass unheralded?

Summary and Conclusion: Epitaph for a Fauna?

The last of the major North American ichthyofaunas to be formally described, that of the intermountain and hot desert regions (Fig. 1-

1), will likely be the first to disappear. All major streams in the western United States are dammed, controlled, and overallocated; waters of the Colorado River basin, for example, are used several times during their passage from the Rocky Mountains to the sea (Fradkin 1984). Groundwaters from deep beneath the floors of desert basins are pumped at rates greatly exceeding those at which aquifers can be recharged, and recharge areas are (in addition) beheaded by dams. Enormous interbasin water transfers are planned and implemented as local supplies for domestic and agricultural uses are exceeded in California and the intermountain West. Water projects that both anticipate and encourage human populations to expand proceed largely unabated throughout the rapidly developing region.

The situation with western North American fishes is grim, and it may soon be replayed elsewhere. Major rivers of the North American plains are failing as groundwaters are extracted for agriculture. Problems of point-source and general pollution continue in the Great Lakes, rivers of the eastern seaboard, and elsewhere. Acid rain is destroying lakes and streams in eastern Canada, the United States, and northern Europe. Drought conditions, coupled with ever-greater water uses, are demanding additional modifications in navigable streams of the Mississippi River valley. Economic woes in Latin America, Africa, and other Third World regions are forcing vast, formerly undeveloped areas of desert, savannah, and rain forest to be opened up for agriculture and human occupation.

---

other lists that dealt with the status of species or subspecies throughout their natural ranges; extinct species were included in the compilations of imperiled taxa, where given. McAllister (1970), McAllister et al. (1985), and Campbell (1984, 1985, 1987, 1988) reviewed rare or endangered Canadian fishes, as did Contreras Balderas (1975, 1978a, 1987, in press) for Mexico, and should also be consulted for details on those faunas.
[2]This summary includes species listed as threatened or endangered (USFWS 1989c), plus those in the *Animal Notice of Review* (USFWS 1989d) in categories 1 and 2 ("substantial evidence toward listing" and "possibly appropriate to list," respectively).
[3]This category includes taxa listed in category 3A (USFWS 1989d), for which "persuasive evidence of extinction" exists.

Table 1-2. North American freshwater fishes that have become extinct in the past century (modified from R. R. Miller et al. [1989]).

Common name	Scientific name	Original distribution[1]
Miller Lake lamprey	*Lampetra minima*	OR
Longjaw cisco	*Coregonus alpenae*	IL, IN, MI, NY, OH, PA, WI, ON
Deepwater cisco	*C. johannae*	IL, IN, MI, MN, WI, ON
Lake Ontario kiyi	*C. kiyi orientalis*	NY, ON
Blackfin cisco	*C. n. nigripinnis*	IL, IN, MI, WI, ON
Yellowfin cutthroat trout	*Oncorhynchus clarki macdonaldi*	CO
Alvord cutthroat trout	*O. clarki* ssp.	NV, OR
Silver trout	*Salvelinus agassizi*	NH
Maravillas red shiner	*Cyprinella lutrensis blairi*	TX
Mexican dace	*Evarra bustamantei*	DFE
Mexican dace	*E. eigenmanni*	DFE
Mexican dace	*E. tlahuacensis*	DFE
Independence Valley tui chub	*Gila bicolor isolata*	NV
Thicktail chub	*G. crassicauda*	CA
Pahranagat spinedace	*Lepidomeda altivelis*	NV
Ameca shiner	*Notropis amecae*	JAL
Durango shiner	*N. aulidion*	DGO
Phantom shiner	*N. orca*	NM, TX, CHI, COA, TAM
Rio Grande blunt-nose shiner	*N. s. simus*	NM, TX(?)
Clear Lake split-tail	*Pogonichthys ciscoides*	CA
Banff longnose dace	*Rhinichthys cataractae smithi*	AB
Las Vegas dace	*R. deaconi*	NV
Grass Valley speckled dace	*R. osculus reliquus*	NV
Stumptooth minnow	*Stypodon signifer*	COA
First June sucker	*Chasmistes l. liorus*	UT
Snake River sucker	*C. muriei*	WY
Harelip sucker	*Lagochila lacera*	AL, AR, GA, IN, KY, OH, TN, VA
Parras pupfish	*Cyprinodon latifasciatus*	COA
Tecopa pupfish	*C. nevadensis calidae*	NV
Monkey Spring pupfish	*Cyprinodon* sp.	AZ
Whiteline topminnow	*Fundulus albolineatus*	AL
Raycraft Ranch poolfish	*Empetrichthys latos concavus*	NV
Pahrump Ranch poolfish	*E. l. pahrump*	NV
Ash Meadows poolfish	*E. merriami*	NV
Opal allotoca	*Allotoca maculata*	JAL
Parras characodon	*Characodon garmani*	COA
Amistad gambusia	*Gambusia amistadensis*	TX
San Marcos gambusia	*G. georgei*	TX
Blue pike	*Stizostedion vitreum glaucum*	MI, NY, OH, PA, ON
Utah Lake sculpin	*Cottus echinatus*	UT

[1]Abbreviations for Mexican states: CHI, Chihuahua; COA, Coahuila; DFE, Districto Federal; DGO, Durango; JAL, Jalisco; and TAM, Tamaulipas. Abbreviations for Canadian provinces: AB, Alberta; ON, Ontario.

Competition, predation, and other detrimental interactions with non-native species are increasingly recognized as important in the decline of native fishes, and western North America may be more seriously affected by the introduction of alien fishes than any other part of the continent (see contributions in Courtenay and Stauffer 1984). Many native western species, because of their antiquity and isolation, appear to lack the competitive abilities and predator defenses developed by fishes in more species-rich areas. As a result, the western fauna is highly susceptible to planned and impromptu introductions of exotics (Moyle et al. 1986). Elsewhere, tropical and subtropical areas such as Florida and parts of Mexico are especially prone to the establishment of aggressive alien forms such as African cichlids that escape from fish farms or are directly stocked to enhance human protein supplies. The spread of non-native fishes in temperate zones also is well under way. As Robins (1986) pointed out, "man can shut off chemical abuse and undo physical damage . . . , but an aquatic organism, once established, is indeed a permanent resident, for better or worse." The presence of non-native fishes may prove a far greater problem to native fish survival than all our other environmental abuses combined.

Slow and progressive development of water resources through minor diversions, hand-built dams, and windmills, as occurred in western Canada and United States over the past century, is not going to happen again. Time, albeit very little, was available to visualize and predict losses of the aquatic fauna in the western United States. The predictions were made and have proven to be accurate.

Today, electric pumps and the hydroelectric dams to power them, followed by a flood of alien species that colonize artificial habitats, are decimating natural aquatic systems and faunas in Mexico more rapidly than has been experienced elsewhere or ever before (Contreras Balderas, *this volume*, chap. 12). There is almost no time available to save that fauna, or even to document its demise.

Technological humans and native freshwater fishes are locked in mortal competition for water, and indigenous fishes will not prevail unless we plan and dictate a scenario that ignores short-term economic concerns. Increased world concern for losses in biodiversity, and the realization that biodiversity and the maintenance of humankind's economic and other systems are inexorably linked (see papers in E. O. Wilson 1988), may be developing too late for the aquatic biota of western North America.

Worldwide clearing of rain forests has had a parallel in the destruction of natural freshwater habitats in the American West for decades. Despite the emergence of an enlightened public, determined efforts by agency and academic biologists, favorable legislation, and other efforts, a wave of native fish extinction is continuing in arid lands. Implementation of conservation plans must be rapid; reaction time is short to avert impending catastrophe. With the application of modern technology to water development, extinction of the native fish fauna will be realized over a period of time far shorter than was needed to discover, collect, and describe it, adding only an indiscernible twitch to a blink of the eye in geologic time.

Chapter 2

Ichthyological Exploration of the American West: The Hubbs-Miller Era, 1915-1950

Robert Rush Miller, Clark Hubbs, and Frances H. Miller[1]

Introduction

These are exciting times, with many new, sophisticated tools and innovative approaches that provide new ideas about zoogeography and classification. But there is also much interest and value in the past, and much of what we learn from history is both fun and rewarding.

In every scientific discipline, the prevalent ideas and even the questions asked are the products of historical development. We progress by successive building blocks, some of which inevitably crumble and are replaced. No single hypothesis can explain all observed patterns; thus it behooves us to be receptive to new ideas, but not to become overzealous champions of any single approach—no matter how novel—since we may later find such novelties to be only passing fads (e.g., chromatography). The most recent "bandwagon" fields of research for determining phylogeny are plate tectonics and molecular systematics. Regarding the latter—which has undergone remarkable change in the last twenty years— there has developed "a commonly held misconception that all evolutionary problems are solvable with molecular data" (Hillis and Moritz 1990:502). Newcomers to a field have also been known to adopt the view that any work older than about a decade or so is irrelevant. The days of such people are numbered.

Arid-region ecosystems (including those of the Arctic and Antarctic) are extremely fragile because of the great daily and seasonal environmental fluctuations, and the biota is typically depauperate (relictual) and usually specialized to the extent that it has lost its ability to compete with generalist organisms. In the American West, these generalists are often exotic species introduced from other places. Thus extinction and decimation of native faunas result as desert marsh, spring, and stream habitats are drained, dredged, impounded, or otherwise altered in the seemingly endless human quest for more water, and as non-native fishes and other organisms are introduced, become established, and begin to exert their competitive and predatory pressures (G. R. Smith 1978; Deacon 1979; R. R. Miller et al. 1989).

The geographic areas we deal with here are the arid and semiarid regions of this continent that include the four great North American deserts (Fig. 2-1): Great Basin, Mohave, Sonoran, and Chihuahuan, as mapped by R. R. Miller (1981). Each desert has its own unique characteristics, and their biotas have distinctive elements and separate derivations. These deserts are not ancient, as some have claimed; they are mostly the products of interglacial and especially postglacial times, approximately covering the last ten to twenty thou-

---

[1]Died 17 October 1987.

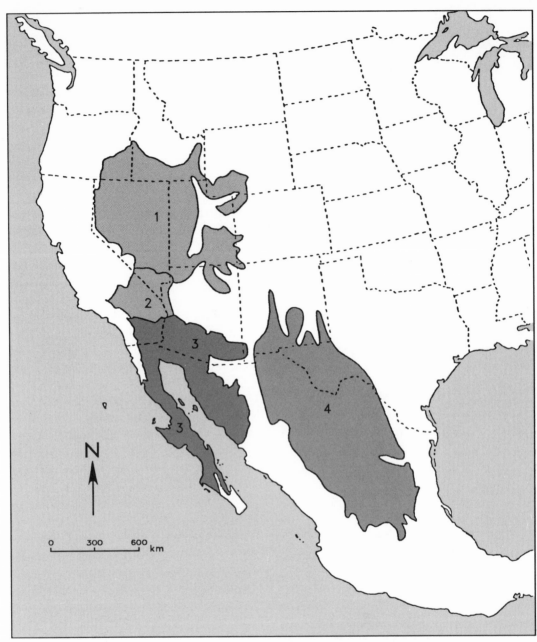

Fig. 2-1. North American deserts: (1) Great
Basin, (2) Mohave, (3) Sonoran, and (4)
Chihuahuan (modified from R. R. Miller 1981).

sand years. In Plio-Pleistocene times (around the last one to five million years) these areas were well watered by many lakes and streams, which became reduced in size and extent or dried as recent climates became warmer and less humid than ever before (Axelrod 1979). As the climate deteriorated and aquatic habitats became reduced in size and disappeared, fishes and other water-dependent animals became more and more restricted. They persisted, however, in many places, and sometimes in unexpected numbers. In the Great Basin, only twenty-nine of seventy-three independent pluvial-lake basins lacked native fish life in the period of exploration covered here (C. L. Hubbs et al. 1974, table 4). What we wish to show is how some of these deserts looked a few short decades ago, before the severe habitat alterations and the introduction of exotic organisms. Our collecting stations are shown in Figure 2-2.

We are grateful to Earl Hubbs for historical insights, Beth Reid for processing the manuscript, Deborah Day for archival material, and Margaret van Bolt for rendering the map for Figure 2-2.

First Visit

Conditions were almost pristine when Carl L. Hubbs was first introduced to the arid West in 1915, on a survey of the fish fauna of the Bonneville Basin of Utah. He was serving as field assistant to John Otterbein Snyder of Stanford University. Only common carp and goldfish (*Cyprinus carpio* and *Carassius auratus*) were mentioned as exotics in Carl Hubbs's account. This trip, in a Model-T Ford, commenced on 22 June and consumed seventy days of camp life, which Hubbs described in a never-published presidential talk to the Zoology Club at Stanford University on 9 February 1916. The role of postpluvial disjunction of Bonneville waters following the desiccation of that great lake was the primary focus of the

expedition. During this trip the first seeds were sown for the Great Basin monograph published thirty-three years later by C. L. Hubbs and Robert R. Miller (1948b).

Getting to Utah was rather eventless, though Hubbs's notes are highlighted now and then by remarks such as, the Truckee River "transfers water from [Lake] Tahoe to Pyramid *and Winnemucca* lakes" (emphasis added; Winnemucca Lake dried in 1938 as Truckee River waters were diverted for agricultural and domestic use; LaRivers 1962). They took the northern route across Nevada (via the Humboldt River valley) rather than the southern route on the official Lincoln Highway, with its scattered mining towns struggling for existence. Their return trip was by the Lincoln Highway. The verbal report of accidentally killing a chicken at a ranch, a chase by the rancher out of the valley, and the final fifty-cent payment for the chicken at perhaps the loss of a horse ridden to overextension during the pursuit was a highlight of later recollections. After plowing through miles of the fine, glaring-white, alkaline lake deposits of the Bonneville Flats, our travelers agreed with an old Kansas farmer they had met earlier at Elko, Nevada, that this part of the road was Blue Hell—in contrast to the Humboldt Valley, which was pure Heaven—"so we passed through both Heaven and Hell in our Ford, and being such, we came through both safely."

After ten days they reached Provo, Utah; the odometer registered over 1000 miles (1600 km) from Palo Alto, California. Utah Lake in midsummer was very silty; the local people said the lake was much clearer before the introduction of carp, which kept the silt in suspension. They remained there over a week and "put up a wash boiler full of fishes," mostly chubs (*Gila atraria*) and suckers; the larger of these were locally called mullet (Utah sucker, *Catostomus ardens*) in contrast to the smooth-mouthed June suckers (*Chasmistes l. liorus*). One haul of their forty-foot (12.2-m) seine in

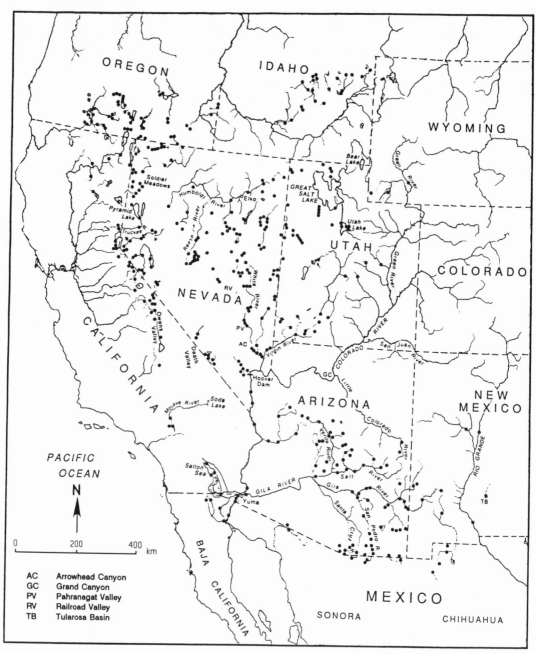

Fig. 2-2. Collecting sites used by Hubbs and Miller and their parties in the American West, 1934–1950.

the Weber River yielded "758 suckers averaging about a pound [0.45 kg] apiece." A memorable week was spent collecting in the clear waters of Bear Lake, on the Utah-Idaho line, described by the Mormons as the most beautiful lake in the world. By the time they returned to Palo Alto they had driven 3500 miles (5600 km), having crossed California and Nevada twice and Utah twice in both its length and breadth, with forays into both Wyoming and Idaho. So far as we can determine, the only publication directly resulting from this survey was J. O. Snyder's (1919) descriptions of three new species of whitefishes (*Prosopium abyssicola*, *P. gemmiferum*, and *P. spilonotus*) from Bear Lake.

In his Zoology Club talk, Hubbs observed that the major disjunct basins of the West

> are often characterized by two types of fishes, one group being known only within the basin (like the endemic *Chasmistes liorus* and the sculpin, *Cottus echinatus* [in Utah Lake]); the other more widely spread with closely related or geminate species (of common origin), found in other basins. These two groups may represent faunas of different ages, the first group being composed of an old fauna which has remained so long within the basin as to become genetically distinct (e.g., *Iotichthys*) from all other fishes, and the second group being composed of a new fauna, which has spread from basin to basin by stream capture (e.g., *Rhinichthys osculus*) or similar modes of transference.

Evidences of Change: Data from 1934 and 1938

Eleven years after the Bonneville survey, as they were returning to Michigan from an expedition to the Pacific Coast in 1926, Carl Hubbs and Leonard P. Schultz collected in the Gila River 3.2 km downstream from Dome, Arizona, and secured seven species. Four of them were non-native (carp, brown bullhead [*Ameiurus nebulosus*], mosquitofish [*Gambusia affinis*], and green sunfish [*Lepomis*

cyanellus]), so the invasion of exotics was already well under way more than sixty years ago. Even in 1904, F. M. Chamberlain, a perceptive observer of the natural scene (Jennings 1987), found five exotic fishes already established in the Gila River basin (R. R. Miller 1961). The drastic habitat degradation of the Colorado River began in the lower part of the system. Documentation for the death of the Gila River is given by Rea (1983), who treats in depth the history of the middle reaches of that stream from prehistoric to Anglo-American perturbations.

The first serious effort to conduct an ichthyological survey of Nevada was the Hubbs expedition of 1934—the great drought year virtually matched in 1988. Their party entered the Red Desert basin, Wyoming, on 11 July, and the last collection was made on 16 September near Green River, Wyoming; forty-six days were spent in the Great Basin itself. An apparently endless sagebrush desert punctuated by spectacular north-south mountain ranges (Fenneman 1931), the Great Basin of western United States, a cool, high desert, would seem an unlikely place to find living fishes. Nevertheless, valley-bottom springs, marked out on the stark landscape by groves of verdant (although introduced) Lombardy trees (*Populus nigra*) and native Frémont poplars (cottonwoods, *P. fremontii*; Fig. 2-3), contain the relics of a wetter (Plio-Pleistocene) regime and form an aquatic archipelago—isolated steppingstones in a sea of desert—similar to the Galápagos Archipelago that Charles Darwin made famous. Early on, traversing the first dirt road in Wyoming, their heavily laden Chevrolet sedan (Fig. 2-4) blew two tires. It was then that Carl discovered that the tires were four-ply (rather than six-ply, as ordered). He wired the dealer in Ann Arbor, Michigan, and six days later he picked up six six-ply tires at Spencer, Idaho, sent by Railway Express.

Everything was in or on the car, which, of course, had running boards (an extinct species today). The back seat had been left at home, and rows of canned goods, utensils, and bedding filled the space so vacated; the three children used these materials for seats. The two tire wells held the spare tires, and all the seines were wedged between these and the hood. There was a special telescoping trunk that made the space two and a half times its normal length. By the driver's side, a luggage rack passing from running board to windows was loaded with collecting gear. Thus the driver could not get out of his side of the car! The party had a tent that was also stored along the running board.

Five times the battery was jarred loose onto the ground by the rough roads, and each time the brakes locked they were beaten loose with a hammer. Chewing gum was used to seal holes that appeared from time to time in the gasoline tank.

Supplies lasted from three weeks to a month before it was necessary to stop and replenish

Fig. 2-3. Blind Spring Valley, Mono County, California, from above Benton. Frémont cottonwoods mark the natural wetlands associated with a spring outflow, and pond margins are lined by introduced lombardy trees planted by humans near aquatic habitats through much of the American West. Photograph by the Hubbs party, 27 July 1938.

Fig. 2-4. Exploring Nevada in Hot Creek Valley, Nye County. Photograph by the Hubbs party, 1934.

them—towns with grocery stores were few and far between. During one period they traveled continuously for twenty-two days off pavement (some of these roads are now paved). Survival under these conditions, with dust and heat often overpowering, required resourcefulness and constant improvisation. Even towels were used on occasion to seine fish.

Except for the first-ever use of commercial powdered derris root (rotenone) to collect fishes in Nevada on 2 August at Italian Camp Springs, Humboldt County (field no. M34-105), only seines (also hands) were used (Fig. 2-5). In order to keep the three children happy under often trying conditions, Carl Hubbs established an "allowance" for them based on the number of species collected (five cents apiece), with special awards for new species or subspecies (a dollar) or new genera or sub-

genera (five dollars). Fortunately for them, he was a taxonomic "splitter," and thus Frances, Clark, and Earl frequently obtained special awards!

Important accomplishments during the 1934 expedition were: (1) collection of a series of sculpins during a six-day examination of the isolated streams of the Snake River Lava Plain, Idaho, that helped to clarify the taxonomy of western *Cottus*—six subspecies considered in the field were later described as the new species *C. confusus* by Bailey and Bond (1963); (2) collection of *Salvelinus confluentus* (Cavender 1978) in this disjunct drainage system, originally identified in the field as *S. malma parkei*; (3) securing large numbers of minnow and sucker hybrids (C. L. Hubbs 1955); (4) collections of series of disjunct, local forms of *Gila bicolor* from diverse habitats of the Railroad Valley system of eastern Nevada (Fig. 2-6A,

Fig. 2-5. Carl L. and Laura C. Hubbs in Nevada.
Photograph by the Hubbs party, 1934.

B), involving six new subspecies (yet to be ana-lyzed; see C. D. Williams and Williams [1981] for population status); and (5) securing excel-lent material for a detailed study of many dis-junct and isolated populations (Fig. 2-7) of speckled dace (*Rhinichthys osculus*; about ten nameable populations), some of which were treated by C. L. Hubbs et al. (1974). Carl Hubbs's extraordinary taxonomic eye detected numerous natural hybrids, an observation that was in direct conflict with the teachings of his mentor, David Starr Jordan.

The year 1934 marked the initiation of the "taming of the Colorado River" with con-struction of Hoover (then Boulder, and now Hoover again) Dam, built in order to trans-form a "natural menace" into a "national re-source" (U.S. Bureau of Reclamation 1946). The dam was closed in 1935. Originally an unimpeded, swift-flowing, silt-laden river with widely fluctuating flows, the Colorado River has become the most regulated and used river in the Western Hemisphere, if not in the world

(Fradkin 1984; Graf 1985). Its depauperate but highly endemic fish fauna is clearly threat-ened with extinction.

New taxa collected in 1934 were: Pit-Kla-math lamprey (*Lampetra lethophaga*), Alvord chub (*Gila alvordensis*), Borax Lake chub (*G. boraxobius*), relict dace (*Relictus solitarius*) (the three children each got five dollars for this one!), the Wall Canyon sucker (*Catosto-mus* n. sp.), shorthead sculpin (*Cottus con-fusus*), Pit sculpin (*C. pitensis*), and at least twenty subspecies.

The 1938 expedition that concentrated again on Nevada lasted two months (sixty-two days) and yielded 115 collections (Fig. 2-8). Carl Hubbs, inveterate collector of many varieties of organisms, obtained thirty-six bats with a single charge of .22 caliber dust shot at Phantom Lake Cave, Texas. He also shot a diamondback rattlesnake (*Crotalus atrox*) at mid-morning on the main, dusty street of Van Horn, Texas, without raising an eyebrow amongst the populace. During the

A

B

Fig. 2-6. Springs in Railroad Valley, Nye County,
Nevada: (A) unnamed springs on the Locke
Ranch; and (B) Green Spring, looking southwest.
Photographs by the Hubbs party, 1934.

Fig. 2-7. Box canyon at Thousand Creek, leading onto the Alvord Desert, Humboldt County, Nevada. Such habitats commonly support populations of fishes, especially speckled dace, isolated by many miles of arid lands from other such stocks (field no. M34-106). Photograph by the Hubbs party, 1934.

work in Nevada, a great deal of time was spent studying the paleohydrology of that state and parts of adjacent ones. This led to publication of a detailed map of pluvial lakes and streams (C. L. Hubbs and Miller 1948b) that remained undiscovered by geologists for more than a decade; the geologists were greatly impressed by the expertise of mere "fish students" (Feth 1961).

It was during this trip that a collection of chubs from springs on the Murphy Ranch between Cherry Creek and Currie, Steptoe Valley, Elko County, Nevada (field no. M38-161) was identified initially as an undescribed form of tui chub (*Gila bicolor*). Chance examina-

tion of the pharyngeal dentition (5-4 in subgenus *Siphateles* and 2, 5-4, 2 in subgenus *Gila*), however, showed that it represented an introduced stock of the Utah chub. The original minnow in these springs, evidently eaten or outcompeted by trout and other exotics, was the relict dace (C. L. Hubbs et al. 1974).

Early fish transplants are difficult to document, and no doubt many have gone unrecorded. Thus, the zoogeographer may try to devise a logical explanation to account for odd displaced stocks, or may believe that a new distribution record is at hand. The history of one early planting in Nevada was fortunately acquired during the 1938 survey, when pioneer resident George Schmidtlein of Big Smoky Valley, Nye County, was interviewed at his ranch. In August 1873 George and his brother Henry, with a neighbor named Smiley, hired an Indian and his wife and pack train to stock Kingston Creek with cutthroat trout (*Oncorhynchus clarki henshawi*) from the Reese River system just over the Toiyabe Mountains to the west. The trout was native to Reese River and its tributaries. They diverted a little stream and caught 139 fish, none more than six inches (150 mm) long, placing them in syrup and vinegar kegs. Those in the syrup kegs died, but those in the vinegar kegs "thrived." The trout were packed across three summits over four days with changes of water and retention of the fish overnight in small dammed ponds. Finally, 39 fish were brought through and put in a dammed pond in Big Smoky Valley, where they were fed for a week. The planting was successful, for despite some dynamiting two years later and diversion of Kingston Creek quite regularly, the trout multiplied rapidly and there was good fishing within three years. From Kingston Creek the trout were soon planted into nearby streams on the eastern side of the Toiyabes and also into a stream on the southern end of that range (Miller and Alcorn 1946). This example may reflect numerous fish transplants by early ranchers or even by Native Americans. Other, later, transfers of native Great Basin fishes are discussed by C. L. Hubbs et al. (1974).

Fig. 2-8. Robert R. Miller (left), Earl Hubbs (right center), and Carl L. Hubbs collecting on the Polkinghorne Ranch, Pleasant Valley, Pershing County, Nevada (field no. M38-80). Photograph by the Hubbs party, 1938.

Fig. 2-9. Frances Voorhies Hubbs at Big Shipley Springs, northwest of Eureka, Nevada. Photograph by the Hubbs party, 13 August 1938.

Fifteen of the 1938 stations (field nos. M38-82–96) yielded lizards, frogs, snakes, toads, and insects. Frances Hubbs (Fig. 2-9) wrote in her diary that by mid-morning on a July day in the desert, "all the lizards came out of their holes to warm up. Dad had dibs on all the lizards on one side of the car and Earl those on the other side. We stopped every 50 feet for a long time. It was really funny." Aquatic habitats were frequently so few and far between that Carl often stopped to obtain large series of "herps" to ease his frustration at going so long between fishing stations! Snakes, amphibians, turtles, aquatic insects,

dragonflies, and butterflies (Fran's job) were also collected whenever they were seen at regular fishing sites.

This collecting habit recently helped to solve the question, in some youthful minds, of whether a frog discovered at Duckwater in Railroad Valley, Nevada, was a disjunct population of *Rana pretiosa*, a hybrid, or even a new species; when the Hubbs party visited there in 1934 and 1938, no frogs were seen. Allozyme analysis of a recent collection demonstrated that the frog is indistinguishable from *R. aurora draytoni* of California's Sierra Nevada (Green 1985), from where it was most likely introduced. Archaeological materials were likewise collected and carefully recorded. Since many sites were remote, much valuable material accrued for disciplines other than ichthyology. Carl Hubbs was, indeed, the "complete collector."

As an impression of collecting in Nevada, Fran wrote on 31 July 1938: "After lunch we went down into Grass Valley and then Pleasant Valley. The first didn't have any grass and the second wasn't pleasant!! But that's typical of Nevada valleys. And you can be sure if it says fish that there won't be any there!! Or else they'll be carp." Again, paraphrasing: "We drove over the longest, bumpiest road I've ever been on—about 40 miles [64 km]. I was positively seasick by the time we got done. We camped at a spring meadow about 2½ miles [4 km] long. There were an awful lot of mosquitoes, but they didn't quite kill us." Diaries kept by Miller and Frances Hubbs recorded flat tires with monotonous frequency—perhaps one every three to five days! Travel conditions were unchanged from 1934, and the canned foods were supplemented by jackrabbits shot with a .22 rifle usually used to obtain lizards with dust shot. Earl was the expert rabbit shooter.

Many of the 1938 collections focused on springfish (*Crenichthys*) habitats (Fig. 2-10A–C). The family visited Sunnyside, Nevada, on

A

B

C

Fig. 2-10. Spring-fed habitats of various
subspecies of White River springfish along the
course of pluvial White River, eastern Nevada:
(A) Hiko Spring, Lincoln County (*Crenichthys
baileyi grandis*); (B) Moromon Spring, Nye

County (*C. b. thermophilus*); and (C) Preston Big
Spring, White Pine County (*C. b. albivalis*; see
J. E. Williams and Wilde 1981). Photographs by
the Hubbs party, 1938.

28 August to sample springfish in Hot Creek, accompanied by a ranch boy. The area was revisited on 4 June 1964 (C. Hubbs and Hettler 1964), and the ranch boy, now the owner, recalled the prior visit almost twenty-six years before as one of the highlights of his life.

On the first night out from Ann Arbor to Nevada, the Hubbs family camped near Indianapolis at an old farmhouse with access to hot water at a cost of only fifty cents! They were up the next morning at 0415 and off by 0515. Carl Hubbs thought nothing of collecting until 0100, and sometimes it was after that hour that they ate "dinner." Lunch frequently was eaten between 1400 and 1530.

Temperature contrasts, so typical of deserts, were noted at a camp south of Caliente, Nevada, on 10 July, when the washbowl left out all night was solid ice at sunrise and the air temperature was 24°C by 0900. The desert's nighttime coolness after a blazing summer day is an important factor in the survival of plants and animals. Rarely were more than two or three collections made in a day because of poor roads and the great distances between water holes, but on 12 July, five fish collections were made, with Fran's comments: "Isn't that marvelous!! And what a lot of good kinds—3 new species and a new genus. What a happy poppa we have." The new genus was *Moapa* (C. L. Hubbs and Miller 1948a), as shown on the data sheet (Fig. 2-11).

The hike up Arrowhead Canyon (Fig. 2-12) on 14 July 1938 to explore this now-extinct waterway of the pluvial White River (C. L. Hubbs and Miller 1948b, item 61 on map) was a memorable one. Fran wrote:

Well inside the canyon the dry river bed was no more than 10 m wide, and the walls rose almost vertically for 100 m! We got to the base of a check dam, nearly at the head of the canyon, with a muddy, one-m-deep pool at its base that was full of *Bufo punctatus*. No water to drink and stifling heat—131°F [55°C] in the sun and no shade. Nearly everyone was sick from heat prostration by the time the car (left well outside the canyon when the "road" became impassable) was reached. To make matters worse, the car got stuck twice on the way back to the Phillips (Home) Ranch near Moapa.

Miller, who had barely survived a trek into the 51°C heat of Salt Creek, Death Valley, less than three weeks earlier, was now a "hardened desert rat" little affected by the Arrowhead Canyon experience and was able to pull the Hubbs's vehicle out. Core body temperatures must have been extreme because the Home Ranch pool (a constant 31.7°C) felt like ice water after the return.

The following day the Hubbs family visited a dusty ranch center and county seat, Las Vegas. After sampling the creek to collect type material of the now-extinct *Rhinichthys deaconi* (Miller 1984), they camped in the city park. During the night a violent windstorm blew large limbs from a cottonwood tree; one hit the tent and just missed permanently immobilizing Frances.

In the Mohave[2] Desert (R. R. Miller 1981), a warm-temperate desert sometimes considered part of the Great Basin (as by C. L. Hubbs and Miller 1943), the Mohave tui chub (*Gila bicolor mohavensis*), with its numerous gill rakers for filtering microorganisms, was adapted for life in pluvial lakes (R. R. Miller

[2]The change in spelling of this name to the Spanish, Mojave, by the U.S. Board on Geographic Names (who reversed their original acceptance of Mohave), is unwarranted and inconsistent. Named after the native aboriginals by Frémont, who spelled the word phonetically as *Mohahve*, the board later changed the name in California but left it as Mohave in Arizona (viz., Mohave County), although there it is derived from the same Indian tribe (Granger 1960:200) whose range included much of the Mohave Desert.

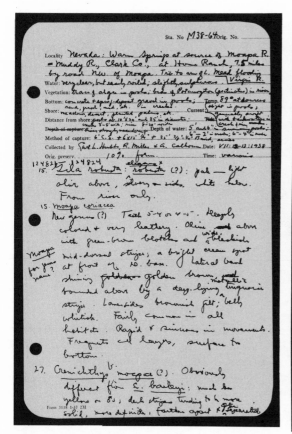

Fig. 2-11. Data sheet for field no. M38-64, where the new genus *Moapa* was collected on 12–13 July 1938. Reproduced from Carl L. Hubbs's field notebooks on file in the University of Michigan Museum of Zoology.

1973). When postpluvial aridity set in, the chub was forced to live in the remnant reaches of the Mohave River from the headwaters to a spring-fed pool on the western edge of Soda Lake, a remnant of pluvial Lake Mohave. Although it was originally believed that the arroyo chub (*G. orcutti*), admirably adapted for stream life, was also native to the Mohave River, it was later learned that bait introductions about 1930 by trout fishermen accounted for its presence. Mass hybridization followed (C. L. Hubbs and Miller 1943). At the time that report appeared, about 9% of the 5604 fishes sampled from the Mohave River basin were hybrids. Several surveys of the river have been made since 1940 (1942,

1955, 1967, and subsequently), and the incidence of apparent hybrids has increased. Although it still persisted in the river and its tributaries in 1967 (R. R. Miller 1969a), the Mohave tui chub is now believed to exist as a pure stock only in the spring-fed lake at Zzyzx Resort on the west side of Soda (Dry) Lake, south of Baker, California. This habitat is described in the recovery plan for this chub (U.S. Fish and Wildlife Service [USFWS] 1984g).

Payoffs: The Values of Data Accumulation

In 1942 Carl Hubbs and Robert Miller each traveled to the Great Basin separately, from 6 June to 1 August and from 10 August to 6

October, respectively, making a total of 191 collections. The purpose of these trips was to tie up loose ends and to explore areas not included on previous trips, as well as to make more observations on pluvial lake deposits and (for Miller) to secure additional evidence on the paleohydrology of the Death Valley system. The Hubbs party included Carl, Laura, and Earl; and the Miller party comprised Robert R., Ralph G., Lucy B. (both to 29 August), and Frances (by now Mrs. R. R. Miller). They were allotted a sum of five hundred dollars (five cents per mile for five thousand miles each) for the two trips, excluding subsistence, which was borne separately by each party, and a collective 110 days in the field yielded more than fifty thousand specimens. In June, at Salt Lake City, Carl presented the Hubbs and Miller contribution to the *Symposium on the Great Basin, with Emphasis on Glacial and Postglacial Times*. The other contributors were Eliot Blackwelder, "The Geological Background," and Ernst Antevs, "Climatic Changes and Pre-white Man." The results of our 1942 fieldwork augmented, confirmed, and sometimes resulted in revision of our conclusions on the taxonomy and history of Great Basin fishes, and this and other trips allowed in-depth examination of a number of taxa and taxonomic problems that would otherwise have been impossible.

The monotypic genus *Eremichthys*, first collected by R. R. and R. G. Miller in 1939, was captured in large numbers at several localities in Soldier Meadows, Nevada, in 1942 (C. L. Hubbs and Miller 1948a), resulting in combined series totaling 1661 specimens for the description of *Eremichthys acros*. This fish is currently regarded as threatened (USFWS

1989c). Local residents supplied testimony on fish populations and valley drainages. An increase in exotic fishes was noted (e.g., yellow perch [*Perca flavescens*], Sacramento perch [*Archoplites interruptus*], and pumpkinseed sunfish [*Lepomis gibbosus*]). Reptiles and amphibians were collected at 83 stations, and insects (excluding aquatic forms) from 114 stations.

One of the controversial taxonomic and nomenclatural problems of Great Basin fishes that could only be clarified by the accumulation of large numbers of specimens over a period of time involves chubs of the subgenus *Siphateles* of *Gila*. Two schools of thought developed on whether the lacustrine form described as *Leucidius pectinifer* by J. O. Snyder (1917) from Pyramid Lake, Nevada, was a full species or only a subspecies of *G. bicolor* (Girard). Snyder based his genus on the numerous gill rakers (29–36 total on the first arch) and its pharyngeal tooth formula (0, 5-5, 0, versus the typical 0, 5-4, 0 in *Gila bicolor obesa*, formerly *Siphateles bicolor obesus*). Although Snyder (1917) never identified any individuals with gill-raker counts intermediate between the lacustrine and fluviatile forms, study (by Miller of eleven paratypes, USNM 93828) of *L. pectinifer* revealed ten individuals with 29–38 rakers, and one with 22 (the holotype of *L. pectinifer* [USNM 76304] has 35 rakers). We have found complete gradation in gill-raker numbers between topotypic "*obesa*," 11–19, and "*pectinifer*," 27–40. Furthermore, isolated demes in other basins show all levels of raker numbers. Moreover, some individuals of "*pectinifer*" have 0, 5-4, 0 teeth, and some "*obesa*" have 0, 5-5, 0 teeth (C. L. Hubbs et al. 1974:148–149).

We regard "*obesa*" and "*pectinifer*" as trophic adaptations with, typically, ecological segregation, but we also found numerous places (including Pyramid Lake, well before the turn of the century) where complete intergradation in raker number and general mor-

Fig. 2-12. Hubbs party entering Arrowhead Canyon, formerly occupied by the White River, Clark County, eastern Nevada, on its way to the Colorado River mainstream in pluvial times. Photograph by the Hubbs party, 13 July 1938.

phology occurs. A typical example is from Little Soda Lake, Churchill County, Nevada, an isolated water body in the central part of the Lahontan Basin: *G. b. obesa*, 158 young to adult with 12–20 rakers; *G. b. pectinifer*, 13 young to half-grown with 30–34 rakers; and intergrades (*G. b. obesa* × *pectinifer*), 86 young to adult with 20–29 rakers. Additional evidence for intergradation or hybridization of the two forms is given by Hubbs et al. (1974), and there remains much additional unpublished data to support this view, which was adopted in the most recent treatment of Nevada fishes (Deacon and Williams 1984). Contrasting views—that "*obesa*" and "*pectinifer*" are valid species or that "*pectinifer*" is not worthy of taxonomic recognition at any level—have been given by Hopkirk and Behnke (1966) and by LaRivers and Trelease (1952), respectively.

Later Studies in Hotter Places

Some accounting of a 1950 trip by R. R. and F. H. Miller and H. E. Winn ends our narrative. This expedition was designed to survey the fish fauna of the lower Colorado River basin in Arizona, adjacent parts of California, and Baja California and Sonora, Mexico. Early in the trip we collected in the Tularosa Basin, New Mexico, and spent three days in northern Chihuahua, Mexico. Our fieldwork, from 4 March to 18 June, resulted in 126 collections totaling 35,378 cataloged specimens. Since almost all earlier expeditions to the West from the Museum of Zoology were during the hot summer months, long after most fishes had spawned, special efforts were made to obtain early life history stages of as many of the native species as possible. This resulted in the first contribution to such biological knowledge for six minnows and six suckers in the Colorado River basin (Winn and Miller 1954). Another major aim of this trip was to evaluate the declining status of native fishes and to determine reasons for their decimation and extinction (R. R. Miller 1961). The role of humans in establishing exotic species in the lower Colorado basin by release of bait fishes was also largely determined during this survey (R. R. Miller 1952). This proved to be only the tip of the iceberg, for the Colorado River fish fauna has since become overwhelmed by exotics (Minckley 1973, 1982).

The total budget for three and a half months of work and 22,830 km of travel, including car rental and salary for the assistant, was $1016.65. Noteworthy discoveries were the first and last collection of desert pupfish (*Cyprinodon macularius*) in the Río San Pedro in Mexico since its description from that area (likely in Arizona) in 1851; first and last records for loach minnow (*Tiaroga cobitis*) from Mexico (R. R. Miller and Winn 1951) and the next-to-the-last record of spikedace (*Meda fulgida*) from the upper San Pedro River (Fig. 2-13) in Arizona; two new species of trouts (R. R. Miller 1950, 1972b); collection of 6100 newborn razorback suckers (*Xyrauchen texanus*; W. F. Sigler and Miller 1963); a new subspecies of *Cyprinodon* (R. R. Miller and Fuiman 1987) from southern Arizona; the last wild Colorado squawfish (*Ptychocheilus lucius*) from the Gila River basin (the Salt River record of Branson et al. [1966] is *Gila robusta*); a new species of spiny-rayed minnow, the Virgin spinedace (*Lepidomeda mollispinis*; R. R. Miller and Hubbs 1960); and observations and data on morphological responses of cyprinids and catostomids to strong currents (R. R. Miller and Webb 1986). Elongated, large-finned, swift-water types are found in torrential, canyon-bound rivers, whereas their thick-bodied ("chubby") counterparts inhabit the quieter tributary waters.

The following descriptions of parts of the trip are from Fran's typewritten account (labeled "Life in a Carryall"):

Fig. 2-13. San Pedro River (field no. M50-55) just south of Charleston, Cochise County, Arizona. Note cut bank at right, the top of which represents the original (early 1800s) valley level. Arroyo cutting, begun about 1880 in the South-west, was probably due to a combination of severe overgrazing and trampling by livestock and an arid weather cycle (Hastings and Turner 1965). Photograph by the Miller party, 24 April 1950.

Everyone was in gay spirits as we left the subzero, snow-covered north and headed south and west to sunny Arizona. This was the start of our four months of travelling and living in a Chevrolet Suburban Carryall, piled high with camping gear and scientific equipment. "Doc" headed our small expedition. Howard (H.) was along as assistant, and a mighty good one he was. I was chief cook and bottle-washer, and occasional puller-of-the-seine. Our two small children Francie [now, heaven forbid, an ichthyologist's wife, which her mother vowed she would never be] and Gifford [now associate professor of geology and director of the Center for Geochronological Research at the University of Colorado] came along for the ride and they loved every minute of the trip.

These Carryalls have an amazing capacity. We started off with clothing for four months of camping, a two-week's food supply, dishes, sleeping bags, two-burner gasoline stove, plus collecting equipment to catch and preserve these elusive fish. On top of all this we placed a two-by-five-foot wooden platform, on which the children sat, played, fought, and often slept. It was a major operation to load and unload every morning and night. No matter how the Carryall was packed, what was needed at the moment was usually at the bottom of the load!

The search was for any remaining native fish in southern Arizona, and desiccated adjoining parts of Mexico. The country was so dry and the water so hard to find that many hours were spent questioning local inhabitants. Our maps told us where water *might* be found. But finding it was not an easy matter. When Doc asked one rancher, "Are there any little fish in that river?" he answered. "Fish?, why, it's dry as a bone. I have a ranch in that valley and there's not a drop of water anywhere." At the slight show of disbelief on my husband's face, the man said, "Go over and see for yourself. You're welcome to stop at the ranch, but there isn't any water in the river." This was a good indication of their attitude toward us: Crazy, but you are welcome to everything we've got, nevertheless. After three hours and a very rough trail later, we rounded a knoll and there it was: a lovely little creek. Doc and H. went hiking upstream, loaded down with nets and formalin in a pail to preserve what fish they might catch.

The children and I decided to take a much-needed bath at this remote spot. As we washed, a strong wind came up and the water began to evaporate right in front of our eyes and our soapy bodies. By the time the last toe was rid of Arizona dust, there wasn't a drop of water left. The rancher was correct now! But he was wrong about the fish. Not one, but three kinds rewarded the trek by Doc and H. up the box canyon. The hospitality of the people of Arizona, like that in other western states, is wonderful. Often when Doc went in to talk to a rancher about the possibility of fishes in the area, they asked me to come in, too. Many a wonderful cup of coffee we had that way. And many tall fish tales resulted.

One additional objective of the 1950 trip was to try to verify or refute a hot spring record of 128°F (53.3°C) for the pupfish *Lucania browni* (Jordan and Richardson 1907; R. R. Miller 1949; = *Cyprinodon macularius*; R. R. Miller 1943a), obtained in the Laguna Salada basin of Baja California del Norte (see photograph in Bailey 1951). A record of the cichlid *Tilapia grahami* from Lake Magadi, Kenya, in water varying from 80° to 120°F (Norman 1931) was later revised to 80°–112°F (Norman and Greenwood 1963), and finally to 70°–104°F (21.1°–40°C) by Norman and Greenwood (1975). In this connection, information from L. C. Beadle (pers. comm. to Miller, September 1962) about this species stated that the pH of the hot springs was about 11.0, and temperatures varied to

42°C. Such a temperature is in line with research conducted on natural temperature tolerances (ca. 45°C) for fishes in North America (R. R. Miller 1981, and see below).

On 14 March 1950 Miller flew with Carl Hubbs and G. E. Kirkpatrick (pilot) in a Super-Cub from La Jolla, California, into the Laguna Salada basin, where we spent about four hours checking for open water around the entire (then dry) lake basin. This was the third and final attempt to locate the hot spring and its fish. The first attempt was made on 9 June 1939 when Miller and John Davis reached the hot spring and found the highest temperature to be 44.4°C and the locale markedly changed from a description given by the original collectors. Thus we thought we had not found the right locality. The second attempt was carried out on 6–7 January 1940, when R. G. Miller (R. R. Miller's father) was led to the spring by Manuel Demara (Fig. 2-14A), a longtime resident in the region who knew the hot spring and its little fish. It was the same spot visited earlier by Miller and Davis. In 1950, our thorough one-week aerial and ground survey showed that the only spring from which *Lucania browni* could have been collected was this one, since it is the only hot spring on the east side of Laguna Salada, lying near the base of the Sierra de los Cucapas (also spelled Cucopas, Cucapahs, Cucopahs, and Cocopahs). The locality is 20.5 km by road along the eastern edge of Laguna Salada from its northern end, at the base of the sierra.

Some twelve to eighteen hot-spring seepages radiated down onto the playa to form a shallow sheet of water (Fig. 2-14B). In a small pool in the course of one spring outflow the temperature was 36.1°C, but only 30 cm away beneath an unbroken green algae and peat muck, and just 2.5 cm below the surface, the temperature was only 27.5°C. The highest temperature in any of the sources was very near 42.8°C (44.4°C in 1939). Evidently the springs have cooled gradually since 1927–

A

B

Fig. 2-14. (A) Manuel Demara at Laguna Salada, Baja California del Norte, Mexico. He helped guide R. R. Miller and party in their search for the hot-spring pupfish. Also shown (B) is the area for the highest thermal record for any known pupfish at the northeastern end of Laguna Salada, Baja California del Norte. Note that the spring outflow (center) lies well below the highwater mark of the lake; this locale was covered by more than 10 m of water by floods in 1983. Photographs by R. R. Miller, 20 March 1950.

1928, when their temperatures varied from 44.4° to 53.3°C (Kniffen 1932); I was assured by Stanley Sykes, who took the original temperatures with standardized instruments, that the water he checked (using several thermometers) was 53.3°C. However, he was unable to say that this temperature was taken at the level where the fish were living. Obviously it was not. M. L. Smith and Chernoff (1981) and Minckley and Minckley (1986) recorded *Cyprinodon pachycephalus* consistently swimming and reproducing in 43.8°C water at San Diego, Chihuahua, Mexico. Thus ends the saga of the remarkable hot-spring fish.

Conclusion

Not until after the end of World War II, when burgeoning human populations began to overflow into the North American deserts, did the biological integrity of desert wildlife begin to be seriously threatened on a broad scale (Sears 1989). When Miller initiated his studies on the pupfish genus *Cyprinodon* in the Mohave Desert–Death Valley–Ash Meadows region in 1934, there were but tiny oases of human invasion, and most aquatic ecosystems were still intact. By the 1970s, however, vast areas of the desert had become dotted with human habitations and cultural trappings. The once-clear desert air was already invaded by smog and other forms of pollution from major cities and industrial chimneys (as at Four Corners) in the American Southwest, even into Death Valley between 1967 and 1971 (smog was first observed in Death Valley in 1971 but was not seen there in 1967 during a survey of the fishes; R. R. Miller 1969a).

The clarity of the crystal-clear air once produced amazing vistas. From the summit of any given mountain range one could easily see for 120 to 160 km. In December 1976, while on a flight from Detroit, Michigan, to Tucson, Arizona, the pilot, on nearing our destination, announced visibility as excellent at 16 km! In July 1934, when Miller flew from Phoenix to Tucson, he could see the Santa Catalina Mountains just north of Tucson, which stood out clearly more than 160 km distant. The present generation thus lacks the historical perspective to evaluate what pristine conditions were like.

New wells, new irrigation ditches, poorly conceived ranching and farming ventures, swimming pools, fancy resorts, air-conditioned homes, and, to be sure, a good many shacks appeared on desert lands out of nowhere. All this activity does not come without cost; one serious price of the so-called miracle on the desert is the inexorable lowering of water tables, the draining away of irreplaceable fossil water. We saw this in subsequent fieldwork of the 1950s, 1960s, and 1970s. Fish populations that were still abundant and relatively undisturbed during our surveys have declined drastically or disappeared completely, such as those in the pluvial White River valley of eastern Nevada (Courtenay et al. 1985) and elsewhere (Deacon 1979).

A matter of greater concern to many has been the rapid disappearance of solitude and privacy, long the primary attributes of deserts. The extreme ecological impacts of off-road vehicles have shattered these arid lands, especially in the Mohave and Sonoran deserts. Only in recent years have conservationists succeeded in convincing managers of desert lands such as the Bureau of Land Management that it is sensible and necessary to establish good zoning principles and declare substantial desert areas as wilderness. In 1986 Great Basin National Park (smaller than it should be to establish its raison d'être) was set up in eastern Nevada (*National Geographic Magazine* 175: 21–75, January 1989).

There are sound biological reasons for holding samples of virgin country in perpetuity. In time, most of the desert will be used by humans for one purpose or another—irrigation, grazing, mining, or playgrounds—but, inevitably, scientists will need to know the original situation. A control is a basic part of every scientific experiment. Retention of wilderness for the maintenance of biological diversity is a necessity for the survival of humans themselves (Wilson 1988). What we must avoid is a domesticated, homogeneous earth; for many, it would be a far less fascinating place in which to live.

Many changes have occurred since onset of the Hubbs-Miller field studies. Between 1934 and 1942 few fishery scientists focused their studies on Great Basin fishes and those of the hotter southern deserts. Virtually all of the workers were from the University of Michigan—the Hubbs-Miller family group. Presently, more research groups have a major interest in western fishes than individuals (regardless of age) who participated in those earlier studies. Our information pool has thus increased proportionally, but the factors threatening these fishes have increased exponentially. Unless we take better care of our regional deserts, nothing will remain but a barren terrain like the largely human-made desert that now stretches uninterrupted from the Atlantic shore of North Africa to the Thar (or Great Indian) Desert of western India (Axelrod 1983).

We are fortunate to have seen the Great Basin and more southern desert lands prior to major perturbations. On occasion, that insight may help resolve (successfully for the biota, we hope) many conflicts. Our fieldwork was enjoyable and our recollections are positive.

SECTION II

Spirals Toward Extinction:
Actions and Reactions

Prior to the 1960s, only a few specialists were alarmed by mankind's clearly detrimental effects on aquatic habitats and fishes in the West. Water development was the only answer to "opening up" the deserts for human use, and early projects enjoyed success in that regard. Derby Dam on the Truckee River, Nevada (1903), Roosevelt Dam on the Salt River, Arizona (1913), and Hoover Dam on the Colorado River (1935) had operated for decades, with tangible benefits varying from flood control and reliable irrigation supplies to abundant hydroelectric power to valuable recreational resources.

It was the last factor that stimulated the poisoning of the Green River in Utah-Wyoming in 1962. The controversy surrounding that operation, which had impacts on fisheries management that persist to this day, is treated in chapter 3. Its intent was to remove "undesirable" fishes and replace them with trout that sportsmen could catch in the newly built Flaming Gorge Reservoir. Included in the "undesirable" category were a number of species now listed as endangered. In theory, at least, such a poisoning will not occur again; by law, such actions must now be evaluated before and after the fact. Consideration must be given to all resources rather than just single species or interests, and endangered species have become part of the guiding protocol for exploitation of water and other natural resources.

Shortly after the Green River incident, land development began to affect native fishes in the Death Valley region. The reaction of conservation-oriented people was swift and decisive, and a diverse group of dedicated individuals formed what was later formalized as the Desert Fishes Council. Their goal was to fight for perpetuation of these animals, whose existence depended on a few hundred liters of water in a few desert springs. Chapter 4 details development of this council, delineates its activities to ensure the continuation of fishes of the desert West, and outlines some of its other accomplishments in promoting communication, education, and conservation of regional, national, and international aquatic resources.

Details on a principal in the Death Valley controversy, the Devils Hole pupfish (*Cyprinodon diabolis*), including a description of its plight, reactions of individuals and agencies to its endangerment, and actions by which the species was saved from extinction, are provided in chapter 5. Litigation over the potential drying of Devil's Hole, Nevada, a part of Death Valley National Monument and the only habitat of this pupfish, went all the way to the Supreme Court of the United States, where its fate was decided in the affirmative. The legacy of this tiny fish in its isolated desert spring, the remarkable and complex sequence of events surrounding its salvation, and the significance of this single species and efforts in its behalf to the conservation movement form the conclusion.

Participation in the Green River poisoning operation as an observer with her husband, W. Jackson Davis, then of Western Michigan University, stimulated Jane Davis to produce this card for Christmas 1962. Part of the text accompanying the artwork was as follows:

> To all of the usual Christmas Greetings and Good Wishes, we add our own special hope that the wilderness creatures of the earth be allowed to multiply in their season, and that their world may persist.

> We could not entertain a more heartfelt wish than that all of you will be touched in some way by the places and lives which lie beyond the boundaries of human habitation, and that those places and lives will continue to exist for the refreshment of your children and your children's children.

> [signed] Jane and Jack Davis

Chapter 3

Ghosts of the Green River: Impacts of Green River Poisoning on Management of Native Fishes

Paul B. Holden

Introduction

In the late 1950s, managing native nongame fish was unheard of in the American West, even among fishery professionals. Fish managers in the western states were in a position very different from their counterparts in the East, Midwest, and South; they had far fewer native fishes to work with. In addition, salmonids, primarily trout in inland areas, were the only native fishes worth managing, at least in the view of the fishing public. Unlike fishery managers in other parts of the country, where centrarchids, ictalurids, percids, and other families provide abundant and diverse sport and commercial opportunities, the western manager was stuck with trouts, minnows, and suckers.

Water development was the practice, and at times the religion, of ranching and agricultural political forces in the West (Reisner 1986). Along with motherhood and apple pie, water development was "American," and pity the poor soul who thought otherwise. Consequently, another western phenomenon faced by fishery managers was the artificial lake, or reservoir.

Along with reservoirs came a new experience. Many local "trash" fishes found them ideal homes, to the dismay of biologists attempting to produce salmonids for the angling public. While some problem fishes were native western cyprinids and catostomids, many were introduced species such as common carp (*Cyprinus carpio*) and yellow perch (*Perca flavescens*). What does one do with a lake full of chubs, suckers, carp, and perch that are eating everything in sight and outcompeting the favored trout? The answer soon became "poison them."

Poisoning fishes with rotenone, a derivative of a number of plants found in South America and elsewhere, was a well-established practice in the fishery profession (Krumholz 1948; Rinne and Turner, *this volume*, chap. 14). It was used to reduce or eliminate fish populations in a variety of habitats, usually manmade systems or those taken over by nonnative species. It was also used as a sampling tool, especially in streams and small lake embayments. The practice of fish poisoning was certainly well known, and it became a logical choice of the western fish manager to reduce trash fish in reservoirs. They seemed undiscouraged to discover that after each operation the target fish rebounded in a few short years, eventually taking over the habitat once again and leading to another round of eradication followed by restocking of salmonids.

This was not the way it began. Native Americans and early settlers found many of the nonsalmonid native species useful (Cope and Yarrow 1875; D. S. Jordan 1891; Rostlund 1952; C. L. Hubbs and Miller 1953). Lakesuckers

43

(fishes of the genera *Chasmistes* and *Deltistes*) were staple foods of some Indians (Scoppettone and Vinyard, *this volume*, chap. 18). Later, commercial fisheries for suckers, including the presently endangered June sucker (*Chasmistes liorus*), were located in Utah Lake near Provo, Utah, in the middle to late 1800s (Carter 1969). The razorback sucker (*Xyrauchen texanus*), another rare fish, was used by lower Colorado River Indians as food and was later marketed in parts of Arizona, where it was commercially fished until the 1940s (C. L. Hubbs and Miller 1953). For one reason or another, however, these native fishes never sustained the commercial or sporting interest of the public, and by the mid-1900s they were no longer prized.

This, then, was the legacy that precipitated the events on the upper Green River. The U.S. Bureau of Reclamation (USBR) planned a massive new reservoir, Flaming Gorge, one of several authorized under the Colorado River Storage Project Act of 1956. The reservoir was to be 145 km long, about 160 km² in area (Utah Department of Fish and Game [UTDFG] 1959), and was supposed to provide excellent habitat for rainbow trout (*Oncorhynchus mykiss*), the principal sport fish stocked in cold-water reservoirs of that region. Excellent sport fishing and a related economic boom accompanying it were forecast by the involved agencies. To ensure that the trout had a head start on trash fish, the Wyoming and Utah fish and game departments decided it was necessary to poison the Green River before the new dam was closed. This paper discusses what happened on the Green River, why it happened, and why it helped ensure that a similar project might never happen again.

Rationale for the Project

In anticipation of construction of Flaming Gorge Reservoir, the two state agencies began planning for the fishery in 1957. The UTDFG conducted preimpoundment surveys of the reservoir basin and the Green River to the Colorado border, 46.5 km below the dam site (Fig. 3-1), in 1959 (McDonald and Dotson 1960). The Wyoming Game and Fish Department (WYGFD) conducted a similar study from the uppermost Green River and some of its tributaries to near the dam site from 1958 to 1960 (Bosley 1960). This study covered a large area because Wyoming was also planning for the USBR Seedskadee Project, which included construction of Fontenelle Reservoir on the upper Green River, to be built about the same time as Flaming Gorge. The swift, deep, and turbid flows of the Green River proved difficult to sample (McDonald and Dotson 1960). Gill netting was the primary sampling method for both studies, but electrofishing, explosives, fyke nets, spot-poisoning with rotenone, and seining were also used.

University of Utah personnel also surveyed the proposed reservoir area (Gaufin et al. 1960; G. R. Smith 1960) in conjunction with the McDonald and Dotson study. The same collections were discussed in both reports, although discrepancies existed in the numbers and species of fishes reported. Taxonomic differences pivoted around identification of chubs of the genus *Gila*.

In the 1950s, two forms of bonytails (a common name then generally applied to Colorado River chubs) were taxonomically recognized as subspecies—roundtail chub (*Gila r. robusta*) and bonytail (*G. r. elegans*). A third form, the humpback chub (*G. cypha*), had only recently been described (R. R. Miller 1946b) and did not yet enjoy general acceptance as a valid taxon. Thus, McDonald and Dotson (1960) reported collections from the Green River in Hideout Canyon in July 1959 that included forty-six "bonytails," but Smith (1960) reported these same collections to include thirteen humpback chubs and thirty-six bonytails. Smith also reported two other

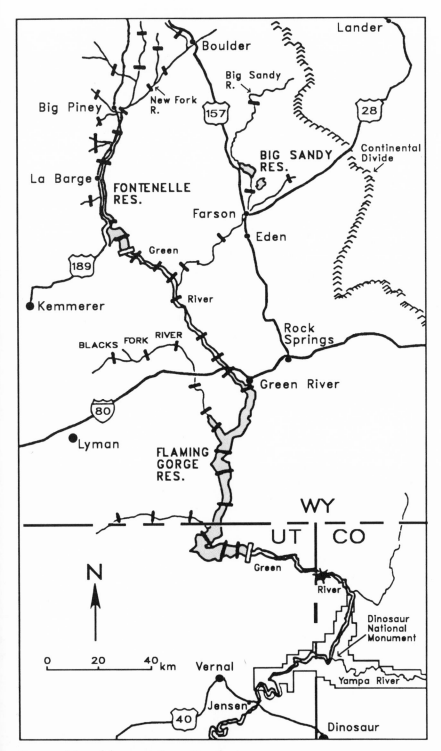

Fig. 3-1. Map of the upper Green River basin,
Colorado-Utah-Wyoming, showing location of
rotenone drip stations (bars) and detoxification
station (asterisk) used in the 1962 poisoning
operation.

humpback chubs and three bonytails from another site in the same general area. Bosley (1960) recorded all the chubs he collected as "bonytails," to which he applied the name *G. r. robusta*, even though he mentioned that "there appears to be change in the physical characteristics of this fish in the extreme lower section of the study area" and presented photographs that clearly included all three forms. McDonald and Dotson (1960) also acknowledged that several morphological variants existed, yet these researchers grouped all the fish into a single taxon since they, and many other state biologists, could not see the clearcut differences suggested by taxonomists.

Several taxonomic studies since that time have recognized roundtail (*G. robusta*), bonytail (*G. elegans*), and humpback (*G. cypha*) chubs as full species (Holden and Stalnaker 1970; Suttkus and Clemmer 1977; G. R. Smith et al. 1979). These studies have not yet resolved the taxonomic status to the satisfaction of many biologists and agencies, and their identification continues to be a problem (Tyus et al. 1987; M. E. Douglas et al. 1989). Most concern today revolves around whether or not the humpback chub has hybridized with one of the other two, or both, rather than its status as a separate species. As we shall see later, the presence of humpback chubs became an important point in controversies over the Green River project.

Both of the state-sponsored studies demonstrated that the Green River in the area of concern contained primarily what were then called "rough," or "trash," fishes, mostly native cyprinids and catostomids. The studies also found numerous introduced species, especially common carp, channel catfish (*Ictalurus punctatus*), redside shiner (*Richardsonius balteatus*), and yellow perch. Of notable interest was the fact that only a few Colorado squawfish (*Ptychocheilus lucius*) and razorback suckers were taken, and no humpback chubs were recognized, but a fair number of "bonytails" were recorded.

The squawfish, razorback sucker, humpback chub, and bonytail were the four large-river, endemic species that elicited much of the controversy over the proposed project. In the early 1960s they were not protected by any governmental entity. At present, all but the razorback sucker are listed as endangered by the U.S. Department of the Interior (USDI) (U.S. Fish and Wildlife Service [USFWS] 1989c), and the razorback sucker has been proposed for listing (USFWS 1990). Furthermore, all four are protected by the Colorado River basin states in which they naturally occur (J. E. Johnson 1987a); all are considered extinct in Wyoming (Baxter and Simon 1970).

As noted above, the same samples recorded in the Utah and Wyoming studies as bonytails were identified by G. R. Smith (1960) as including fifteen humpback chubs from Hideout Canyon, a few miles upstream from the Flaming Gorge dam site. Binns (1967a), who reviewed earlier agency studies and conducted pre- and postimpoundment studies from 1962 to 1964 for the WYGFD, did not cite or report this discrepancy, although he did reidentify one fish collected by Bosley (1960) as a humpback chub.

This situation exemplifies a fairly common phenomenon of the time. Many fishery biologists either could not or would not identify the more difficult nongame fishes to species, especially closely related minnows and suckers, or else they misidentified them. Because these species had little prominence in management decisions, they were often identified only to family even though a multitude of kinds were represented. Taxonomic concerns were the realm of taxonomists, usually housed in universities rather than fish and game departments. In this instance, however, biologists involved with the two state-sponsored studies attempted to identify all the fishes collected to species, and even to subspecies at times (Bosley 1960), and apparently they did an excellent job of identifying difficult taxa except for the three forms of *Gila*. Another criticism of the preimpoundment investigations, raised later, was a failure to sample intensively for small fishes. Although seines were available, they apparently were seldom used. This may relate to the discussion above; identification of small, hard-to-determine, "unimportant" species was not a standard practice for management biologists.

Both state-sponsored studies recommended that the river system be treated with rotenone to rid the new reservoir of rough fish, at least initially, in order to allow planted salmonids a chance to become established. It appears, however, that this conclusion had essentially been reached by the respective departments well before preimpoundment studies were completed. Binns et al. (1963) stated, "A rehabilitation program was informally discussed as a possible step . . . as early as 1957." Very little justification for poisoning was presented, except the broad generalization that populations of undesirable species would "explode" in the new reservoir. A summary report (Binns et al. 1963) that noted common carp as the main concern more clearly stated the reasons for the project after it was completed. Boysen Reservoir in Wyoming was represented as an example of a system in which carp became dominant because poison had not been used, but it was further noted that they did not become abundant until two years after the reservoir was filled, and that a good trout fishery existed during the intervening period. Two other Wyoming examples, Glendo and Pathfinder reservoirs, where poisoning had removed rough fish and provided for excellent trout fisheries, were also cited.

Therefore, biological justification for the massive project primarily involved precon-ceived notions about what would happen in the reservoir. No ecological projections were attempted to estimate which fish could become detrimental to the stocked trout, or what the real chances were for such an event. No apparent thought was given to the fate of native fishes, because they had no value to sportsmen. But that was a prevailing attitude of the fishery profession at the time.

The project was also justified on economic grounds. Several reports and memoranda discussed projected use of the new reservoir by fishermen and potential economic benefits of that use (D. Andriano, UTDFG [letter to R. H. Stroud, Sport Fishing Institute], 26 January 1962; Binns et al. 1963; UTDFG 1959; USDI 1963). In almost every case the entire economic benefit was related to the fish eradication program. Nowhere were estimates made of economic effects of not poisoning. It appears reasonable to assume that some sort of a trout fishery, perhaps a very good one, could have existed without poisoning. This fishery would have generated economic benefits perhaps nearly as great as those estimated for the fishery after poisoning. For some reason that potential apparently was never discussed, an indication of deep-seated convictions that poisoning had to be conducted, and that the possibility of not doing so had been discarded several years earlier.

The Opposition

Opposition began to emerge soon after the proposed project was made public. The primary opponent was Robert Rush Miller of the University of Michigan Museum of Zoology, who had been studying the Colorado River system and its fishes for nearly twenty-five years, carrying on some of the earlier investigations of his mentor and father-in-law, Carl Leavitt Hubbs. Both men were respected ichthyologists with considerable knowledge of and appreciation for native fishes. Miller's research was of particular interest; he was assembling information on the demise of native fishes in the Colorado River system. His publication (Miller 1961), "Man and the Changing Fish Fauna of the American Southwest," documented major losses sustained by Colorado squawfish, bonytail, and razorback sucker, among others. He had recently described the humpback chub (Miller 1946b) and was attempting to locate populations and obtain specimens to clarify its taxonomic position. Miller, more than anyone, knew what the loss of additional native fishes meant.

The major organized opposition came from the American Society of Ichthyologists and Herpetologists (ASIH). A resolution from their 1961 meeting that opposed the project because of concern about losses of rare native fishes aroused additional concern from both scientists and politicians. At the same time, it tended to solidify the resolve of involved state agencies and their supporters to proceed. The USFWS (then the Bureau of Sport Fisheries and Wildlife [USBSFW]), a participant because federal funding was used, became more involved. The Sport Fishing Institute published supportive articles (R. H. Stroud, Sport Fishing Institute [letter to D. Andriano, UTDFG], 30 January 1962) and promised more support if needed. The opposing sides thus became better defined, and the controversy over the project became known over much of the nation.

Although he was concerned about all the native species in the river, it is apparent from letters and comments that Miller was especially alarmed for the humpback chub. The location of the largest known concentration of this rare species in Hideout Canyon, an area to be poisoned and then inundated by the reservoir, was a major point that the agencies could not deny; nor did they acknowledge it.

Flaming Gorge Dam was planned for completion in 1962, and poisoning was originally scheduled to coincide with its closing, thus holding the rotenone until it detoxified. Late in 1961 it became obvious that the dam would not be completed until late 1962, at a time of year when cold water would reduce the effectiveness of the toxicant (Binns et al. 1963). It was therefore necessary to conduct the operation before the dam closed, which presented a major problem—the poison could potentially move 40 km downstream into Dinosaur National Monument (NM). This forced the agencies to consider detoxification, a relatively untried approach, especially in a stream as large and complex as the Green River. The opposition seized this opportunity to point out potential dangers to the national park system, and the National Park Service (USNPS) quickly became more involved.

After considerable investigation and debate, the states decided to employ a detoxification program designed to neutralize rotenone below the dam site. The detoxification station would be placed at a bridge in Brown's Park, just south of the Utah-Colorado border (Fig. 3-1), but only 26 km upstream from the boundary of Dinosaur NM (Binns et al. 1963). This site was chosen after extensive study and research into detoxification. One reason for choosing a location so far below the last planned rotenone station at the dam site was to allow for natural oxidation of the rotenone, thereby reducing the need for detoxification. Potassium permanganate ($KMnO_4$) was the selected oxidant, and several studies were conducted to identify application methods and dosage (Binns et al. 1963).

The controversy engendered additional sampling (especially within Dinosaur NM) in 1962 by the USFWS (Azevedo 1962a, b, c) and USNPS (Hagen and Banks 1963). These efforts concentrated on the four large-river species, with emphasis on humpback chub. The state of Wyoming initiated a Federal Aid to Fisheries study of impacts of the poisoning operation, scheduled to run from 1962 to 1964 (Binns 1967a). Opposition to the project had caused substantial consternation within various federal and state agencies and stimulated additional efforts to ensure the protection of rare native fishes.

In June 1962 upper-level personnel from the USFWS and the Sport Fishing Institute met with the ASIH to discuss the project. Carl Hubbs was spokesman for the society and pointed out their general concern with use of rotenone on such a massive scale. He also pointed out that poisoning could create conditions for rapid expansion of undesirable resistant species such as common carp and bullheads (*Ameiurus* spp.) (R. E. Johnson, Division of Sport Fisheries, USBSFW [meeting notes on meeting with ASIH, 14 June, Washington], 19 June 1962). He suggested that the agencies try to protect species such as humpback chub by transferring them to suitable unpoisoned areas, and that dead specimens of rare species be salvaged during the operation and preserved. The USFWS pointed out "that encroachments of civilization's demand on fish environments, the modifications of flowing water for power, irrigation, navigation, flood control, and the changes in water quality brought about by many forms of pollution were soon going to renovate our fish populations for us." The meeting provided both sides with the opportunity to discuss their basic differences and reach a common accord on some points, but major differences remained in their philosophies concerning native fish management. The ASIH, composed primarily of academic biologists, considered native fish management an important issue. The agencies, composed of management biologists, were trying to manage man-made environments to benefit the angling public. Native fishes did not fit into the latter scenario.

In addition to the planned poisoning, another battle emerged from discussions surrounding the project. This involved the native chubs, especially the humpback chub, and whether it was in fact rare, or even a valid species. Although scarcely mentioned in written reports, interagency correspondence and papers presented at meetings clearly suggest agency concern about this issue. Following sampling in 1961 and 1962, agency letters and memoranda attempted to show that humpback chub were common in Dinosaur NM (D. Andriano, UTDFG [letter to R. H. Stroud, Sport Fishing Institute], 26 January 1962; J. E. Hemphill, Fisheries Service Branch, USBSFW [memorandum to regional director, Region 2, USBSFW], 26 July 1962), but when the studies were completed, only two specimens identified as humpback chubs had been collected (Azevedo 1962c). The two fish were later reidentified as bonytails (Vanicek et al. 1970). The agencies still did not acknowledge that at least some fish caught in 1959 in Hideout Canyon were humpback chubs, and although the misidentification in 1962 was undoubtedly an honest mistake, it emphasizes the continuing confusion over identification of the chubs.

The Operation

Rotenone application began at 0800 on 4 September 1962 at upstream stations on the Green and upper New Fork rivers (Fig. 3-1). Downstream stations were programmed to begin releasing rotenone based on computed time of travel so that the entire area was covered in one long, sweeping treatment. About 715 km of the Green River and its tributaries were treated using fifty-five drip stations. The operation lasted three days, used 81,350 l of rotenone, and required more than a hundred people, numerous vehicles, airboats, a helicopter, and a remarkable amount of logistic preparation and support (Binns et al. 1963). The enormous amount of preparation and planning paid off in a well-run and relatively effective program. No major problems occurred with the poisoning portion of the project, and the goal of eliminating fish in the river was generally achieved.

Unfortunately, such was not true for the planned detoxification. Measurements of rotenone concentrations above the dam site indicated that the toxicant was reaching far higher concentrations than predicted. This meant detoxification would take longer and require more oxidant than anticipated. Rotenone application at remaining stations was reduced, all available $KMnO_4$ was located and purchased, and amounts applied were reduced to the minimum required as determined by in situ tests with live fish.

Rotenone was first detected at the detoxification station on 8 September, and $KMnO_4$ application was initiated within two hours. Application lasted for eighty-three hours and used 7800 kg of $KMnO_4$, essentially all that was locally available. The amount of detoxicant was continuously adjusted as apparent concentrations of rotenone changed, although some lag occurred between rotenone concentration detection and the application of appropriate detoxicant levels. Other problems that plagued detoxification included inclement weather, difficulty in applying the crystalline agent, declining river discharges, and analytical inaccuracies in determining rotenone concentrations (Binns et al. 1963). The result of these factors was incomplete detoxification. All the $KMnO_4$ had been used by late in the day on 11 September.

Binns et al. (1963) suggested that detoxification worked for 95% of the operation, and that rotenone concentrations during the remainder of the time were very low. Nonetheless, dead and dying fishes were reported by USNPS personnel in Dinosaur NM on 13, 14, and 15 September. Rotenone probably reached the mouth of the Yampa River early on 13 September and the lower end of Split

Mountain Canyon on the evening of 14 September. USFWS and UTDFG biologists arrived on the fifteenth and sampled at Split Mountain (USDI 1963). Many live fishes were found, but there were many dead fishes along the shorelines as well. Additional sampling was conducted by state agencies and the USFWS (Azevedo 1962a, b, c) later in 1962 and into 1963 to assess effects on fish populations in the monument.

Charges and Countercharges

What occurred in the following days and months could be described as a nightmare for both proponents and opponents of the Green River project. The major bone of contention was the level of impact on the fishes in Dinosaur NM. The USNPS and R. R. Miller immediately charged that the project had caused major fish kills due to inadequacies of the detoxification program (J. C. Gatlin, USBSFW, Albuquerque [memorandum to regional director, USNPS, Omaha], 20 December 1962; R. E. Johnson, Division of Sport Fisheries, USBSFW [letter to R. R. Miller], 4 December 1962; R. R. Miller [letter to Anthony Smith, National Parks Association, Washington], 26 November 1962 [letter to Rachel Carson], 19 December 1962). The state agencies and the USFWS countered by pointing out that post-treatment studies indicated good populations of most species of concern, and that habitat changes caused by the dam would certainly alter native fish populations anyway. This did little to slow the campaign to report the operation as a disaster. By early 1963, Miller and other ASIH members had taken the matter all the way to the U.S. Congress, and pressures were being applied at high levels within the USFWS (I. LaRivers, University of Nevada, Reno [letter to Senator A. Bible], 29 November 1962; A. Bible, U.S. Senate [letter to D. H. Janzen, director, USBSFW], 20 December 1962). Claims by the involved agencies that detoxification was a success, as well as the inability to find any humpback chubs during posttreatment sampling, were major concerns of the Washington office of the USFWS (W. King, Fisheries Management Services, USBSFW [letter to regional director, Region 2, USBSFW], 2 January 1963).

On 25 March 1963 Secretary of the Interior Stewart L. Udall released a directive dealing with the project (Udall 1963). Although he neither favored nor opposed the operation, he directed the following:

> That adequate research be undertaken on the effects of rotenone, potassium permanganate, or other fish controlling agents, under varying environmental conditions, before additional management programs are undertaken, and that when such programs are carried out, research results are applied in a way that is relevant.
>
> Whenever there is a question of danger to a unique species, the potential loss to the pool of genes of living material is of such significance that this must be a dominant consideration in evaluating the advisability of the total project.
>
> I am taking measures to assure that future projects are reviewed to assure that experimental work is taken into consideration, and that possible deleterious effects are evaluated by competent and disinterested parties.
>
> As a follow-up of last September's operation on the Green River, I am asking the National Park Service and the Bureau of Sport Fisheries and Wildlife to undertake fish population studies this summer in Dinosaur National Monument to determine the extent of species and population impairment. I am also asking them to plan a longer range research project which will assess the changes on habitat and populations in Dinosaur National Monument brought about by the closing of Flaming Gorge Dam.

In May 1963 Miller published a paper charging that the Green River project had gotten "out of control," resulting in "heavy losses" of native fishes in Dinosaur NM (R. R. Miller 1963). He called for consideration of the value of native fishes in management decisions and the use of caution when poisons were used for management purposes. The involved agencies tended to see the negative comments only, however, and their resolve to show that the project was not detrimental increased.

In early June 1963 Congressman Robert E. Jones, chairman of the Natural Resources and Power Subcommittee, wrote Secretary Udall (R. E. Jones, U.S. Congress [letter to Stewart Udall], 13 June 1963) and, using the Miller (1963) paper as a blueprint, questioned the poisoning project and the USFWS role as follows:

> The subcommittee would appreciate your informing us:
>
> 1. When, and by whom, was the Fish and Wildlife Service warned of the dangers of the Green River extermination project?
>
> 2. Why did the Fish and Wildlife Service continue to support this poisoning project?
>
> 3. At what point along the River were efforts made to detoxify the poison?
>
> 4. How far downstream did the poison have killing effects?
>
> 5. When the poison was introduced into the River, 8 miles [12.9 km] above Ashley [= Flaming Gorge] Dam (between September 4 and 8, 1962), was it known that the Dam would not be closed by the time the poison reached the upstream face of the Dam?
>
> 6. What efforts were made, prior to introducing the poison into the River, to determine whether the toxicant could be effectively detoxified at a particular point in the stretch of the River?
>
> 7. Please state the Department's responsibilities (and statutory base) with respect to the preservation of aquatic life in Dinosaur National Monument.
>
> 8. Does the Department believe, in light of results in this poisoning project, that the Department has fully complied with its responsibilities relating to (a) the aquatic life of Dinosaur National Monument and (b) the fishery resources in navigable waters of the United States?

Udall answered with an eight-page letter (S. L. Udall, secretary of the interior [to R. E. Jones, U.S. House of Representatives], 23 July 1963), noting his directive (Udall 1963) and pointing out the following: the warnings of the ASIH; attempts that were made to ensure that no fish were killed in the monument; that major impacts to native fish due to habitat changes caused by the dam and reservoir were to be expected anyway; and that the treatment did not eliminate any species from the monument. In essence, he supported the poisoning operation and the involvement of his department, but it was also obvious that his directive would make it difficult for such an operation to occur in the future.

Additional concern and debate continued into 1963, as did studies to determine what had happened. Studies during that summer located humpback chub and other native fishes in Dinosaur NM (D. R. Franklin, Utah State University [memorandum to superintendent, Dinosaur National Monument], 29 October 1963). Both young and adults of most species were found, strengthening the conclusion that the accidental kill in Dinosaur NM was probably fairly small.

The other point of contention between Miller and the agencies—the humpback chub and its taxonomic validity (R. R. Miller [letter to W. King, USBSFW, Washington], 29 October 1963; W. King, Fisheries Management Services, USBSFW [letter to regional director, USBSFW], 26 June 1963)—continued to stimulate debate. Apparently, some agency people thought the humpback chub might be a male form of the bonytail (see King letter above), and such a revelation could discredit Miller's criticisms of the project. Donald Franklin, leader of the Utah Cooperative Fishery Unit at Utah State University, was appointed to lead studies assessing effects of the dam on fishes in Dinosaur NM, as directed by Secretary Udall. He recommended collecting chubs from throughout the upper Colorado River basin and, using this specimen base, resolving the question once and for all (D. R. Franklin, Utah State University [memorandum to superintendent, Dinosaur National Monument], 29 October 1963). That study was completed in 1968 by myself (Holden and Stalnaker 1970); I wandered into the controversy with no idea of its beginnings.

The Aftermath

Flaming Gorge Reservoir was stocked with rainbow trout and kokanee salmon (*Oncorhynchus nerka*) in 1963. Surveys of the lower Green River in Wyoming in spring 1963 found no fish. By summer, native flannelmouth sucker (*Catostomus latipinnis*) and mountain whitefish (*Prosopium williamsoni*) had recolonized much of the treated area, and common carp were there by 1964. Young-of-year carp and Utah chub (*Gila atraria*) appeared in the reservoir in 1964. Rainbow trout planted in the river above the reservoir did not feed on nongame fishes, which were becoming abundant (Binns 1967a).

Growth of rainbow trout in the reservoir, the upper Green River, and tailwaters of Flaming Gorge Dam was excellent, and a substantial fishery developed. Flaming Gorge was living up to the expectations of the agencies that had planned so long and invested so much in the poisoning operation (Eiserman et al. 1964). Binns (1967a) conducted postimpoundment studies beginning in 1962, and his was the last official report dealing with the poisoning. The very last sentence read: "nongame fish populations in the lower treated area appear to be increasing and may again reach problem status at some time in the future." Little did he know how prophetic that statement would prove to be.

Introduced Utah chub became the major management concern in Flaming Gorge Reservoir by the late 1960s (Schmidt 1979). The species was poorly utilized by rainbow trout, and not only appeared to compete with trout for planktonic foods but established growth records of its own. It is interesting that Schmidt (1979), who was in charge of fish management at Flaming Gorge for the UTDFG (by now the Utah Division of Wildlife Resources), suggested that the Utah chub had had a head start. Although not intended as such, this could be taken as an admission by the very agency that fought so hard and worked so diligently to accomplish it, that the poisoning operation opened up the reservoir for that species!

The words of Miller and Hubbs thus had a ring of truth; poisoning without consideration for ecological consequences caused concerns for both native fishes and non-native salmonids. In recent years, management of Flaming Gorge has deemphasized rainbow trout, as kokanee salmon (a planktivore) and lake trout (*Salvelinus namaycush*, a large piscivore) have attracted most of the fishermen's attention. The lake trout, although never predicted in early plans, has developed a trophy fishery (9–15 kg fish and larger) of world-class dimensions. Common carp, a species targeted in the rotenone operation, never became a problem, mostly because of the deep, cold nature of the reservoir. As should have been obvious during the planning stage, Flaming Gorge is not good habitat for this species. Utah chub populations declined in numbers as the reservoir matured, and other non-native species, such as white suckers (*Catostomus commersoni*), became more abundant. Present management policy considers nongame fishes to be part of the reservoir fauna, and the game fish program has been adjusted accordingly.

Reservoirs in Wyoming, presented as supporting evidence for poisoning Green River, have also changed. Both Boysen and Glendo reservoirs are now cool-water habitats, with walleye (*Stizostedion vitreum*) the principal game species (T. Annear, WYGFD [letter to P. Holden], 12 February 1988). Neither is considered a good trout reservoir. Pathfinder Reservoir remains an excellent rainbow trout fishery. This, along with what happened at Flaming Gorge, indicates that successful management in large reservoirs in the West is likely more dependent on overall ecological conditions than on which fish has a head start.

Although hindsight always has a distinct advantage, it would seem that agencies planning the fishery at Flaming Gorge should have been able to foresee that the reservoir would be poor habitat for common carp. Also, none of the native Colorado River fishes were known to be problems in reservoirs elsewhere, and in fact might have acted as a buffer against population explosions of other nongame species in the early years following impoundment. Unfortunately, this type of ecological thinking was not part of management planning at the time.

Ghosts of the Green River

It is difficult to view the Green River poisoning and draw a direct line to changes in fishery management. Nevertheless, it is evident that this single event attracted so much notoriety and involved so many people from so many walks of life that it could have been a turning point. Miller, Hubbs, and their colleagues continued to push for native and rare fish protection and study, and continued to use the Green River poisoning as an example when making a point about their concerns. State agencies slowly began to manage for native fishes as the Endangered Species Act of 1973 became the law of the land, and nongame biologists were hired and administrative units were established. Ironically, many biologists who worked on the Green River project eventually held positions in which they were responsible for protecting the very species they had once worked to destroy.

During 1962 and 1963, amid confrontation and controversy, the USDI and scientists began discussing lists of rare species (R. E. Johnson, Division of Sports Fisheries, USBSFW [meeting notes on meeting with ASIH, 14 June, Washington], 19 June 1962; S. L. Udall, secretary of the interior [to R. E. Jones, U.S. House of Representatives], 23 July 1963). These lists undoubtedly were the beginnings of the first compilation of endangered species endorsed by the federal government. Prominent among names of fishes of concern would be Colorado squawfish and humpback chub.

In retrospect, it can be said that neither side was correct in their appraisals of the project. The agencies were incorrect in their belief that fish eradication in such a large system would be successful enough to be worth the effort. They "opened up" the new reservoir not only for trout but also for other fishes that could take advantage of such a large, unfilled ecosystem. In this case, Utah chub, another species introduced into the Colorado River basin, found a home, even though rainbow trout were given the "head start." Simply stocking rainbow trout was a far cry from developing the kind of ecosystem needed at such a site. Consideration of a potential food base, as well as temperature patterns and habitat types, would have aided in identifying the game and nongame species that would likely do well. This was also borne out at Boysen and Glendo reservoirs in Wyoming.

The opposition was similarly incorrect in their claims that the operation was poorly handled and that major fish kills had occurred in Dinosaur NM. Clearly, fishes in the Green River were much more affected by the dam and changes in flow regime and temperature than by the amount of undetoxified rotenone (Vanicek and Kramer 1969; Vanicek et al. 1970; Holden 1979).

A major negative impact on the now-rare native fishes was that remnant populations did not remain in the river or reservoir above the dam. Remnant populations of bonytails and razorback suckers in Lake Mohave on the lower Colorado River are today providing important brood stocks for recovery of these species. It appears likely that bonytails, and perhaps humpback chubs, could have survived for a time in the reservoir and might have provided endangered species biologists with additional brood stocks. Unfortunately, they never had this chance in the upper Green River.

Problems brought out by the Green River project are still with us. Major differences of opinion still exist between sport and nongame fish managers, even though they work for the same state or federal organizations. The value of native fishes is not universally accepted, and many species remain threatened by our actions on their environments. The major difference between the situation in 1962 and now is that many native fishes are now protected by law, and state and federal agencies are working to learn more about them and how best to manage them. There is no question that the Green River project helped speed the process that brought about awareness not only of the native fishes but also of the natural ecosystems on which they depend.

Chapter 4

The Desert Fishes Council: Catalyst for Change

Edwin Philip Pister

> *One of the penalties of an ecological education is that one lives alone in a world of wounds. Much of the damage inflicted on land is quite invisible to laymen. An ecologist must either harden his shell and make believe that the consequences of science are none of his business, or he must be the doctor who sees the marks of death in a community that believes itself well and does not want to be told otherwise.*
>
> — Aldo Leopold *Round River* (1953)

Introduction

For more than a century isolated individuals have expressed concern over a vanishing flora and fauna (Wallace 1863), but until enough of them were brought together by their common concern, little remedial action was taken. Although voices in the wilderness might have protested extinction of the Ash Meadows poolfish (*Empetrichthys merriami*), Tecopa pupfish (*Cyprinodon nevadensis calidae*), and thicktail chub (*Gila crassicauda*), such losses through societal oversight were largely passed off as isolated, albeit unfortunate, incidents, with the only known casualties (besides the fishes) being those few who witnessed and were saddened by their passing. Then, in September 1962, came the infamous Green River poisoning incident, which proceeded despite strong objections from the nation's ichthyologists and resulted in major losses to native fishes and invertebrates in more than 700 km of the Green River, a significant portion within Dinosaur National Monument. So great was the protest following this event that a letter of apology was sent by Secretary of the Interior Udall to Carl L. Hubbs, who at that time was serving as chairman of the Committee on Fish Conservation of the American Society of Ichthyologists and Herpetologists. Secretary Udall emphasized:

> Whenever there is a question of danger to a unique species, the potential loss to the pool of genes of living material is of such significance that this must be a dominant consideration in evaluating the advisability of the total project. I am taking measures to assure that future projects are reviewed to assure that experimental work is taken into consideration, and that possible deleterious effects are evaluated by competent and disinterested parties. (C. L. Hubbs 1963)

The Green River incident occurred, of course, before implementation of either the Endangered Species Preservation Act of 1966, the National Environmental Policy Act (NEPA) of 1969, or the Endangered Species Act (ESA) of 1973. So, until enactment of the NEPA, followed by similar laws in certain states, little could be done legally to prevent such occurrences. Almost no opposition was forthcoming from either federal or state conservation agencies, especially since it was a consortium of the U.S. Fish and Wildlife Service (USFWS) and counterpart state agencies within Wyoming and Utah that conceived and conducted the Green River fish eradication project. No opposition was expressed, either legal or philosophical, from the downstream states of Arizona, Colorado, Nevada, or California—tacit admission of the highly utilitarian philosophies of the time.

Green River was a seed that was slow to germinate but ultimately created an aura of awareness that would turn the remainder of the 1960s into a period of sharply increased environmental concern. Early "prophets" of the time were Robert Rush Miller (1961), Wendell L. Minckley (1965), Martin R. Brittan (1967), and James E. Deacon (1968b, 1969), whose writings warned of things to come.

Established conservation groups (Audubon Society, Defenders of Wildlife, Sierra Club, The Nature Conservancy [TNC], etc.) had not yet evolved to a point where fishes were of major concern. They were something to be caught and eaten, and their ecological significance was yet to be recognized. Nongame fishes were viewed primarily as competitors for game species.

My own, essentially dormant, concern over such matters was sharply awakened when field research uncovered the sorry state of native fishes within my eastern California purview (Pister 1974, 1985a, b, 1987a). Thus began a shift, in both philosophy and program direction, resulting in remedial actions. The Owens Valley Native Fish Sanctuary, the first such facility of its kind within California, was proposed in 1967 as a program of the California Department of Fish and Game (CADFG) to preserve the four native fishes of the Owens River portion of the Death Valley hydrographic area; it was formally adopted at the April 1968 meeting of the California Fish and Game Commission. The newness of this type of program was reflected by the fact that a paper describing the project published in the July 1971 issue of *Transactions of the American Fisheries Society* (Miller and Pister 1971) was the first ever included in that journal concerning management of a nongame, or noncommercial, fish species.

So, when U.S. National Park Service (USNPS) naturalist Dwight T. Warren telephoned in early March 1968 and reported that bad things were happening in the Amargosa River drainage of Death Valley National Monument (NM) that the state of California should be made aware of, his concern fell on sympathetic ears. A field trip on 12–14 March revealed major habitat disruption throughout Nevada's Ash Meadows, an area of enormous biological importance (Pister 1971, 1974; Cook and Williams 1982; Sada and Mozejko 1984). Agricultural development, especially groundwater pumping, threatened to accelerate destruction of a series of desert springs and fishes, and with them an evolutionary drama in progress since the Pleistocene or before.

Various agencies were notified of the problem, but the bureaucracy, equipped only with untested laws and no budget with which to implement them, found itself almost helpless when faced with an unprecedented emergency. Adding to the problem were archaic philosophies pervading upper echelons of government, which gave little value to nongame components of the biota despite the Endangered Species Preservation Act of 1966. A year passed while habitat destruction and alteration proceeded unabated.

Origin of the Desert Fishes Council

The week of 20 April 1969 proved to be a historic milestone. Following three days of visits to key desert fish habitats, both within Death Valley NM and in the Amargosa River drainage to the east and south, an informal field meeting was held in Ash Meadows to begin a preservation plan. In attendance were A. Edward Smith, Robert L. Borovicka, James D. Yoakum, and Lewis H. Myers, U.S. Bureau of Land Management [USBLM]; Clinton H.

Lostetter, USFWS; Dwight T. Warren and Superintendent Robert J. Murphy, USNPS; James E. Deacon, University of Nevada, Las Vegas; Dale V. Lockard, Nevada Department of Fish and Game (now the Nevada Department of Wildlife); and Leonard O. Fisk and Edwin P. Pister, CADFG. A busy field season was upon us, and neither time nor funds were budgeted for such work; another more general meeting of interested and concerned parties was scheduled for 18–19 November 1969 at Death Valley NM headquarters at Furnace Creek, California.

Although we did not realize it at the time, this April meeting constituted the beginning of an insurrection against established and conventional fisheries management procedures and philosophies. We *did* realize, however, that species extinction would precede implementation of any agency-sponsored preservation program if we were to follow the normal bureaucratic course. It was clear that new ground had to be broken.

The November meeting was viewed as the only conceivable means of circumventing agency inertia and thereby starting a movement to preserve endangered desert ecosystems and their associated life-forms. We were keenly aware of the inherent values involved, but our problem was mainly one of convincing agency administrators, and ultimately the public, of these values. Ironically, the latter generally proved simpler; bureaucratic intransigence is not easily overcome (Pister 1985a, b, 1987a). I agreed to serve as general coordinator for the November meeting, and Superintendent Murphy volunteered the auditorium and full cooperation of the USNPS staff at Death Valley toward making our meeting a success.

Much of the intervening time between April and November was devoted to related activities. My own work was directed primarily toward expediting construction of the Owens Valley Sanctuary and monitoring the isolated population of Owens pupfish (*Cyprinodon radiosus*) designated to be its initial inhabitants. Despite careful surveillance, an unusual combination of circumstances almost caused the loss of the population (and the species) on a hot August afternoon in 1969 (Miller and Pister 1971). Fortunately, we were able to keep a viable stock alive until June 1970, at which time they were introduced into the newly constructed sanctuary.

Although construction of the sanctuary and the protection afforded it in Fish Slough made things look brighter for the Owens pupfish, the scene was definitely less optimistic for Ash Meadows species and their habitats. Reports from Deacon, Lockard, and others in Nevada revealed that land development, groundwater pumping, and habitat destruction were accelerating (Pister 1974).

Development of a Common Effort

It was with a mixture of hope and despair that forty-four individuals concerned over the well-being of desert fishes and their habitats met at Death Valley in November 1969 for a symposium relating to their protection and preservation (Pister 1970; Fig. 4-1). In contrast to the more sophisticated symposia of later years, the first meeting was strictly a "brass tacks" affair devoted to an assessment of the resource and what might be done to save it. We had no direction, either legal or practical, and were guided only by our incomplete knowledge of the biology of the fishes. As I review my notes, I can see that we devised rough recovery plans and appointed recovery teams to accomplish their objectives, utilizing procedures not too different, although far less elaborate, than those in effect today under the ESA.

We left the meeting as we entered it, with mixed emotions. We were encouraged by the fact that a roughly organized effort was at least under way, and that in some cases we could see light at the end of the tunnel in the form of refugia and land acquisition programs. However, looming ever more ominously was the specter of continuing development in Ash Meadows, a result of basic interagency conflicts.

This first meeting was conducted amidst obviously conflicting philosophies and programs of several Interior Department agencies in attendance: the Bureaus of Land Management and Reclamation (USBR) creating the basic problem through promotion of land development and irrigation; the USNPS wondering what to do about declining water levels in Devil's Hole (a disjunct part of Death Valley NM); the USFWS (then the Bureau of Sport Fisheries and Wildlife) perplexed over how to administer provisions of the Endangered Species Acts of 1966 and 1969 when the basic problem was caused by sister agencies; and the Geological Survey (USGS) warning that the USBR's Amargosa River Basin Development Project might well spell the doom of aquatic resources within Death Valley proper, to say nothing of the aquatic habitats and endemic fish species to the east in Ash Meadows.

Two statements made during the first symposium perhaps provide an accurate philosophical perspective for that time. At one point, Carl L. Hubbs, one of the world's great ichthyologists and a pioneer in the study of western fishes (Pister 1979c), rose from his chair and, with great emotion, stated, "I can't tell you what this means to us. Bob Miller [of comparable stature as a scientist] and I thought that those of you in government would never see what we have seen for so long!" The other statement was by Robert E. Brown, then a graduate student at the University of California, Los Angeles, studying *C. radiosus*. Standing toward the front and against the west wall of the Death Valley auditorium, he stated with emotion similar to Hubbs's: "There just *has*

to be something we can do!" Indeed there was, but we were yet unaware that Providence would soon smile upon us.

At this point I feel it appropriate to enter an observation that has been the key to much of the effectiveness of action groups and recovery actions performed to date. When the announcement was made of the proposed meeting, response to a common problem was equally enthusiastic from within both government and academe. Yet, when I looked down from the rostrum, I noted that agency biologists and resource managers were clustered together in one part of the auditorium while university professors and their graduate students occupied another (Pister 1985a, b). Very little communication between the groups was evident; one could sense a feeling of mutual distrust and misunderstanding. When the meeting ended the following evening, these barriers had largely dissolved. The remaining barrier was the one presented by conflicts in goals among various Department of the Interior agencies, which, irrespective of personal feelings of agency representatives, still determined management directions.

Following the meeting, we disbanded, at least physically, to do what we could to accomplish our basic goals and solve our common problems. Not long afterward, I received calls from Sierra Club attorneys asking detailed questions about the problem we were encountering in Ash Meadows. Inasmuch as Devil's Hole was being affected and no one in the upper echelons of the Department of the Interior seemed concerned, the Sierra Club legal staff was preparing a writ of mandamus to force government action to preserve the area's biological integrity. The threat was posed but fortunately never had to be implemented.

Breakthrough

I must now relate a delightfully fortunate circumstance that constituted a turning point in

Fig. 4-1. Participants in the first symposium on desert fishes (a group that ultimately formed the Desert Fishes Council) at Furnace Creek, Death Valley National Monument, California. U.S. National Park Service photograph, 19 November 1968. First row (kneeling): Tina Nappe, Tom Jenkins, Bob Brown, Jim St. Amant, Frances Clark, Frances Miller, Jim Blaisdell, Dale Lockard, Paul Fodor, Bill Richardson, Pete Sanchez, and Phil Pister; middle row: Leonard Fisk, Bob Miller, Vlad Walters, Bob Liu, Laura Hubbs, W. L. Minckley, Dave Greenfield, Wayne Alley, Clint Lostetter, Jim Deacon, Sterling Bunnell, Bob Borovicka, Jim Yoakum, and Ward Gillilan; back row: Jim LaBounty, Bob Jennings, Carl Hubbs, Bill Templeton, Lew Myers, Wally Wallis, Vern Burandt, Ed Smith, Martin Litton, Bill Newman, and Don Cain.

agency involvement to preserve the native western aquatic fauna. During the early part of my career, while I was involved in anadromous salmonid research along California's northern coast, I worked with fellow CADFG biologist Charles H. (Chuck) Meacham. By great coincidence, he was a native of the small, eastern Sierra Nevada town of Bishop, near which I had conducted my earlier graduate research, and where I still live. We obviously had much in common.

Meacham had great interest in salmon research and management, and his pioneering spirit eventually took him northward to fill a position with the territory of Alaska. Alaska achieved statehood not long after that (1959), and when Walter J. Hickel became governor, Meacham worked with him as an adviser in matters relating to Alaska's vitally important salmon fisheries.

Following a long-standing tradition of selecting a westerner as secretary of the interior, President Richard M. Nixon brought Hickel down from Juneau during the early part of his administration. It was not surprising, then, that Hickel should choose certain of his Juneau staff to fill key positions in Washington. Meacham was confirmed as commissioner of the USFWS in June 1969 and given a second hat to wear with a secretarial appointment as deputy assistant secretary for fish, wildlife, and parks. The timing could not have been better.

Following the 1969 symposium, a series of articles was published in the spring 1970 issue of *Cry California* that generally described the natural history of pupfishes (Bunnell 1970), the destructive processes at work (Deacon and Bunnell 1970), and a general plan of action for saving the pupfish (Litton 1970). Despite this publicity—and an increasing awareness within the media regarding endangered species—conditions in Ash Meadows grew progressively worse. The full influence of the media was yet to be felt.

In early 1970 we were in dire need of help, especially in the matter of coordination of effort within the Department of the Interior. Some incredible inconsistencies existed. Although the USBLM's Jim Yoakum and Lew Myers had constructed a refugium at School Spring to protect the Warm Springs pupfish (*C. nevadensis pectoralis*), on USBLM land at Jackrabbit Spring a short distance away, a developer was using a gasoline-powered pump to remove all the water, including the entire population of Ash Meadows pupfish (*C. n. mionectes*). And while our major problem was the decreasing groundwater levels, the USBLM and USBR persisted in trying to put more land into irrigated agriculture. Something had to be done, and the time was ripe to do it.

On the morning of 4 May 1970, I phoned Commissioner Meacham's office in Washington, D.C., and was informed that he was "out west," but would return the call as soon as possible. Later that day I heard his familiar voice ask: "How are things in Bishop, Phil?" I outlined our problem throughout the Death Valley hydrographic area, ending with the seriously depleted populations of Owens pupfish in their sole remaining habitat at Fish Slough. His next query both amazed and delighted me. "Which spring is involved, the northeast or northwest?" Fish Slough, it turned out, was one of his favorite boyhood haunts.

Things happened quickly after that. Within a matter of hours Meacham had started the process of establishing a preservation program, wisely choosing as its leader James T. McBroom, assistant director for cooperative services in the Bureau of Sport Fisheries and Wildlife. Reporting on this at the Third Annual Symposium of the Desert Fishes Council, McBroom (1983:21–22), wrote, "About May 5 last year (1970), Chuck Meacham, then deputy Assistant Secretary for Fish and Wildlife, Parks and Marine Services, called me away from a meeting I was attending. He said to me, 'Jim, the Interior Department has got to organize an effort to protect the desert pupfish.' Then he said, 'When Secretary Hickel told me to do this, I knew I needed the roughest, toughest man I could think of to lead the program. It turned out to be you!'" McBroom's report is fascinating and should be read in its entirety by anyone interested in the early days of the preservation effort, or by anyone who doubts the ability of government to move quickly and effectively *when* the correct avenue of communication is discovered and individuals in key positions espouse philosophical agreement (Pister 1974).

McBroom's first move was to establish the Pupfish Task Force (U.S. Department of the Interior 1970, 1971) comprising representatives of the USBLM, USBR, USFWS, USGS, USNPS, Office of Water Resources Research, and (very wisely) Office of the Solicitor. A corresponding field advisory group included representatives of the California and Nevada departments of fish and game, the Nevada Department of Conservation and Natural Resources, and the universities of Nevada (Las Vegas), Michigan, and California (Scripps Institution of Oceanography). A Department of the Interior press release dated 14 June 1970 described the task force program and concluded with a comforting statement quoting Secretary Hickel: "The Interior Department will vigorously oppose adverse water use which would endanger the continued existence of these surviving species of fish."

The primary effort of the task force during its seventeen-month existence was to establish programs for hydrologic studies, surveillance of the resource, investigation of transplant sites and refugia, aquarium culture (if feasible), reclassification of 29.3 km^2 of public domain in the Ash Meadows area as not appropriate for disposal or exchange, and an investigation of the legal status of water withdrawals (an action that ultimately led to a favorable judg-

ment by the U.S. Supreme Court; Pister 1985a, b; Deacon and Williams, *this volume*, chap. 5).

It was stimulating to participate in activities of the task force, both locally and in Washington, D.C. Possibly the most remarkable occurrence was how, by secretarial directive, the various Department of the Interior agencies were changed from antagonists to cooperators, each contributing equally to financing a hydrologic study by the USGS and the University of Nevada Desert Research Institute that later would form the basis for the government's legal action against the land developers and the state of Nevada (Pister 1985a, b).

The only known exception to this cooperative atmosphere occurred when, in a discussion with Death Valley NM Superintendent Murphy, the USBLM district manager at Las Vegas staunchly defended the Ash Meadows operation and alleged that it was a shame to interfere with it "just to save a few worthless fish." Such a statement should never be made to a park service superintendent named Murphy!

The task force continued to provide a flow of information and suggestions concerning the common goal of saving the pupfish. The unofficial group that had been meeting regularly since April 1969, and which indirectly caused the task force to be formed, needed a more official status and designation in order to achieve maximum effectiveness. In a letter to me dated 24 August 1970, McBroom stated: "We believe it would be a splendid idea if you and others involved with desert fishes established a council, similar to the Desert Bighorn Council, to coordinate efforts toward the mutual objective. Such a council could speak with more authority and could add more strength to the effort than the independent effort of individuals. You may wish to consider this proposal at the November meeting." This was done, as noted in the summary of the second annual symposium (Pister 1971:12): "*Session III*—Consideration of proposal to establish a 'Desert Fishes Protective Council' and a Pupfish Advisory Committee to work with Interior's Pupfish task force." Considerable group discussion was held concerning this item of business, and the Desert Fishes Council (DFC) was formed by unanimous vote. I was selected as chairman and given the assignment of preparing a constitution and bylaws, as well as appointing committees and generally getting the council off the ground. Also involved were setting up advisory groups, designating procedures requisite to administering affairs of the DFC, and assisting the task force in solving technical problems. Formal discussion and adoption of the constitution were set for the 16–17 November 1971 symposium (Pister 1984).

The technical advisory group, already in existence for several months, was formalized, and the task of drafting a constitution was given to Peter G. Sanchez of Death Valley NM. Sanchez, a geologist familiar with the program and procedures of the Geological Society of America, drafted the new constitution on the format of that organization. His initial draft, with only minor changes, has served the council well.

Initial Success: An Impetus for the Future

Press releases from the task force and state fish and game agencies, combined with the inherent newsworthiness of anomalous "fish in the desert" threatened by "big business," soon gained the attention of major news media. The early 1970s saw burgeoning publicity in the form of television documentation, feature articles in major newspapers such as the *Wall Street Journal* and *Los Angeles Times*, and several articles in national magazines (*Audubon, Scientific American, Smithsonian*). Strong support was received from more specialized publications of the American Killifish Association, *California Caver, Defenders of Wildlife News, Desert, Dodge News Magazine, National Parks and Conservation Magazine*, the National Speleological Society, *The Nature Conservancy News, Outdoor California, Westways*, and others. These were accompanied by numerous technical papers published in agency reports and scientific journals (R. R. Miller et al. 1985).

In August 1970 a Columbia Broadcasting System television news team was flown by CADFG aircraft from Los Angeles to Ash Meadows to film a transplant of Devils Hole pupfish (*Cyprinodon diabolis*) to a site in Saline Valley, Inyo County, California. The ensuing telecast, shown on the evening news, resulted in a rash of letters directed to political figures ranging from California governor Ronald Reagan to Vice President Spiro Agnew, and to a variety of state and federal elected representatives. There seems little question that the publicity given desert fishes in the early 1970s also played a key role in providing a favorable atmosphere for passage by Congress of the new ESA, and its signing by President Nixon in December 1973. Such publicity also provided a favorable political climate for legal action by the Interior Department, acting through the Department of Justice and carried successfully to the U.S. Supreme Court, to protect Devil's Hole from pumping (Pister 1979b, 1985a, b; Deacon and Williams, *this volume*, chap. 5).

Reflections

As I review events leading to formation of the DFC, and to its evolution into an effective action group, the following points appear to have been instrumental. Concerned and competent persons must first become aware of a critical problem, and then must be willing to make virtually all other considerations secondary to solving it. The initial division between government and academe lasted only a matter of days and was overcome through a process of communication built around an atmosphere of interdependence, respect, and mutual trust (Pister 1985a, b). Very simply, fishes need water to exist; further information regarding biological requirements was a logical function of academic researchers. Research data were used by concerned government scientists and administrators to devise strategies to acquire or build habitats fulfilling the needs of the species.

Passage of the ESA, despite the funding it provided for state, federal, and international programs, did not result in a mad rush by agencies to involve themselves deeply in such matters. The utilitarian philosophies that directed agency programs at that time persist to a large extent even to this day. Laws and programs are no more effective than the motivation of those who implement them. I am reminded of an observation attributed to the German philosopher Goethe, that "man has only enough strength to accomplish those things of which he is fully convinced of their importance."

Death Valley's problems provided a valuable testing ground as the DFC moved into areas of concern throughout other parts of the southwestern United States and northern Mexico. Agency intransigence was so firmly entrenched during the late 1960s and early 1970s that the DFC was compelled to implement a program involving appointment of area coordinators, one for each of the twelve major hydrographic areas within the Basin and Range Province, to be watchdogs over the specific areas of concern and to work with sympathetic agency employees to ensure that no species was lost. The area coordinator program remains in effect at this writing.

Fiat Lux

One cannot help but wonder why such a program was necessary, why professional biologists employed by western fish and wildlife management agencies lacked the motivation to inventory their native faunas and devise programs to ensure their perpetuation. Having spent most of the past two decades pondering this question, and based on my own experiences as a state fish and game agency biologist (Pister 1985, 1987a), I would lay most of the blame at the feet of a bureaucracy rooted in tradition, an almost-universal program direction and professional ethic built around sport and commercial fishing, and university curricula devoid of courses in environmental ethics and ecological and evolutionary principles.

Lack of funding is often presented as an excuse. However, even with funding provided through the ESA, a large percentage of fishery biologists, state and federal alike, show little interest in the nongame component of the resource and are more than willing to leave such responsibilities to woefully inadequate staffs (or an individual) retained specifically for that purpose. Again, Aldo Leopold (1949) expressed it succinctly: "We fancy that game species support us, forgetting what supports game species."

Many persons, and most agencies, are simply not prepared to deal with the complexities and challenges of nonutilitarian aspects of ecosystem management. Their academic preparation and philosophical orientation are insufficient in breadth, content, or perspective for them to embrace more than a cost-benefit approach (D. A. Brown 1987). Dependence on competent and dedicated persons with knowledge of ecological principles to provide insight into future needs of a biological resource is often rejected as a viable option in decision making. This occurs even though most science, technology, and (for that matter) human cultural development are based on historical projections.

It thus seems appropriate and not at all surprising that the philosophies of great scientists of the past laid the groundwork for conservation efforts for western fishes. The lineage they started has carried on their tradition with admirable dedication. Carl Hubbs was a student of David Starr Jordan, Charles Henry Gilbert, and John Otterbein Snyder at Stanford University, and Hubbs's students, and their students in turn, continue to provide much of the philosophical and academic direction necessary for such work to succeed (Pister 1979c, 1987a). Hubbs made it clear that utilitarian management practices were of limited long-term value and would do little to accommodate future needs, which can best be met by preserving and managing as complete a native fauna as possible (Pister 1976). Native species constitute a dictionary from which words may be chosen to compose management prescriptions for the future.

Progress in conservation is being made, although leadership emanates not from agency administrators but from academe, isolated field biologists, and the private sector (Pister 1976, 1979a). TNC Natural Heritage Programs have been exemplary in this last regard (Pister and Unkel 1989) and deserve much of the credit for existing state programs. Bureaucratic intransigence still exists in federal agencies, as discussed by Williams and Deacon in chapter 7 of this volume, and is even more pervasive in some state conservation departments.

Table 4-1 summarizes the current degree of responsibility and involvement exhibited by western states regarding the protection of nongame fish faunas. It is evident that, after a substantial delay, most have begun to participate.

I am reminded of two incidents that eloquently express the sentiments frequently displayed by the leaders of western state fish and wildlife agencies in the early 1970s. In the first instance, I responded favorably in public to a keynote address on endangered species problems and solutions delivered in 1972 by Assistant Secretary of the Interior Nathaniel Reed at a meeting of the Western Division, American Fisheries Society (AFS), in Portland, Oregon (Pister 1979a). Shortly thereafter I was soundly admonished by the CADFG top leadership for my "embarrassing behavior" in supporting federal involvement in nongame and endangered species programs. This fit exactly with the second incident I encountered during summer 1974. As a member of the Endangered Species Committee of the Western Division, AFS, I attended a meeting of the International Association of Game, Fish, and Conservation Commissioners in Las Vegas, Nevada. During one session the newly implemented ESA was discussed from

Table 4-1. Status of nongame and endangered fish programs in the western states.

Category	AZ	CA	CO	ID	NV	NM
Date of cooperative USFWS animal agreement	1985	1976	1976	1979	1979	1976
Date of cooperative USFWS plant agreement	1979	1980	1987	1985	1985	1985
Full-time nongame fish biologist?	yes	yes	yes	no	yes	yes
Nongame division or branch?	yes	yes	no	no	no	no
Statutory authority for nongame protection?	yes	yes	yes	yes	yes	yes
Funding for nongame fish programs[1]	1, 2, 3	1, 2, 3	1, 2	2, 3	1	1, 2, 3
State endangered species act?	no[2]	yes	yes	no[2]	no	yes

[1]1, Section 6 (ESA) funding; 2, state income tax check-off monies; 3, funds from general agency budget.
[2]Has commission-approved threatened and endangered species lists.

the perspective of how it might be applied within the states. One commissioner rose and blurted out with considerable emotion: "The Feds better stay out of *my* state. I'd rather have our species become extinct than have the federal government become involved."

Our aquatic resources owe much to private individuals who saw a need in the early years and responded to it. Pasadena, California, schoolteacher Miriam Romero, working with the Sierra Club, organized and published a superb environmental inventory of the Amargosa River Gorge biota (Inyo and San Bernardino counties, California; Romero 1972) and paved the way for establishment of a USBLM Area of Critical Environmental Concern, which protects flora and fauna alike. Barbara Kelly Sada and Cindy Deacon Williams, representing the DFC's Ash Meadows Education Committee, literally accomplished miracles in working with TNC and top Nevada politicians, up to and including Senator Paul Laxalt, to acquire private holdings within Ash Meadows (Cook and Williams 1982; Sada and Mozejko 1984), now a national wildlife refuge (Deacon and Williams, *this volume*, chap. 5).

Garland R. (Bob) Love, a Union Oil Company chemist and longtime weekend resident

of Ash Meadows, laid much of the groundwork for preservation of that area and now represents TNC there. Tasker and Beula Edmiston, well-known conservationists from Los Angeles, brought a tradition of political expertise to the grass-roots efforts of the preservation effort. And Peter B. Moyle, a University of California (Davis) professor, served as chairman of the CADFG's Citizens' Nongame Advisory Committee and did much to get California's nongame conservation program off the ground. Tina Nappe did likewise as a member of the Nevada Fish and Game Commission.

These are but a few of the concerned citizens who have donated their talents and funds toward the preservation of western fishes. For the most part unacquainted with the need for preserving biological diversity that provides compelling motivation to professional biologists, these lay citizens, working through the democratic process, have provided the political support that ultimately underlies any significant conservation effort.

Looking Forward

It is not surprising that all the above-named individuals are members of the DFC, which provides a medium to enhance their effectiveness. The same is true of virtually every agency fishery biologist in the desert Southwest who possesses a strong interest in native life-forms and recognizes the need for an ecosystem approach to species preservation (J. E. Williams et al. 1985).

These individuals are separated from their agency peers by the way in which they define and practice the land ethic. Aldo Leopold (1949) defined the difference: "Conservationists are known for their dissensions. In each field one group (A) regards the land as soil and its function as commodity-production; another group (B) regards the land as a biota, and its function as something broader." Although the ranks of the B group in the various agencies are growing, they are still outnum-

OR	TX	UT	WY
1986	1987	1979	1981
1985	1987	1979	1981
no	part	yes	no
no	yes	yes	no
yes	yes	yes	yes
1, 2, 3	1, 3	1, 2, 3	1, 3
yes	yes	no	no

bered by the A faction. The transition from A to B will occur gradually through attrition, and by the ultimate and inescapable realization that human populations will increase in inverse proportion to the availability of fish and wildlife habitat. It will then become apparent that fish and wildlife–oriented recreation can no longer be fully met through contemporary and conventional programs (Pister 1976, 1987a). "To promote perception is the only truly creative part of recreational engineering, [and] . . . the only true development in American recreational resources is the development of the perceptive faculty of Americans. All of the other acts we grace by that name are, at best, attempts to retard or mask the process of dilution" (Leopold 1949). Ironically, these observations apply equally to American outdoorsmen and to fish and wildlife agency staffs dedicated to providing meaningful outdoor experiences. As I view native fishes in a context of ever-increasing human populations in an increasingly arid Southwest, it occurs to me that current management procedures (generally with non-native species) only touch on the potentials inherent within our native fauna. Already angler demand exceeds fish supply in most areas of the Southwest, and there are no signs that point to any real improvement of the situation (Pister 1976). However, although maximum sustained yield obviously has its limits, the infinite recreational and scientific resources provided by natural ecosystems are limited only by our ability to comprehend and appreciate them.

Major areas of emphasis for the DFC have evolved. Today the council is concerned with the following: (1) general coordination of ecosystem and species preservation in the Southwest, (2) encouraging agency biologists and university scientists and students to conduct species inventories and life history research useful in recovery efforts, (3) performing pioneering work in the fields of species

and habitat restoration, (4) furthering communication between government and academe, and (5) continuing to schedule annual symposia for the presentation and discussion of relevant data and programs. The general philosophy and procedures of the newly recognized field of conservation biology (Soulé 1986) have long been subscribed to and practiced by the DFC.

Inasmuch as the council's area of concern extends well below the border into the desert areas of Mexico, Mexican scientists and students have been full partners in the DFC's efforts. The council met in Mexico in 1980 at La Universidad Autónoma de Nuevo León (Pister 1981a), in 1984 at El Instituto de Zonas Desiérticas de San Luis Potosí (Pister 1987b), in 1987 at El Centro Ecológico de Sonora in Hermosillo (Pister 1990), and in 1990 at La Universidad Autonoma de Baja California in Ensenada.

Recovery efforts for Mexican fishes are also under way, but more slowly than in the United States. Mexican colleagues are restricted in their efforts by a lack of strong environmental legislation and money, and a conservation ethic within the Mexican public that is understandably tied more closely to immediate human needs than to a concern for genetic diversity or ecosystem integrity. The International Program of TNC is combining with support from within Mexican academia to make some headway here, with one of the first and most important target areas being the biologically rich Cuatro Ciénegas basin of Coahuila (Marsh 1984; Contreras Balderas 1977, 1984, in press, *this volume*, chap. 12).

An example of international cooperation is provided by the 1987 Nineteenth DFC Symposium hosted by El Centro Ecológico de Sonora. Of eighty-four registrants, twenty-eight were Mexican scientists or students; and of twenty-seven universities and research institutions represented, ten were from Mexico. Fifty technical papers were presented, seventeen from Mexican students and researchers traveling from as far as Cancún (Puerto Morelos) on the Yucatán peninsula, Mexico City, Baja California, and Nuevo León. Both student paper awards were won by Mexican nationals, one of whom is pursuing an advanced degree in the United States (Pister 1990). Although language barriers posed a problem, they were minimized by presentation of English abstracts for papers given in Spanish, and vice versa. An increasing bilingual capability within the membership is also of great assistance. The formation of the Mexican Ichthyological Society in 1987, which had its first meeting in La Paz, Baja California Sur, in November 1988, will do much to further the cause of native fish preservation in Mexico.

The DFC has now grown from a handful of persons who first met in Ash Meadows in April 1969 to an international representation of more than 500 agency and university scientists and resource managers, members of conservation organizations, and private citizens, all concerned with the preservation of aquatic ecosystem integrity throughout the deserts of the United States and Mexico. Related concerns are being expressed, and research conducted, by council members in all three nations of North America. The broad function of the DFC is to detect weak areas within the field of desert ecosystem preservation and provide the full strength of its membership to compensate for bureaucratic inadequacies, and to enhance governmental preservation programs.

The DFC provides a means—unimpeded by constraints of bureaucracy and politics—to meet specific conservation needs and express relevant concerns. In recent years it has used its expertise and influences not only to acquire major land areas within Ash Meadows (working with TNC) but also to influence and work with the USFWS in rehabilitating the area, establishing a national wildlife refuge, and assisting in pioneering efforts to develop new management strategies directed toward preservation of native life-forms. This is in sharp contrast to the consumptive harvest philosophies underlying the acquisition and management of most such areas.

In 1986 the DFC formed a special committee of experts to review a plan drafted by USFWS biologists to preserve native fishes of the upper Colorado River basin above Glen Canyon Dam (USFWS 1987a). Several areas of concern were detected, and although full agreement was not reached, the plan ultimately will be a better one because of council input. Speaking as a representative of a state fish and wildlife management agency, I can see obvious and enormous value in being in a position to state facts without concern for politics. Such information gives agency administrators a vastly superior negotiating stance when compared with a starting position already weakened by inevitable concern over political considerations. Entrenched bureaucracy does not welcome a rebel.

To Promote Perception

In spring 1979 a new journal appeared, devoted to a field touched on occasionally by environmental biologists but still far enough afield from the mainstream of biological thought as to belong to a distant but related discipline. Centered originally within the Philosophy Department of the University of New Mexico (later moving to the University of Georgia and in 1990 to the University of North Texas) and led by Eugene C. Hargrove, *Environmental Ethics* is described on the publication cover as "an interdisciplinary journal devoted to the philosophical aspects of environmental problems." It brings together writings of philosophers and biologists who detect an obvious need for the application of ethical considerations to the work in which environmentalists have been engaged for decades, frequently without giving the subject of ethics more than a passing thought.

There is little question that this new partner in the cause of conservation biology will grow and likely assume a role of importance comparable to the philosophies of Stephen Mather and Gifford Pinchot at the turn of the century (Rolston 1988). Concurrent with development of the journal has been the appearance of courses devoted to environmental philosophy and ethics at certain universities. With the philosophy expressed in Aldo Leopold's "The Land Ethic" (*A Sand County Almanac*) receiving major emphasis, such courses give badly needed maturity and direction to students who otherwise all too often become missiles without guidance systems. In an era fraught with technology and technological advances, this new emphasis on environmental ethics will likely become the most important innovation of the twentieth century as it gradually (yet inexorably) pervades both agency and academe with a philosophy more in line with what we can do for our fish and wildlife resources, rather than what our fish and wildlife resources can do for us. Holmes Rolston (*this volume*, chap. 6) addresses this subject in depth.

It is ironic that when Europeans first landed in North America they encountered native populations who recognized their dependence on the land and lived accordingly. It is only in recent years (particularly since World War II) that our quest for an ever-higher standard of living has resulted in the accelerating habitat loss and extinction rates that characterize the Western Hemisphere. So now, with increasing perception and understanding, we begin to build a new ethic (Pister 1981b, 1985a, b, 1987a). There is still time to act if we are only willing to sacrifice blind economic expediency to achieve long-term survival. It is also ironic that two of our most basic motivations, greed and self-preservation, should come into conflict in such a terrifying way. It is only through a new awareness of habitat and resource dependence by industrialized nations, and our rejection of the humanistic philosophies that have placed us in our current dilemma, that we can hope to survive over the long term. To paraphrase the theme of David Ehrenfeld's (1978) superb treatment of the subject in his book *The Arrogance of Humanism*: "Just who do we think we are?" It seems appropriate to close this chapter with the same scripture from Isaiah 47:10 (Jewish Publication Society 1985) that Ehrenfeld used to end his book.

> It was your skill and science
>   that led you astray
> And you thought to yourself,
>   I am, and there is none but me.

Chapter 5

Ash Meadows and the Legacy of the Devils Hole Pupfish

James E. Deacon and Cynthia Deacon Williams

Introduction

"We arrived at a beautiful valley [Ash Meadows] considerably lower than we had been before and quite a warm region so that we encountered flies, butterflies, beetles, etc. At the entrance to the valley to the right is a hole in the rocks [Devil's Hole] which contains magnificent warm water and in which Hadapp and I enjoyed an extremely refreshing bath." So wrote Louis Nusbaumer in his diary on the evening of 23 December 1849 (L. Johnson and Johnson 1987), just before continuing on into Death Valley on the arduous and, for some, fatal trek of the forty-niners that gave the valley its name. Thirty-nine years later, remembering that journey, William Manly wrote, "On the second or third night we camped near a hole of clear water which was quite deep and had some little minus [sic] in" (L. Johnson and Johnson 1987). The Devils Hole pupfish (*Cyprinodon diabolis*) thus began its earliest recorded syntopic existence with *Homo sapiens* by playing a role in significant historical events. That ecological relationship continues with increased intensity today.

Devil's Hole has held an enduring fascination for travelers, residents, and scientists who have enjoyed refreshing baths, marveled at the magical appearance of its caverns, and been enthralled by the insights into natural history it has yielded. The pool, a skylight into the aquifer (Fig. 5-1), was known locally as a "miners' bathtub" (Halliday 1955); it was designated a disjunct part of Death Valley National Monument (NM) on 17 January 1952.

Devil's Hole is located on the edge of a 94.7-km^2 oasis that incorporates most of the remaining habitat for at least twenty-three endemic taxa of plants and animals, the largest such assemblage in so small an area in the United States (Cook and Williams 1982; Hershler and Sada 1987). In June 1984 the area became the Ash Meadows National Wildlife Refuge (NWR). Had it been so designated forty years earlier, two of these taxa, the Ash Meadows poolfish (*Empetrichthys merriami*) and the Ash Meadows vole (*Microtus montanus nevadensis*), might not have become extinct.

In 1969, with a corporate farm in the early stages of development in Ash Meadows, the U.S. Bureau of Land Management (USBLM) selling land to them and classifying more for ultimate disposal, the state of Nevada issuing water rights that would guarantee lowering the water table, and local politicians encouraging development as the area's best hope for an expanded economy, the plight of the largest oasis in the Mohave Desert seemed hopeless. The Devils Hole pupfish, a colorful and improbable inhabitant, became a focal point around which the conservation community developed a groundswell of public interest, leading to the establishment of a legal framework making it difficult to develop the area in ways that might destroy the ecosystem's integrity. Concern for the pupfish became a rallying point for a diverse group of scientists, resource managers, and lay citizens, who organized the Desert Fishes Council, which subsequently has had more influence on conservation and management of native western fishes and their habitats than any of its founders dreamed possible (Pister, *this volume*, chap. 4).

These were essential precursors to a congressional appropriation of $5 million for purchase of Ash Meadows NWR in 1984. Creation of the refuge is thus a story in microcosm of a society wrestling with conflicting values and finally making a decision compatible with its developing environmental ethic. Litigation, legislation, agency leadership, agency obstruction, public controversy, and catalytic actions by conservation and scientific organizations all were involved in the process. But, as one can imagine, this story started much earlier, and could not have begun without the philosophical, legislative, and legal ferment of the time (Pister, Rolston, Williams and Deacon, *this volume*, chaps. 4, 6, and 7, respectively).

The Devils Hole pupfish played a pivotal role in establishing Ash Meadows NWR, influenced development of an environmental ethic in the United States, and, under authority of a decision by the U.S. Supreme Court, guided our society's expression of that ethic. Thus, the battle for survival of a pupfish served as a catalyst for change—an incident that helped define and delimit values—and heralded the dawn of the environmental epoch in America. The present paper deals with the human interactions that formed the battle to perpetuate the Devils Hole pupfish and Ash Meadows, and their implications for the conservation of western fishes and aquatic habitats in general. A review of its biology and the specific actions taken to perpetuate the Devils Hole pupfish while development, litigation, and negotia-

tions proceeded is left for other times and places.

We thank Robert Blesse and the Special Collections Department of the Getchell Library, University of Nevada, Reno, for providing access to the Laxalt Papers, and for kind and most welcome assistance. The following colleagues provided thoughtful critiques of this paper, for which we are deeply indebted: David J. Brown, Mary Dale Deacon, Maxine S. Deacon, David Livermore, Paul C. Marsh, E. P. Pister, Holmes Rolston III, Barbara Sada, Donald W. Sada, and Jack E. Williams. Carl L. Hubbs had the insight to recognize the special nature of Devil's Hole, the persistence to get it added to Death Valley NM, and the enthusiasm to transmit its wonder to many of his followers on this globe. Judge Roger D. Foley had the patience to listen to days of testimony, the interest to probe for a clear definition of the philosophical and scientific as well as the legal issues involved, the courage to write a precedent-setting opinion, and the scholarship to win unanimous concurrence from the Supreme Court. These persons have won the admiration and respect of all who strive to maintain the integrity, stability, and beauty of our world.

We dedicate our efforts here to the memory of David E. Deacon, whose awe-inspired response to diving in Devil's Hole was always unmistakable evidence of the intrinsic value of this magical cavern.

Devil's Hole and Ash Meadows

Since its discovery, Devil's Hole has been recognized as a special place. Nusbaumer noted that "the saline cavity itself presents a magical appearance" (L. Johnson and Johnson 1987). Halliday (1966) described the first series of dives into the clear, warm waters of the hole in the early 1950s. Hoffman (1988) compiled a record of all known organized dives into the

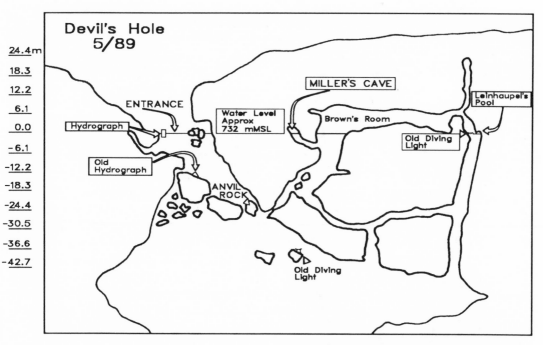

Fig. 5-1. Diagrammatic cross-section of Devil's Hole cavern system, Nye County, Nevada, as presently known.

hole, demonstrating the continuing fascination it holds for divers and scientists alike.

Establishment of the Monument

The campaign for protection of the Devils Hole pupfish and its cavern habitat was begun by Carl Hubbs in the late 1940s and scored its first success on 17 January 1952, when President Harry S Truman issued a proclamation declaring Devil's Hole part of Death Valley NM. Urging this protection was difficult because Hubbs had to overcome the national impression of the time that all the U.S. natural treasures had been cared for already (Williams and Deacon, *this volume*, chap. 7). Arguments for protecting the pupfish as well as the unusual geological characteristics of Devil's Hole ultimately proved adequate. Hubbs's recognition of the geological and hydrological features as exceptional proved unusually pro-

phetic when it was subsequently shown (Winograd et al. 1988) that a calcitic vein on the subaquatic cavern walls contains a continuous record of oxygen-18 variations that permit the interpretation of climatic history of the area from 310,000 to 50,000 years before present. This detailed record may require a major revision of the hypotheses used to explain the causes of glaciation (Monastersky 1988; Winograd et al. 1988).

In 1956, four years after its inclusion in Death Valley NM, the U.S. Geological Survey (USGS) installed a water-stage recorder in Devil's Hole, and shortly thereafter public access was inhibited by a locked gate placed by the National Park Service (USNPS) in a rock fissure at the entrance. Although regular trespass occurred, little damage to either the fish or habitat was evident until 1969 (Dudley and Larson 1976).

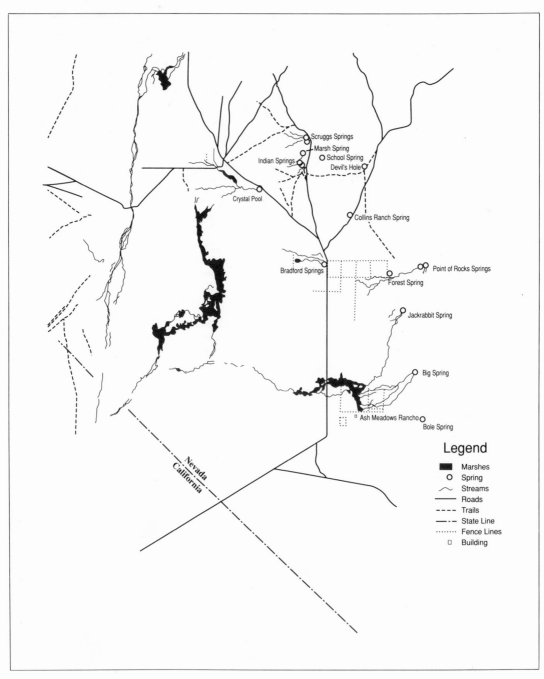

A

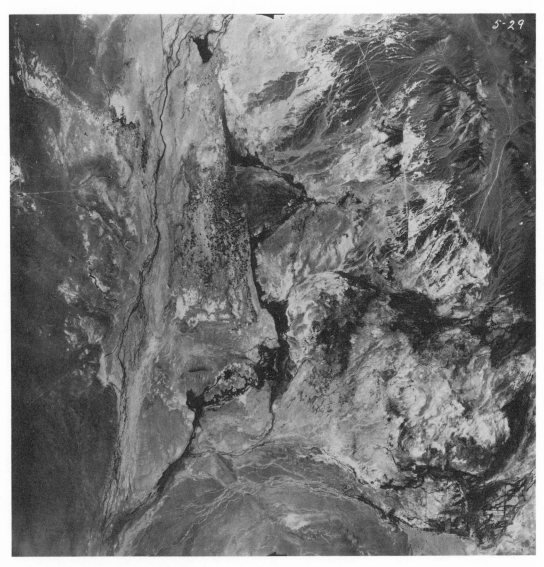

B

Fig. 5-2. Line drawings and aerial photographs of
Ash Meadows, Nevada, before and after major
modification of the valley for agriculture and a
projected subdivided community. Sketch maps
paired with aerial photographs to the same scale
(1 cm = 555 m) depict conditions in 1947 (A and
B) and 1988 (C and D). The 1947 photograph
was provided by the Fairchild Aerial Photography
Collection at Whittier College (flight number
C-13100, frame 5:29), and that for 1988 by
D. J. Brown, USFWS, Ash Meadows National
Wildlife Refuge.

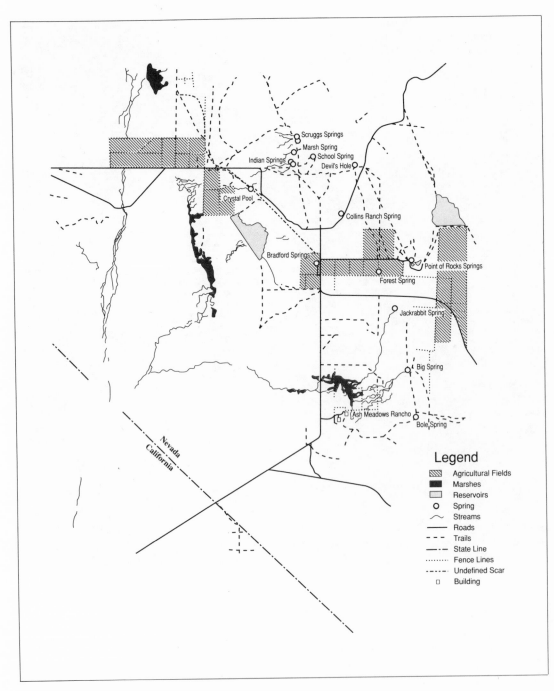

Scruggs Springs
Marsh Spring
Indian Springs   School Spring
Devil's Hole

Crystal Pool

Collins Ranch Spring

Bradford Springs   Point of Rocks Springs
Forest Spring

Jackrabbit Spring

Big Spring

Nevada
California

Ash Meadows Rancho
Bole Spring

Legend

▨ Agricultural Fields
■ Marshes
▢ Reservoirs
○ Spring
⌒ Streams
— Roads
- - - Trails
-·-· State Line
······ Fence Lines
-··-·· Undefined Scar
▫ Building

C

D

Early Human Impacts

From the time Devil's Hole was added to Death Valley NM, the USNPS was concerned about being able to maintain its natural character in the face of local development. Negative influences on the Ash Meadows area by humans had begun in the first half of the twen-

tieth century with subsistence farming and the apparently innocuous stocking of mosquito-fish (*Gambusia affinis*) and crayfish (*Procambarus clarki*). Introduction of sailfin mollies (*Poecilia latipinna*) into springs near Devil's Hole in the 1950s unfortunately must have combined with earlier introductions to contribute to the extinction of the Ash Meadows poolfish (Minckley and Deacon 1968).

A second major assault on the region's environmental integrity occurred in the early 1960s, when George Swink, a local rancher, drained Carson Slough in Ash Meadows and for three years mined peat from the area. The land was then sold to Spring Meadows, Inc., who reorganized watercourses and roads, and converted the peat areas into agricultural fields by bulldozing nearby sand dunes into them and tilling the land (Fig. 5-2). At the same time, a commercial aquarium fish-rearing facility was installed at Forest Spring, temporarily adding green swordtail (*Xiphophorus helleri*; Deacon, unpub. data) and arawana (*Osteoglossum bicirrhosum*; Soltz and Naiman 1978) to the exotic aquatic fauna. These developers were not unaware of the unique natural values of the area, but they were encouraged to proceed by prodevelopment government policies and prevailing social attitudes.

In 1963 the USGS was asked by the USNPS to determine whether increasing groundwater development would cause a decline in pool level in Devil's Hole. Worts (1963) concluded that groundwater pumping 19 km to the northwest had had no effect on Devil's Hole, and probably would have none for at least the next several decades. He did caution, however, that pumping within 1.6 km would probably affect pool level within a year.

Devil's Hole and the Courts

In 1967 Spring Meadows began acquiring large tracts in Ash Meadows, many of them from the USBLM. Spring Meadows, on the basis of an evaluation by their hydrological consultant, anticipated developing about 48.6 km^2 for irrigated cropland, to be used for growing cattle feed. Intensive development of a well field was undertaken from 1967 to 1970 to support this enterprise (Dudley and Larson 1976). By winter 1968, this latest assault was causing a decline in pool level in Devil's Hole (Fig. 5-3). The decline intensified in 1969–1972, as did the concern of conservationists and federal and state agencies. Not only would a decline in pool level alter the character of the place, but a recently completed study of the ecology of the pupfish (C. James 1969) showed that the shallow shelf in Devil's Hole was essentially the only breeding and feeding area for the species.

Developing public awareness

A public constituency to support and stimulate government action to preserve the pupfish and other species and habitats in Ash Meadows began to develop in earnest in April 1969. Concerned scientists and conservationists from state and federal agencies, universities, and the private sector reviewed the situation and agreed to hold a symposium on 18–19 November to formulate strategies for protection and preservation of fishes of the Death Valley system (Pister 1970, 1971, *this volume*, chap. 4). The media also sounded an alarm. The March 1970 issue of *Cry California* was devoted to an in-depth description of the Ash Meadows crisis: "We know what must be done to save the pupfish. What we need is the commitment to see to it that they continue to thrive. For if by our inaction we allow the desert pupfish, and perhaps with them the wonders of Death Valley, to be exterminated for the short-term economic advantage of a few, we will have committed a crime comparable to bombing the Louvre to make way for a parking lot" (Deacon and Bunnell 1970).

In May 1970 National Broadcasting Corporation television filmed at Devil's Hole and Ash Meadows for a documentary on water pollution titled *Timetable for Disaster*. The hour-long program, narrated by Jack Lemmon, featured a fifteen-minute segment on the Devils Hole pupfish. It aired in southern California in July and in the northeastern United States in September, and won an Emmy Award as the best television documentary of 1970. In

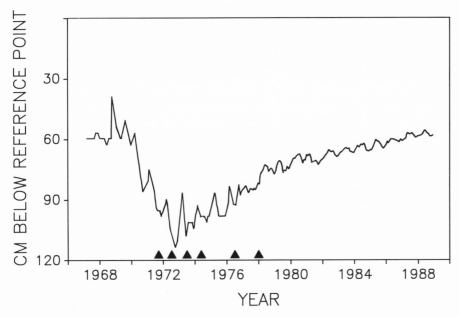

Fig. 5-3. Monthly mean maximum levels in Devil's Hole, Nye County, Nevada, 1967–1988. Triangles designate (sequentially, from left to right) time of occurrence of the following events: (1) suit filed in district court and three wells shut down; (2) suit reactivated; (3) preliminary injunction; (4) permanent injunction; (5) U.S. Supreme Court ruling; and (6) final water level ordered by district court.

September, American Broadcasting Company television featured pupfish on "Bill Burrud's Animal World." This national exposure informed a huge audience and sensitized both politicians and bureaucrats to the plight of the pupfish. A dramatic twist was added to the process when Don Widener, the producer/director for *Timetable for Disaster*, visited a sympathetic Walter Hickel, then secretary of the interior, and informed him that he would do another program blasting the inadequate government response to the crisis if the Devils Hole pupfish were allowed to become extinct.

In addition to national television coverage, other media exhibited interest in the issue, with major newspapers (*New York Times, Wall Street Journal, San Francisco Chronicle, Los Angeles Times,* and *Christian Science Monitor*) carrying at least one story each during the height of the controversy. Stanton Films produced an educational movie titled *The Desert Pupfish*. Local newspapers and television provided frequent coverage, and popular articles appeared in a number of magazines (Deacon 1969; Litton 1969; Findley 1970; McLane 1971a, b; Nappe 1972a, b; Trusso 1972; McNulty 1973; Sharpe et al. 1973). One consequence of this extensive coverage was to make citizens of southern Nevada and adjacent California, as well as a substantial segment of the U.S. public, aware of the plight of the pupfish. This created a climate that allowed, in some cases forced, public agencies to take action.

The November 1969 symposium developed a priority listing of the most endangered fishes of the area, formulated an immediate action plan for each taxon, and assigned specific responsibility for carrying out each plan (Pister 1970). The success of this approach depended on acceptance of responsibilities by individuals with the ability and authority to accomplish them. The magic of the group, which became the Desert Fishes Council (DFC), was its coalition of private conservationists, state and federal agency people, and academicians—someone for every task (Pister 1981b, 1985a, b, *this volume*, chap. 4).

Shortly afterward, Donald Harris, cochairman of the Sierra Club's legal committee, called the undersecretary of the interior three times a day until his calls were returned. He outlined the problem in Ash Meadows, received a sympathetic response, and then announced that the Sierra Club was considering bringing a mandamus action against Secretary Hickel to compel him to take action to protect the Devils Hole pupfish (R. F. Fisher 1971). Added to the media attention, scientific planning, and personal contacts described by Pister (*this volume*, chap. 4), this legal threat stimulated a number of positive and reasonably effective steps by the U.S. Department of the Interior.

Pupfish Task Force

In May 1970 the secretary of the interior formed the Pupfish Task Force, with representation from seven agencies within the Department of the Interior. The task force was directed to devise a plan and take immediate action to save the remaining Death Valley fishes. Their first step was to contract for preliminary hydrologic studies. Within a month, Fiero and Maxey (1970) analyzed existing hy-

drogeologic data and concluded that continued pumping of groundwater at 1969 rates would result in continued decline in pool level at Devil's Hole, reduce flows from the major springs in Ash Meadows, and perhaps dry many of the springs. Natural discharge was estimated to be sufficient, in conjunction with surface storage facilities, to irrigate about 12.1 (but certainly not 48.6) km^2. They recommended that the task force begin monitoring wells and springs, conduct long-term pumping tests on selected wells, attempt to stop pumping, and encourage agricultural interests to depend on surface discharge. Construction of a predictive optimization model was recommended to guide the use of groundwater so as to minimize or avoid adverse effects on pool level in Devil's Hole and discharges from major springs. They pointed out that even with such a model there would be no hope of irrigating more than 20.2 km^2. It was evident that Spring Meadows was engaged in a destructive and ultimately doomed endeavor, perhaps because of advice based on a maximum short-term use ethic provided by their hydrologists.

In addition, Secretary Hickel issued an order on 3 September 1970, at the direction of the task force, "declassifying" approximately 29.5 km^2 of public land previously classified for exchange to private ownership. This was the first positive action by the federal government taken on behalf of the ecosystem in the face of imminent development.

Heeding the recommendation of Fiero and Maxey (1970), the task force coordinated funding from five government agencies for a study by the USGS to determine causes of water-level decline in Devil's Hole and decreased discharge of Ash Meadows springs. Preliminary work was begun in September 1970. With considerably more detailed data on spring discharge and effects of pumping, Dudley and Larson (1976) concluded that substantial removal of groundwater, except perhaps in a northern sector, would result in adverse effects on spring discharge, water level in Devil's Hole, and pupfish habitats.

A second symposium was held on 17–19 November 1970 (Pister 1971). Reports of progress on tasks assigned at the first meeting, actions to date by the Pupfish Task Force, and other actions taken by federal and state agencies, universities, and conservationists were presented and discussed, and additional assignments were made.

In February 1971 a third groundwater study was initiated to develop a predictive model and formulate a water development and management plan for the Ash Meadows area to ensure the safety of the pupfish (Bateman et al. 1974). The model demonstrated that a reconfiguration of the well field could reduce adverse effects on Devil's Hole, but only at the cost of decreased discharges elsewhere. Altering the pumping schedule of existing wells also showed little promise that irrigated agriculture and pupfish (or natural aquatic habitats) could be compatible.

Litigation

On 17 August 1971, with all reasonable areas of compromise eliminated, and at the request of the Department of the Interior, the U.S. Department of Justice filed a complaint in U.S. District Court in Las Vegas seeking to enjoin Spring Meadows from pumping three wells identified by William Dudley of the USGS as most strongly influencing the water level in Devil's Hole (McBroom 1983). The complaint was based primarily on the federal property power and the USNPS Organic Act, contending that when Devil's Hole was incorporated into Death Valley NM, sufficient water also was reserved to serve the requirements and purposes of the monument. While it was noted that the Devils Hole pupfish was federally listed as endangered, no violation of the 1966 Endangered Species Preservation Act was alleged. Recall that at that time the more stringent pro-

visions of the 1973 Endangered Species Act had yet to be enacted. On 31 August 1971 the federal government and Spring Meadows agreed to a stipulation of continuance that provided for the three most offending wells to be shut down, with no compensatory increase in pumping elsewhere.

Following a brief recovery, the water level in Devil's Hole resumed its decline under the influence of continued pumping from other wells (Dudley and Larson 1976; Pister 1983). Therefore, in June 1972, the government reactivated its suit in an effort to have the court enjoin Spring Meadows (now Cappaert Enterprises) from using any water from any wells, aquifers, and springs within 4 km of Devil's Hole for other than domestic purposes. The government asserted its claim to both surface and groundwater on the basis of the implied reservation doctrine, first established in *Winters v. United States* (207 U.S. 564, 28 S. Ct. 207, 52 L. Ed. 340 [1908]). Federal rights to groundwater under that doctrine had not previously been established; however, the government was granted a preliminary injunction on 5 June 1973.

Displeased with the district court's decision, Cappaert Enterprises appealed it, and on 9 April 1974 the Ninth U.S. Circuit Court of Appeals upheld the preliminary injunction and issued a permanent injunction that allowed slightly more water to be pumped. The circuit court decision was appealed to the U.S. Supreme Court, which, on 7 June 1976, upheld the permanent injunction, returned the allowable water level to that originally set by the district court, and directed the district court to review the facts and establish a final minimum water level that would tend to ensure survival of the pupfish. The final district court judgment, handed down on 22 December 1977, established a minimum water level of 0.82 m below a reference point on the rock wall, a level guaranteeing a slightly higher water level than had been specified by the original injunction. Justification for the higher water level came from studies supported by Death Valley NM throughout the litigation (Deacon 1979; Deacon and Deacon 1979; Baugh and Deacon 1983; Chernoff 1985).

President Truman had used the acquisition and management tool to incorporate Devil's Hole into Death Valley NM. The USNPS in the 1970s defended the species in court based on the federal property power combined with their directed purpose to protect lands under their jurisdiction for the enjoyment of present and future generations. The U.S. Supreme Court came down decidedly in favor of the pupfish in *Cappaert v. United States* (426 U.S. 128 [1976]) when it ruled that in reserving a tract of land in Nevada as part of Death Valley NM in order to protect a rare species of fish, the federal government also implicitly reserved the groundwater appurtenant to the land, "lest the purpose of preserving the monument 'unimpaired for the enjoyment of future generations' be frustrated." The Court rejected the contention that the Antiquities Act, under which Devil's Hole was added to Death Valley NM, was intended to protect only archaeological sites, and held that the pupfish inhabiting the land in question were "objects of historic and scientific interest." Significantly, the Court held that the reservation of unappropriated groundwater need not comply with state law, once again dismissing the argument of state supremacy articulated only once in 1912, in the *Abby Dodge* case (Williams and Deacon, *this volume*, chap. 7), a concept that, while never explicitly overturned, has been given a quiet burial. To all appearances, the battle for Devil's Hole and Ash Meadows had been won; it certainly should have been!

The Ash Meadows National Wildlife Refuge

During the protracted litigation, local support for continued development of farming in Ash Meadows was frequently expressed. Nye County Commissioner Robert Rudd produced a KILL THE PUPFISH bumper sticker in response to a SAVE THE PUPFISH sticker produced as a joke by California Assemblyman Eugene Chappi and distributed by the DFC (Fig. 5-4). The Nevada State Legislature passed a resolution opposing a bill introduced in Congress by Senator Alan Cranston (D., California) to create a Pupfish National Monument in Ash Meadows. Local newspaper coverage was consistently and strongly supportive of the farming interests, although the Las Vegas press provided a more balanced perspective. Conflicting values were being discussed, defended, and evaluated. An irresponsibly extreme position was taken by the editor of the *Elko* (Nevada) *Daily Free Press* on 8 March 1976:

> There is an insecticide on the market called rotenone which has been used successfully to eradicate "problem" fish on many other occasions. This substance holds the key to resolving the "Pupfish Caper" before any more governmental time and money are wasted on this fraudulent attempt to establish federal authority as being greater than Nevada's jurisdiction of its own state water rights. An appropriate quantity of rotenone dumped into that desert sinkhole would effectively and abruptly halt the federal attempt at usurpation.

This illustrates the moral fervor with which a segment of society holds to a belief in their right to unfettered use of property (even dedicated public property) and their right to exercise dominion unrestricted by responsibilities of stewardship (Williams and Deacon, *this volume*, chap. 7).

The DFC had focused attention on the plight of all Death Valley area fishes, and the Pupfish Task Force had accepted responsibility for them. Without the council's efforts to organize an evolving action plan to save the fishes and stimulate agencies to accept and discharge their responsibilities, and The Nature Conser-

Fig. 5-4. Bumper stickers displayed during the times of heated controversy over the fate of Ash Meadows and the Devils Hole pupfish. Photographs by E. P. Pister.

vancy's (TNC) persistent efforts to acquire land in Ash Meadows, it is doubtful that the final battle would have been joined. Most assuredly, it would not have been won. Phase 2 of the battle for the region can be attributed to the failure of the Portland Regional Office of the USFWS to follow through with responsibilities assumed from the disbanded task force.

The great mistake

Following final action by the Supreme Court, Cappaert Enterprises determined they would be unable to develop a viable farming operation with the water available and offered to sell to the USFWS. The service's regional director, Kahler Martinson, acting on a recommendation from his endangered species adviser,

Philip Lehenbauer, determined that the USFWS had no interest in acquiring Ash Meadows. He apparently was unaware of the responsibility transferred to him from the Pupfish Task Force to save *all* remaining Death Valley fishes, and declined an offer from the Nevada Department of Wildlife to participate in a joint purchase. Martinson felt the safety of the Devils Hole pupfish was assured by court decisions and that other endemic species were not USFWS responsibility (verbal communication to J. E. Deacon, 1978).

Scientists and conservationists were stunned. The USFWS had walked away from a victory earned by a decade of hard work, refusing to claim the prize! Cappaert Enterprises sold its water rights and 51 km^2 of land to Preferred Equities Corporation in 1980, and phase 2 battle lines were drawn.

The recovery

Almost immediately, TNC opened discussions with Jack Soules, president of Preferred Equities, to negotiate full purchase of the land, and explored other options to ensure protection of endemic species of the region. These discussions required tireless and persistent efforts by David Livermore of TNC from 1980 through 1983. Soules's original offer to sell to TNC was made partly in jest, when he claimed the property would be worth $20–$25 million fully developed. That figure was well above the approximately $5 million land value estimated by TNC. For the next three years TNC was in more or less monthly, and sometimes weekly, contact with Soules.

During this time, Preferred Equities added to their original Ash Meadows investment with the purchase of an additional 20.2 km^2 of private lands and proceeded with plans for an urban/suburban/agricultural complex known as Calvada Lakes. The development proposed 33,636 residential parcels on 36.7 km^2, and commercial/agricultural/industrial development on an additional 11.8 km^2. Esti-

mates of water use for this type of development in southern Nevada indicated there would be a demand for 368% of the total water discharged by springs in Ash Meadows (Cook and Williams 1982). Ironically, there was some concern that the proposal for a new town of more than thirty thousand people would not be bound by the decision that had forced Cappaert Enterprises out of the region, because the injunction originally sought by the federal government in 1971 was directed to limiting water use for "other than domestic purposes."

Preferred Equities continued to farm some agricultural lands developed by Spring Meadows, built new roads, redirected spring outflows, and altered spring pools in preparation for marketing the project (Fig. 5-2). These activities inevitably encroached on habitats supporting endemic species. As late as 9 April 1982, Soules was quoted as saying: "There's no way the development can be stopped now. They might want to buy the project. But that would take something like $25 million for the land and works that has [*sic*] been done so far" (*Death Valley Gateway Gazette*, Beatty, Nevada ["Last Minute Efforts Seek to Stop Preferred Equity's Project in Amargosa"]).

It was time for conservationists to change their legal tools. Reserved federal rights, coupled with directed agency purpose, had been used to defend against threats from Cappaert Enterprises. However, in intervening years, the Endangered Species Act (ESA) of 1973 had been passed by the U.S. Congress. Prior to 1980, when Preferred Equities began preparations for urban development, federally listed endangered species in the area included Devils Hole pupfish and the Warm Springs pupfish (*Cyprinodon nevadensis pectoralis*). State-listed species included the federally listed ones plus the Ash Meadows pupfish (*C. n. mionectes*), Ash Meadows milkvetch (*Astragalus phoenix*), and the Ash Meadows blazing star (*Mentzelia leucophylla*). In the face of threats from the proposed development, the USFWS prepared a notice of review for two more fishes, nine plants, an insect, and twelve snails believed to be jeopardized.

On 10 May 1982, in a bizarre and totally unexpected twist of fate, Secretary of the Interior James Watt, the nation's self-proclaimed chief development officer, authorized emergency listing of the Ash Meadows pupfish and Ash Meadows speckled dace (*Rhinichthys osculus nevadensis*) as endangered. This action extended protection of the ESA for 240 days and made it obvious to everyone that development of the kind proposed would be impossible without destroying individuals and reducing populations of listed species, and thus violating the "take" provisions of the act. Shortly thereafter, Preferred Equities temporarily suspended its on-the-ground activities, while TNC and others sought a compromise.

Following extensive negotiations by David Livermore of TNC and Arch Mehrhoff and Donald Sada of the USFWS, Preferred Equities agreed to explore seriously an exchange of its Ash Meadows holdings for USBLM property in the Las Vegas basin. TNC proposed such an exchange to USBLM director Robert Burford on 10 August 1982, and at the same time solicited assistance from Senator Paul Laxalt (R., Nevada). The USBLM initially appeared willing to consider an exchange for all Preferred Equities's lands in Ash Meadows; however, by 5 November 1982, the agency decided that its asset management policy (which called for sale of federal lands to raise money to help reduce the national debt) and other problems precluded an exchange for more than about 81 ha immediately adjacent to springs containing endangered fishes. An attempt to exchange for land in Pahrump Valley, Nevada, also failed.

With the 240-day emergency protection running out and solutions for removing the threat disappearing, a second emergency listing for the two fish species was published on 5 January 1983. In mid-February Preferred Equities called TNC with an offer to sell at a price that would cover their investments. The conservancy, unable to purchase, restore, and manage Ash Meadows using their own resources, approached Senator Laxalt with an offer to purchase and hold the property until a federal appropriation could be made that allowed the USFWS to purchase it for a national wildlife refuge. By early March 1983, Senator Laxalt, with support from Senator Chic Hecht (R., Nevada), Congressman Harry Reid (D., Nevada), Nevada governor Richard Bryan, and the Nevada Department of Wildlife, had agreed to request an add-on appropriation from the Land and Water Conservation Fund that would permit such a purchase. The Nevada delegation made it clear that the appropriation would not be pursued unless TNC held an option to purchase all of Preferred Equities's property in Ash Meadows prior to markup of the bill, scheduled for early June. The conservancy immediately contracted for an appraisal of the property and entered negotiations with Preferred Equities for an option to purchase.

In early May 1983 Preferred Equities, in an apparent effort to establish land values favorable to their negotiating position and demonstrate their ability to sell without "taking" endangered species, sold ten lots. This action appeared to make it nearly impossible to meet the conditions essential for continued support of the appropriation by the Nevada delegation. Jack Soules, however, assured TNC that if a reasonable appraisal was obtained, the federal government took appropriate action, and TNC offered a reasonable option to purchase, the corporation would be able to deliver the entire property by repurchasing the lots they had sold (Soules [letter to Livermore, TNC], 13 May 1983).

By late May, with Senator Laxalt's support, the appraisal value of $5,128,000 was added to the fiscal 1983 Supplemental Appropriations Bill by the Interior Appropriations Subcommittee for purchase of Preferred Equities's property. An appropriation of $5 million was approved by Congress in July 1983. During this time, TNC was becoming acquainted with the owners of Preferred Equities, Leonard and Ron Rosen, and the new corporate president, Clark Wysong. These individuals had become the negotiators upon Soules's sudden death. Despite having expectations of government reimbursement of only $5 million, TNC gambled and made an offer of $5.5 million to Ron Rosen on 12 July 1983.

The corporation insisted on a sale price of $6 million and, in spite of having offered to sell, continued to encourage local residents and politicians to apply as much pressure as possible to promote development. Corporate representatives attended meetings of local town boards to explain the advantages of development, wrote letters to the congressional delegation, state and local politicians, and newspapers (Wysong [letter to Livermore, TNC], 22 June 1983; Robison [memorandum to Laxalt, U.S. Senate], 8 September 1983; *Death Valley Gateway Gazette*, Beatty, Nevada ["Amargosa Residents Petition Feds to Save Ash Meadows"], 12 September 1983, Laxalt letter to the editor, 7 October 1983).

Apparently convinced that the development would proceed on schedule, and based on anticipated tax revenues, the Amargosa Town Board went into bonded indebtedness to build a medical clinic, multipurpose room, and town hall. Concern for repayment of that debt put local politicians in a near panic (Robison [memorandum to Laxalt, U.S. Senate], 8 September 1983). With this background, Preferred Equities informed TNC that a purchase agreement must be consummated by the end of August or they would begin to seriously promote and sell lots.

While the corporation and many local politicians and citizens attempted to promote the development, some local residents, the Ash Meadows Education Fund of the DFC, and the Sierra Club Legal Defense Fund (SCLDF) were doing everything possible to convince Preferred Equities of the futility of their cause. A report to the Nevada state attorney general sponsored by the Ash Meadows Education Fund (Cook and Williams 1982) resulted in the U.S. Department of Housing and Urban Development informing the corporation on 8 June 1983 of a large number of apparent violations of the Interstate Land Sales Full Disclosure Act (Peterson [letter to Soules, Preferred Equities], 8 June 1983). The corporation was given fifteen days to file amendments and suspend their statement of record, or be subject to a notice of proceedings and an opportunity for hearing.

On 2 August 1983 the SCLDF, on behalf of Defenders of Wildlife, Natural Resources Defense Council, Nevada Outdoor Recreation Association, Sierra Club–Toiyabe Chapter, Humane Societies of the United States, Nevada, and southern Nevada, and the International Humane Society, filed with the Department of the Interior and Preferred Equities written notice of violation of the ESA and informed them of their intent to file suit. On the same date the corporation also was informed of their violations of the Federal Water Pollution Control Act and the SCLDF's intent to bring suit for those violations.

The corporation immediately asked TNC to request that the SCLDF take no further action and agreed to extend the end-of-August deadline, thereby permitting a second appraisal as required by the USFWS. In their 9 August response to the corporation, TNC noted that it was reasonable to attempt intervention with the SCLDF only if a signed agreement for purchase existed, reaffirmed its offer to purchase Ash Meadows for $5.5 million (at least until the results of the second appraisal were available), and noted that the Nevada congressional delegation was prepared to reprogram the $5 million appropriation if they became disillusioned with the progress of negotiations.

On 26 August 1983 the SCLDF asked the U.S. Army Corps of Engineers to regulate discharges of dredge-and-fill material into the waters of Ash Meadows, as required by section 404 of the Federal Water Pollution Control Act. They specified points of law, definitions, and court decisions that obligated the corps to do so. The question of corps jurisdiction over the waters and wetlands had been discussed by the Environmental Protection Agency (EPA), the USFWS, and the corps over the past year. The EPA informed the corps (Wise [letter to Williams, U.S. Army Corps of Engineers], 9 September 1983) that "the evidence clearly and overwhelmingly supports the assertion that the waters and wetlands of Ash Meadows are waters of the United States." The corps therefore had to exercise jurisdiction and was required under the ESA to prevent adverse modification of habitat critical to the survival of listed species.

A second appraisal of $5.4 million was completed on 28 September 1983. TNC's negotiating position was, to a certain extent, defined by the two appraisals, and they were unable to go above the offered $5.5 million. The corporation, however, held fast to a demand for $6 million. The impasse was broken by the Richard King Mellon Foundation and their National Wetlands Grant program. On 30 November 1983 the Mellon trustees toured and were impressed by the area. Their willingness to make available a $1 million interest-free loan allowed TNC to negotiate the purchase agreement.

One small problem remained. TNC's cash offer to Preferred Equities was $500,000 more than was available for reimbursement from the federal appropriation. This difficulty was overcome when the Goodhill Foundation made that amount available to TNC for Ash Meadows as part of their national commitment to acquire critically endangered natural areas throughout America.

Acquisition

Thus, thirty-two years after Devil's Hole was included in Death Valley NM, and nearly fifteen years after the beginning of intense efforts to save Ash Meadows from unsustainable development, TNC consummated its purchase. On 7 February 1984 Ash Meadows was acquired for $5.5 million cash plus a five-year loan of $1 million at 5% interest. The refuge was assured! However, difficulties with the Portland Regional Office of the USFWS forced a delay in actual transfer of most of the property until June 1984 and led to an additional cost to the conservancy of nearly $1 million. A supplemental appropriation orchestrated by Senator Laxalt allowed final transfer.

Enlargement of the refuge by about 182.1 km^2 to include most of the area containing endemic plants then was proposed by USFWS staff. The proposal was immediately stopped by Senator Laxalt and his staff, who were acutely conscious of local Sagebrush Rebellion sentiments (Robison and Abbott [memorandum to Laxalt, U.S. Senate], 13 December 1983; Abbott [memorandum to Laxalt], 16 March 1984; Robison [memorandum to Laxalt], 26 March 1984). After a series of meetings between USFWS staff and Senator Laxalt's office, a final proposal was drafted with a goal of acquiring "the minimum [land] required to protect presently listed species." The approved refuge thus consists of 51.1 km^2 purchased from TNC, 10.9 km^2 of USBLM land to be protectively withdrawn, 6.3 km^2 of privately owned land to be purchased as available, and 26.6 km^2 of public domain to be cooperatively managed by the USFWS and USBLM.

Although some controversies regarding management policies have surfaced, the first refuge manager, David J. Brown, is attempting to develop a plan focused on maintaining ecosystem integrity as the only way to ensure survival of the numerous endemic species. The next phase of human activity will, for the first time, be an attempt to restore and then maintain the integrity of the Ash Meadows ecosystem.

The Legacy

Hope for the Devils Hole pupfish was spawned by the environmental consciousness of the 1960s. It was a dream strong enough to stimulate formation of the Desert Fishes Council. The practice of resource managers, conservation organizations, scientists, and lay conservationists collaborating through a professional society to influence state and federal conservation management was developed, or at least extensively refined by this group (Pister, *this volume*, chap. 4). At least four similar organizations (Southeastern Fishes Council, Desert Tortoise Council, Arizona Native Plant Society, and Arizona Riparian Council) have since been patterned after the DFC.

Anchored by the Supreme Court decision on Devil's Hole, promoted by the DFC, supported by a vocal public, bolstered by the ESA, and negotiated by TNC, Ash Meadows National Wildlife Refuge was created. Events leading to its establishment doubtless also aided the creation of at least two other national refuges for endangered fishes (Moapa NWR, Nevada, in 1979, and San Bernardino NWR, Arizona, in 1980). In addition, TNC, beginning with its commitment in Ash Meadows, developed an effective, expanding program of preservation for desert aquatic systems (Dodge 1984).

The Supreme Court decision on Devil's Hole in June 1976 reaffirmed USNPS responsibility to protect species living on their lands, and reaffirmed the federal property right to wildlife. For the first time, the implied reserva-

tion doctrine of water rights was extended to groundwaters.

One problem the Supreme Court opinion left with us is the apparent precedent established to provide only minimum water rights. The Court stated that when the U.S. government withdraws land, it

> by implication, reserves appurtenant water then unappropriated to the extent needed to accomplish the purpose of the reservation. . . . The purpose of reserving Devil's Hole being the preservation of the underground pool, the District Court appropriately tailored its injunction to the minimal need, curtailing pumping only to the extent necessary to preserve a water level adequate to protect the pool's scientific value as the natural habitat of the fish species sought to be preserved. (*Cappaert v. United States*, 426 U.S. 128[1976])

With regard to other federal lands, there is concern over the *Cappaert* Court's requirement to define minimal need. Maintenance of minimum environmental conditions is not considered desirable for most species, nor was it for the pupfish. The Court focused on reestablishing habitat conditions most likely to ensure survival of the species. That is the standard that justifiably should be taken from the decision. Decisions that tend "to preserve the integrity, stability, and beauty of the biotic community" (Williams and Deacon, *this volume*, chap. 7), in Devil's Hole, can only provide for a minimum viable population of pupfish, since that is all there was under pristine conditions. In nearly all other ecosystems, higher goals are desirable.

In environmental law there is a developing trend toward giving legal standing to natural objects, species, and ecosystems. The Devils Hole pupfish was not named as a plaintiff in this case. Nevertheless, in the minds of many proponents, opponents, and the general public, it was regarded as the injured party. The pupfish therefore contributed to the concept that natural objects could and perhaps should have legal standing. Legal standing for natural objects has yet to be specifically affirmed by the courts or granted by Congress, although the Endangered Species Act comes close (Varner 1987).

The pupfish became both a beneficiary and a shaper of society's worldview. The plight of a tiny fish in a tiny pool in the seemingly desolate desert caused a large number of people to evaluate seriously the relative importance of farms, subdivisions, jobs, broadening the tax base, species, and ecosystems. The pupfish case was based on values supported by an environmental ethic clearly distinguished from an animal rights ethic. The position was established eloquently at the onset of the battle by Martin Litton (1970) when he called on the president and the secretary of the interior to take bold, aggressive steps "to carry out whatever management programs prove to be necessary—not to guarantee survival of the pupfish for all time but to make sure that civilized man is not their killer."

Death Valley NM's resource specialist, Peter Sanchez, noted early in the battle that we were fighting to permit the Devils Hole pupfish to "program its own extinction." From the beginning, the focus has been on maintaining a relatively natural ecosystem in which all native species are capable of continuing the creative evolutionary process. We all recognize that that process ultimately will result in extinction, most probably by evolution of new forms. Never has there been the suggestion, characteristic of an animal rights ethic, that civilized man must not kill some individuals. The species and its supporting ecosystem, not the individual, have been the units of concern.

The fact that in this instance society decided in favor of ecosystem integrity does not imply that it always will. But at least sometimes we can express an environmental ethic that ennobles the human spirit by acknowledging that humans participate in, rather than determine, ecosystem function. In Ash Meadows, an effort is being organized to restore and perpetuate an ecological setting in which species, as products of evolution, may continue their creative process undiminished through time. We are witness to a symbiotic evolution of species and ethics. Such is the legacy of the Devils Hole pupfish.

SECTION III

Swimming Against the Current: Ethical, Political, and Other Conservation Tools

What should be the goal of a management program for native fishes—species survival? ecosystem integrity? maintenance of the evolutionary process? How do we reconcile these goals with the often conflicting goals of creating jobs, maintaining an expanding economy, and developing recreational opportunities for increasing human populations? How do values held by individuals influence decisions affecting native fishes? Before approaching answers to such questions, the value of native fishes to humans must be defined, both to the manager and to the public. The process of definition results in a change in moral standards of the individual that must be translated to and assimilated by society to succeed.

This collection of papers addresses these kinds of questions. It commences, in chapter 6, with the paradox of fishes in deserts and examines ethical reasons for perpetuating them. The development of an ethical foundation requires a deep understanding of the interplay between biology and ethics. Anthropocentric rationale (although impressive) is ultimately insufficient either to provide protection or to adequately respond to humanity's stewardship responsibilities. Desert fishes are dynamic historical lineages in which speciation has been impressively influenced by ecosystem change. Both the process and its products deserve respect. Humans, as sapient creatures, have duties not only to themselves and other humans but to all life, including desert fishes—instance, increment, and symbol of respect for life on earth.

From here, chapter 7 moves to a combination of ethics, outlines of available federal legislation, and an examination of the role of litigation in conservation of western fishes. Religious views on conservation are briefly reviewed, along with a consideration of Leopold's land ethic and the need to develop a worldview for natural systems. Federal legislation available for use in conservation and preservation efforts is outlined, both in general and for specific agencies. The substantial discretionary latitude afforded agency administrators is identified as a major factor in the application of conservation legislation in the real world of management and the recovery of natural habitats and endangered species. Examples of litigation important in solidifying legal boundaries in conservation are also cited, where appropriate.

This section ends with a case history of the development of the most ambitious implementation plan for recovery of endangered fishes yet conceived. Chapter 8 includes a brief history of human development of the arid Colorado River drainage, plus an outline of projected needs for an ever-expanding population of humans, which even now includes more than fifteen million people. The first legal protection for most native nongame fishes and their natural habitats in the 1960s was followed by a surge of research. This led in the 1970s to the formal Colorado River Fish Project, which was designed to increase knowledge of both the fishes and their imperiled ecosystem in the upper Colorado River basin. Conflict between water development and endangered fishes continued, however, stimulating the initiation of a special project to find an administrative solution to the problem. This process ended in formation of a fifteen-year Recovery Implementation Program to coordinate federal, state, and private actions to conserve rare fishes in a manner thought compatible with existing states' water rights as well as interstate pacts and agreements that guide water allocation, development, and management. If it works, both fishes and developers will benefit. If not, the earth will lose more species. Only time will tell.

Condition of Jackrabbit Spring, Ash Meadows, Nye County, Nevada (A and B), during active pumping for irrigated agriculture in the Death Valley/Ash Meadows region; photograph by L. Meyers, 1969. Jackrabbit Spring (C) after recovery following its incorporation into Ash Meadows National Wildlife Refuge; photograph by E. P. Pister, November 1988.

A

B

C

Chapter 6

Fishes in the Desert: Paradox and Responsibility

Holmes Rolston III

Puzzles about Fishes

Fishes in the desert—that can seem almost a contradiction in terms since by definition fishes inhabit water and deserts have little. Adverse even for terrestrial life, deserts are impossible environments for aquatic species. But, conflicting expectations to the contrary, there is water in the desert and fishes do live there—no contradiction but quite a marvel. Having discovered fishes in the desert, to ask next whether there can be duties to them seems incongruous again—a category mistake; neither a particular fish nor its species is a possible object of duty. Persons count morally, but fishes do not. Again, the prevailing expectations are wrong. Humans do have responsibilities to these marvelous fishes.

Admittedly, though, there is something odd about taking ethics underwater into the desert. Even if fishes do live there, that is only a biological description of anomalous life in arid lands. Can one conclude that humans ought to save such fishes—a prescription for conduct—without committing what philosophers call the naturalistic fallacy, which forbids logical passage from the *is* to the *ought*? Can we be more specific about what is really a double difficulty: the biological difficulty of being a fish in the desert and how this connects with an ethical difficulty? What are the challenging human responsibilities we en-counter with such exceptional fishes in troubled waters?

Fishes in the North American Deserts

The North American deserts were not always dry. Pleistocene times were pluvial; lakes and streams were abundant. But the waters left, and the lands have been arid for more than ten thousand years, resulting in a dry climate presently more severe than at any earlier time (Axelrod 1979; M. L. Smith 1981). Though the fishes largely vanished, relicts managed to survive in oases—springs, pools, and seeps, often fed by underground aquifers with waters that rained in the ancient past; that is, "fossil" water. They also survived in rivers, especially those that crossed the deserts but had their headwaters in wetter, mountainous terrain.

The isolation and duress produced some remarkable fishes. They were subject to extremes: shifting water supplies—cold torrents during spring floods, followed by dry-up during summer heat—shifting streambeds, salinity across a spectrum from the fresh waters of melting snow to briny seeps, to playa lakes on alkali flats, more than three times as salty as the sea. Desert life demands unique fish. Although such fishes are often endemic to local areas, the desert has regularly produced such endemics—about two hundred such species in the American West (J. E. Williams et al. 1985; Minckley and Douglas, *this volume*, chap. 1).

Ash Meadows, a sprawling oasis in the Nevada desert about 100 km² in extent, contains more than twenty springs, numerous lime-encrusted pools, small streams that flow year-round, and seepage and swampy areas. One of the most unusual places in the United States, it supports a unique flora and fauna— twenty-six plants and animals found nowhere else in the world—more endemics for its size than any other place in the continental United States (Beatley 1971, 1977; Schwartz 1984). Ash Meadows is named for the endemic velvet ash, *Fraxinus velutina* var. *coriacea*. At least eleven species of invertebrates, including insects and snails, are found only there. Lying in one of the most arid areas of the world, all its native fish are endemics: three desert pupfishes (*Cyprinodon* spp.) in twenty distinct populations, the Ash Meadows speckled dace (*Rhinichthys osculus nevadensis*), and the Ash Meadows poolfish (*Empetrichthys merriami* [extinct]; Soltz and Naiman 1978; J. E. Williams et al. 1985).

The Colorado River basin similarly contains a higher percentage of endemic species than does any other river in North America. Sixty-four percent of its native freshwater fish species (35% of native genera) are found nowhere else (R. R. Miller 1963; Carlson and Muth 1989). As a contrasting extreme in size to the tiny pupfish, the big mainstream rivers flowing from the Rocky Mountains through the desert have here shaped the Colorado squawfish (*Ptychocheilus lucius*), the largest member of the minnow family in this hemisphere and one of the largest in the world, once attaining lengths approaching 1.8 m and weights of 40 kg (Deacon 1979). The humpback chub (*Gila cypha*) is one of the most bizarre fishes of this continent, extraordinarily specialized for life in torrential waters. Of all fishes, it has the most extreme stabilizing nu-

chal hump. The related bonytail (*G. elegans*) has the most fusiform body. All these fishes have expansive fins for maximum power in swift currents.

We could go on cataloging the queer, the rare, the curious twists and turns that life has taken underwater in the desert. But a lengthening list of the aberrant and weird does not imply increasing duty. The blue whale is the largest animal that has ever lived on Earth, with 3600 l of blood and a heart big enough for a person to crawl around inside. Should one therefore save whales? Does bigness generate duty? Ptiliid beetles are smaller than the periods on this page, yet each has six legs, a pair of wings, a digestive tract, reproductive organs, a nervous system, and genetic information that, translated into a code of English words printed in letters of standard size, would stretch 1600 km. Should one save ptiliid beetles because they are so small? The facts are striking, but they do not as such yield obligations. Extremes have no evident logical connection with value.

Some will argue that such odd facts as premises really yield another conclusion: that desert fishes are a fluke, there by luck already—living Tertiary fossils, detritus from the past. What survives or perishes does so by chance; it has nothing to do with value or human duties. Nature has no standards of value, and there is no reason to think that she has been protecting treasures in the desert, conserving them either because they had value in themselves or because humans were coming. So there is no cause for humans to care particularly about this scanty, chance selection of desert fishes. Precariously situated by a whim of nature, they are going to become extinct by natural causes sooner or later anyway. Most of the other fauna that were there in the Pliocene and Pleistocene are long since extinct—mastodons, ground sloths, saber-toothed tigers, dire wolves, camels, horses, to say nothing of other species of fishes. If these desert fishes come to an end, they have already gone on long enough, and they are all an accident in the first place. So why should we care?

What is happening to these fishes now is nothing different from what happened in geological times; desert streams dry up all the time, not just when humans draw them down; floods inundate channels and backwaters annually; landslides and lava flows dammed rivers before the U.S. Bureau of Reclamation arrived. Always, the fittest survive, and the rules do not change when humans arrive, modify habitats, and introduce exotic parasites, predators, or game species that deplete the natives. It does not seem far out of line with evolutionary natural history that humans should drive a few more species to extinction.

Human Duties to Desert Fishes

Faced with these difficulties, those who argue for preservation along the commonest, easiest path take an anthropocentric turn. Regarding humans, we do have duties. Regarding the fish, we need only deal with the present and be pragmatic about that. The first biological premise is the descriptive fact that fishes exist in desert waters; a second biological fact is the anomaly of their existence. But no conclusion needs to be drawn about duties to fishes. The conclusions have rather to do with human benefits, such as sport fishing and food. The resulting ethic is one of resource management. With this redirecting of the argument, there is no problem with the premise about anomalous luck, because humans often value resources that they have obtained by whim of nature. Whatever humans desire is, ipso facto, valuable; the natural history of the origin of the object of desire—chance or necessity, common or rare, typical or anomalous—is irrelevant.

In the Endangered Species Act, Congress lamented the lack of "adequate concern [for] and conservation [of]" species, and insisted that endangered species are of "esthetic, eco-logical, educational, historical, recreational, and scientific value to the Nation and its people (U.S. Congress 1973, sec. 2[a])." On the masthead of the journal *Fisheries*, after notice that the American Fisheries Society is, since 1870, the oldest and largest professional society representing fisheries scientists, we read that "AFS promotes scientific research and enlightened management of aquatic resources for optimum use and enjoyment by the public." There is similar language in the enabling legislation of every federal and state agency charged with managing fishery resources.

Confronting directly the question of why we should bother to save desert fishes, James Everett Deacon (1979:56), a pioneer in desert fish preservation, gave two answers: "Because it is in our self-interest to do so, and because our society's values, expressed through federal law, require us to bother. . . ." The first, he says, is "really the core of the endangered species debate."

The question is one of human class self-interest, and any duties are embedded in that class self-interest. "The preservation of species," by the usual utilitarian account reported by Hampshire (1972:3–4), is "to be aimed at and commended only in so far as human beings are, or will be emotionally and sentimentally interested." Feinberg (1974:56) says, "We do have duties to protect threatened species, not duties to the species themselves as such, but rather duties to future human beings, duties derived from our housekeeping role as temporary inhabitants of this planet." All this simplifies the logic and the ethics. It enables philosophers to concur with the arguments of legislators and resource managers. Within the collective human self-interest there are no duties *to* endangered species, only duties *to* persons. The relation is threefold. Person A has a duty *to* person B that *concerns* species C, but is not *to* C. A's duty is to promote benefits deriving from C that satisfy B's preferences.

Human Benefits from Desert Fishes

A third tacit premise must be made explicit before this anthropocentric argument can succeed: that desert fishes do yield human benefits—aesthetic, ecological, scientific, and so on—in excess of any benefits to be gained by their extinction. Can we be more specific about how the preservation of desert fishes is in our self-interest? How do these odd fishes satisfy our preferences?

Persons have a strong duty of nonmaleficence not to harm other humans, and a weaker, though important, duty of beneficence to help other humans. Humans will be harmed if their ecosystems are degraded, and diverse species are critical to our life-support systems. Arguing the threat of harm, Paul and Anne Ehrlich (1981) maintained that the myriad species are rivets in the airplane in which humans are flying. Extinctions are maleficent rivet popping. On the earthship on which we ride there is redundancy, but humans cannot safely lose 1.5 million species/rivets, and any loss of redundancy is to be deplored. Ecosystems have no useless parts, and we are foolish to think they do. Species, including endangered ones, are stabilizers. What the hump is to the humpback chub, endangered species are to humans.

Once this premise is made explicit, it is not always convincing. *Astragalus detritalis*, an uncommon milk vetch and one of the few legumes that grow on shale in the Uinta Basin of eastern Utah, fixes nitrogen and might be important in that ecosystem. But if this or that desert fish goes extinct, everything else going on in the West—ecologically and culturally—will continue about as usual. Just because they are relict species, these fishes form no significant part of our human life-support system. They are not rivets in spaceship earth. They are not even rivets in California or Nevada or Arizona. If they have any ecological value, it must be of some other kind.

Some argue that this value lies in their role as ecological indicators. The rare species are the first to show environmental stresses; they are a red flag indicating that even common species, including humans, will soon be in trouble if trends go unreversed. It is not just the fishes in the desert that need water; every living thing there needs water—from plants and invertebrates through bobcats and bighorn sheep. Fishes are but early indicators of the water quality, and the quality of life in the desert. Still, perhaps we can read that signal of trouble and take remedial action; after we get the warning it does not matter whether the indicator fish is protected or goes extinct—unless we need it as an ongoing indicator. Also, given increased expertise in building instruments, we can eventually make better monitors and will no longer need indicator fish. Once miners used canaries to detect foul air; now they use electronic meters.

Congress also expects "recreational" benefits from conserving endangered species. One whooping crane in a flock of sandhills perks up a bird-watcher's day. People go on field trips to see the endangered Arizona hedgehog cactus (*Echinocereus triglochidiatus* var. *arizonicus*), known only from small populations in central Arizona. Others take cruises to watch whales and dolphins. There is fish-watching at Virgin Islands National Park and the Great Barrier Reef. Does this work with desert fish? Recreators come to visit Devil's Hole, and these odd fish can fascinate enthusiastic ichthyologists.

But let us be frank. These fish are underwater, not part of the scenery. They are out of sight and largely out of mind. Recreators overlooking a marsh or a spring may experience a bit of excitement at viewing the sole habitat for an endemic fish, but there is not and cannot be widespread, recreational desert fish-watching analogous to bird-watching. According to surveys, one American in four takes at least occasional time each year to watch birds—in backyard, field, or woods. But not one American in four million watches desert fish.

Anglers are as numerous as bird-watchers. But most of these desert fishes are disliked by anglers; either there are not enough to catch, or they are not desirable, or they are protected by law and cannot be caught. From an angler's point of view the western fish fauna is depauperate; that is why fishes have been introduced into every major stream in the West: to provide recreation that the native fishes did not. These introduced fishes outcompete the natives, yield more fish per kilometer of stream, and—to recall the goal of the American Fisheries Society—we thus have "enlightened management of aquatic resources for optimum use and enjoyment by the public." With this management objective in mind, in September 1962 more than 81,000 l of rotenone were applied to 700 km of the Green River in Wyoming, Utah, and Colorado to rid the river of nine species of native "trash" fish such as squawfish and bonytail (as well as some introduced trash fish), so that, after the poison had passed or been neutralized, Flaming Gorge Reservoir, which was soon to fill, could be stocked with rainbow trout (*Oncorhynchus mykiss*) for quality fishing (R. R. Miller 1963; Holden, *this volume*, chap. 3). Proponents of this project alleged that the waters had to be made safe for sport fishing by killing the native species.

Anglers like to catch golden trout (*O. aguabonita*) endemic to three California creeks—the South Fork of the Kern River, Golden Trout Creek, and the Little Kern River. When the golden trout became threatened by introduced brown trout (*Salmo trutta*), the California Department of Fish and Game spent $300,000 over eighteen years (1966–1984) in a campaign to eliminate the browns and restore the goldens in their native habitat (E. P. Pister, pers. comm.). This time the poisoning was applied to remove the introduced fish and restore the endemic. A major justification was so that anglers could have their prized, flashy catch. The Colorado River cutthroat trout (*O. clarki pleuriticus*), the only native trout in the upper Colorado River drainage, is another desirable catch. The Gila trout (*O. gilae*) and Apache trout (*O. apache*) are also game species. Sometimes desert fish have recreational value; but a major problem for conservation is that usually they do not.

"Economic" is not on the list of endangered species benefits specified by Congress. Congress seems to have omitted it deliberately in order to suggest that the noneconomic benefits of conservation will override thoughtless human-caused extinctions in the name of development. At least in later amendments of the law, the burden of proof lies with those who think economic benefits justify extinction. Nevertheless, the most pragmatic argument for conserving endangered species is that some of them—which ones we do not know—will have agricultural, industrial, or medical uses in the future. The International Union for the Conservation of Nature and Natural Resources says, "The ultimate protection of nature, . . . and all its endangered forms of life, demands . . . an enlightened exploitation of its wild resources" (J. Fisher et al. 1969:19). Myers (1979a:56) says, "If species can prove their worth through their contributions to agriculture, technology, and other down-to-earth activities, they can stake a strong claim to survival space in a crowded world." He urges "conserving our global stock" (Myers 1979b).

Those species that are neither rivets nor indicators nor recreationally desirable may be raw materials. They may provide medicines or chemicals or genetic breeding materials. This argument works on occasion. Most species of *Aloe*, succulent plants, grow in deserts. The juice of *Aloe vera* promotes rapid healing of burns; rare species of *Aloe* may be destroyed before they can be examined for this effect.

But it seems unlikely that desert fishes are going to be good for anything agriculturally, medically, or industrially. Exploit desert fishes! That advice is not even pragmatically persuasive, and it seems somewhat demeaning for humans to regard all nonhuman species as "stock."

Congress anticipated that endangered species will have "scientific value." Indeed, they sometimes are key study species for both applied and theoretical science. A National Science Foundation report (NSF 1977:28) advocated saving the Devils Hole pupfish (*Cyprinodon diabolis*) because it and its relatives thrive in hot or salty water.

> Such extreme conditions tell us something about the creatures' extraordinary thermoregulatory system and kidney function—but not enough as yet. . . . They can serve as useful biological models for future research on the human kidney—and on survival in a seemingly hostile environment. . . . Man, in the opinion of many ecologists, will need all the help he can get in understanding and adapting to the expansion of arid areas over the Earth.

The pupfish has a sort of medical use after all; it is a survival study tool.

Where applied scientific value fails, there still remains theoretical scientific value. Species are clues to natural history; desert fishes, like fossils, help us to decode the past. Paleogeographers can figure out where the rivers formerly ran, where the lakes once were. Paleobiologists can figure out how fast speciation takes place and learn how dispersal occurs across wide ranges.

Some of these fishes are genetic anomalies because of their small population sizes. "The Devil's Hole pupfish . . . has apparently existed for thousands of generations with populations hovering near several hundred individuals. Classical genetic models predict that continual inbreeding should probably have already led to the extinction of this species, yet it still thrives in its single locality" (Meffe 1986:21).

The even smaller population of *C. nevadensis pectoralis* (twenty to forty fish) in Mexican Spring (the size of a bathtub) should not have been there—in theory (Soltz and Naiman 1978). But there it was, and had been for thousands of years (J. H. Brown 1971). Until very recently, before humans interfered, there it still was; geneticists cannot yet say how. From the viewpoint of pure theory, it would be interesting to know—even if this knowledge had no trickle-down benefit in applied genetics. It might help us to understand founder effects in evolutionary natural history, where accidental events in small, early populations may have large consequences later on.

Destroying species is like tearing pages out of an unread book, written in a language humans hardly know how to read, about the place where they live. No sensible person would destroy the Rosetta Stone, and no self-respecting person will destroy desert fishes. Humans need insight into the full text of natural history. They need to understand the evolving world in which they are placed, and scientific study of these fishes is likely to reveal something presently unknown about the pre-human history of the lands we now possess as the American West. Following this logic, humans do not have duties to the book, the stone, or the species, but to themselves—duties both of prudence and education. Fishes have, as Congress expected, "educational," "scientific," and "historical" values.

These arguments, sometimes sound, can quickly become overstated. No one can be sure that the pupfish will not teach us something vital about human kidneys or how to survive in arid lands, but it seems unlikely that these lessons can be learned only or best with *Cyprinodon diabolis*, and not—if that species should be lost—with *C. nevadensis*, or even some plentiful anadromous fish like salmon, which migrate from salt to fresh water. If certain information that scientists need to revise genetic theory can be obtained only from *C.*

diabolis, what happens after we have obtained it? We can discard the fish as we please, like laboratory rats after an experiment is over—unless a new argument is brought forth that *C. diabolis* might hold further theoretical or practical secrets.

All these utilitarian reasons will not work all the time; no single one will work in every case. Still, as a collective set some will work nearly all the time. It is a versatile tool kit; there is something handy for almost every job, even though, rarely, one may not be able to find a suitable tool. Most of the desert fishes can be conserved by one or another of these pragmatic justifications, although for a few rare fishes we can anticipate no likely benefits. That will get us 95% conservation.

Duties and Human Excellence

We can preserve the remaining, nonresource, fishes (the 5%) with a final, double-sided humanistic argument—so continues this anthropocentric environmental ethics. On the positive side, an admirable trait in persons is their capacity to appreciate things outside themselves, things that have no economic, medical, or industrial uses, perhaps even no ordinary recreational, aesthetic, or scientific value. An interest in natural history ennobles persons. It stretches them out into bigger persons. Humans must inevitably be consumers of nature; but they can and ought to be more—admirers of nature—and that redounds to their excellence. A condition necessary for humans to flourish is that humans enjoy natural things in as much diversity as possible—and enjoy them at times because such creatures flourish in themselves.

On the negative side, there is something philistine and small-spirited about the inveterate exploiter of nature. There is always something wrong with callous destruction. Vandals destroying art objects cheapen their own character. Humans of decent character will refrain

from needless destruction of all kinds, including destruction of even unimportant species. Americans are ashamed of having destroyed the passenger pigeon. They will be ashamed if they destroy these desert fishes; they will be more excellent persons if they conserve them. Destruction of these desert fishes is "uncalled for." Short of overriding justifications, humans really ought to save them all—including those few species from which we can gain no conceivable pragmatic, economic, ecological, aesthetic, recreational, scientific, educational, historical, or other benefits. We can always gain excellence of character from acts of conservation. We have a duty to our higher selves to save these fishes.

In another version of this argument, humans ought to preserve an environment adequate to match their capacity to wonder. Human life is often routine and boring, especially in town and on the job, and the great outdoors stimulates wonder that enriches human life. The desert evokes the sense of the sublime, and these curious desert fishes can certainly serve as objects of wonder. We have a duty to our higher selves to keep life wonderful.

At this point, however, we have pushed the anthropic arguments to the breaking point. Straining to develop a conservation ethic that is in our enlightened, highest human self-interest, the argument has become increasingly refined, only—alas—to become increasingly hollow. The logic of the utilitarian arguments was sometimes hard, but often soft. The promised benefits were real enough on some occasions, but on other occasions probabilistic and iffy. The loftiest preservationist argument is to preserve human excellence, to stretch humans out of themeselves in wonder. But let us be frank again. It seems unexcellent—cheap and philistine, in fact—to say that excellence of human character is what we are after when we preserve these fishes. We want virtue in the beholder; is value in the fishes only tributary to that? If a person made a large donation to the Desert Fishes Council, and, being asked what motivated his charity, replied that he was cultivating his excellence of character, we should rightly react that, small of spirit, he had a long way to go!

Why is callous destruction of desert fishes uncalled for if not because there is something in the fish that calls for a more appropriate attitude? Excellence of human character does indeed result from a concern for these fishes, but if this excellence of character really comes from appreciating otherness, then why not value that otherness in wild nature first? Let the human virtue come tributary to that. It is hard to gain much excellence of character from appreciating an otherwise worthless thing. One does not gain nobility just from respecting curios. Prohibiting needless destruction of fish species seems to depend on some value in the species as such, for there need be no prohibition against destroying a valueless thing. The excellence of human character depends on a sensitivity to excellence in these marvelous fishes flourishing in the desert.

The human mind grows toward the realization of its possibilities (excellences) by appropriate respect for nature (fishes), but that respect is the end, and the growth the by-product. It is even true that realizing this excellent humanity in *Homo sapiens* is a greater value than the flourishing of fish life in *Cyprinodon diabolis*, but the realization of excellent humanity here is exactly the *expansion* of human life into a concern for fish life for what it is in itself, past concern for utility, resource conservation, or self-development. Here humans are higher than fishes only as and because humans, moving outside their own immediate sector of interest, can and ought to be morally concerned for fishes, while fishes have no moral capacities at all and can neither cognitively entertain a concept of humans nor evaluate the worth of humans. "Higher" means here having the capacity to be concerned for

the "lower." Humans are subjectively enriched in their experience as and because they love the other, nonhuman species for what they objectively are.

Excellence is intrinsically a good state for the self, but there are various intrinsic goods that the self desires and pursues in its relation to others (for example, welfare of another human, or of desert pupfish) that are not self-states of the person who is desiring and pursuing. The preservation of the pupfish is not covertly the cultivation of human excellences; the life of the pupfish is the overt value defended. An enriched humanity results, with values in the fishes and values in persons compounded—but only if the loci of value are not confounded.

One does indeed want to keep life wonderful, but the logic is topsy-turvy if we only value the *experience* of wonder, and not the *objects* of that wonder. Merely valuing the experience commits a fallacy of misplaced wonder; it puts the virtue in the beholder, not in the species beheld. Earth's five to ten million species are among the marvels of the universe, and fishes tenaciously speciating in the desert are exceptional even on earth. Valuing species and speciation directly, however, seems to attach value to the long-standing evolutionary products and processes (the wonders, the wonderland), not merely to subjective experiences that arise when latecoming humans reflect over events (the felt wonder).

Evolutionary development in these fishes runs to quantitative extremes, and human awareness of this can enrich our quality of life. But what is objectively there, before human subjective experience, is already quality in life, something remarkable because it is exceptional. If you like, humans *need* to admire and respect these fishes more than they need bluegrass lawns, or an overpopulated Arizona, or a few more beef cattle, or introduced game fish. That is a moral need. Humans need moral development more than they need water development; they need a moral development that constrains any water development that endangers species.

Authorities are to be commended because, on the Virgin River drainage in Utah in 1980, they abandoned the Warner Valley Project lest it jeopardize the woundfin (*Plagopterus argentissimus*) and built the Quail Creek Project instead (Deacon 1988). Humans needed to do that. But the focus of this *need* cannot be simply a matter of human excellences. The alternate dam was not built to generate noble human character, or to preserve experiences of wonder. The alternative was chosen to preserve notable fishes and their natural excellences.

It is safe to say that, in the decades ahead, the quality of life in the American West will decline in proportion to the loss of biotic diversity, though it is usually thought that we are sacrificing biotic diversity to improve human life. So there is a sense in which humans will not be losers if we save endangered fishes, cactuses, snakes, toads, and butterflies. There is a sense in which those who do the right thing never lose, even when they respect values other than their own. Slave owners do not really lose when they free their slaves, since the slave owners become better persons by freeing people to whom they can thereafter relate person to person. Subsequent human relationships will be richer. After we get the deepest values clear in morality, only the immoral lose. Similarly, humans who protect endangered fishes will, if and when they change their value priorities, be better persons for their admiring respect for other forms of life.

But this should not obscure the fact that humans can and sometimes should be short-term losers. Sometimes we ought to make sacrifices, at least in terms of what we presently value, to preserve species. On such occasions humans might be duty-bound to be losers in the sense that they sacrificed values and adopted an altered set of values, although they would still be winners for doing the right thing. Ethics is

not merely about what humans love, enjoy, and find rewarding, nor about what they find wonderful, ennobling, or want as souvenirs. It is sometimes a matter of what humans *ought* to do, like it or not, and these *oughts* may not always rest on the likes of other humans or on what ennobles character.

Sometimes we ought to consider worth beyond that within ourselves. It would be better, in addition to our strategies, our loves, our self-development, our class self-interest, to know the full truth of the human obligation— to have the best reasons as well as the good ones. If one insists on putting it this way— emphasizing a paradox in responsibility— concern for nonhumans can ennoble humans (although this concern short-circuits if the concern is explicitly or tacitly just for noble humans). Genuine concern for nonhumans could humanize our race all the more. That is what the argument about human excellence is trying to say, only it confuses a desirable result with the primary locus of value.

Where the preceding arguments work, we have an ethic concerning the environment, but we have not yet reached an environmental ethic in a primary sense. The deeper problem with the anthropocentric rationale, beyond overstatement, is that its justifications are submoral and fundamentally exploitive and self-serving, even if subtly so. This need not be true intraspecifically, when out of a sense of duty one human altruistically defers to the values of fellow humans. But it is true interspecifically, since *Homo sapiens* treats all other species as rivets, resources, study materials, entertainments, curios, or occasions for wonder and character building. Ethics has always been about partners with entwined destinies. But it has never been very convincing when argued as enlightened self-interest (that one ought always to do only what is in one's intelligent self-interest), including class self-interest, even though in practice altruistic ethics often needs to be reinforced psychologically by self-interest. Some humans—scientists who have learned to be disinterested, ethicists who have learned to consider the interests of others, naturalists exceptionally concerned for these odd fishes—ought to be able to see further. Humans have learned some intraspecific altruism. The challenge now is to learn interspecific altruism.

Species as Historical Lineages

There are many barriers to thinking of duties between and to species, however, and scientific ones precede ethical ones. It is difficult enough to argue from an *is* (that a species exists) to an *ought* (that a species ought to exist). If the concept of species is flawed to begin with, it will be impossible to get the right ethical conclusion from a flawed biological premise. Perhaps the species concept is arbitrary, conventional, a mapping device that is only theoretical. Perhaps species do not exist. Individual fish exist, but *Cyprinodon milleri*, the Cottonball Marsh pupfish, once described as a full species from Death Valley (LaBounty and Deacon 1972), became just a subspecies (R. R. Miller 1981) when ichthyologists changed their minds. If species do not exist except embedded in a theory in the minds of classifiers, it is hard to see how there can be duties to save them. Duties to them would be as imaginary as duties to contour lines, or to lines of latitude and longitude. Is there enough factual reality in species to base duty there?

If a species is only a category or class, boundary lines may be arbitrarily drawn because the class is nothing more than a convenient grouping of its members. Darwin (1968 [1859]:108) wrote, "I look at the term species, as one arbitrarily given for the sake of convenience to a set of individuals closely resembling each other." Which natural properties are used for classification—reproductive structures, fins, or scales—and where the lines are drawn are decisions that vary with taxonomists. Indeed,

biologists routinely put after a species the name of the "author" who, they say, "erected" the taxon.

But a biological "species" is not just a class. A species is a living historical form (Latin *species*), propagated in individual organisms, that flows dynamically over generations. Simpson (1961:153) concluded that "an evolutionary species is a lineage (an ancestral-descendant sequence of populations) evolving separately from others and with its own unitary evolutionary role and tendencies."

Eldredge and Cracraft (1980:92) found that "a species is a diagnosable cluster of individuals within which there is a parental pattern of ancestry and descent, beyond which there is not, and which exhibits a pattern of phylogenetic ancestry and descent among units of like kind." Species, they insisted, are "discrete entities in time as well as space." Grene (1987:508) claimed, "species . . . can be thought of as definite historical entities playing a role in the evolutionary process. Lineages, chunks of a genealogical nexus, can count as real, just as genes or organisms do."

It is difficult to pinpoint precisely what a species is, and there may be no single, quintessential way to define species; a polythetic or polytypic gestalt of features may be required. All we need to raise the issue of duty, however, is that species be objectively there as living processes in the evolutionary ecosystem; the varied criteria for defining them (descent, reproductive isolation, morphology, gene pool) come together at least in providing evidence that species are really there. In this sense, species are dynamic natural kinds. A species is a coherent, ongoing form of life expressed in organisms, encoded in gene flow, and shaped by the environment.

The claim that there are specific forms of life historically maintained in their environments over time does not seem arbitrary or fictitious at all, but rather is as certain as anything else we believe about the empirical world. After all, the fishes are objectively there in Ash Meadows, and the reason we are concerned about them is that they are unlike fishes anywhere else. Species are not so much like lines of latitude and longitude as they are like mountains and rivers—phenomena objectively there to be mapped. What we want to protect is kinds of desert fishes, not taxa that taxonomists have made up to classify them. Humans do not want to protect the labels they use but the living process in the environment.

Taxonomists from time to time revise the theories and taxa with which they map these forms. They make mistakes and improve their phylogenetic knowledge. They successfully map numerous species that are distinctively different. Beyond that, we can expect that one species will slide into another over evolutionary time. But the fact that speciation is sometimes in progress does not mean that species are merely made up, instead of being found as evolutionary lines articulated into diverse forms, each with its more or less distinct integrity, breeding population, gene pool, and role in its ecosystem. That one river flows into another, and that we make some choices about what names to apply where, does not disprove the existence of rivers.

We can begin to see how there can be duties to species. What humans ought to respect are dynamic life-forms preserved in historical lines, vital informational processes that persist genetically over millions of years, overleaping short-lived individuals. It is not *form* (species) as mere morphology, but the *formative* (speciating) process that humans ought to preserve, although the process cannot be preserved without its products. Endangered "species" is a convenient and realistic way of tagging this process, but protection can be interpreted (as the Endangered Species Act permits) in terms of subspecies, varieties, or other taxa or categories that point out the diversity of life.

Our concern is with the products of the pro-

cess, but it is just as much with the process itself—as much with speciation as with species. Here fishes in the desert are of concern whether or not the edges between species are sharp. Where the edges are clear, we have a well-defined product of the evolutionary process. Where the edges are transitional, we have the process under way. As we have already noted, *Cyprinodon milleri*, the Cottonball Marsh pupfish, was first described as a species separate from *C. salinus*, the Salt Creek pupfish in nearby Salt Creek. It is smaller and more slender, with a more posterior dorsal fin and a marked reduction or complete absence of pelvic fins (Soltz and Naiman 1978). But LaBounty and Deacon (1972) found evidence that at high water they may mix and interbreed; R. R. Miller (1981) found similarities in tooth structures; and *C. milleri* is now considered a subspecies of *C. salinus*. Still, the discovery that this fish is only a subspecies is no reason for less concern; it is reason for concern that speciation under way be allowed to continue.

The Death Valley area, including Ash Meadows, is a good place to see what is wrong with a proposal sometimes made that—while we do want to preserve all the species of mammals—with fishes, especially nongame fishes, it is enough to save at the genus level. Perhaps one *Cyprinodon* will do; they are all pretty much alike. Extending the same logic to insects, unless they have special economic or ecosystemic importance, saving beetles at the family level is enough. One member of the Ptiliidae will do. But that kind of representative saving does nothing to save the speciating process. Species are most similar where the speciating process is fecund; there the dynamic lineages are profuse and procreative. This speciating fertility would be reduced to nothing if but one such species were preserved.

Even saving a species in a hatchery stops speciation. A species removed from the full set of interactions with its competitors and neighbors no longer works as it formerly did in the biotic community. A species is what it is where it is. Wild fish brought into hatcheries are soon selected for hatchery conditions, and the genome deteriorates, sometimes within a few generations (Meffe 1986). This is especially true with groups of fishes that speciate rapidly, groups that often include the endemics. Ex situ preservation, at times vital to the survival of a species whose habitat humans have radically disturbed, can never be more than an interim means to gain in situ preservation. We want to protect endangered speciation as well as endangered species.

Vanishing Desert Fishes and Human Development

These speciating processes and their product species will come to a stop—if present development trends go unreversed. The Endangered Species Committee of the Desert Fishes Council identified 164 fishes in North American deserts as endangered, vulnerable, rare, or of indeterminate status and suspected to be of concern. In addition, 18 fishes have already become extinct (J. E. Williams et al. 1985). In the West, Deacon (1979) listed 55 taxa (species and subspecies in 26 genera) of fishes that are extinct, endangered, threatened, or of concern. Four species and 6 subspecies in 6 genera have become extinct in recent decades. A fifth species feared extinct has been rediscovered (Pister 1981a). In Arizona, 81% of the native fish fauna is presently classified or proposed as threatened or endangered by state or federal agencies. In New Mexico, 42% are in trouble; and California, Nevada, and Texas fishes are in no better shape (J. E. Johnson and Rinne 1982; Rinne et al. 1986).

Most of the big-river fishes endemic to the Colorado River basin are in grave danger; three (Colorado squawfish, humpback chub, and bonytail) are listed as endangered, the fourth, the razorback sucker (*Xyrauchen texanus*), is reduced to scattered individuals in all but Lake Mohave, where adult fish are of great age (thirty years or more) and are not being replaced. Unless there are sustained recovery efforts, the sucker is predicted to be extinct in the lake by the year 2000 (Minckley 1983; McCarthy and Minckley 1987). The bonytail is functionally extinct; only a few rare individuals exist. Behnke and Benson (1980:20) said of the bonytail's demise, "If it were not for the stark example provided by the passenger pigeon, such rapid disappearance of a species once so abundant would be almost beyond belief."

The cui-ui (*Chasmistes cujus*) is endemic to Pyramid Lake, Nevada, a deep, large Pleistocene remnant. Withdrawal of upstream water has reduced the lake level more than 20 m and endangered the lacustrine sucker, which is now maintained in the lake by hatchery reintroductions and by providing assistance to the spawning run (Scoppettone and Vinyard, *this volume*, chap. 18). The U.S. Fish and Wildlife Service estimates that more than thirty-five species of southwestern fishes will need some type of artificial propagation if they are to survive (J. E. Johnson and Rinne 1982; Rinne et al. 1986).

The native fish fauna of North America has been tampered with possibly as extensively as, and certainly more rapidly than, the fish fauna of any other continent—by introductions of "game" and elimination of "trash" fish, by dams, pollution, and erosional sedimentation, and by thoughtless development, together with the accidental results of development such as introduced parasites and diseases. Of the endangered and threatened fishes of the world, about 70% are in North America (Ono et al. 1983). Of fish species in the United States and Canada, 56% are receiving some degree of protection (J. E. Johnson 1987a). The fishes in the United States have been as disturbed as any other faunal component, more so in the West than the East, and most of all in the Southwest (Moyle et al. 1986). Sixty-seven non-native fishes have been introduced into the Colorado River basin (Carlson and Muth 1989, in press).

The fishes of the West are like the birds of Hawaii. Both have a unique past natural history; both have been disastrously upset by the arrival of modern culture; both have a doubtful future. Desert fishes evolved in oases in an ocean of sand; Hawaiian birds evolved on islands in the sea. Both are bellwethers, casualties of explosive development. Of sixty-eight species of birds unique to Hawaii, forty-one are extinct or virtually so (Ehrlich and Ehrlich 1981). If there is any place in the United States that today approaches and even exceeds the catastrophic extinction rates of the geological past, it is in Hawaii and the West. Extinction rates rise with development rates.

Development seems like a good thing, but we cannot really know what we are doing in the West until we know what we are undoing. What is evident in the West is its development—condominiums, dams, highways, shopping centers, mushrooming cities. Less evident is how this cultural development is bringing about a tragedy—the catastrophic collapse of evolutionary developments there since the Pleistocene and earlier, a collapse unprecedented in scale since Tertiary times. Irreversible destruction of the generative and regenerative powers on earth cannot be the positive "development" that humans want.

This is why arguing the matter in terms of sport fishing versus trash fish (as was done in the Green River poisoning) is blind to what is really going on. Sport fishing does not justify the extinction of fish species that offer humans no fun. That pits trivial, short-range, nonbasic human pleasures against long-range evolutionary vitality. The deeper issue is respect for life, not "optimum enjoyment by the public." No non-native fish should be stocked in desert waters unless it has been determined that this practice does not adversely affect (officially or unofficially) threatened or endangered species. (This is the U.S. Fish and Wildlife Service policy for listed species in the Colorado River basin.) Non-native fish presently adversely affecting such species ought to be eliminated. The reintroduction of vanished fishes into their historic ranges ought to have priority over sport fishing.

Even to argue the matter in terms of water development requires caution. Not all water use is vital. Often one is trading bluegrass lawns, new golf courses, and two showers a day for shutting down evolutionary history. The Devils Hole pupfish was threatened by irrigation drawdown so that a few thousand cattle could be raised on land clearly marginal for that purpose (Deacon and Deacon 1979; Deacon and Williams, *this volume*, chap. 5). After that, until Preferred Equities sold its holdings in Ash Meadows to The Nature Conservancy, the threat to Ash Meadows was water development for a pleasure city (Adler 1984). Not even a pleasure city justifies tragedy in natural history!

Some who claim to be forward-looking will reply that the American West is in a post-evolutionary stage; the current story there is culture, and the latest chapter is the twentieth-century boom. The old rules do not apply. For millennia development took place through natural selection; development today takes place through real estate agencies and state legislatures. Nature must give way to culture. You cannot allow a few relict fish to hold up progress. Or, if you like, the old rules do apply even after the advent of culture: the fittest survive, and these archaic fishes cannot compete. Culture triumphs. That is the way it is, and that is the way it ought to be!

But before humans undo the natural history of the desert, we ought to ask whether cultural development compatible with a respect for developments going on independently of our presence is possible. In the first decade of the Endangered Species Act there were 1632 consultations on possible adverse effects to endangered species by federally sponsored projects in Arizona, New Mexico, Texas, and Oklahoma. Only 13 resulted in jeopardy opinions, and in all 13 cases alternatives were found to alleviate the impact (J. E. Johnson and Rinne 1982). That does not mean that development will never be seriously constrained by efforts to preserve species, but it does indicate that forms of development compatible with preservation are possible.

Is it not the time to reconsider whether the "enlightened management of aquatic resources for optimum use and enjoyment by the public" is all there is to be said? Is it only a matter of exploiting resources, or is it also one of admiring the sources, the creative powers that wrought the land we would now manage entirely in our self-interest? From that perspective, the deepest reason to deplore the loss of these fishes is not senseless destabilizing, not the loss of resources and rivets, but the maelstrom of killing and insensitivity to forms of life and the sources producing them. This final imperative does not urge optimal human use and pleasure, or prudent reclamation, but principled responsibility to the biospheric Earth.

Duties to Desert Fishes

These fishes are objectively there! That primary, long-standing biological fact is one premise of the argument. After that, we go astray if we emphasize anomalous luck as a second premise, or inevitable natural extinction as a third premise, or if we treat human-caused extinction as equivalent—biologically or morally—to natural extinction. The argument begins to move toward another conclusion if, for instance, after the primary biological fact that the fishes are still there, we posit a remarkable biological competence (instead of luck) as a second premise. Then we put as a third premise that speciation is still going on in the desert (along with inevitable extinction) and, fourth, we distinguish between natural and human-caused extinction rather like we do between death from old age and murder.

We initially suppose that desert fishes are dead ends in the evolutionary process; active speciation is being shut down, and the few remaining fishes are anomalous relicts. But that is to misjudge the story. Fishes speciate extensively; there are more species of fishes in the world than of all other vertebrates (mammals, birds, reptiles, and amphibians) combined. Fishes can speciate explosively. In fishes, speciation has taken place spontaneously during recorded human history (Greenwood 1981); fishes are the highest phylogenetic category— the only vertebrate taxon—in which this is known to have happened. In less than five thousand years, since ancestral Lake Manly in Death Valley dried up with the retreat of the glaciers, different *Cyprinodon* species learned to survive in remarkably different environments—in shallow streams and marshes, in groundwater springs, in water as salty as the sea, in thermal springs, in springs where water levels fluctuate widely, in hot artesian wells dug by humans. Some survive in environments as constant as any known in the temperate zone; others live in environments that fluctuate widely from cold winter rains to summer heat. About all *Cyprinodon* seems to need is water—any kind, place, or amount— and a little time to adapt to circumstances.

Though a place like Ash Meadows is a freakish anomaly, the life that prospers there has extraordinary vigor forced to ingenious modes of adaptation. Accidental life is matched with tenacity of life. The hardy, sprightly *Cyprinodon diabolis* has been clinging to life on a small shelf of rock for ten thousand years or more. No other vertebrate species is known to exist in so small a habitat (Pister 1981b; Deacon 1979). This species "has evolved in probably the most restricted and isolated habitat of any fish in the world" (Soltz and Naiman 1978:35). We begin to wonder if there is not something admirable taking place as well as something accidental, something excellent because it is extreme.

Although the West is as dry as it has ever been in geologic history, and its fishes are as stressed as they have been in millennia, there are no signs of incompetence in the remaining fishes or of the slowing down of speciation. Death Valley *Cyprinodon* evolved into four species in at least twenty-eight populations (twenty remaining, eight exterminated by humans), with almost every population of *C. nevadensis* exhibiting evident differences. That shows an unusual capacity for rapid evolution (McNulty 1973). Desert fishes "present one of the clearest illustrations of the evolutionary process in North America, rivaling in diversity the finches of the Galápagos Islands which first caused Charles Darwin to crystallize his ideas on the evolutionary process" (Soltz and Naiman 1978:1). Relicts of the past, these fishes also live on the cutting edge of adaptability. They are endemics, and—far from being evidence of any biological incompetence—that attests to their specialized achievements in harsh habitats.

The same is true with hundreds of endemic fishes, reptiles, amphibians, invertebrates, and plants throughout the desert West. Even though fishes have been less common in the increasingly arid environment in recent times than in earlier eras (fishes in the United States as a whole were not), these desert fishes persisted more than ten thousand years in hundreds of endemic species. Before Europeans arrived in Arizona, California, and New Mexico, there was no end in sight for the fish.

Pushing on at the edge of perishing, in their struggle for life they offer a moment of perennial truth.

In terms of conservation biology, the humanist scientist thinks that conservation biology begins with human concern. But conservation biology has been going on in the desert since before Pleistocene times. The pupfish, the squawfish, the woundfin—these are projects in biological conservation; these species have been conserving their kind for ten thousand years; they have been passing into transformed species tracking fitness in their environments. What human conservation biologists should do, arriving in this dramatic natural history, is admire and respect biological conservation taking place objectively to their conservation goals.

The wrong that humans are doing, or allowing to happen through carelessness or apathy, is stopping the historical flow of the vitality of life. One generation of one species is stopping all generation. Every extinction is an incremental decay in this stopping of life—no small thing. Every extinction is a kind of superkilling. It kills forms (*species*), beyond individuals. It kills "essences" beyond "existences," the "soul" as well as the "body." It kills birth as well as death. It kills collectively, not just distributively. It is not merely the loss of potential human information that we lament, but the loss of biological information, present independent of instrumental human uses for it. At stake is something vital, beyond something biological.

This superkilling is unprecedented in either natural history or human experience, and it is happening now in Arizona, New Mexico, Colorado, and Nevada. European Americans arrived in the West a few hundred years ago and gained the technological power to become a serious threat to fishes only a few decades ago. True, the issue faced here—desert fish—is not the whole global story. But it is an increment in it. "Ought desert fish to exist?" is a distributive element in a collective question, "Ought life on Earth to exist?" The answer to the local question is not identical with that of the global question, but the two are sufficiently related that the burden of proof lies with those who wish to superkill the fishes and simultaneously to care for life on Earth. If these fishes become extinct, that event alone will not stop evolutionary development elsewhere on the globe. But it will stop the story underwater in the desert. Life is a many-splendored thing; fishes sparkle in desert waters. Extinction dims that lustre.

Can humans reside in the desert West with a respect for place, fauna, and flora? Is there not something morally naïve about one species taking itself as absolute and regarding everything else relative to its utility? Though we have to make tradeoffs, do not these exceptional fishes claim our responsible care? They are right (fit) for life, right where they are, and that biological fact generates an ethical duty: it is right for humans to let them be, to let them evolve.

A Developing Ethic

Nature has equipped *Homo sapiens*, the wise species, with a conscience to direct the fearful power of the brain and hand. Only the human species contains moral agents, but perhaps conscience is less wisely used than it ought to be when it exempts every other form of life from consideration, with the resulting paradox that the sole moral species acts only in its collective self-interest toward all the rest. Among the remarkable developments on Earth with which we have to reckon, there is the long-standing ingenuity of these fishes, underwater in the desert; there is the recent, explosive human development in the American West; and there ought to be, and is, a developing environmental ethic. This is the biology of ultimate concern.

The author appreciates critical comments from R. J. Behnke, C. A. Carlson, D. A. Crosby, and E. P. Pister.

Chapter 7

Ethics, Federal Legislation, and Litigation in the Battle Against Extinction

Cynthia Deacon Williams and James E. Deacon

Introduction

The most critical threat to survival of civilization, barring all-out nuclear war, is human-induced ecosystem disruption resulting in worldwide extinction of plants and animals (Ehrlich 1988; E. O. Wilson 1988). The present wave of extinction, for the first time in earth's history caused by activities of a single species, carries the potential of reducing diversity to such an extent that the free ecosystem maintenance services provided by microorganisms, plants, and animals could become inadequate to serve the requirements of civilization (Ehrlich and Ehrlich 1981). Similar convulsions of extinction in the past have been followed by millennia of depauperate biotas, but eventually evolutionary radiation carried diversity to new heights (Raup and Sepkoski 1982; Raup 1986, 1988). Public awareness of the implications of losing 20% to 50% of the earth's biota over the next few decades is developing slowly.

While the problem is critical in the tropics, it is also of immediate concern in the arid American West, where fishes dramatically illustrate its magnitude. For example, the state of Arizona recognizes 111 species (27 fishes) needing special protection, California recognizes 254 (14), Colorado 74 (15), Nevada 78 (29), New Mexico 106 (24), Oregon 30 (4), Texas 126 (30), Utah 90 (18), and Washington 26 (0). More than 531 taxa (88 fishes) are recognized by the federal government as threatened or endangered in the United States alone. The inadequacy of the federal list is illustrated by recent American Fisheries Society reviews of the status of fishes of North America which listed 254 freshwater taxa in the United States as endangered, threatened, or of special concern (J. E. Williams et al. 1989), and 214 Pacific Coast anadromous salmonid populations at risk (Nehlsen et al. 1991).

We believe it is of critical importance for everyone in the United States, and indeed the world, to understand the essential linkages between human population size, resource exploitation, ecosystem disruption, and biotic diversity. History has shown that public understanding precedes attempts to correct major environmental abuse. Conservation legislation in the United States grew out of moral outrage over environmental abuses so obvious that even a Congress dedicated to economic development and pork-barrel politics could not ignore them. The national wildlife refuge system began as a response to extinction of the passenger pigeon (*Ectopistes migratorius*). Our national forests and parks were created following the gross land abuses associated with settlement of the West. The Bureau of Land Management emerged from the Dust Bowl. The Bureau of Reclamation's water-moving projects were initiated in response to settlers streaming back eastward, abandoning the parched western lands on which farming should never have been attempted. In the deeper sense, the need for these agencies was generated by anthropocentric moral standards that directed human activities in ways inevitably incompatible with nature. The agencies, in effect, represent our attempts to reconcile deep-seated moral conflicts.

The various federal agencies have differing obligations and responsibilities based on their enabling legislation, as well as the philosophies under which they were founded. Recent environmental legislation places obligations or provides discretionary authority that may be discharged differently depending on the philosophy of the administrator or the current administration in Washington, D.C. We hope to show that an understanding of both legislation and agency viewpoints is essential for success in the continuing efforts to maintain and improve aquatic habitats and native fish populations.

The following colleagues have provided thoughtful critiques of this paper, for which we are deeply indebted: David J. Brown, Mary Dale Deacon, Maxine S. Deacon, David Livermore, Paul C. Marsh, E. P. Pister, Holmes Rolston III, Alan Riggs, Barbara Sada, Donald W. Sada, and Jack E. Williams. We owe a special debt of gratitude to John Muir for so eloquently championing intrinsic values in nature, to Aldo Leopold for developing a simple, yet profound, land ethic, and to Edward Abbey for reminding us that "philosophy without action is the death of the soul."

Bases for Environmentally Destructive Behavior

There seems little doubt that the pervasive environmental abuse we are experiencing today has its roots in our fundamental worldview of humans as separate from and independent of nature, and in our concept of land as property rather than community. Western philosophic and religious worldviews predominate in to-

day's more developed countries, and therefore have a more prominent, although not exclusive, role in ecosystem disruption.

Religion

In North America, Christianity is by far the predominant religion, and the Christian church and its Bible are frequently blamed for encouraging environmental abuse by placing humankind in a special position of dominion over nature, relegating the remainder of created life to the category of organisms without souls (White 1967; Nash 1970). Christianity's emphasis on miracles, original sin, the corruption inherent in things of the world, and the supremacy of the afterlife make development of an appropriate Christian environmental ethic difficult (Ehrenfeld and Ehrenfeld 1985). Accepting environmental difficulties inherent in the classical Christian worldview, Cobb (1979), a Christian theologian, indicated that there were historical and naturalistic reasons to doubt that the human species had the requisite capacity to change its worldview, but he believed that Christians, through the grace of God, could do so.

By contrast, while accepting many of the criticisms regarding the Christian environmental ethic as correct interpretations of the Old Testament, Bratton (1984) contended that the belief that nature praises God, and that the entirety of creation is the work of God and under God's continuing care, gives nature intrinsic value that should be recognized by Christians. "Unnecessary" extinction at the hand of humankind is therefore viewed by at least some theologians as abrogating the stewardship responsibility inherent in the dominion over earth given to humans by God (Nibley 1978; Cobb 1988).

Judaism is similarly blamed and exonerated. Schwartzschild (1984) maintained that Judaism and Jewish culture have operated with a fundamental dichotomy between nature and ethics, in which nature remains subject to humanly enacted ends. Ehrenfeld and Ehrenfeld (1985) represented Judaism as containing the most detailed and ecologically sensitive code of ethics among world religions. The code includes laws prohibiting cruelty to animals, mandates the protection of nature during war, and introduces the idea of stewardship. Marxism and Taoism also deliver conflicting messages on environmental ethics (Goodman 1980; D. C. Lee 1980), and we expect that elsewhere, West or East, similar problems exist with other philosophical and religious worldviews.

Properly understood, it may be true that none of the major philosophic and religious teachings support environmental abuse. On the other hand, none has developed an ethic capable of preventing the pervasive abuses of our times. Humans almost uniformly have viewed themselves as separate from and independent of nature. Today, in the United States, for instance, Christian fundamentalists provide the political Right with some of its staunchest support for decreased environmental regulation, instead espousing "freedom" for businesses and individuals to pursue personal goals unfettered by governmental restrictions or concern for ecosystem integrity. Military security is viewed as the central responsibility of government, while the government's responsibility in providing for security or sustainability of ecosystems is largely ignored. In effect, we have abrogated our responsibilities of stewardship while exercising the privileges of dominion.

Property

Another frequently identified cause of environmental problems is the American concept of land as property. Historically and legally, the landowner has rights of ownership based on an infusion of labor that conveys a right to uncontrolled use. This follows the tradition of the seminomadic Saxon freemen of Germany and England. During and after the American

Revolution, while land-use legislation was being drafted, Thomas Jefferson's eloquence led a deliberate effort to adopt the Saxon concept of a freehold tenure system, modified by the doctrine of John Locke. According to that doctrine, landholding was not subject to obligations to a superior (the king) but was authorized by the infusion of labor and given to men for their support and comfort. In fact, Locke went so far as to claim that society was formed primarily so that individuals could enjoy property rights more safely and securely. Jefferson ultimately stopped trying to justify his position in terms of historical precedent and instead argued that small landholders were the most morally responsible individuals any state could hope to have (Hargrove 1989). Consequently, modern landowners inherited a belief in their legal and moral right to unfettered use of their property.

This contrasts with the Norman tradition, in which property rights were vested in the king (state) and granted, with restrictions, to the people. Locke took the divine right of kings to property and vested it in individual property owners, but he did not transfer the divine obligation of kings to act in accordance with the highest moral standard of right and wrong and to consider the welfare of the entire state and the common good.

The landowners' belief in their right to do what they individually pleased with their land was a concept developed when the human population size was below carrying capacity. The error of this concept was strikingly exemplified during the rush to settle the arid West, when vast areas were destroyed by poor land-use practices. Many unsuspecting settlers were doomed from the start by false claims of opportunity and assurances that "rain follows the plow." This absurd notion, ironically and briefly substantiated by a wet climatic cycle during the last decades of the 1800s, induced thousands to occupy land incapable of supporting sustained agriculture (Stegner 1954).

An almost continual stream of wagons moving *east* with failed, starving settlers displaying banners proclaiming "In God We Trusted, in the West We Busted" was the result (Terrell 1969).

Much of the destruction of western lands along with the dreams of countless intended settlers could have been avoided had Congress heeded John Wesley Powell's (1878) *Report on the Lands of the Arid Region.* Unfortunately, Powell's recommendations for orderly development within the confines of the region's carrying capacity were ignored. Passage of the 1902 Reclamation Act, the 1933 Tennessee Valley Authority, and the 1934 Taylor Grazing Act were partial, belated steps toward implementing Powell's concepts of basinwide water resource and arid land management. Reisner (1986) documented the degree to which even those timid steps can be perverted by greed and blind adherence to traditional concepts of land as property. Sustainable, ecologically sensitive land management for the common good is an idea explored with difficulty by our society. To our credit, we have made some overtures in that direction.

Reserving Lands for the Common Good

Near the end of the nineteenth century, evolving land-use attitudes led to formation of the country's first national parks. Bratton (1985) labeled the period including establishment of Yellowstone National Park in 1872, Yosemite, Sequoia, and Kings Canyon national parks in 1890, and most of our geologic national monuments between 1900 and 1940 as the "scenic wonders era." The first park management strategy, which involved setting boundaries and establishing antipoaching patrols, he called the "paradise garden mode." Congressional debates leading to establishment of the first national parks prominently discussed concerns that creating parks amounted to stealing unimproved lands, potentially usable for agriculture or industry, from citizens of the country. The Lockean view of land as a repository of raw materials awaiting improvements did not prevail, but it was partially responsible for refocusing American environmental discussions from issues of intrinsic value to utilitarianism. For example, facilities are provided to promote use by tourists, and park value is measured partially by "visitor days."

Gifford Pinchot, the country's leading proponent of utilitarianism, founding light of the U.S. Forest Service (USFS), and environmental adviser to President Theodore Roosevelt, urged the rational management and use of natural resources for present and future generations (Hargrove 1979, 1989). By the end of the nineteenth century the intrinsic values promoted by John Muir were largely replaced within the federal bureaucracy by the utilitarian values of parks and forests promoted by Pinchot.

Beginning in the early twentieth century, foresters and naturalists working within a Pinchot-style framework began a slow movement back toward intrinsic value through a doctrine of multiple use. This led to an expanded view of what could be included in a national park and resulted in a second phase of acquisition labeled by Bratton (1985) the "biological and landscape era." Management strategies shifted to a "succession and restoration" mode, in which the protection of wildlife and flora became important and the perceived need to manipulate parks to change or restore specific biotic communities increased.

In a third phase, from about 1940 to 1959, only five natural areas were added to the national park system in the United States. Management during this "we think we've done it all era" focused on improved visitor access in a "people plus" management scheme (Bratton 1985). A report by A. Starker Leopold (1963) summarized increasing concerns over excessive human impact on parks and culminated in a call for more ecologically sensitive management. This report stimulated a shift toward ecological concerns in the 1960s and early 1970s.

The most recent phase of acquisition (Bratton 1985) is the "coastal and recreational era," in which remnant landscapes and representative ecosystems and plant communities are protected from the threats posed by summer homes and other human modifications. Much of the justification for providing protection rests on claims that species, ecosystems, and other natural objects have intrinsic value (A. Leopold 1949; Rolston 1988, *this volume*, chap. 6), which, in the view of some, results in our according them rights (C. Stone 1974, 1987; Varner 1987).

Ethical, Philosophical, and Artistic Influences

Despite what appears to be strong historic and philosophic support for attitudes toward nature that lead to environmental abuse, recent polls show that Americans consistently support wildlife, land protection, and improved maintenance of ecosystem integrity. Such attitudes also have a significant, if not as lengthy, history.

Aesthetics, Science, and Philosophy

Attitudes appreciative of nature first appeared in poetry and landscape gardening between 1725 and 1730, and about 30 years later in landscape painting, fiction, and travel literature (Reynolds 1966). Brewer (1904) and Hargrove (1979, 1989) claimed that science created conditions permitting development of attitudes appreciative of nature, and Geike (1962) asserted that these attitudes arose from an interplay between science and art. During the eighteenth and nineteenth centuries, before photography was developed, naturalists had to take artists into the field with them or develop artistic talents of their own. Consequently, an appreciation of en-

vironmental beauty was incorporated into scientific investigations, while the artistic world gained a detailed understanding of nature. Audubon, Wilson, and others accurately interpreted natural beauty in their reports and illustrated its intrinsic value on canvas, while Thoreau, Muir, Emerson, and others translated it into words. Science, art, and literature were thus combined to strongly influence the American population's feelings toward nature over the past 150 years (Hargrove 1979, 1989; Ronald 1989).

In this context, after extensive experience in the wilds of the American West and elsewhere, Aldo Leopold (1949; L. B. Leopold 1953) combined a profound understanding of the science of ecology with a powerful appreciation of the beauty of nature to develop a land ethic that is only now being assimilated by a small but significant fraction of our society (Callicott 1987a, 1991). The land ethic provides a philosophical basis for building a sustainable relationship with our natural world and could be incorporated into nearly any worldview. It requires a deep appreciation of nature and a comparable understanding of ecological relationships. Once that is achieved, it becomes evident that "a thing is right when it tends to preserve the integrity, stability, and beauty of the biotic community. It is wrong when it tends otherwise" (A. Leopold 1949). Earlier in the same work Leopold wrote: "In short, a land ethic changes *Homo sapiens* from conqueror of the land community to plain member and citizen of it. It implies respect for his fellow members, and also respect for the community as such."

The land ethic generated a subdiscipline within philosophy, called environmental ethics (Callicott 1987b), that provides the philosophical underpinnings for the recently defined subdiscipline of conservation biology. Soulé (1986) emphasized that conservation biologists have an obligation to communicate not only their science but also their love of nature derived from a lifetime of study. Naess (1986) demonstrated that the apparent gulf between attitudes and public statements or actions by powerful politicians in Norway could in part be attributed to experts who gave practical, professional recommendations for a problem without considering its broader philosophical or ethical context. He noted that even sympathetic managers, politicians, and decision makers must have the benefit of an expert's "ecosophy," or wisdom of household, rather than merely their ecology, or knowledge of household, if environmentally responsible actions and attitudes are to be woven into social and political fabrics.

Conflicting Values

The dichotomy between perceptions based on our Saxon heritage and religious/philosophic teachings, and perceptions based on appreciation of natural beauty and our understanding of ecology, leads to confrontation. As ecosystem disruption proceeds, the necessity of incorporating the land ethic into society's normal functions becomes increasingly obvious. This necessity is frequently advocated by a majority of the population before social and political institutions are forced to change their traditional practices.

Value change in society requires structural adjustments in social institutions (Petulla 1980), a process frequently delayed by bureaucrats holding other values. Attempts by the Reagan administration to circumvent, subvert, or ignore values expressed by environmental legislation (Pope 1988) provide frustratingly ample evidence of this point. For example, Frank Dunkle, former director of the U.S. Fish and Wildlife Service (USFWS), in a speech to the Montana Ski Area Association on 13 April 1988, left no doubt about his differences with the Endangered Species Act (ESA). Dunkle pointed out that the act "is the most precise, detailed piece of legislation Congress has ever written telling an agency what

to do. The act dominated the U.S. Fish and Wildlife Service." He then proceeded to explain that he had moved the agency out from under the domination of the ESA to make it more responsive to public needs: "I have tried to make our approach to endangered species protection more balanced and realistic. Helping rare species is fine—in fact, it is the law. But punishing people and preventing them from going about normal and legitimate business pursuits in the absence of any solid indications [that] it will help wildlife is not a credible approach to conservation" (Morgan, *The Daily Interlake*, Kalispell, Montana ["Dunkle: USFWS Responds to Needs of the People"], 14 April 1988).

Congress, by contrast, had found earlier that "various species . . . have been rendered extinct as a consequence of economic growth and development untempered by adequate concern and conservation" (sec. 2[a] ESA, 1973). It is abundantly clear that Congress found "the normal and legitimate business pursuits" (which Dunkle wished to protect) one of the major justifications for passing that legislation.

In February 1988 a similar agency intransigence caused the California/Nevada chapter of the American Fisheries Society to sue the National Marine Fisheries Service (NMFS) to list Sacramento River winter chinook salmon (*Oncorhynchus tshawytscha*) as a threatened species. The case, one of a long list of suits against the NMFS for ESA violations, ultimately was settled by an agreement to designate the salmon as threatened through an emergency regulation.

After James Watt became secretary of the interior in 1980, 90% of the regulations from the Federal Office of Surface Mining were rewritten to reflect looser standards. Even then, by 1985, 6000 sites had been illegally strip-mined. Unbelievably, only 8% of the civil penalties assessed by the agency were collected (Pope 1988). From 1981 to 1988 more than

$6 billion designated for park acquisition and recreational development was left unspent by the Reagan administration (*New York Times*, 15 March 1988). This occurred despite documentation in the 1987 *Report of the President's Commission* that the USFS in 1985 considered only 29% of its recreational facilities adequately maintained (Pope 1988).

Difficulties with agency implementation demonstrate that both legislation and support by an administration are essential to the process of institutionalizing environmental values. Over the past hundred years, incorporation of a new environmental ethic into bureaucracies has been a result of the slow evolution of wildlife law. This evolution has seen "wildlife" come to mean "all fauna and flora," management goals turn from single species toward ecosystems, and legally protected uses of wildlife extend far beyond those associated with food or other commercial products (Bean 1983). During the 1970s, many federal agencies were attempting to deal with the trauma of internalizing new, legislatively mandated directives. By focusing single-mindedly on economic growth during the 1980s, the Reagan administration ripped most governmental policies compatible with an environmental ethic from the painstakingly woven fabric of our federal bureaucracies (Pope 1988). Public opinion, environmental law, and the ecological requirements of maintaining a sustainable earth demand reversal of that trend.

Constitutional Basis of Federal Law

Federal wildlife law is largely statutory, and therefore more recent than the body of common law that guides many of our fundamental legal relationships. During the nineteenth century, the doctrine of state ownership of wildlife was developed, based on the transfer of the king's public trust responsibilities to the states, but subject to the rights given by the Constitution to the federal government (*Martin v. Waddell*, 41 U.S. [16 Pet.] 234 [1842]). Because of a tendency to ignore the constitutional qualifier, the state's public trust responsibilities and rights, as outlined in more detail in *Geer v. Connecticut* (161 U.S. 519 [1896]), were used by many to argue that the federal government was excluded from developing wildlife law. That position was strengthened in 1912 when the courts, expanding on *Geer v. Connecticut*, explicitly confirmed the state supremacy argument by ruling that state ownership precluded federal wildlife regulation (*The Abby Dodge*, 223 U.S. 166 [1912]). This was the first and last time the courts so ruled. Since then, at least three separate constitutional powers (commerce, treaty making, and property) have been identified that confer authority for federal wildlife regulation. These provide the basis for the development of federal wildlife law (Bean 1983). Congress typically has availed itself of these powers by regulating commerce in wildlife and wildlife products, regulating take, ratifying treaties negotiated by the executive branch of government, acquiring and managing wildlife habitat, and requiring consideration of impacts on wildlife of various forms of development.

Commerce

The Lacy Act of 1900, controlling interstate transportation of wildlife and birds and motivated in part by the demise of the passenger pigeon, was the first federal wildlife legislation. The Black Bass Act of 1926 regulated interstate transportation of largemouth bass (*Micropterus salmoides*). These two acts, expanded upon and ultimately combined in 1981, formed the cornerstone of federal efforts to protect wildlife through commerce power (Bean 1983). It was not until 1977 that the U.S. Supreme Court spelled out the scope of federal wildlife regulatory power conferred by the commerce clause. In a series of cases (*Douglas v. Seacoast Products, Inc.*, 431 U.S. [1977]; *Andrus v. Allard*, 444 U.S. 51 [1977];

Palila v. Hawaii Department of Land and Natural Resources, 471 F. Supp. 985 [D.Ha. 1979], aff'd on other grounds, 639 F.3d 495 [9th Cir. 1981]), the Supreme Court ruled that the commerce clause confers wide powers indeed.

Treaties

Federal regulation of wildlife under the treaty-making power was initiated with the Migratory Bird Treaty signed with Great Britain on behalf of Canada in 1916. Signature was rushed because of government concern regarding constitutional support for the Migratory Bird Act of 1913 in the absence of such a treaty. In fact, a legal challenge of that act was dismissed by the Supreme Court following the passage of implementing legislation for the treaty in 1918. Further litigation challenging the federal government's authority to regulate migratory wildlife established beyond a doubt the supremacy of the federal treaty-making power (*Missouri v. Holland*, 252 U.S. 416 [1920]).

Treaties signed with various American Indian nations have considerable potential for establishing a basis for the protection of western fishes. Courts, especially recently, have attempted to interpret Indian treaties in a way that is consistent with the probable understanding of them by the Indians at the time they were signed (Bean 1983). In a number of instances the degree to which treaty Indians are freed from the regulatory authority of state and federal governments has been specified. Judge George H. Boldt, in a series of decisions (*Puyallup Tribe v. Washington Department of Game*, 391 U.S. 392 [1968]; 414 U.S. 44 [1973]; 433 U.S. 165 [1977]) suggested that not only are treaty Indians freed from state regulatory authority, but that the treaties may impose affirmative duties on state and federal governments to protect the fish and wildlife resources to which tribal rights are granted. Perhaps this principle could be tested

further with respect to government responsibilities to protect the endangered cui-ui (*Chasmistes cujus*) on the Pyramid Lake Indian Reservation, and in other situations where Indians were granted fish and wildlife rights by treaty.

Property

The federal property power is most often asserted through hunting or fishing restrictions or governmental removal of wildlife on property acquired and managed by a federal agency. In 1894 hunting was prohibited in Yellowstone National Park. The national wildlife refuge system was started in 1903 to protect wildlife from year-round market hunting and habitat destruction. Despite challenges of the federal property authority via the Migratory Bird Act, hunting and fishing restrictions on national parks and wildlife refuges were not challenged, undoubtedly because everyone thought the federal government had the same authority as any other landowner to restrict actions on their property (Bean 1983).

A number of cases have, however, challenged the federal government's authority to regulate wildlife, asserting that some connections to property protection must be demonstrated. The courts generally ruled in favor of broad discretion being given to federal regulation of wildlife pursuant to the property clause, and in a definitive ruling (*Kleppe v. New Mexico*, 426 U.S. [1976]), the Supreme Court held that the protection of federal land is a sufficient, but not a necessary, basis for action under the property clause, and asserted that the property power necessarily includes the authority to regulate and protect wildlife living there. The courts consistently seem to uphold the principle that any reasonable federal property right, wholly without regard to state law, may be recognized (*U.S. v. Albrecht*, 496 F. 2d [8th Cir. 1974]; *North Dakota v. U.S.*, 103 S. Ct. [1985]). This line of reasoning also has been accepted with regard to implied reservation of a federal water right on lands owned by the government (*Winters v. U.S.*, 128 [1976]; Deacon and Williams, *this volume*, chap. 5).

Inherent Agency Obligations

Despite the fact that during the eighteenth and nineteenth centuries laws were made to promote quick settlement, not conservation (Petulla 1980), much land remains in federal ownership. This is partly because congressional reaction to abuse of settled lands resulted in removing certain especially valuable lands from laws promoting disposal (Bean 1983). The specific mechanisms used to remove the lands strongly influence the strategies available for preventing reductions in species diversity.

Forest Service

The first systematic withdrawal of federal lands was made by President Theodore Roosevelt pursuant to the Forest Reserve Act of 1891. Because he acted so extensively, Congress subsequently limited the lands a president could withdraw by passing the USFS Organic Administration Act of 1897, specifying that only lands needed to improve and protect the forest within the reservation, secure favorable conditions of water flows, or furnish a continuous supply of timber could be withdrawn. The USFS pursued a multiple-use strategy on their lands, through which the needs of fish and wildlife could be considered even prior to specific conveyance of that authority by the Multiple Use–Sustained Yield Act of 1960.

Unfortunately, in a 1978 Supreme Court case involving the Organic Act (*U.S. v. New Mexico*, 438 U.S. 696 [1978]), the Court ruled that reservation of water rights needed for protection of in-stream needs of fish and wildlife was not implicit in establishment of the Gila National Forest, New Mexico. Further blows to the interests of fish and wildlife conservation on national forests have been struck by court rulings that "due consideration" means some, but not equal, consideration of competing values. USFS hydrologists are increasingly securing favorable conditions of flow by attempting to protect the hydraulic, rather than the biotic, integrity of streams. Many argue that water rights for that purpose fall under the implied reservation doctrine. It appears that preservation of native fishes on USFS lands, except in cases involving threatened or endangered species, will be mostly a by-product of efforts to maintain the hydraulic integrity of stream channels.

National Park Service (USNPS)

On the other hand, the U.S. National Park Service Act of 1916 established a system of parks, monuments, seashores, lakeshores, wild rivers, and other preserves for the recreational enjoyment of present and future generations. In contrast to authorities in the USFS Organic Act, the conservation of wildlife was explicitly recognized as a purpose for parks. Several court cases have given the secretary of the interior considerable discretionary authority about when or whether to act on behalf of the resource. When the secretary takes action to leave the parks "unimpaired for the enjoyment of future generations," the courts find ample support for protection of fish and wildlife (*Cappaert v. U.S.*, 426 U.S. 128 [1976]; *U.S. v. Brown*, 552 F.2d 817 [8th Cir. 1977]).

Fish and Wildlife Service

The acquisition of national wildlife refuges was stimulated by the Migratory Bird Treaty Act of 1918 and the Migratory Bird Conservation Act of 1929. In 1966 refuges were consolidated as dominant-use lands under management of the USFWS pursuant to the National Wildlife Refuge System Administration Act of 1929, which provided that refuges acquired pursuant to its authority be operated as "inviolate sanctuaries" (16 U.S. C sec. 715d

[1976]). This was amended in 1949 to author- ize public hunting on up to 25% of an area, in 1958 on up to 40%, in 1966 to restrict the 40% to waterfowl and allow compatible hunting on 100%, and in 1978 to authorize waterfowl hunting on more than 40% of an area, if "beneficial to waterfowl." The Refuge Recreation Act of 1962 allows public recre- ation as an incidental or secondary use, and the Refuge Revenue Sharing Act of 1964 re- quires payments to counties from receipts of revenue-generating activities on refuges. Both acts provide pressures and incentives to man- age refuges for purposes other than fish and wildlife needs.

The right of the secretary of the interior to prohibit incompatible secondary uses on ref- uges was confirmed, however, in *Coupland v. Morton* (5 Envt. L. Rep. [Envt. L. Inst.] 20507 (4th Cir. 7 July 1975]), and established as a judicially enforceable limitation on the secre- tary's discretion in *Defenders of Wildlife v. An- drus* (455 F. Supp. 446, 449 [D.D.C. 1978]). The protection of western fishes on national wildlife refuges appears to have adequate statutory and legal support, but as with the USNPS, the secretary has considerable dis- cretionary authority.

Bureau of Land Management (USBLM)

The Dust Bowl of the 1930s stimulated Con- gress to provide for management of unre- served and unappropriated public domain in the western United States with passage of the Taylor Grazing Act of 1934. Until the mid- 1970s this vast area was managed almost ex- clusively to serve the livestock industry. In 1975 the USBLM reported 83% of the range in fair or worse condition. They estimated that forage conditions were improving on 19%, declining on 16%, and indefinite or stable on 65% of the range in the western United States (Braun 1986). Appalled by continued degra- dation of vast tracts of land, Congress passed

the Federal Land Policy Management Act of 1976 and the Public Rangeland Improvement Act of 1978, which provided a mandate to "restore a viable ecological system that bene- fits both range users and the wildlife habitat." This represented an effort to broaden the con- stituency served by the USBLM and force a shift from single-use toward multiple-use management.

In response, the "Sagebrush Rebellion" was mounted by the livestock industry, with the support of many western states. With continu- ing pressure from their traditional constitu- ency and administrative direction from avowed sagebrush rebels (Robert Burford, Donald Hodel, Ronald Reagan, and James Watt, for example), the USBLM demonstrated its ability to circumvent congressional intent. When the "rebellion" was challenged in court (*Natural Resources Defense Council v. Hodel* (624 F. Supp. 1045 [D. Nev. 1985]), the court stated: "Although I might privately agree with plaintiffs that a more aggressive approach to range improvement would be environmentally preferable, or might even be closer to what Congress had in mind, . . . the courts are not at liberty to break the tie choosing one theory of range management as superior to another." In effect, the court ruled that the USBLM need not make more than a convincing pretense at improving range conditions. This situation promises little hope for widespread restora- tion of ecosystem integrity, or systematic pro- tection of native fishes on USBLM lands in the absence of administrative desire to do so.

External Agency Obligations

A number of laws passed by Congress impose on *all* federal agencies environmental man- dates to be considered in addition to require- ments enunciated in specific acts governing the agency. Many of these "external" acts involve requirements that fish and wildlife needs be considered during project planning. The imposition of wildlife values on federal agencies whose primary purpose focuses elsewhere is one of the most significant recent trends in national environmental law (Bean 1978, 1983).

Fish and Wildlife Coordination Act

The Fish and Wildlife Coordination Act of 1934 was a remarkably forward-looking piece of legislation (Bean 1978, 1983). It called for investigation into the effects of pollution on wildlife, encouraged the development of a program to maintain wildlife on federal lands, and promised a national program of conservation. Congress attempted to add enforceable provisions to the act in 1946, and again in 1958, by adding language conferring "equal consideration" to wildlife and directing construction agencies to give "full consideration" to recommendations of wildlife agencies.

The independent impact of this law was never fully felt, however, because the courts ruled in 1972 that good-faith compliance with the National Environmental Policy Act (NEPA) automatically takes into consideration all the factors required by the Coordination Act (*Environmental Defense Fund v. U.S. Army Corps of Engineers*, 325 F. Supp. 749 [E.D. Ark, 1971], injunction dissolved, 342 F. Supp. 1211 [1972], aff'd, 470 F.2d 289 [8th Cir. 1972]). Hopes for a synergistic relationship between the Coordination Act and the NEPA (Guilbert 1974; Shipley 1974) were dashed when the court found that the USFWS did not need to respond to requests for consultation in the face of funding and personnel shortages (*Sun Enterprises, Ltd. v. Train*, 394 F. Supp. 211 (S.D.N.Y. 1975], aff'd, 532 F.2d 280 [2d Cir. 1976]), nor could private plaintiffs assert causes of action under the Coordination Act (*Sierra Club v. Morton*, 400 F. Supp. 610 [N.D. Cal. 1975]). The Fish and Wildlife Coordination Act was slow in achieving clear enunciation, and when its provisions finally were complete, it was overshadowed by the National Environmental Policy Act.

National Environmental Policy Act

Although the National Environmental Policy Act of 1969 never mentions the word *wildlife*, great benefits accrue to wildlife as an effect of implementing this act's broad environmental policies. The NEPA requires development of environmental impact statements that attempt to anticipate and, where possible, avoid adverse environmental effects. Application of the act's requirement for environmental impact analysis has been extensively litigated, with the answers turning on whether the action is "major," "federal," or "significant" (Bean 1983), ambiguous terms that defy definition. A related impact of the act is the enlargement of federal agencies' statutory authority for environmental protection (*Public Service Company of New Hampshire v. U.S. Nuclear Regulatory Commission*, 582 F.2d 77 [1st Cir. 1978]; *Detroit Edison Company v. U.S. Nuclear Regulatory Commission*, 630 F.2d 450 [6th Cir. 1980]).

Although many early interpretations of the NEPA discussed substantive and judicially enforceable standards of environmental quality, the courts have found the statute to be procedural in nature (Bean 1983). In *Vermont Yankee Nuclear Power Corporation v. National Resources Defense Council* (435 U.S. 519 [1978]), the Supreme Court held that although the act established "significant goals for the Nation," its obligations were essentially procedural. This concept of the act also was enunciated in *Stryker's Bay Neighborhood Council, Inc. v. Karlen* (444 U.S. 223 [1980]), although the Court also implied that an agency's considerations of environmental consequences can be struck down if they are arbitrary and capricious. At base, though, the NEPA simply is a requirement for full disclosure.

Federal Water Pollution Control (Clean Water) Act

Braun (1986) noted that the Federal Water Pollution Control Act of 1972 provides the

Bureau of Land Management and all other federal agencies with a legally enforceable duty to restore riparian zones. This act has been successfully applied to constrain Forest Service road construction and timber sale proposals because of water quality concerns (*Northwestern Indian Cemetery Protective Association v. Peterson*, 795 F.2d 688 [9th Cir. 1986]). Action taken to compel compliance may be one of the most effective strategies available to restore viable populations of native fishes in many western streams, especially on public domain. In fact, sometimes the simple threat of a "Clean Water" Act suit can be effective, e.g., threat of a lawsuit by the Sierra Club Legal Defense Fund and others in 1983 appeared instrumental in bringing the developers of Ash Meadows, Nevada, to the bargaining table (Deacon and Williams, *this volume*, chap. 5).

Endangered Species Act

The Endangered Species Act of 1973 was the first federal statute to embody a comprehensive effort at wildlife preservation and, as amended, is one of the strongest pieces of environmental legislation ever passed. The 1966 Endangered Species Preservation Act gave authority to the USFWS to "protect those native wildlife threatened with extinction," and provided up to $15 million annually from the Land and Water Conservation Fund for that purpose. The act's provisions involving listing and coordination with other federal and state agencies were ambiguous. Expansion of the scope, clarification of some of the ambiguities, and modification of the 1966 act were accomplished in 1969, 1973, 1978, 1982, and 1988, resulting in a gradual strengthening of legal rights and protections for vulnerable species.

Species are brought under the protection of the ESA by being placed on a list of threatened or endangered wildlife and plants. Amendments in 1982 made it clear that inclusion on the list must be based solely on an evaluation of biological status. While there are numerous provisions, the core of the ESA directs federal agencies to conserve listed species and prohibits them from undertaking, funding, or permitting any action likely to compromise the continued existence of threatened or endangered species. The act also prohibits any person from taking or harming a listed species.

For a time there was uncertainty about whether indirect effects resulting from habitat modifications constituted violations of the ESA. In a suit also noteworthy because one of the plaintiffs was the listed species itself (*Palila v. Hawaii Department of Land and Water Resources*, 4761 F. Supp. 985 [D. Ha. 1979], aff'd, 639 F.2d 495 [9th Cir. 1981]), the courts ruled that habitat modification constituted "harm" and was therefore prohibited by the ESA. An administrative attempt by the USFWS to overrule the *Palila* decision by eliminating habitat modification from its regulatory definition of "harm" drew a storm of protest, and ultimately the USFWS withdrew its proposal.

Summary and Conclusions

Federal wildlife law was originally a patchwork designed to fill gaps left by state regulations or address specific environmental contingencies. Initially, it was directed only toward game species or those with commercial worth. That scope expanded slowly until, under the ESA, almost all members of the animal and plant kingdoms are included. Legal recognition of the value of wildlife for food or commercial products has expanded to include symbolic (bald eagle), aesthetic, ecological, educational, historical, recreational, and scientific value to the nation and its people. This comes close to legal recognition of intrinsic values. Management goals also are changing, a trend most obviously embodied in law by the Marine Mammal Protection Act (MMPA), which signaled a trend away from species- and harvest-oriented management toward ecosystem management.

The ESA, MMPA, and Fishery Conservation and Management Act of 1976 permit and encourage active participation by the public. They bring wildlife law into the domain of administrative law, and have led to an expanded oversight role for the judiciary. The courts have a long history of adjudicating disputes between states and the federal government over the constitutional limits of their respective wildlife authorities. Judicial oversight of the actual implementation of wildlife policies has been much more recent and is the result of legislation imposing increasingly detailed standards for wildlife administrators and federal agencies, and providing opportunities for citizen involvement. In turn, active court intervention has thrown back to Congress the task of fine-tuning the originally broad regulatory standards, heretofore the exclusive province of federal wildlife agencies.

Used as intended, the ESA provides enforceable tools for attempting to prevent the extinction of any species, subspecies, or (for vertebrates only) distinct population. Values expressed by the ESA and now being more fully enunciated in the emerging Biodiversity Act (which calls for management to preserve genetic, species, and ecosystem diversity) are compatible with Aldo Leopold's land ethic and the developing field of environmental ethics discussed by Rolston (1988, *this volume*, chap. 6), Callicott (1987a), and others. To at least some degree, Stone's (1974) revolutionary idea of according legal standing to natural objects has happened. Because of the ESA, species, and perhaps ecosystems, do indeed have legal rights (Varner 1987). It seems a small step to give them formal legal standing, thereby conferring on them the authority to seek enforcement actions brought entirely in their own names!

A variety of legislative and judicial tools are available to aid in the battle against extinction of western fishes. Many battles will continue to be waged over differing perceptions of ethical responsibilities under the law. Success will be measured not by numbers of species saved from extinction but by how successfully we are able to shift society's values and institutions toward building a sustainable earth. As Rolston (*this volume*, chap. 6) notes, we must work to sustain the creative evolutionary process, not just its products. We must learn to behave as plain members and citizens of the biotic community, to live within the carrying capacity of the earth for the sake of all members of the community, and to exercise dominion as stewards rather than as a destructive force.

Chapter 8

Evolution of a Cooperative Recovery Program for Endangered Fishes in the Upper Colorado River Basin

Richard S. Wydoski and John Hamill

Introduction

The Colorado River originates in clear, cold streams of the Rocky, Uinta, and Wind River mountains of the western United States, flows through high deserts, and has carved spectacular canyons by erosion of soft sandstones and other rocks. Historically, the river and its larger tributaries were warm and turbid in summer and characterized by large changes in water volumes and velocities. Several unique fishes evolved in this distinctive riverine environment, where 74% of the native fish fauna was endemic (R. R. Miller 1959). Three of the endemic large-river fishes—the Colorado squawfish (*Ptychocheilus lucius*), humpback chub (*Gila cypha*), and bonytail (*G. elegans*)—are federally listed as endangered (U.S. Fish and Wildlife Service [USFWS] 1990). A fourth, the razorback sucker (*Xyrauchen texanus*), has been proposed for listing as endangered. These four species are collectively referred to as the endangered Colorado River fishes, and this paper describes and discusses the evolution of a program toward their recovery in the upper Colorado River basin (hereinafter referred to as the upper basin).

We thank the many persons who reviewed and provided suggestions for improvement of the manuscript, especially J. Bennett, L. Kaeding, W. Miller, H. Tyus, and M. Zallen.

Setting the Stage

An understanding of recovery efforts for endangered fishes of the upper basin requires an appreciation of the importance of the rivers as a source of water for municipal, industrial, and agricultural purposes. Settlement of the arid West began more than a hundred years ago and emphasized "mastery over nature." The Colorado River was thus altered to develop and control its waters to such an extent that it has been described as the most heavily used, controlled, and fought-over river in the world (Crawford and Petersen 1974). Although the basin receives less precipitation per square kilometer of drainage than any other major watershed in the United States, it provides more than fifteen million people with water (Utah Water Research Laboratory 1975). Further alteration of this already over-allocated resource is considered by some people as necessary to supply water to an expanding human population and to develop some of the largest fuel deposits (coal, oil, oil shales, and uranium) in the nation (Bishop et al. 1975).

Legal control of the river began with the Colorado River Compact of 1922. This compact divided the water between the seven states comprising the upper and lower basins, anticipated demands for water in Mexico that were eventually agreed upon, and imposed certain restrictions on quantities and scheduling of flows (Harris et al. 1982). The 1948 Upper Colorado River Compact provided consumptive water rights for Arizona, California, and Nevada, and apportioned the remainder to the upper basin states of Colorado, New Mexico, Utah, and Wyoming. Division of water among the states cleared the way for development of several major water projects in the upper basin. In 1956 the Colorado River Storage Project Act authorized construction of large main-stem dams on the upper Colorado and its tributaries. The project included six reservoirs: Blue Mesa, Crystal, Flaming Gorge, Morrow Point, Navajo, and Powell. The last to be constructed, Lake Powell, was completed in 1962 and filled by 1980.

In the early 1960s the endangered Colorado River fishes were considered undesirable "rough" fish by conservation agencies. As a result, the largest rotenone treatment ever applied in the United States until that time was conducted to remove them and other unwanted species such as common carp (*Cyprinus carpio*) from the Green River above Flaming Gorge Dam. The goal was to create a sport fishery for introduced salmonids (Binns 1965; Holden, *this volume*, chap. 3). Before the rotenone could be completely detoxified, it continued downstream into Dinosaur National Monument (NM). Reductions in native fish numbers were greatest in the uppermost part of the monument, at the Gates of Lodore, and diminished as the chemical moved downstream. Although the poison adversely affected native fishes and invertebrates, the eventual ecological changes to riverine environments caused by Flaming Gorge Dam were believed more important in ultimately reducing their populations (Holden, *this volume*, chap. 3).

The Colorado River Storage Project dams and reservoirs, along with private and local water developments, all combined to alter nat-

ural flow, water temperature, and sediment transport in much of the upper basin. Peak spring flows were diminished and low stable flows at other times of year were elevated by reservoir releases that changed the natural hydrograph (Vanicek et al. 1970). These projects also resulted in direct losses of stream habitat through inundation by reservoirs and blockage of migration routes.

Concurrently, various non-native fishes were intentionally or accidentally introduced. In 1976 the Colorado River Wildlife Council listed twenty species (40%) as native to the system and thirty (60%) as introduced (W. M. Richardson 1976). At the same time, Holden and Stalnaker (1975a) reported ten native fishes (34.5%) and nineteen non-native (65.5%) in the upper basin. By 1982 the nonnative species had increased to 76% of the fifty-five fishes known to occur in the upper basin (Tyus et al. 1982a). It is generally believed that predation and competition by introduced fishes are major factors adversely affecting the endemic species.

Competition for Water

Drought in the late 1800s followed by prolonged flooding in the early 1900s in the lower Colorado River basin stimulated demands for control of the Colorado River (Fradkin 1984). Construction of Roosevelt Dam on the Salt River in 1913 and Hoover Dam in 1935, followed by other main-stem dams, changed much of the free-flowing river to a lacustrine environment. Stream flow and temperature regimes in the remaining river were greatly altered. These events were closely followed by declines in native fishes in the lower basin. Colorado squawfish were extirpated, and bonytails, humpback chubs, and razorback suckers were drastically reduced in numbers and distribution. Only large, old bonytails and razorback suckers were still found in reservoirs of the lower basin by the

1980s (Minckley et al., *this volume*, chap. 17). A viable population of humpback chubs remained in the Little Colorado River near its confluence with the Colorado. Few early studies on the biology and ecological requirements of these fishes were made in the lower basin, but their declines were documented by R. R. Miller (1961), Minckley and Deacon (1968), and others.

The four endangered fishes were still found in the unaltered upper basin during the early 1960s (Fig. 8-1). Limited preimpoundment studies indicated that numbers of some species were low, and further suggested that these fishes may never have been abundant. Competition for water resources intensified in the upper basin during the 1960s. Various reports summarized supply and demand of water and evaluated alternative uses of the water resources (e.g., National Research Council 1968). The U.S. Water Resources Council (1968) completed a comprehensive appraisal of water resources and their geographic distributions, made projections of future requirements, defined problems and needs, and presented a program for water development and conservation to the year 2020. The council concluded that ample resources were available to meet fishing needs in the upper basin if minimum stream flows and adequate conservation pools in reservoirs were maintained for game fish. The report did not include conservation of endangered Colorado River fishes because their status was not then recognized or appreciated.

In the mid-1970s the Western U.S. Water Plan (known as the Westwide Study), conceived under authority of the Colorado River Basin Project Act of 1968, proposed development of adequate information as a basis for decisions on water and related resources in the eleven western states (U.S. Bureau of Reclamation [USBR] 1975). The plan focused primarily on the quantity and quality of water. In 1976 the USFWS funded a symposium through Resources for the Future to summarize probable impacts of potential energy developments on water, fish (including endangered species), and wildlife in the upper Colorado River (Spofford et al. 1980).

Future demands and allocations of Colorado River water were concisely summarized by Weatherford and Jacoby (1975):

> In broad terms, the problem of managing the Colorado River is the problem of allocating a flow resource in such a way as to satisfy legally preferred current demands without foreclosing the satisfaction of a different set or configuration of demands in the future. When so viewed, it is clear that there will be no single or final solution to the problems of allocation and management in the Colorado River basin. The time for seriously addressing emerging generation of problems, however, is now.

Key Federal Environmental Legislation

The American public's concern about environmental issues has been marked by surges and declines during the past century. This concern was interrupted by two world wars and an economic depression, but a major revival occurred (McEvoy 1973) when the Fish and Wildlife Coordination Act of 1934 was amended in 1958 to confer "equal consideration" to wildlife and directed development programs to give "full consideration" to recommendations of wildlife agencies (Williams and Deacon, *this volume*, chap. 7). It was not until the 1960s, however, that an "environmental movement" took shape in the United States.

During this period the federal government acknowledged a national responsibility to save endangered species through the Endangered Species Preservation Act of 1966, amended as the Endangered Species Conservation Act of 1969. The National Environmental Policy Act (NEPA) also became law that same year, and regulations to implement it were published by the Council on Environmental Quality. The NEPA requires that environmental impacts be described, alternative actions be considered, and public input be sought for all federal development projects.

The most significant federal legislation providing protection for endangered fauna and flora, however, was the Endangered Species Act (ESA) of 1973. Section 7 of this act is particularly significant because it states that all federal agencies must "insure that actions authorized, funded, or carried out by them do not jeopardize the continued existence of such endangered species and threatened species or result in the destruction or modification of habitat of such species." Section 4 provides for the listing and recovery of threatened or endangered species and directs the secretary of the interior to develop and implement recovery plans. Section 6 encourages the federal government to cooperate with the states in conservation of threatened or endangered species and provides funds to states to conduct studies on such species. Section 9 prohibits the taking (including activities from harassment to capture) of listed species without proper federal and state permits. The 1973 act, and later amendments of 1978, 1982, and 1988, provided the foundation for the recovery efforts now under way for endangered Colorado River fishes. This act, in concert with the NEPA and the Fish and Wildlife Coordination Act, provides the major legal mandate for recovery efforts in the upper basin.

Section 7 Consultation Procedures

Section 7 of the ESA requires federal agencies to determine whether their proposed action may affect a threatened or endangered species. If so, formal consultation with the USFWS is required. One of the USFWS's principal concerns raised by the section 7 consultation process was the cumulative effect of water depletions on habitats of endangered Colorado River fishes. The USFWS maintained that water depletions:

1. reduced the quantity and quality of backwater habitats formed by high runoff during spring—habitats that are used extensively during migration and spawning;

2. reduced the availability of nursery areas or rearing habitat essential for survival of young;

3. reduced sediment transport capacity of the river, which in turn affected basic productivity and availability of important habitats used by the endangered fishes;

4. created river habitats favoring non-native fishes that compete or prey upon endangered fishes; and

5. reduced the future flexibility to manage stream flows to benefit endangered fishes (i.e., the water that is consumptively used cannot be managed, appropriated, or acquired to benefit endangered fishes).

Beginning in 1977 and continuing through 1981, the USFWS wrote "jeopardy" biological opinions for all major water-depletion projects in the upper basin. However, none of these projects was cancelled, because each opinion contained "reasonable and prudent alternatives" that, if implemented, would offset adverse impacts to endangered fishes. The most common alternative was the commitment by the project sponsor (usually the USBR) to provide releases from existing storage reservoirs (e.g., Flaming Gorge and Blue Mesa) to offset water depletions. Major consultations completed using this approach included the Dolores, Dallas Creek, and Central Utah projects.

In 1981 the USFWS reviewed the Windy Gap and Moon Lake projects, which would deplete water from the Colorado, Green, and White rivers. These projects were not capable of guaranteeing releases to offset depletions. They were privately funded, and the sponsors wished to avoid jeopardy opinions because, they argued, such opinions would affect their ability to obtain financing and necessary con-

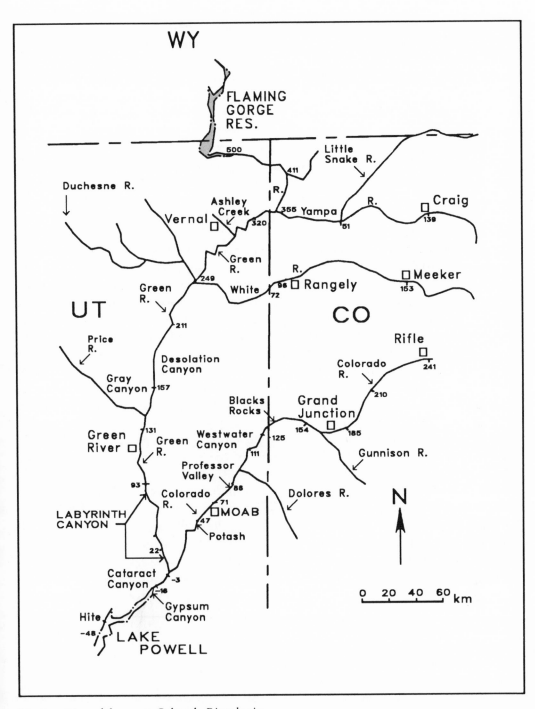

Fig. 8-1. Map of the upper Colorado River basin, western United States, showing locations of major cities, rivers, and canyon reaches. Numbers refer to river kilometers, with 0.0 at the confluence of the Colorado and Green rivers.

struction permits. In response to this issue, the USFWS developed an approach known as the Windy Gap process, which allowed projects to proceed with a "no jeopardy" finding assuming that the sponsor contributed monetarily to a comprehensive recovery effort for endangered fishes.

The Windy Gap process was based on an estimate that it would cost $25 million to implement a comprehensive fishery conservation effort for the upper basin. Under this process, sponsors of a private development project were required to provide monetary compensation for actions that would result in jeopardy to threatened or endangered species, based on a formula that considered quantity of water to be depleted and the volume of water remaining after interstate compact flows to the lower basin were delivered. Consultation under the Windy Gap process resulted in nearly $1.3 million being provided to the USFWS between 1981 and 1987. These funds were used for studies of ecological requirements, propagation and stocking, habitat improvements, and other recovery efforts. The Windy Gap process did not apply to USBR projects. That agency instead agreed to set aside water in its reservoirs for later release to habitats occupied by endangered fishes.

The Windy Gap process proved controversial. In testimony before Congress, several environmental groups alleged that it entailed "excessive and unnecessary risks of extinction for these species" (Environmental Defense Fund et. al. 1985). The USFWS discontinued the process on large water projects after 1985 and formed the Interagency Coordinating Committee that developed a section 7 consultation process acceptable to both environmental and water-development interests.

Ironically, the new process that is described in the Recovery Program for Endangered Fish Species in the Upper Colorado River Basin (USFWS 1987a) is patterned after the Windy Gap process. Under the recovery program di-

rect project impacts, such as obstruction of migration routes or physical alteration of occupied habitat, will be evaluated on a case-by-case basis by the USFWS during the section 7 process. Whenever possible, the USFWS will suggest reasonable and prudent alternatives to offset direct project impacts. However, the recovery program identifies 159 km of the Colorado, Green, Yampa, and White rivers as extremely important to the protection and recovery of the endangered fishes. Direct impacts to these areas would likely result in a "jeopardy" opinion without any reasonable and prudent alternatives to avoid jeopardy.

Indirect impacts caused by water depletions will be offset by a onetime contribution of ten dollars per acre-foot of a project's average annual depletion and used to fund recovery activities under the recovery program. This amount is adjusted for inflation; for 1991 it was set at $10.91 per acre-foot. However, such contributions are subject to a determination by the USFWS that progress under the recovery program has been sufficient to offset the impacts of a proposed project, especially in the protection of in-stream flows. To date, the USFWS has maintained that there is significant uncertainty that legal protection of in-stream flows will be achieved under the recovery program in a timely manner. As a result, the USFWS has required proponents of large depletion projects to agree to additional conservation measures such as dedicating a quantity of water for in-stream flows required by the endangered fishes in order to receive a "no jeopardy" opinion.

Early Studies

Early detailed investigations of the biota of the upper Colorado River basin were directed toward pre- and postimpoundment studies to answer questions about water quality (W. F. Sigler et al. 1966; Tsivoglov et al. 1959) or the probable effects of future alterations in water quality and stream flow on game fish (G. C. Powell 1958; Weber 1959; Coon 1965).

In March 1963, Secretary of the Interior Stewart Udall requested the Utah Cooperative Fishery Research Unit to investigate changes in habitat and fish populations of the Green River in Dinosaur NM, Colorado-Utah, resulting from the 1962 poisoning operation and closure of Flaming Gorge Dam. These studies, conducted between 1963 and 1966, concluded that cold-water releases caused the disappearance of Colorado squawfish, bonytail, humpback chub, and razorback sucker from the 105-km reach below the dam (Vanicek 1967; Vanicek et al,. 1970). The studies also found that the Green River below its confluence with the Yampa River still contained the endangered Colorado River fishes, a presumed consequence of the Yampa's ameliorating effect on water temperature.

Work by the Cooperative Fishery Unit during 1963–1966 focused on basic life history of the Colorado squawfish and bonytail (Vanicek 1967; Vanicek and Kramer 1969; Vanicek et al. 1970) and on macroinvertebrate abundance and distribution (Pearson 1967; Pearson et al. 1968). These studies suggested that year classes of Colorado squawfish were strong in 1959, 1961, 1963, 1964, and possibly 1966 (Vanicek 1967), but weak in 1962 and 1965. Formation of strong year classes during years when stream flow and water temperatures are more suitable for survival of recruits could be an evolutionary life-history strategy of this species (Tyus 1986, *this volume*, chap. 19). Vanicek (1967) also reported difficulty in separating the various species of *Gila*, especially during early life stages. This led to a study of *Gila* taxonomy by Holden (1968; Holden and Stalnaker 1970), but the taxonomic questions were not fully answered. In 1988 the USFWS contracted with the Smithsonian Institution to review and develop a program to resolve the problem.

Holden (1973) studied relative abundance and distribution of native fishes in the upper basin and documented problems with recruitment of squawfish in Echo Park (Dinosaur NM), where juveniles were abundant in 1968, scarce in 1969, and nonexistent in 1970. However, he reported young-of-year in Desolation Canyon in 1971, and in the Green River in Canyonlands National Park in 1970 and 1971. During 1974–1976, studies focused on obtaining further life-history information (McAda 1977; Seethaler 1978). McAda (1977; McAda and Wydoski 1980) provided new biological information and synthesized all known life-history data on razorback suckers in the upper basin. Seethaler (1978) provided a comprehensive summary of the life history of Colorado squawfish, including the first data on reproduction, fecundity, maturity, and early development. An annotated bibliography assembled in 1976 and updated in 1980 listed studies of the native fishes and macroinvertebrates, and provided selected references that discussed economic, political, and sociological factors confounding management of Colorado River fishes (Wydoski et al. 1980). Studies of macroinvertebrates were completed by Carlson et al. (1979).

Colorado River Fish Project

The requirements of section 7 of the ESA had potentially serious ramifications for new water projects proposed by the USBR, and for the operation of several of their existing facilities such as Flaming Gorge and Blue Mesa reservoirs. However, lack of data on in-stream flow and other habitat requirements of endangered Colorado River fishes was recognized by both the USBR and USFWS as a serious impediment to the development of reasonable and prudent alternatives for operating existing, and designing proposed, projects. As a result, the USBR agreed to provide funding to gather essential information on ecological requirements of the rare fishes.

In June 1979 the USFWS and USBR signed a memorandum of understanding (MOU) whereby the USBR agreed to fund a comprehensive investigation of the endangered Colorado River fishes in the upper basin. The USFWS agreed to obtain the information essential to providing biological opinions on impacts of existing and proposed water-development projects. The primary objective of the USFWS's effort, named the Colorado River Fish Project (CRFP), was to acquire information needed to recover endangered fishes while allowing the USBR to operate existing projects as well as to plan and construct proposed projects.

Initially, the CRFP extended over 965 km of the Colorado and Green rivers and their tributaries in Colorado and Utah. Studies focused on identification of spawning requirements, habitat requirements of young and adults, migratory behavior, interspecific competition, predation and food habits, effects of temperature, salinity, and chemicals, development of culture techniques, disease and parasite diagnostics, and taxonomy (W. H. Miller et al. 1982d). Fieldwork from 1979 to 1981 emphasized sampling the upper Colorado River from Lake Powell to Rifle, Colorado (Valdez et al. 1982b), and the Green River from its confluence with the Colorado upstream to Split Mountain Gorge (Tyus et al. 1982b), to determine distribution, relative abundance, movements, and habitats of various life stages.

The 1979 MOU was amended in 1981 to include an investigation of humpback chubs in the Little Colorado River, Arizona (Kaeding and Zimmerman 1983), and to expand field studies in the upper basin to include the Dolores and Gunnison rivers, Colorado-Utah (Valdez et al. 1982a). Additional funds provided by the U.S. Bureau of Land Management to include White River, Colorado-Utah (W. H. Miller et al. 1982b), and funds from Congress and the U.S. National Park Service (USNPS) supported field research on the Yampa

and Green rivers in Dinosaur NM (W. H. Miller et al. 1982c). Funding provided through the Windy Gap process was used for a three-year habitat use and radiotelemetry study of Colorado squawfish and humpback chubs on the upper Colorado River, and to investigate use of "nonflow alternatives" (habitat development, fish passage, and stocking) as means of maintaining and ensuring recovery of the fishes of concern (Archer et al. 1985).

Laboratory research for Colorado squawfish outlined in the MOU included swimming stamina, bioassays of potentially toxic trace element tolerances, and determination of preferences for temperature and total dissolved solids. In addition, various contracted studies examined physicochemical habitat conditions in the Green and Colorado rivers, culture of rare fishes, diseases, movements of Colorado squawfish in the inlet to Lake Powell, and stomach contents of fishes (W. H. Miller et al. 1982a).

The accumulated knowledge of the endangered Colorado River fishes was summarized in a 1981 symposium that emphasized studies conducted after 1975 (W. H. Miller et al. 1982c). The CRFP continued its work after 1982, emphasizing filling gaps in knowledge of ecological requirements of the four rare fishes; for example, delineating necessary stream flows, describing movements (especially those associated with reproduction), and identifying factors limiting recruitment (Archer et al. 1984; Kaeding et al. 1986; Tyus et al. 1987).

Other Studies

Studies of the ecology of rare fishes were also conducted by biologists from the Colorado Division of Wildlife (CODOW) and the Utah Division of Wildlife Resources (UTDWR) from the early 1980s to the present (e.g., Haynes and Muth 1982; Radant 1982, 1986; Wick et al. 1985). This work, funded through section

6 of the ESA and by individual states, has provided much of the basic information used to make decisions on efforts in behalf of the rare species. CODOW studies documented the importance of the Yampa River to the Colorado squawfish. The importance of the Green River to squawfish and humpback chub was verified by the UTDWR. Distribution of endangered Colorado River fishes in Cataract Canyon was examined by private contractors with USBR support (Valdez and Williams 1986; Valdez 1990). Culture and propagation techniques for rare Colorado River fishes were developed at Willow Beach (Arizona) and Dexter (New Mexico) national fish hatcheries (Hamman 1982a, et seq.: Inslee 1982a, b). Description of the larval stages of these fishes was completed at Colorado State University by Muth (1988).

San Juan River—The Forgotten Basin

The San Juan River originates in the mountains of southwestern Colorado and flows South into Navajo Reservoir and then east through the deserts of New Mexico and Utah to eventually join the Colorado River in Lake Powell. Historically, the San Juan provided habitat for Colorado squawfish, razorback sucker, and perhaps bonytail, although historical data indicate that none of these species was common. An effort to eradicate native and introduced rough fish before closing the Navajo Reservoir in the early 1960s was described as effective, but only four large squawfish were documented as being killed by the treatment (Olson 1962a, b). A limited survey in the 1970s resulted in the capture of a single juvenile Colorado squawfish near Aneth, Utah (VTN Consolidated 1978). Razorback suckers were rare, and records were based more on local testimony than on actual specimens (Minckley et al., *this volume*, chap. 17). Finally, the name "bonytail" was used for other *Gila* species, particularly roundtail chub (*G.*

robusta), so the presence of bonytail in the San Juan remains questionable.

Until recently the San Juan was relegated to a relatively low priority in upper-basin recovery efforts for endangered fishes. This low priority is mostly attributable to the 1979 USFWS biological opinion on the proposed Animas–La Plata Project, which concluded: "because of the apparent small size of the San Juan River squawfish population and its already tenuous hold on survival, its possible loss will have little impact on the successfully reproducing Green and Colorado River squawfish populations and therefore on the species itself." The opinion also concluded that construction of the Animas–La Plata Project would further change the San Juan River to a point where the Colorado squawfish population in the river would likely be lost.

The Animas–La Plata biological opinion also recommended that the USBR conduct a thorough study of the native fishes in the San Juan River. These studies were initiated by the states of Utah and New Mexico in 1987 under contract with the USBR. Salient findings of these studies (Platania and Young 1990) include the following: (1) Colorado squawfish were collected from the San Juan River at several locations from Shiprock, New Mexico, to Lake Powell, Utah; (2) successful reproduction of Colorado squawfish was confirmed in 1987 and 1988 by the capture of eighteen young-of-year fish; and (3) suitable Colorado squawfish habitat appears present throughout the San Juan River, and this species appears to occupy the river on a year-round basis.

As a result of these findings, the USBR reinitiated section 7 consultation on the Animas–La Plata Project in early 1990. Based on results of the San Juan fisheries surveys, and the belief that Colorado squawfish populations were not stable or were showing additional signs of decline in other parts of the upper basin, the USFWS issued a draft biological opinion in May 1990 concluding that the project was likely to jeopardize the continued existence of the species. Furthermore, the USFWS concluded that no reasonable or prudent alternatives were available to avoid jeopardy and recommended additional study to develop a more complete data base, especially on stream-flow needs for the fish in the San Juan River.

The USFWS opinion placed the project on hold and evoked a major outcry from Colorado and New Mexico congressional delegations, the governor of Colorado, a variety of state and local officials, Indian tribes, and other project supporters. Even Secretary of the Interior Manuel Lujan suggested that the ESA was too restrictive and recommended that Congress amend it to allow consideration of economic factors in the section 7 consultation process.

As a result of the draft biological opinion, the Animas–La Plata project was reexamined. A revised draft opinion was issued by the service in May 1991 that would approve development of a scaled-down version of the project subject to several conditions, including:

1. Operation of Navajo Reservoir to provide releases that mimic a natural hydrograph in the San Juan River;
2. Conducting a seven year research program to better assess the habitat requirements of the endangered fishes in the San Juan River;
3. Implementing a long term recovery program for the endangered fishes in the San Juan River basin.

Whether this opinion will ultimately be accepted and the Animas–La Plata project constructed remains to be determined. However, it now appears that the San Juan River will become a major focal point for recovery efforts in the next decade.

Coordination and Cooperation between Agencies

Colorado River Fishes Recovery Team (CRFRT)

The ESA directs the secretary of the interior to develop and implement recovery plans for threatened and endangered species with the aid of appropriate public and private agencies, institutions, and qualified individuals. By this authority the USFWS invited various agencies interested in management of the Colorado River fishes to participate on the CRFRT. Formed in December 1975 as the Colorado Squawfish Recovery Team, the effort was expanded in 1976 to include all endangered Colorado River fishes in the upper basin (K. D. Miller 1982). The recovery team included representatives from the states of Arizona, California, Colorado, Nevada, New Mexico, and Utah, and from the USBR, USNPS, and USFWS. Team members have written recovery plans for bonytail, humpback chub, and Colorado squawfish. The humpback chub and bonytail plans were revised and approved in 1990. The Colorado squawfish plan is expected to be approved in 1991.

Colorado River Endangered Fishes Researchers Meetings

Since 1983, fishery biologists and other researchers from state and federal agencies, universities, and private consulting firms in the upper basin have held an annual meeting sponsored by the states of Colorado and Utah. The open communication and coordination provided by this meeting have been effective at integrating research efforts among biologists. Communication among biologists from the upper and lower basins has been enhanced as well through annual meetings of the Desert Fishes Council and through meetings of the CRFRT.

American Fisheries Society (AFS)

Members of the Endangered Species Committee of the Bonneville Chapter of the AFS wrote a position paper in 1974 that strongly supported protection of natural habitats and native species that are threatened, endangered, or of special concern in Utah (Holden et al. 1974). Members of the Threatened and Endangered Species Committee of the AFS developed systematic guidelines and policies for introductions of threatened and endangered fishes to supplement existing populations or to establish new populations (J. E. Williams et al. 1988). These recommended guidelines focus on planning, implementation, and evaluation of introductions in ways intended to increase the probability of success in recovery efforts. Further communications about endangered fishes occur at meetings of the Western Division of the AFS (W. H. Miller et al. 1982c) and at meetings of the Bonneville and Colorado-Wyoming chapters of that organization.

Interagency Coordinating Committee

The 1982 amendments to the ESA declare that "the policy of Congress is that Federal agencies shall coordinate with State and local agencies to resolve water resource issues in concert with the conservation of endangered species." This amendment was added to the ESA to address specific conflicts concerning water development and conservation of the endangered species in the upper Colorado and Platte river basins.

By 1984 the USFWS had issued nearly a hundred biological opinions, concluding that the site-specific cumulative effect of water developments and depletions was likely to jeopardize the continued existence of endangered Colorado River fishes. Also in 1984, the USFWS issued a draft conservation plan that specified minimum stream flows needed by endangered fishes for all major streams in the upper basin. This plan drew harsh reactions from the upper-basin states because stream-flow recommendations were based on historic conditions rather than on the specifically documented biological needs of the species themselves (Zallen 1986). The plan was interpreted as a threat to future water development and state water-rights systems.

In response to this controversy and a failure to weaken the ESA (Tarlock 1984), water-development interests became more actively involved in trying to resolve growing concerns over endangered species versus water development in the upper basin. For example, the directors of the Colorado Water Congress established a Special Project on Threatened and Endangered Species in December 1983 (Pitts 1988). Its goal was to find an administrative solution acceptable to water-development interests, the federal government, states, and environmental organizations that would allow water development to continue in the upper Colorado and Platte river basins while avoiding conflicts with the ESA.

Also in response to the growing controversy, the USFWS began discussions among representatives of the USBR and the states of Colorado, Utah, and Wyoming, private water-development interests, and environmental groups. These led to the formation of the Upper Colorado River Coordinating Committee (UCRCC) in March 1984. The committee's primary goal was to develop a recovery program for endangered fishes in the upper basin within the framework of the ESA, existing states' water rights, and terms of the Colorado River compacts. The San Juan River was excluded from discussions because it had not been identified as a priority recovery area for endangered fishes.

One of the first UCRCC activities was the formation of biology and hydrology subcommittees to review and synthesize technical information on the fishes and their stream-flow requirements. Biological and hydrological data thus summarized were used as a basis for drafting a recovery program for the endangered species. After nearly four years of intense discussions, data analyses, and negotiations, a recovery program was finalized in September 1987 (USFWS 1987a). After completion of an environmental assessment of the recovery program (USFWS 1987c), the secretary of the interior joined the governors of Colorado, Utah, and Wyoming, and the administrator for the Western Area Power Administration (WAPA), in executing a cooperative agreement to formally implement the recovery program.

The agreement created the ten-member Upper Colorado River Implementation Committee (UCRIC) to oversee the USFWS recovery efforts (Rose and Hamill 1988). Voting members included the USFWS, USBR, WAPA, the states of Colorado, Utah, and Wyoming, one representative of environmental organizations (e.g., the Environmental Defense Fund, Audubon Society), and private water-development interests in the three states. The UCRCC, including its biology and hydrology subcommittees, was replaced by three technical committees (biology, water acquisition, and information and education) and a management committee to oversee ongoing activities. The signing of this agreement and implementation of the recovery program were made possible, in part, by legislation in Colorado that recognizes in-stream flows for fishes as a beneficial use of water. The goal of the recovery program is to recover, delist, and manage the three endangered fishes and to free the razorback sucker of a need for protection under the ESA by the year 2002.

The recovery program outlines a fifteen-year, $60 million effort consisting of five elements:

1. *Provision of in-stream flows.* The USFWS will quantify in-stream flows needed for recovery of the four rare fishes in the upper Colorado and Green River sub-basins. The UCRIC, in cooperation with state agencies, will iden-

tify and recommend alternatives to the secretary of the interior for implementing USFWS flow recommendations. It is anticipated that in-stream flow needs of rare fishes in major reaches of the Colorado and Green rivers can be provided through program refinement and protection of releases from federal reservoirs such as Flaming Gorge and Blue Mesa. Specifically, the recovery program provides for the acquisition and appropriation of water rights in relatively unregulated systems such as the Yampa and White rivers, conversion of these rights into in-steam flows for fishes, and administration of these rights for in-stream flow for fish pursuant to state water laws. More than half of the recovery program's budget was targeted for this purpose. In 1988 Congress appropriated $1 million to initiate acquisition of water for in-stream flow. Efforts to date have focused on the Yampa River and a 24-km reach of the Colorado River near Grand Junction, Colorado.

2. *Habitat development and maintenance.* The goal of this element is to enhance populations of fishes through habitat development or management measures such as creation of backwaters and construction of jetties, fish ladders, and so on. The effectiveness of these measures has never been demonstrated, and consequently, experimental research and demonstration projects will be conducted before large-scale implementation is attempted.

3. *Native fish stocking.* The goal of this recovery element is to produce a sufficient supply of hatchery-reared fish to support research and recovery efforts, and to preserve the genetic diversity present in wild fish. Consideration will be given to supplementing existing populations where studies conclude it would help promote self-sustaining populations.

4. *Management of non-native species and sport fishing.* Some introduced fish species are known to prey on, compete with, or limit recruitment of endangered fishes. Angling may also increase mortality in the Colorado squaw-

fish and humpback chub because of their high vulnerability. The recovery program prescribes that stocking of non-native species be confined to areas where absence of conflict with the endangered fishes can be demonstrated. These potential recovery problems will be closely monitored by CRFP personnel, the CODOW, and the UTDWR. Where necessary and feasible, state conservation agencies and the USFWS will cooperatively plan and implement controls.

5. *Research, monitoring, and data management.* The UCRIC provides a forum to guide and coordinate research, management, and recovery activities for endangered fishes. Its management and technical committees meet four to six times annually, and the entire UCRIC meets semiannually to review progress on recovery activities and research on life history, ecology, and habitat requirements of endangered fishes, and to assess the effectiveness of recovery and management procedures.

Funding is also a cooperative responsibility. The projected annual budget of $2.3 million, adjusted for inflation, is to be provided by federal and state governments, power and water users, and private donations. Two capital funds will be requested from Congress. A minimum of $10 million will be requested for purchase of water rights to protect in-stream flows. In addition, a $5-million fund will be requested for construction of fish passageways and rearing facilities.

Summary and Epilogue

The recovery effort for the rare fishes in the upper Colorado River basin has been the largest and most comprehensive project of its kind in the United States. It evolved as an effort to resolve conflicts between water use for municipal, industrial, and agricultural purposes and water requirements for endangered fishes. The USFWS Colorado River Fishery Project, the USBR, and the states of Colorado and Utah conducted and sponsored intensive biological studies on endangered fishes. Technical information provided by these studies was used by the Upper Colorado River Coordinating Committee to develop a recovery program that has been formally endorsed by water users, federal and state governments, and conservation groups (USFWS 1987a).

For the first time there is now a long-term, cooperative commitment to fund and implement a comprehensive effort aimed at recovery of endangered fishes. The challenge will be to ensure that water can be managed and allocated to meet existing and new municipal, industrial, and agricultural uses, while at the same time providing adequate in-stream flows required by the fishes of concern. In addition, effective measures to control non-native fishes that prey on or compete with endangered fishes must be found and implemented.

Recovery and section 7 actions must be based on sound biological principles so that the primary objective of the ESA is achieved; that is, to preserve the ecosystem on which the endangered Colorado River fishes depend. The success of the recovery program requires a strong and continued commitment by participants on the Upper Colorado River Implementation Committee to balance the needs of all parties (water developers, power users, environmentalists, anglers, etc.) who have a concern and interest in water and fishery resources of the upper Colorado River. Constraints of knowledge, time, politics, and available funds require creative thinking and actions for effective multiple use of this important resource. The recovery program provides an unprecedented opportunity to demonstrate that realistic management of endangered species is possible.

SECTION IV

Some Concerns, Facilities, and Methods of Management

Desert fishes often exist in small numbers, either normally or due to episodic natural phenomena or, more often, as a result of human activities. A small population is genetically dangerous because a few individuals simply cannot contain as much genetic variability as a population of thousands. Furthermore, each time a large population is reduced to a few individuals (termed bottlenecking), it is likely that more variability will be lost. If fitness to survive and reproduce is more likely for genetically variable individuals than for nonvarying ones, then repeated bottlenecking will force a species or population toward greater and greater homozygosity, and presumably to lower levels of fitness. Western fishes face such dangers now that they are significantly reduced in numbers and their distributions are fragmented. In many cases they have become so rare that refuges and preserves must serve as enclaves against extinction, with considerations of genetic variability having critical importance.

The first contribution in this section examines genetics and genic diversity in fishes of the West. Only a few species have been studied, and those have shown considerable variation, so few generalizations can be made. As expected, stocks isolated in small, constant springs tend to be more homozygous than large populations in more extensive and variable habitats. Individual samples from species with widespread, continuous populations carry a larger proportion of the total heterogeneity than those from isolated species. Thus, range fragmentation and lack of migration have promoted local uniqueness. Hybridization also increases local heterozygosity, and both natural and anthropogenic hybridization and introgression are identified as major factors affecting western fish populations. The danger of direct and indirect effects of introgressive hybridization to native forms is evident. An assessment of the myriad genetic problems facing species subjected to artificial propagation or placed under refuge conditions forms an end for the chapter.

Selection and design of preserves and refuges for western fishes are examined in chapter 10. Theoretical considerations are followed by proposed design criteria. A classification of preserves is proposed, along with a strategy for development of regional systems for maintaining the native western aquatic fauna. To be effective and successful such places must be capable of maintaining the complex interactions among organisms as well as among the organisms and their environment. Naturalness, appropriate size, and buffers against invasion or other perturbations, among other factors, are deemed important to protect natural diversity.

This treatment of the proposed and theoretical is followed in chapter 11 by a historical accounting of preserves and refuges already in place, and an evaluation of their effectiveness in protecting fishes in the American West. The argument is forwarded that aquatic communities, including their native fishes, should be identified and protected before it is necessary to invoke legal action under the Endangered Species Act. Emphasis on habitat preservation and enhancement is clearly the key to species salvation, and the sooner we identify and set aside areas with high native species richness, the sooner we can claim progress toward a viable program of conservation.

Chapter 12 first lists and describes some refuges where fishes and aquatic habitats have been preserved in Mexico. Springs, wherever they occur, hold a fascination all their own, and in arid lands they provide welcome oases for rest, recreation, and therapeutic values (real or imaginary), as well as reliable water supplies. As a result, springs in Mexico (and in arid parts of the United States, as covered in later chapters), and their fishes with them, are often set aside and protected. The second part of this chapter announces new legislation in Mexico that deals in part with the conservation of natural systems, providing the first major recognition in that country of the pressing needs to ameliorate utilitarian uses of land and water by a degree of environmental concern.

Stopgap measures in species preservation often include captive propagation. Thus, many well-known endangered species, such as whooping cranes and black-footed ferrets from North America, and numerous large mammals and birds from other lands, are subject to such measures. Efforts for western fishes are almost unheralded but have been under way for several decades, and especially since 1974 at Dexter National Fish Hatchery, New Mexico. Chapter 13 describes and analyzes the Dexter experiment, its successes and failures, and the underlying philosophies, challenges, and paradoxes that have served to shape and direct its development and progress.

The contributions of the Dexter hatchery, acting as a refuge and research center as well as a production facility for reintroducing imperiled native fishes back into nature, have been substantial. It is nonetheless only an interim solution to immediate problems of species disappearing in nature, a means of perpetuating them until suitable habitat can be found, rebuilt, or created, rather than an end in itself.

Recovery efforts through manipulations of natural and artificial habitats are reviewed in chapter 14. Again, habitat integrity is the factor in question, and the emphasis is on improvement, alteration, or renovation of streams, springs, and other waters for conservation of native fishes. A long history of habitat manipulation exists for western streams, mostly for enhancement of sport fisheries. This literature is used to assess potential damages suffered by native faunas through such actions in the past, and the information accumulated may now be applied to management of native fishes. It seems clear that physical modifications were often a shortsighted approach to rectify longer-term problems of watershed abuse through overgrazing or logging. Emphasis was, and largely remains, limited in scope, attempting to enhance individual species rather than the ecosystem. Use of poisons to remove undesired species was also a short-term solution to perceived problems in game fish production. Most renovations failed or attained the desired result for only a few years, and the same has been realized in some, but not all, recent attempts to renovate aquatic systems for nongame native fishes. Chemical renovation nonetheless remains a valuable management option, applicable under certain circumstances to reclaim habitat and ensure survival of native species.

Dexter National Fish Hatchery, Chaves County, New Mexico. This U.S. Fish and Wildlife Service facility, converted from a warm-water game fish hatchery, has played a major role in recovery efforts for native western freshwater fishes. Photograph by B. L. Jensen, 1990.

Chapter 9

Conservation Genetics and Genic Diversity in Freshwater Fishes of Western North America

Anthony A. Echelle

Genetic wildlife conservation makes sense only in terms of an evolutionary time scale. Its sights must reach into the distant future. — Frankel 1974

Introduction

Frankel's view may seem too idealistic to conservationists dealing with Soulé's (1986) "real world: ... the world of politics and economics, and all the vagaries of human nature that we associate with these areas." Nevertheless, concern for evolutionary potential in the distant, unforeseeable future should underlie every action in conservation biology. At the level of individual taxa, this translates into management of genetic resources within the framework of evolutionary theory (Frankel and Soulé 1981; Schonewald-Cox et al. 1983; Soulé 1986).

Once a taxon or some geographic subdivision of a taxon has been targeted for conservation efforts, management needs and priorities should depend primarily on knowledge of the geographic pattern of genetic variation (Allendorf and Phelps 1981; Chambers and Bayless 1983; Vrijenhoek et al. 1985). Which populations merit the most attention? How many populations should be established in artificial refugia? Which populations should provide founding stock for establishing new captive or natural populations? Without knowledge of geographic patterns of genetic variation, decisions will be determined more by chance or short-term costs and convenience than by what is best for the evolutionary potential of the taxon.

In this paper I review electrophoretic studies of genetic variation in fishes of concern to conservationists in western North America. I also offer suggestions for management that stem largely from the results of such studies. Protein electrophoresis represents the major source of data on genetic variation in natural populations. Beginning in the 1960s, protein electrophoresis enabled analysis of distinct alleles at large numbers of defined gene loci. This provided a powerful and as-yet-unparalleled tool for studies of population structure, including estimates of local genetic variation, divergence among populations, and levels of hybridization and introgression. The advantages of protein electrophoresis over other approaches were recently discussed by Allendorf et al. (1987) and Campton (1987).

I thank Alice F. Echelle for her editorial and laboratory help; P. J. Conner and G. R. Wilde for information on Pecos pupfish; B. L. Jensen, J. E. Brooks, D. C. Hales, R. L. Hamman, and T. Winham for helpful discussions on management of native fishes of the Southwest; and F. W. Allendorf, D. G. Buth, R. J. Behnke, T. E. Dowling, E. J. Loudenslager, G. K. Meffe, D. C. Morizot, and J. N. Rinne for helpful comments or for allowing me to use their unpublished manuscripts. This paper was written while I was supported by an Intergovernmental Personnel Act Agreement between the U.S. Fish and Wildlife Service (USFWS) and Oklahoma State University at Dexter National Fish Hatchery, New Mexico. Partial support was also provided by a grant from the National Science Foundation (BSR 88-18004).

General Perspectives

Behnke (1968, 1972) was an early advocate of the position that conservationists should not be concerned with simply preserving species or subspecies. He emphasized the diversity of biotypes often embedded in a single taxon. For example, the population of Lahontan cutthroat trout (*Oncorhynchus clarki henshawi*, now extinct) that once occupied Pyramid Lake, Nevada, were large predators attaining weights of 20 kg. Although not taxonomically distinguished from other populations of the subspecies, the Pyramid Lake population apparently was genetically different in body size. Individuals of the same subspecies introduced from elsewhere in the Lahontan Basin into Pyramid Lake "do not approach even one half the maximum size attained by the original population" (Behnke 1972).

Human activities are causing losses of genetic diversity through attrition in population sizes, extinction of local populations, and, perhaps not so obviously, by encouraging hybridization (primarily through introductions of non-native forms). It might be argued that the effect of hybridization is relatively trivial. As I discuss more fully later on, however, hybridization and its resulting genetic contamination can quickly cause losses of native fishes over large geographic areas. Loss of a genetically distinctive biotype, whether by extinction or by hybridization, reduces the present and future options that the biotype represents

for conservation, management for recreation and aesthetic appeal, and experimentation in applied and basic research.

The best strategy for preserving both management options and evolutionary flexibility of taxa is to maintain as many populations as possible while retaining natural patterns of genetic diversity within and between populations. Protein electrophoresis can provide relatively sensitive indexes of such patterns of diversity, thereby giving insight into what needs preserving.

Biochemical indexes of diversity should not be used to the exclusion of morphological data and life-history information. In general, the small subset of genes assayed biochemically will not be responsible for observed developmental differences. Striking intraspecific polymorphism, and even speciation, can occur with little or no detectable divergence at protein-coding loci (B. J. Turner 1974; Kornfield et al. 1982; A. A. Echelle and Kornfield 1984). On the other hand, many recently evolved forms may exhibit low levels of biochemical differentiation.

Electrophoretic Studies of Genetic Diversity

Two measures of genetic variation in local populations typically are reported from electrophoretic data: heterozygosity (H), the average frequency of heterozygotes per locus per individual, and polymorphism (P), the proportion of loci having two or more alleles. Of the two, H is less arbitrary and more precise because it is relatively insensitive to the number of individuals sampled (Nei 1975).

The distribution of genetic variation in a population typically exhibits a hierarchical structure in which the number of levels depends on patterns of gene flow and other evolutionary factors (Nei 1977). In a population composed of subpopulations, total gene diversity is computed as $H_t = H_s + D_{st}$, where H_t is expected heterozygosity in a pan-mictic population with allele frequencies equal to the unweighted mean over all sub-populations, H_s is unweighted average hetero-zygosity among populations, and D_{st} is aver-age gene diversity due to differences between subpopulations. D_{st} can be further broken down depending on the number of hierarchi-cal levels. For example, if groups of subpopu-lations exists, $H_t = H_s + D_{sg} + D_{gt}$, where D_{sg} is gene diversity due to differences be-tween subpopulations within groups, and D_{gt} is gene diversity due to differences between groups. In assessing biases associated with sampling and alternative methods of comput-ing hierarchical gene diversity, Chakraborty and Leimar (1987) considered effects of sam-pling to be more critical than methods of anal-ysis. Allendorf and Phelps (1981) and Chak-raborty and Leimar (1987) provided helpful discussions of sampling designs for such studies.

Table 9-1 summarizes available informa-tion on genetic diversity in the threatened fishes of western North America. These taxa were listed by J. E. Johnson (1987a) and J. E. Williams et al. (1985) as receiving legal pro-tection from state or federal agencies or other-wise of special concern due to low numbers, limited distributions, or recent declines. Gene diversity analyses were either taken directly from published sources or computed with the BIOSYS-I program (Swofford and Selander 1981).

Because of limited geographic sampling, the values for a number of species in Table 9-1 are only crude approximations of the distribu-tion of diversity: *Oncorhynchus apache, O. gilae,* all of the catostomids, and three species (*Gambusia longispinis, G. marshi,* and *Cich-lasoma minckleyi*) represented by one to three samples from Cuatro Ciénegas, and two spe-cies of *Gambusia* (*G. hurtadoi* and *G. krum-holzi*) represented by single samples from relatively large spring populations. For four spring-dwellers effectively restricted to a single location—*Gambusia alvarezi, G. gaigei, G. georgei,* and *G. heterochir*—the available data from one or two samples probably repre-sent reliable estimates of genetic diversity.

In this review I emphasize the hierarchical approach to genetic diversity because it facili-tates summarization of data for diverse spe-cies. This approach indicates the level of the population structure at which diversity in a given taxon is concentrated. Most studies of variation also include a cluster analysis, in which samples are grouped by overall genetic similarity. Hierarchical analysis of two species having the same geographic range may indi-cate that similar proportions of genetic diver-sity in the two are due to "between drainage" differences between populations. However, the specific drainages occupied by homogene-ous subsets of samples may be quite different between the two species. Cluster analysis aids the manager by indicating the geographic limits of genetically defined subpopulations.

Genetic Variation in Local Populations

Heterozygosities in local populations of threatened freshwater fishes of western North America are low relative to averages given by P. J. Smith and Fujio (1982) for 106 species of marine fishes ($H = 0.055$), and by Nevo et al. (1984) for 183 species of fishes in general ($H = 0.051$). Twenty-three (85%) of the 27 threatened taxa in Table 9-1 exhibited H_s lower than 0.050. Nevo et al. (1984) found no significant differences between marine and inland fishes. Gyllensten's (1985) review of fewer, but thoroughly assayed, species found lower heterozygosities in "stationary fresh-water species" than in marine fishes. This was attributed to smaller population sizes and more restricted gene flow—pronounced attri-butes of many threatened species.

Theory predicts low electrophoretically de-tectable variability for many populations of fishes in desert areas. Many desert fishes are

Table 9-1. Genetic diversity in threatened fishes of western North America. Status = area of concern for the species: US, whole taxon in United States; S, one or more states of the United States; MX, whole taxon in Mexico; SM, present in both countries; P/L, number of populations and number of gene loci assayed; H_t, total gene diversity; and H_s, average heterozygosity in local populations. Asterisks signify taxa examined over only a small portion of their total geographic ranges. Parentheses indicate values based on minimal sampling; these are considered crude approximations (see text).

Taxa	Status	P/L	H_t	H_s	Between drainages	Between samples within drainages	Within samples	References[1]
Salmonidae								
Oncorhynchus nerka*	S	18/26	.046	.044	2.5	3.1	94.4	1
O. apache (Salt R.)	US	5/35	(.006)	(.006)	—	9.5	90.5	2
O. clarki bouvieri	S	8/46	.014	.013	—	3.7	96.3	3
O. c. henshawi	US	15/35	.065	.036	—	44.5	55.5	4
O. c. lewisi	S	103/29	.029	.019	16.7	15.7	67.6	3
O. mykiss*	SM	38/16	.069	.058	7.3	7.7	85.0	1
		31/24	.106	.092	—	—	86.8	5
O. gilae (Gila R.)	US	4/35	(.059)	(.051)	—	13.6	86.4	2
Catostomidae								
Catostomus (Pantosteus) discobolus yarroui	US	3/45	(.030)	(.055)	45.2	—	54.8	6
C. (P.) plebeius[2]*	SM	5/45	(.014)	(.013)	7.1	—	92.9	6
		4/27	(.111)	(.013)	88.8	—	11.3	7
Xyrauchen texanus	US	2/21	(.029)	(.029)	—	1.1	98.9	8
Cyprinodontidae								
Cyprinodon bovinus	US	5/28[3]	.046	.045	—	1.4	98.6	9
C. elegans	US	7/28[3]	.056	.050	—	10.8	89.2	9
C. macularius[2]	SM	3/38	.049	.034	29.9	—	70.1	10
C. pecosensis	US	6/28	.029	.027	—	7.7	92.3	9
C. tularosa	US	3/28	.028	.017	—	19.0	81.0	9

Poeciliidae

Gambusia alvarezi	MX	1/23	.000	.000	—	—	—	11
G. gaigei	US	1/23	.000	.000	—	—	—	11
		2/60	.000	.000	—	—	—	12
G. georgei	US	1/23	.060	.060	—	—	—	11
G. heterochir	US	1/23	.093	.093	—	—	—	11
G. hurtadoi	MX	1/23	(.046)	(.046)	—	—	—	11
G. krumbolzi*	MX	1/23	(.021)	(.021)	—	—	—	11
G. longispinis*	MX	2/23	(.000)	(.000)	—	—	—	11
G. marshi (Cuatro Ciénegas)*	MX	1/23	(.000)	(.000)	—	—	—	11
G. nobilis	US	16/24	.063	.030	—	51.6	48.4	13
Poeciliopsis o. occidentalis	SM	10/25	.033	.020	0.0	40.7	59.3	14

Cichlidae

Cichlasoma minckleyi	MX	3/13	(.036)	(.035)	—	2.3	97.7	15
		3/27	(.011)	(.010)	—	5.4	94.6	16

Cottidae

Cottus confusus	S	16/33	.028	.015	—	46.1	53.9	17

Gasterosteidae

Gasterosteus aculeatus williamsoni	US	1/45	—	(.029)	—	—	—	18

[1], Ryman (1983); 2, Loudenslager et al. (1986); 3, Allendorf and Leary (1988); 4, Loudenslager and Gall (1980); 5, Berg and Gall (1988); 6, Crabtree and Buth (1987); 7, Ferris et al. (1982); 8, Buth et al. (1987); 9, A. A. Echelle et al. (1987); 10, Turner (1983); 11, A. A. Echelle et al. (1989); 12, C. Hubbs et al. (1986); 13, A. F. Echelle et al. (1989); 14, Vrijenhoek et al. (1985); 15, Kornfield and Koehn (1975); 16, Sage and Selander (1975); 17, Gyllensten (1985); and 18, Buth et al. (1984).

[2]Multiple samples from one drainage were treated as a single population with the average allele frequencies of the samples from the drainage.

[3]A. A. Echelle et al. (1987) mistakenly reported thirty loci; the two peptidases were not assayed in the two species indicated.

isolated in small bodies of water (e.g., a single spring or spring system). In the absence of gene flow, and assuming selective neutrality or near neutrality of most allozymes (Nei 1983; Chakraborty 1980), variability is determined by the effective size (N_e) of the population. Bottlenecks in N_e probably are more common in fishes of desert regions than elsewhere. Each bottleneck intensifies genetic drift, which then decreases variability. Repeated bottlenecking may have an even greater effect on variability than classical theory generally predicts (Motro and Thomson 1982).

Extreme examples of reduced variability in desert populations include zero variability ($H = 0.000$, $P = 0.000$) in electrophoretic assays of the following: (1) Devils Hole pupfish (*Cyprinodon diabolis*; B. J. Turner 1974); (2) three or four isolated populations of Sonoran topminnow (*Poeciliopsis o. occidentalis*) in Arizona (Vrijenhoek et al. 1985); (3) a sample of *Gambusia marshi* from the Cuatro Ciénegas basin, Coahuila, Mexico; (4) a sample of *G. longispinis* from the Cuatro Ciénegas basin; and (5) Big Bend gambusia (*G. gaigei*) from Big Bend National Park, Texas (Table 9-1).

A comparison of eight species of *Gambusia* restricted to the Chihuahuan Desert with fourteen species of the genus from other physiographic areas (eastern United States and tropical-subtropical areas) found significantly lower variability in the desert dwellers (A. A. Echelle et al. 1989). Instances of no detectable variability were restricted to samples of the four desert species listed above. Most indigenous desert *Gambusia* are restricted to spring environments, but there was no overall association between spring dwelling and level of variability. Thus, selection for an optimal genotype in spring environments apparently does not explain the reduced levels of variability observed. The heightened susceptibility of desert dwellers to repeated population bottle-necking, coupled with extended periods of isolation, seems to be the most likely explanation.

Total Gene Diversity

Extremely low levels of total gene diversity among threatened western fishes are indicated for several species effectively restricted to single aquatic habitats. In such species, heterozygosity in a single sample should closely approximate H_t for the species. Thus the absence of detected variability in assays of *Cyprinodon diabolis*, *Gambusia alvarezi*, and *G. gaigei* suggests minimal genetic variability in these taxa.

The estimate of zero variability in Big Bend gambusia is especially reliable because it is based on independent surveys of twenty-three and sixty loci (Table 9-1). The existing population within its native range is artificially maintained in Big Bend National Park. In 1956 the species passed through an especially severe bottleneck—down to one female and two males (C. Hubbs and Broderick 1963). Recent population bottlenecking probably helps account for the reduced variability. Because brief bottlenecks are not likely to cause a total lack of detectable variability (Nei et al. 1975), low genetic diversity in Big Bend gambusia probably existed before 1956.

Excluding single-sample estimates, H_t in threatened fishes of western North America (mean = 0.04, range = 0.01–0.07) is similar to H_t of nonmigratory freshwater fishes in general (0.04, 0.01–0.08; Gyllensten 1985), and somewhat lower than H_t of marine and anadromous fishes (0.06, 0.03–0.09). Two highly restricted species, the San Marcos gambusia (*G. georgei*) and Clear Creek gambusia (*G. heterochir*), had total gene diversities (0.06 and 0.09, respectively) well above the average, while several wider-ranging taxa were well below the average. The highest diversity (0.11; Table 9-1) was in the Rio Grande sucker (*Catostomus* [*Pantosteus*] *plebeius*) from three

Mexican drainages; this reflects the presence of two forms (possibly different taxa) that were strongly divergent genetically (Ferris et al. 1982). Loudenslager and Gall (1980) reported a similar level of H_t for cutthroat trout (0.13) when computed for the species as a whole rather than for each subspecies separately.

Geographical Distribution of Genetic Diversity

Distribution of genetic diversity within taxa is primarily the result of patterns of gene flow. For example, local populations of marine and migratory fishes typically represent 85% or more of the species' H_t, while local populations of nonmigratory freshwater species generally carry a much smaller proportion (Gyllensten 1985; Allendorf and Leary 1988). These differences presumably reflect differential opportunities for gene flow among local populations.

The effects of migration and gene flow on the distribution of gene diversity are well illustrated by salmonids. Allendorf and Leary (1988) reported that the average local population in nonmigratory forms (e.g., *O. c. henshawi* and *O. c. lewisi*; Table 9-1) generally carries a much smaller proportion of H_t than do migratory taxa (e.g., *O. mykiss*). Exceptions to this rule include the highly restricted, nonmigratory Yellowstone cutthroat (*O. c. lewisi*), Gila (*O. gilae*), and Apache (*O. apache*) trouts, in which 86% to 96% of H_t occurs in the average local population. Both *O. apache* and *O. gilae* need more thorough geographic sampling (J. Rinne, U.S. Forest Service, pers. comm.). An unknown proportion of the genetic diversity of *O. gilae* was lost with extirpation of the presumably large, disjunct population from the Verde River drainage in Arizona (R. R. Miller 1972b).

The effect of range fragmentation is illustrated on microgeographic and macrogeographic bases by three sets of comparisons of fishes endemic to the Pecos River drainage of New Mexico and Texas. First, the Pecos gambusia (*Gambusia nobilis*) is restricted to four isolated, spring-fed areas, while mosquitofish (*G. affinis*) and Pecos pupfish (*Cyprinodon pecosensis*) are more continuously distributed. Presumably because of restricted gene flow, the average sample of *G. nobilis* carries only 48% of H_t (A. F. Echelle et al. 1989). In contrast, the average is 88% for *G. affinis* from three of the areas occupied by *G. nobilis* (A. A. Echelle et al. 1989) and 92% for *C. pecosensis* from riverine habitats (A. A. Echelle et al. 1987).

Second, in the rather complex system of springs and irrigation canals in the Balmorhea area of Texas, *G. nobilis* is restricted to headwaters of several different springs, while in Leon Creek, Texas, and Blue Spring, New Mexico, it is more continuously distributed. Correspondingly, an average sample from the Balmorhea area represents 75.6% of H_t, while average samples from the other two populations represent 98.7% and 99.6% of H_t in their respective areas (A. F. Echelle et al. 1989). In the fourth occupied area, Bitter Lakes National Wildlife Refuge, New Mexico, the pattern is confounded by human intervention.

Third, within the Balmorhea area, the Comanche Springs pupfish (*Cyprinodon elegans*) is more continuously distributed than *G. nobilis* (A. A. Echelle and Echelle 1980). As expected, the average sample of *C. elegans* carries a significantly higher percentage of H_t than does such a sample of *G. nobilis* from that area (89% versus 76%).

Effects of Natural Hybridization and Genetic Introgression

Genetic introgression occurs when hybridization and backcrossing lead to an exchange of genetic material between different species or genetically differentiated populations of the same species. Temporarily, at least, this heightens levels of diversity in the affected populations.

There must be many instances of natural introgression among fishes of western North America. This is to be expected from the complex geographic history of drainage systems in the region (see contributions in Hocutt and Wiley 1986) and the frequent hybridization between sympatric species (C. L. Hubbs 1955; F. J. Schwartz 1981). Few examples have been well documented, however, due in part to a paucity of genetic studies designed to detect introgression. Even when such studies have been done, it is difficult to determine whether similarities are due to introgression, retention of primitive traits, or descent from a recent common ancestor. For example, Leary et al. (1987) found that three of the subspecies of cutthroat trout were electrophoretically less similar to other cutthroat subspecies than to rainbow trout. These three also occur in natural sympatry with rainbow trout, while the other cutthroat subspecies do not. Although zoogeography in this instance is consistent with an introgression hypothesis, other evolutionary explanations cannot be discounted (Leary et al. 1987).

In some instances, geographic patterns of variation almost certainly reflect ongoing or past natural hybridization between taxa. The most thoroughly studied example from the threatened fishes of western North America is the Zuñi bluehead sucker (*Catostomus* [*Pantosteus*] *discobolus yarrowi*), which shows evidence of introgressive hybridization with the Rio Grande sucker (G. R. Smith et al. 1983; Crabtree and Buth 1987). Relative to other populations of *C. d. yarrowi*, the Nutria Creek, New Mexico, population, in an area of probable stream capture between the Rio Grande and Little Colorado River basins, has high heterozygosity ($H = 0.052$ versus $0.016–0.022$ in other populations) and an al-

lelic composition suggestive of introgression (Crabtree and Buth 1987).

Anthropogenic Hybridization and Introgression

Two types of human activities contribute to loss of native populations by encouraging hybridization: habitat alterations and introductions of fishes into places outside their natural areas of occurrence. Habitat changes can heighten hybridization by allowing contact between previously separated populations, by altering environmental features important in reproductive isolation, or by reducing abundance of a species until contact with individuals of another species becomes more likely than contact with conspecifics (C. L. Hubbs 1955; Minckley 1978; G. R. Smith et al. 1979; R. R. Miller and Smith 1981). The persistence of hybridizing taxa as separate evolutionary units indicates that hybrids generally suffer low evolutionary fitness. When a taxon is rare, however, hybridization enhanced by habitat alterations may threaten the genetic integrity of the entire taxon. Examples include the following: (1) hybridization between mosquitofish and Clear Creek gambusia, the latter being restricted to a single springhead situation (C. Hubbs 1971); (2) apparent hybridization of the rare humpback chub (*Gila cypha*) and bonytail (*G. elegans*) with the more abundant roundtail chub (*G. robusta*; G. R. Smith et al. 1979); and (3) hybridization between declining shortnose sucker (*Chasmistes brevirostris*) and other suckers within its limited geographic range (R. R. Miller and Smith 1981; J. E. Williams et al. 1985). In another instance involving June suckers (*Chasmistes l. liorus*), an original population was so influenced by hybridization following population lows after severe drawdown for local irrigation during drought that it apparently assimilated foreign genes from Utah suckers (*Catostomus ardens*) so that the original form became extinct and a new subspecies (*C. l. mictus*) was formed (R. R. Miller and Smith 1981).

Many taxa in the West have existed for long periods of time in the absence of interactions with congeneric species. Consequently, extensive hybridization often occurs when congeners suddenly co-occur as a result of introductions. The best-known examples of this problem among threatened species of the region involve native trouts and pupfishes.

Beginning in the late 1800s and continuing to the present, trouts of western North America have been repeatedly introduced outside their natural ranges, precipitating contact among many formerly allopatric native taxa (R. R. Miller 1950; Behnke 1972; Allendorf and Leary 1988). Introduced rainbow trout have now replaced or genetically introgressed native cutthroats in many areas (Behnke 1972; Busack and Gall 1981; Allendorf and Phelps 1981; Allendorf and Leary 1988). Introgression, replacement by rainbow trout, or both, are also largely responsible for severe reductions in Gila and Apache trouts (Rinne 1985b; Rinne and Minckley 1985; Loudenslager et al. 1986). In some areas native trout have also been introgressed by introduced subspecies of cutthroat trout (Gyllensten et al. 1985; Allendorf and Leary 1988).

With few exceptions (R. R. Miller 1981; Humphries 1984; Minckley and Minckley 1986), the thirty or so species of pupfishes are allopatric in distribution. High levels of reproductive compatibility occur between morphologically distinct pupfish species in laboratory situations (B. J. Turner and Liu 1977; Garrett 1980b; Cokendolpher 1980; Loiselle 1982). In addition, hybridization has been documented in all wild populations where native pupfish species have been brought into contact by human-caused habitat alterations, or where they have been exposed to non-native pupfishes (Stevenson and Buchanan 1973; Minckley 1978; C. Hubbs 1980; A. A. Echelle and Conner 1989).

An electrophoretic study of hybridization between Pecos pupfish and an introduced pupfish, the sheepshead minnow (*C. variegatus*), revealed how rapidly and to what magnitude introgression can occur after introduction of non-native species (A. A. Echelle and Conner 1989). The sheepshead minnow apparently was stocked into the Pecos River sometime between 1980 and 1984, and by 1985 a hybrid swarm had developed throughout 430 river kilometers. The frequency of foreign alleles in 1985 varied from 0.18 to 0.84. The original introduction apparently occurred in a mid-reach of the river and was followed by both upstream and downstream dispersal of foreign genes. The hybrid swarm now occupies about half the original range of the Pecos pupfish.

Catastrophic introgression may occur after only minimal introduction of a foreign biotype. For example, the introduction of *C. variegatus* into the Pecos River probably was an accidental result of bait transport or the stocking of sport fish and presumably did not involve massive numbers. The observed magnitude and rapidity of change apparently reflect intense natural selection favoring the introduced genetic material (A. A. Echelle and Conner 1989).

Introgressive hybridization has both direct and indirect effects in causing declines in native forms. Allendorf and Leary (1988) summarized the cutthroat trout situation as follows: "The presence of numerous introgressed populations throughout the range of cutthroat trout threatens the remaining native populations. If this condition persists, the only native populations that are likely to remain will be those isolated by dispersal barriers. This fragmentation into a number of small, isolated refuges is expected to increase the chances of extinction." Similar circumstances are developing for the Pecos pupfish and for the Guadalupe bass (*Micropterus treculi*), a species endemic to the Edwards Plateau of Texas that has been exposed to introduced smallmouth bass (*M. dolomieui*; R. J. Edwards 1979; Whitmore 1983). Heightened chances of extinction as small genetically uncontaminated populations become even smaller and more isolated are due to both stochastic and deterministic forces, including reduced variability resulting from restricted gene flow and drift, and the inexorable losses of local habitat through human activity.

Management

There is a disturbing lack of data on genetic diversity in threatened fishes of the American West. The twenty-six taxa represented in Table 9-1 are less than 20% of the fishes "of concern" in the region (J. E. Williams et al. 1985; J. E. Johnson 1987a). Furthermore, less than half of these twenty-six taxa have been adequately surveyed for levels of diversity and geographic patterns of variation.

Some species that are poorly understood genetically are nonetheless the objects of intensive management: for example, Colorado squawfish (*Ptychocheilus lucius*), bonytail, woundfin (*Plagopterus argentissimus*), and razorback sucker (*Xyrauchen texanus*). Management of these fishes includes artificial propagation, reintroductions into formerly occupied areas, and augmentation of existing populations. Ideally, such activities should be based on patterns of genetic variation. The USFWS is initiating efforts to obtain such information on some of these species (Buth et al. 1987; Ammerman 1988; Ammerman and Morizot 1989; Minckley et al. 1989), but much more needs to be done. One of the first priorities in plans for recovery of any species should be geographic surveys of genetic variation.

Management of Natural Populations

Once a species, subspecies, or population has been selected for conservation efforts, guide-lines must be established for preserving maximal proportions of the remaining genetic variation. Allendorf and Leary (1988) argued for preserving the diversity of alleles rather than allele frequencies. They pointed out that extinctions of alleles are essentially irrevocable, while changes in frequencies of available alleles are not. When variation is largely due to alleles with narrow geographic distributions, a number of populations will have to be maintained to protect diversity. Such protection will require maintenance of fewer populations if variation is due to frequency differences in alleles present in nearly all populations (Allendorf and Leary 1988).

Although the hierarchical level at which unique alleles are distributed is a useful guide to the preservation of genetic variability, the lack of detected, uniquely distributed alleles does not necessarily mean that a taxon should be treated as a homogeneous unit (Chakraborty and Leimer 1987; Allendorf and Leary 1988). In both of the endangered fishes in springs of the Balmorhea area of Texas, the major source of between-sample diversity is allele frequency differences between Phantom Lake Spring populations and the remainder of the system (A. A. Echelle et al. 1987; A. F. Echelle et al. 1989). In *Cyprinodon elegans*, the genetic pattern is congruent with morphological differentiation (A. A. Echelle et al. 1987). Although no uniquely distributed alleles have been detected, the zoogeographic pattern reflects a potential for local adaptations that should be preserved.

Cutthroat trout provide another example. Forms showing no diagnostic differences electrophoretically (Loudenslager and Gall 1980; Leary et al. 1987) exhibit well-defined differences in pigment pattern and life-history traits (Behnke 1980). As I mentioned previously, no one modality of data should necessarily take precedence over other forms of information in efforts to preserve meaningful genetic diversity.

Gene flow is an important contributor to maintenance of allele variation. Gene flow from outside may, in fact, be the *only* important contributor in very small populations because mutation is much too slow and genetic drift will overcome selection favoring allele diversity (Lacy 1987). Human activities have disrupted gene flow in many taxa. Fishes of the Colorado River system provide striking examples: many native species that once had relatively continuous distributions now consist of disjunct, remnant populations (Minckley 1973; Molles 1980) with little, if any, opportunity for interpopulation gene flow. In such instances gene diversity analysis can provide valuable information for attempts to recreate natural patterns of gene flow through transplantation. Such efforts should take care to avoid swamping the "opportunity for local adaptation" (Meffe and Vrijenhoek 1988) and should be done only when the systematics of the involved populations are thoroughly understood. Although requiring caution, this is a potentially valuable management option for preserving the evolutionary potential of many threatened taxa.

By now, it should be generally recognized that introductions of fishes foreign to an area must be discouraged. The many problems caused by such introductions are well illustrated in the western United States (J. N. Taylor et al. 1984; Moyle et al. 1986), where loss of native populations through introgressive hybridization is a major factor in fish extinctions. In most situations, restoration of an introgressed stock to its original genome is difficult or impossible. Such an effort apparently was successful in a small (3–4 km) spring-fed stream where the Leon Springs pupfish was introgressed by *C. variegatus* (C. Hubbs et al. 1978; C. Hubbs 1980; A. A. Echelle et al. 1987). Usually, the system is too complex or introgressed populations are too widespread for complete removal of the problem; introgressed Pecos pupfish, Guadalupe

bass, Gila and Apache trouts, and subspecies of cutthroat trout are good examples.

In some instances efforts are made to replace introgressed populations with the native taxon in isolated portions of its original geographic range; for example, trout in small headwater lakes and streams (Allendorf and Leary 1988). Such efforts include constructing barriers to dispersal, eradicating contaminated populations, and introducing genetically "pure" stocks of the native taxon (Rinne and Turner, *this volume*, chap. 14). In some situations it may be possible to swamp out contamination by continually introducing genetically pure stocks, especially if the degree of introgression is low.

State and federal agencies should protect uncontaminated populations by prohibiting *all* human transport of live fish (i.e., bait and sport fish) in the area of concern. Such restrictions are being considered for local areas by the New Mexico Department of Fish and Game (NMDFG) and the USFWS in response to threats posed by pupfish transport (D. Propst, NMDFG, and J. Brooks, USFWS, pers. comm.).

Managers handling introgressed populations must decide what frequency of foreign genes necessitates restoration efforts (Campton 1987) and at what gene frequency a population has been effectively restored. Allendorf and Leary (1988) suggested a level of 1% foreign genes. At that frequency introgression is difficult to detect and is unlikely to alter biological characteristics from the native state. It is important to realize that from the standpoint of maintaining genetic diversity, even highly introgressed populations represent a valuable resource if they represent all that remains of a taxon.

Artificial Propagation

A number of threatened fishes are already being maintained and propagated artificially. Most of this activity is centered at Dexter National Fish Hatchery (NFH) in New Mexico

(Rinne et al. 1986; Johnson and Jensen, *this volume*, chap. 13). The primary purposes of such efforts are threefold: (1) to protect genetic resources against catastrophic loss of natural populations, (2) to allow research on needs of threatened species, and (3) to provide stocks for reintroduction into the historic range. The genetic structure of the captive population(s) is critical for all three purposes.

After a taxon has been targeted for artificial propagation, the source of captive stock must be chosen with care. Whenever possible the choice should be consistent with concerns regarding maintenance of the natural pattern(s) of genetic diversity. However, when a species is extremely rare, such as the bonytail (Minckley et al. 1989), management goals must shift from preserving geographic or other patterns of variation to simply preserving as much of the remaining diversity as possible.

Having chosen a population, the manager must decide which site(s) should supply the founding stock of captives. In general, sites showing the highest heterozygosity are preferable (Meffe 1986). The value of this rule is illustrated by the Sonoran topminnow. Based on knowledge of heterozygosity, Vrijenhoek et al. (1985) recommended that the source of topminnows for captive propagation be switched from the Monkey Spring population ($H = 0.000$) to the Sharp Spring population ($H = 0.037$). Other studies indicate the wisdom of this change. Topminnows from Sharp Spring have higher fecundity, growth rates, and other measures of fitness than do those from Monkey Spring (Meffe 1985b; Quattro and Vrijenhoek 1989; Vrijenhoek and Sadowski, unpub. data).

Such choices should be based on thoroughly established patterns of diversity, and not on preliminary data such as those presented for most species in Table 9-1. For example, the USFWS is propagating Apache trout from the East Fork of the White River to provide stock for reintroduction (Rinne, pers. comm.). The

sample from that source of brood stock has a very low heterozygosity value (0.007), two and a half times less than the mean for the species (0.018), and nearly four times less than the maximum (0.026 in Flash Creek; Loudenslager et al. 1986). Thus, a change in the source for captive brood stock may be advisable, although more samples are needed to ensure that the differences are not due to sampling error. Populations can show considerable microgeographic variation in heterogeneity, especially if physical barriers prevent gene flow (A. A. Echelle et al. 1987; A. F. Echelle et al. 1989).

The primary problems to be avoided in preserving the genetic integrity of captive populations are loss of variability due to drift, changes due to selection, and contamination with foreign genes. Management recommendations vary somewhat depending on the species of concern. Captive threatened fishes tend to fall into two groups: "spontaneous breeders," in which large numbers of individuals spawn successfully in environments such as hatchery ponds, and "artificial breeders" requiring artificial inducements to spawn (hormone injections, stripping of gametes, etc.). The spontaneous breeders at Dexter NFH include pupfishes, poeciliids, and most cyprinids (*Gila* and *Cyprinella*); those being spawned artificially include Colorado squawfish and razorback suckers.

Allendorf and Ryman (1987) gave a number of baseline criteria for reducing genetic drift that are especially relevant for artificial breeders. The major recommendations are threefold: (1) the initial (founding) population size should have an "absolute minimum" of twenty-five females and twenty-five males, (2) succeeding generations in captivity should be maintained at no fewer than one hundred females and one hundred males, and (3) efforts should be made to equalize reproduction of all individuals in the founding population and in all succeeding generations. Such equalization can increase the effective population size up to twice that of a random-mating population of the same size (Denniston 1978).

Wild-caught fish of some species (e.g., bonytail) are so rare that it will be extremely difficult to meet the minimum requirements for the founding population. For such taxa, equalization of reproductive contributions of all individuals in the founding population is even more important. When new wild-caught individuals are added to such stocks, every effort should be made to ensure their contribution to the next generation of brood fish. This will help replace the potentially substantial (Allendorf 1986) loss of allele diversity associated with the initial (founding) event.

Although each species needs individual evaluation, management of genetic diversity in the spontaneous breeders generally seems less complicated than management in artificial breeders. Typically, spontaneous breeders are smaller, shorter-lived, locally abundant fishes, and founding populations can comprise hundreds of individuals. Most such species breed in hatchery ponds over a period of several months and quickly attain relatively high densities (e.g., hundreds or thousands; B. L. Jensen, USFWS, pers. comm.). Comparisons involving two pupfish species and a gambusia species held at Dexter NFH for six to eight years revealed no large changes in heterozygosity. Some losses of rare alleles apparently occurred, possibly due to the small founding stocks of thirty to eighty individuals (Edds and Echelle 1989). The present practice at the hatchery is to include two hundred or more fish in the founding stock of each species. An electrophoretic survey of desert pupfish (*Cyprinodon m. macularius*) maintained in several artificial refugia revealed minor changes that appeared attributable to natural selection rather than to drift (B. J. Turner 1983, 1984).

Captive populations are simultaneously subjected to selective pressures unique to their new artificial environment and to a "release" from selection pressures that normally operate in the wild. Allendorf and Ryman (1987) made three major recommendations to help avoid unwanted genetic change due to hatchery effects: (1) avoid selecting for particular traits—like high growth rate and high fecundity, (2) eliminate specimens showing deformities, and (3) periodically introduce new individuals from wild stocks. These authors also provided suggestions for monitoring captive populations to detect unwanted change, whatever the cause.

The USFWS estimates that more than thirty-five species of fish from the southwestern United States alone may need artificial propagation if they are to be recovered (Rinne et al. 1986). Those numbers are likely to increase in the future. Currently, most captive fishes are maintained in hatchery ponds. Given economic realities and the numbers of taxa involved, a high proportion of captive stocks will be maintained as single populations. When the captives represent a major portion of the remaining genetic resources in a taxon, multiple stocks should be maintained to avoid losses due to catastrophic events.

Chapter 10

On the Design of Preserves to Protect Native Fishes

Peter B. Moyle and Georgina M. Sato

Introduction

As long as human populations continue to grow, accompanied by an even more rapidly growing demand for goods and services, native fishes will decline. These declines are caused directly by diversion of water for human use and indirectly by pollution, habitat alteration, and introductions of non-native fishes. The decline of native fishes is particularly severe in western North America, where a limited supply of water is in high demand for irrigated agriculture and urban development. Problems of supply and demand have consistently made water controversies the most important political issues in the western United States (Fradkin 1984; Reisner 1986).

In recent years, growing public concern over the loss of native plants and animals has resulted in laws that protect at least some of the native fish fauna. As other chapters in this volume indicate, this legislation mainly protects species officially listed as endangered or threatened, and usually on an emergency basis. Ideally, however, native fishes should be protected before they become imperiled; this policy protects not only species but also the biotic communities of which they are a part. Proper protection requires an understanding of what communities need to persist indefinitely, and such an understanding is essential for the design of aquatic preserves.

We broadly define a preserve to mean any natural area that is established and managed to maintain its native biotic communities. In this paper we establish criteria for the design of such places and then discuss their implications for management. First, we review the theoretical basis for preserve design in light of the developing field of conservation biology. Second, we discuss design criteria. Third, we present a classification system for aquatic preserves and discuss how this and other classification systems satisfy our design criteria. Fourth, we propose a strategy for development of regional systems of aquatic preserves.

This paper is based on field studies sponsored by many agencies, but most recently by the U.S. National Park Service and the California Department of Fish and Game (CADFG). The ideas outlined here are currently being applied to the design of a statewide aquatic preserve system in California, sponsored by the CADFG. Working with the senior author on this venture are J. E. Williams (U.S. Bureau of Land Management) and C. Swift (Los Angeles County Museum of Natural History). We appreciate the support and encouragement of J. Brode and B. Bolster, CADFG. Various drafts of this paper were debated and dissected by L. Brown, E. Strange, E. Wikramanayake, W. Bennett, T. Hopkins, A. Brasher, D. Castleberry, and M. Parker.

Theoretical Basis for Preserve Design

The design of preserves is an integral part of conservation biology, which has been labeled "the science of scarcity and diversity" (Soulé 1986). As such it can be identified as a crisis-oriented discipline whose focus is on the prevention of extinctions and the protection of natural diversity. Conservation biologists trying to establish a theoretical basis for their new field of endeavor have borrowed and re-synthesized many theories from other disciplines, especially ecology and genetics. Here we review theories most relevant to preserve design from community ecology, biogeography, population dynamics, and population genetics.

Community Ecology

If preserves are to maintain their biotic communities, they must be capable of maintaining all the complex interactions among the organisms as well as interactions among the organisms and their environment. The area of community ecology that seems to provide the best insights into these complex interactions is food-web theory, which focuses on direct and indirect trophic interactions.

Food-web theory suggests that the number and complexity of trophic links within a community (connectivity) contribute to its stability, which translates to persistence of its constituent species. The amount of connectivity depends on the number of generalists or specialists in the food web, and on the number of trophic levels (Pimm 1986). The loss of a single species, depending on its position in the food web, could thus result in loss of other species in the community (Pimm 1980). Particularly important are "keystone" species, such as a predator that keeps populations of a competitively dominant herbivore in check, or an organism that keeps some limiting nutrient in circulation through its activities.

In aquatic systems, the importance of maintaining a high degree of connectivity in food webs has long been recognized. In lakes, for example, stabilizing links have been demonstrated among various trophic levels, including detritus, bacteria, phytoplankton, zooplankton, and fishes (Rich and Wetzel 1978; Cole 1982; Tilman et al. 1982). Thus, introduction into a lake of a new species that directly alters one part of the food web can change the structure of the entire community through indirect effects (Moyle et al. 1986). Often, much of the energy that forms the basis for a food web is produced outside the system. This is particularly true for streams in which leaves fallen from surrounding trees are the largest source of energy, or streams in which decaying carcasses of migrating salmon are a major source of nutrients (Richey et al. 1975). Maintenance of such allochthonous sources of energy must be yet another consideration of any preserve design; if energy sources are altered, communities may be significantly changed.

Community ecology theory supports the idea that other outside influences must be maintained as well if that community is to continue to maintain its structure. This includes abiotic as well as biotic factors. An important role of such factors may be in creating unpredictable disturbances to the community that prevent a small number of species from dominating it completely through competition, predation, or other means. In general, maximum diversity of a system is promoted by intermediate levels of disturbance (Menge and Sutherland 1976; Connell 1978; Sousa 1984). This "intermediate disturbance hypothesis" suggests that it is important to maintain disturbances in preserves at a high enough frequency and intensity to keep competitively dominant species from eliminating competitively inferior ones, yet low enough to prevent elimination of species through environmental stress. In a stream preserve, for example, floods of varying magnitude are probably important for maintaining high species diversity; creation of excessive flooding by poor watershed management or elimination of floods by dams are both likely to decrease natural diversity (R. M. Baxter 1977; Minckley and Meffe 1987).

Island Biogeographic Theory

The earliest theory in community ecology to be applied to the question of maintaining species diversity in a limited area (e.g., a preserve) was that of island biogeography. Preserves were viewed as islands of natural habitat in a sea of human-influenced habitats. It was therefore assumed that this theory could be applied to preserve design, because it was developed to explain how species diversity of islands originated and changed. Unlike newly formed oceanic islands, however, preserves are rarely empty and awaiting colonization. Likewise, unlike mountaintop "islands" resulting from changes in topography over geologic time, preserves usually result from relatively rapid isolation. Because of these differences, the elements of island biogeographic theory most pertinent to preserve design are those which address questions of what parameters best predict species diversity and what effects fragmentation and isolation have on species diversity.

Prediction of species diversity

Island biogeographic models relating numbers of species to land area have played a major role in conservation biology, even if their true value developed indirectly. Species-area models were first used to describe, in part, differences in species diversity on islands of different areas (MacArthur and Wilson 1967). However, applications of such models to empirical studies have not always been successful (Diamond 1976; Simberloff and Abele

1976, 1982; Terborgh 1976; Abele and Connor 1979; Cole 1982; Wilcox and Murphy 1985). Statistically, the robustness and high variance of species-area models make them unreliable as predictors of species diversity (Haas 1975; Diamond and Mayr 1976; Connor and McCoy 1979; Quinn and Hastings 1987). Furthermore, biological interpretation of the parameters used has also come into question (Connor and McCoy 1979; I. Abbott 1983). Overall, area alone is not sufficient for predicting species diversity (J. H. Brown 1978; Abele and Connor 1979; Connor and McCoy 1979; Simberloff and Abele 1982; I. Abbott 1983; Boecklen and Botelli 1984; Quinn and Robinson, in press).

As the controversy over species-area models developed, various workers showed that species diversity could be better predicted by parameters such as elevation (M. P. Johnson and Raven 1983; Diamond and Mayr 1976; Picton 1979), production (Wright 1983), and, especially, habitat complexity (MacArthur 1964; Kohn 1967; Johnson and Raven 1973; Diamond 1976; Picton 1979; Forman and Godron 1981; Simberloff and Abele 1982; Boecklen and Botelli 1984; Soulé and Simberloff 1986). Species diversity may also be dependent on biological factors such as initial numbers of species and individuals, immigration and emigration rates, and presence of dominant predators or competitors (Haas 1975; Diamond 1976; Diamond and May 1976; Simberloff and Abele 1976, 1982; Terborgh 1976; R. I. Miller and Harris 1979; Soulé et al. 1979; Cole 1982; Shaffer and Samson 1985; Wilcove 1985; Soulé and Simberloff 1986).

Parameters other than area may be better predictors of species diversity in aquatic systems. In natural lakes, when area has a clear and positive correlation with species diversity, it also has a positive correlation with habitat diversity (Barbour and Brown 1974; Tonn and Magnuson 1982; Eadie and Keast 1984). Physical parameters related to habitat diversity—such as substrate heterogeneity, nutrient load, natural acidity or alkalinity, and degree of isolation—account for much of the variability among lakes in numbers of species (Keast 1978; Tilman et al. 1982; Rahel 1986; Eadie et al. 1986). Browne (1981) also indicated that limnological factors relating to productivity were the best predictors of fish species diversity in a small set of lakes.

For streams, as for lakes, habitat heterogeneity is presumably a key to fish species diversity (Gorman and Karr 1978; Eadie et al. 1986). Within a region there are typically positive relationships between watershed area and number of species (Thompson and Hunt 1930; Lake 1982; Brönmark et al. 1984; Fausch et al. 1984; Welcomme 1985; Eadie et al. 1986; Sheldon 1988) that result from increased habitat diversity associated with increased length and size of streams. Some Mexican springs follow this same pattern (Minckley 1984). Numbers of species of both fishes and aquatic insects generally increase within a stream or spring system in a downstream direction (Hynes 1970); new species tend to be added while only a few headwater habitat specialists drop out (Moyle and Li 1979; Mahon 1984). Nevertheless, factors such as discharge, gradient, substrate type, temperature, productivity, water chemistry, or even distance from centers of evolution may be more important in promoting such a trend (Sepkoski and Rex 1974; Lake 1982; Brönmark et al. 1984; Eadie et al. 1986; Sheldon 1987).

Fragmentation and isolation

One of the biggest problems in preserve design is compensating for the effects of isolation on persistence of species. As natural habitats become fragmented through expansion of altered habitats, populations become increasingly isolated and confined to smaller areas (Lovejoy et al. 1984). Although fragmentation can occur naturally through geologic or climatic events (for fish examples see Moyle 1976; G. R. Smith 1981a; Minckley et al. 1986; Sheldon 1987), anthropogenic fragmentation occurs on a far shorter time scale, so its effects are likely to be more severe.

Species richness and abundance in non-isolated areas are balanced by the birth, death, emigration, and immigration rates of each taxon (MacArthur and Wilson 1967). In contrast, in isolated places, particularly those isolated by human disturbance, species numbers can decline as a result of extinction or emigration without restorative immigration (J. H. Brown 1978; Pickett and Thompson 1978; R. I. Miller and Harris 1979; Soulé et al. 1979; Frankel and Soulé 1981; Cole 1982; Wilcox and Murphy 1985). Because of this effect, compensating for factors that can cause extinction of populations is a major concern in preserve design. Two important long-term factors are the possibilities that small isolated populations will become extinct either through random population fluctuations or through loss of genetic diversity; these are dealt with below. More immediate factors are the constant external threats to each preserve and its species. Without active management a preserve becomes more and more like the degraded areas surrounding it. The smaller and more isolated a preserve, the more susceptible it becomes to degradation (Janzen 1983). Factors that threaten preserve integrity from the outside are of two basic types: edge effects and external effects.

Edge effects are threats that result from the presence of a steep gradient from good- to poor-quality habitat (Lovejoy et al. 1986). The presence of altered habitat adjacent to preserve edges may create subtle changes such as altered microclimates inside remnant natural areas (Lovejoy et al. 1986) or changes in the behavior of organisms on both sides of the boundary. Because conditions near an edge are transitional, assemblages are likely to contain mostly species that are good colonizers. Also, as a gradient of quality develops across an edge, organisms in areas being degraded may move into the preserve to take advantage of better conditions, causing overpopulation and habitat destruction (Lovejoy et al. 1984; Dobson and May 1986; Janzen 1986). After a gradient is established, organisms within a preserve forced out through flooding in streams, for example, may enter habitats that can no longer sustain them, resulting in population losses that cannot be balanced by immigration or recolonization. Even if none of these things happens, organisms inside the preserve may alter their behavior (e.g., restrict their movements), possibly causing changes in population selection pressures or extinction rates (Forman and Godron 1981; Wilcove 1985).

Edge effects that degrade a preserve are often accompanied by more insidious external effects that do not recognize boundary lines. These "eternal external" threats (Janzen 1986) include pollutants, fires, diseases, and exotic species (Frankel and Soulé 1981; Dobson and May 1986; Janzen 1986; Wilcove 1985). External forces can be particularly damaging to an aquatic preserve because any deleterious effect occurring in a drainage may eventually make its way through the heart of it (Sioli 1986), resulting in habitat degradation and consequent reductions of native species (Moyle et al. 1986). Dams and impoundments in streams can cause changes that influence both upstream and downstream reaches (Baxter 1977; Sioli 1986). The negative effects of chemical pollutants, including fertilizers and pesticides, have been repeatedly demonstrated for all types of aquatic systems and many taxa (e.g., Pryde 1972; Dermott and Spence 1984; Millemann et al. 1984; Sanders 1986; Lewis

and Morris 1986). Turbidity from agriculture, recreation, logging, or road construction can also have severe negative effects (Atapattu and Wickremasinghe 1974; Moss 1977; Bisson and Bailey 1982; J. W. Sigler et al. 1984; Murphy et al. 1986). Competition and predation from non-native invading species can reduce numbers of native species (Gorman and Nielsen 1982; Post and Cucin 1984; Moyle et al 1986).

One of the best ways to protect natural areas from edge and external effects is to create buffer zones around them (Schonewald-Cox and Bayless 1986). Buffer zones provide a gradient from natural to altered conditions and become increasingly effective as barriers to perturbation as their size (width) is increased and their natural qualities are enhanced (Lovejoy et al. 1986). Like the areas they are designed to protect, buffer zones must be tightly regulated (Dobson and May 1986; Soulé and Simberloff 1986).

Well-managed buffer zones of terrestrial habitats are particularly crucial for the protection of aquatic preserves, especially for streams, because their linear shape means they have a high proportion of edge and are therefore exceptionally vulnerable. Thus, buffer strips of riparian vegetation along streams have repeatedly been shown to reduce negative impacts of logging on fishes and other organisms (e.g., Murphy et al. 1986). One of the difficulties in establishing buffer zones around aquatic systems is psychological: the sharp visual boundaries that exist between aquatic and terrestrial habitats are often interpreted, if unconsciously, as barriers to external influences.

Population Dynamics

Within a preserve, population size alone may threaten persistence of a species, even if habitat requirements are met. Basically, the population size must be large enough so it will not be driven to extinction by two types of stochastic events. The first involves preserve-wide environmental catastrophes. Such catastrophes, whether human or natural in origin, can range from the species-specific, such as a disease epidemic (Arai and Mudry 1983), to those encompassing entire regions, as with the effects of acid precipitation. The second event is a random drop in population size below a threshold of recovery for that species (Gilpin and Soulé 1986). Thus, determination of minimum viable population size for each species must be based on an understanding of both its population dynamics and the disturbance regime of the preserve region (Diamond and May 1976; Pickett and Thompson 1978; Shaffer 1981).

Problems of demographic stochasticity can be countered in two ways. First, to guard against extinctions due to local catastrophes there must be a stock of the species in other preserves (Simberloff and Abele 1976, 1982; Frankel and Soulé 1981; Dobson and May 1986). Second, to guard against effects of random population fluctuations, within-preserve populations must be large and widespread enough so that extinction cannot occur even when populations are at their lowest levels (Richter-Dyn and Goel 1972; Frankel 1974; Kushlan 1979; Schonewald-Cox and Bayless 1986; Soulé and Simberloff 1986). In short, replication of populations within and among preserves is necessary to perpetuate species in the event of local extirpations due to random fluctuations or events.

Population Genetics

When isolation of preserves leads to disruption of natural gene flow, the population within each preserve faces possible loss of genetic variation. Frankel and Soulé (1981), Schonewald-Cox et al. (1983), Meffe and Vrijenhoek (1988), and Echelle (*this volume*, chap. 9) review the importance of genetic diversity for fitness and survival of species. All

four works cover the threats to genetic diversity resulting from the creation of preserves that isolate small populations of species that were once more widely distributed. Problems of particular importance include: (1) founder effects or genetic bottlenecks (effect of small initial populations being random subsamples of an entire species' gene pool), (2) inbreeding depression (possible increase in deleterious alleles through increased mating with relatives), (3) outbreeding depression (possible loss of fitness through increased occurrence of maladaptive traits from outside the population), (4) genetic drift (stochastic loss of diversity through reproductive events being equivalent to mere sampling events), and (5) hybridization (possible loss of species identity through introgression with other species).

Threats to a species' genetic diversity can be reduced by having populations replicated elsewhere and in large enough numbers to guard against drift, inbreeding, or hybridization. Loss of genetic diversity due to small population sizes under captive propagation can be decreased through controlled breeding procedures (Frankel and Soulé 1981). It is important to establish preserves based on some knowledge of the natural distribution of genetic diversity in order to begin conservation efforts with as much diversity as possible and with knowledge of the effects isolation is likely to have (Echelle, *this volume*, chap. 9). Although natural gene flow is important for maintenance of genetic diversity (L. M. Cook 1961; Slatkin 1985), human manipulations of gene flow through exchanges of individuals among artificially isolated populations should be done only with caution because of the risks of transferring maladaptive traits or diseases (Kushlan 1979; Boecklen 1986; Dobson and May 1986). With fishes, Meffe and Vrijenhoek (1988) suggest that experiments in mixing genetic stocks to look for potential problems should be conducted before transplants take place.

Criteria for the Design of Natural Preserves

The ultimate goal of preserve design should be protection of entire, naturally functioning, native communities, because this is the way to ensure the survival of species in their evolutionary context. Preserves must not be designed as museums that freeze present conditions (or historic conditions; the "way it was") in place, but as dynamic, evolving ecosystems in which both short-term and long-term natural processes are functioning. Most preserves are unlikely to meet this goal because of human-caused environmental changes. This means they will not be self-maintaining, but will require active management. In most instances, protected natural areas contain only a fraction of the original biota and are maintained mainly to protect species that might otherwise disappear altogether. Areas where design goals are species oriented rather than community oriented are best labeled refuges rather than preserves, although there is really no sharp dividing line between these two types of protected areas. In fact, an ultimate goal of a refuge design should be to convert it into a preserve by restoring lost complexity. Thus, criteria for preserves should be applied as much as possible to all the areas protected for the sake of perpetuating natural diversity.

Many problems involved in designing aquatic preserves are the same as those for terrestrial ones, but aquatic systems have many special management problems. In fact, aquatic habitats are difficult to protect if their surrounding terrestrial environs are not protected as well, so the aquatic components are typically the most sensitive portions of preserves in general. Therefore, we list some of the more important criteria for preserve design in general, derived from our preceding discussions of theory, and outline some problems in applying them to aquatic systems:

1. *A preserve must contain resources and habitat conditions that are known to be necessary to the persistence of all species in the communities it is designed to protect.* This includes migratory species that use the preserve for only part of their life cycles. To satisfy this criterion the requirements of all life-history stages of all species should be known. Because this is clearly impossible, design should be based on the needs of the largest and most mobile species, on the assumption that their needs will encompass those of smaller and less well understood taxa. In most cases these species are the best studied anyway. For aquatic systems, "design" organisms will usually be fishes and macroinvertebrates.

2. *A preserve must be large enough to maintain the range and variability in conditions needed to maintain natural species diversity.* The actual size of a preserve will depend on the communities being protected: a spring may require only a few hundred square meters, whereas a riverine system may require thousands of square kilometers. In general, aquatic preserves need to be large enough that their water sources are included, or at least protected, whether these are aquifers feeding springs, headwaters of streams, or tributaries to lakes. Streams present a special problem because of their linear nature and unidirectional flow. Not only do stream preserves need to include headwaters, but they also should maintain their characteristic longitudinal faunal zonation (Moyle and Li 1979).

3. *A preserve must be protected from edge and external effects in order to maintain good internal quality.* For aquatic preserves this means: (a) protection of water sources and upstream areas to prevent siltation, pollution, and other problems; (b) creation of substantial terrestrial buffer zones; and (c) construction of barriers to preclude invasion by non-native species. Design of barriers is a particular problem because they should be able to pass native migrants while excluding non-native invaders.

For many aquatic systems, but especially for streams, the best barrier to invasion is to have habitat as pristine as possible, thus giving native species adapted to such conditions competitive/predatory superiority.

4. *A preserve must have enough within-boundary replication to avoid problems created by local extirpations.* For lakes and springs this means the entire water body needs protection. For stream systems, tributaries of different orders need replication within the preserve.

5. *A preserve should be replicated by having one or more similar areas protected far enough distant so that replicates will not be affected by the same disaster.* For many aquatic preserves, such as those protecting springs and lakes, replication may not be possible because of highly localized endemism. For streams, replication means protecting separate drainages with similar characteristics and biotas.

6. *A preserve must be able to support populations large enough to be self-sustaining in the face of demographic or genetic stochasticity.* This is a particular problem in the design of riverine preserves because natural streams fluctuate widely and local extirpations are probably common. Under normal conditions, rapid recolonization of organisms from nearby streams would occur. As more of a drainage is protected, colonists are more likely to be available to reoccupy decimated areas.

Classifying Preserves and Other Waters

The ideal preserve should be not an isolated entity but part of a system that provides redundancy for protection of widespread communities as well as those which are local and unique (such as those often found in isolated desert springs). Developing a regional preserve system requires that available natural areas be realistically evaluated in terms of satisfying the criteria just delineated. This,

in turn, requires a realistic classification and ranking system for the waters to be protected.

Most classification systems now available do not adequately integrate the biotic and abiotic requirements of organisms. One type focuses on natural diversity, emphasizing special species or conditions (e.g., Rabe and Savage 1979; Rapport et al. 1986) without considering the ranges of the species, biotic and abiotic interactions that stabilize communities, or potential threats to each natural area. At the other extreme are classification systems that rely mainly on physical environment, an approach that is especially common for aquatic systems (e.g., Savage and Rabe 1979; Lotspeich and Platts 1982). These generally assume that a given type of habitat will support a given array of species regardless of the many factors that affect the ability of the habitat to act as a preserve. Margules and Usher (1981) reviewed selection criteria employed in various attempts to classify natural areas and found that the top-ranking criteria were diversity, rarity, naturalness, size of area, and threat of human interference. However, no study used all these criteria or another important criterion—the likelihood of survival of critical species populations (keystone carnivores, etc.) after isolation.

Our classification system attempts to involve all these factors in classifying and ranking waters according to their ability to maintain natural biotic diversity. Basically, it consists of six classes of protected water that form a continuum of quality from class I (best) to class VI (worst). Ideally, all aquatic organisms should be protected in class I waters, but such places are now extremely rare, especially in the American West. Any practical system will necessarily include some or most of the six classes, although its focus should be on those of the highest quality.

Class I Waters

Completely pristine watersheds or drainage

systems may no longer exist, but those present today that bear the closest resemblance to what we think original watersheds were like are class I waters. Class I waters contain a nearly complete set of the native biota, have suffered comparatively little from human disturbance, and have a high degree of natural or official protection, including ridge-to-ridge protection of the watershed, barriers against invasion of non-native species, and wide buffer zones. Although class I waters can be of any size or area, if they are to protect native fishes (or other aquatic biota) they must (1) contain a substantial percentage of the regional fish fauna, (2) have a high degree of habitat diversity, and (3) be large enough to maintain minimum viable populations of all resident taxa. Unequivocal examples of class I waters are rare, but Aravaipa Creek (Graham and Pinal counties, Arizona), Cottonball Marsh (Riverside County, California), and Elder Creek (Mendocino County, California) have most of the class I attributes.

Aravaipa Creek is a small, canyon-bound, Sonoran Desert stream famous for its native fishes and discussed in more detail by Jack E. Williams (*this volume*, chap. 11). Seven of the original thirteen species native to the Gila River basin are present, most notable of which are the threatened spikedace (*Meda fulgida*) and loach minnow (*Tiaroga cobitis*), along with a substantial percentage of the native macroinvertebrates, riparian vertebrates, and native vegetational components to be expected in such a system. The stream is protected by the Aravaipa Wilderness Area (administered by the U.S. Bureau of Land Management [USBLM]), which includes a central canyon reach, extensive holdings by The Nature Conservancy (TNC) both upstream and downstream and on adjacent mountain slopes, and substantial USBLM and U.S. Forest Service (USFS) holdings in other parts of its watershed (in part, J. E. Williams et al. 1985).

Cottonball Marsh, home of the Cottonball

Marsh pupfish (*Cyprinodon salinus milleri*), is on the floor of Death Valley, an environment too severe and inaccessible to be disturbed by people, and protected by Death Valley National Monument.

Elder Creek is a small third-order stream with its entire drainage contained within the Northern California Coast Range Preserve of TNC. Because of its pristine nature it is used by the U.S. Geological Survey as an official benchmark watershed for comparison with other, more degraded streams in the area. The drainage is heavily vegetated with old-growth Douglas fir. Dominant fishes are juveniles of anadromous rainbow trout (*Oncorhynchus mykiss*), but Pacific lamprey (*Lampetra tridentata*) and coho salmon (*O. kisutch*) also spawn there. Numerous fish analogs (amphibian larvae) also reside in the stream. The fishes present are only a fraction of the species native to the Eel River, a coastal stream to which Elder Creek is tributary and a class II water itself (see below).

Class II Waters

There are no sharp differences between class I and class II waters, but the latter show a greater degree of modification by human activities. Important attributes are: (1) they still contain a native biota, even if reduced in numbers; and (2) they could be restored to class I status without unreasonable efforts (e.g., removal of a major dam or relocation of a city). Although class II waters are capable of being upgraded, restoration is not necessarily a goal of their management. Typically in the West, they are part of a state or national park, forest, or other public land in which multiple use is a major goal. Maintenance of natural diversity in such areas may be part of this goal, but it has to be compatible with activities such as recreation, logging, grazing, and road building. Thus, maintenance of natural diversity often is one of many use-oriented goals. In the best of such preserves, protection of the native biota is the primary goal, all else being secondary, and more incompatible uses (e.g., off-road vehicle use) are eliminated.

The most important preserves in the American West are class II because (1) they are numerous and often large in size; (2) they are likely to be familiar to the public, who can apply pressures to agencies to manage them for native biota (especially fishes like salmonids); and (3) users often provide funds for management. Unfortunately, because class II waters are rarely established specifically as aquatic preserves, upstream and downstream reaches may not have the same protection as designated segments. This places quality at risk to external influences such as increased sedimentation due to road building upstream or invasions of alien species from downstream. Such changes, if allowed to multiply, may become irreversible and alter the status to class III, with many faunal elements lost. Thus, a class II water requires active management just to maintain the status quo, and considerable effort would be necessary to improve its ability to support a native biota. Often a first step is to recognize its importance as a preserve.

Examples of class II waters are numerous. Three from California are Eagle Lake in Lassen County, which approaches class I; Deer Creek, Tehama County, a high-quality class II water; and South Fork of the Eel River, Mendocino County, which could soon deteriorate to class III.

Eagle Lake is a large (11,500 ha) terminal lake located on the edge of the Great Basin. The lake's high alkalinity (*p*H around 9.0) has been a major factor preventing invasion by introduced species (such has occurred in nearby Lake Tahoe), despite numerous attempts to introduce non-native fishes (unpub. records, CADFG). The aquatic biota is not particularly rich in species, but it is entirely native and extraordinarily abundant, as reflected in large populations of grebes, cormorants, pelicans, and other waterfowl. The lake and its limited drainage basin have been protected through geographic isolation, general unsuitability for most forms of aquatic recreation, and ownership (90% of the land is under the jurisdiction of the USFS and USBLM). The lake is not class I because extensive logging and grazing in the basin have caused major degradation of tributary streams. Particularly hard hit was Pine Creek, the largest tributary and once the principal spawning stream of the Eagle Lake trout (*O. mykiss aquilarum*). Pine Creek became unsuitable for spawning, and the endemic trout is now maintained only by trapping ripe adults, rearing young in hatcheries, and planting them in the lake. Fortunately, county, state, and federal agencies have recognized the unique nature of this lake and its drainage and are taking steps to protect it and to restore Pine Creek (E. Ekman, pers. comm.). Eventually, it may merit class I status, if protection and restoration are successful.

Deer Creek, along with a companion stream, Mill Creek, is the least-disturbed large tributary to the Sacramento River (Sato and Moyle 1987) and is noteworthy for its intact native fish assemblage, including one of the last distinct stocks of spring-run chinook salmon (*O. tshawytscha*). Its protection stems from its near inaccessibility as it flows through a rugged canyon, some of which is now within the Ishi Wilderness Area. Despite a high value for native fishes, it is not pristine. The lowermost reaches may dry in summer due to irrigation diversions, while the uppermost reaches receive high recreational use because they are paralleled in part by a highway. Both upper meadow and lower valley regions are heavily grazed. The most altered segments are dominated by introduced species. Land ownership in much of the watershed is split between the USFS and private logging companies, although logging in the basin has generally respected the steep slopes and erosible soils, helping to maintain a natural condition. Most changes to the creek and its biota are reversible or compatible with maintaining it for the native aquatic biota. Nevertheless, because of its varied ownership and multiple uses, it is unlikely that the drainage will ever receive the protection needed to make it class I.

The South Fork of Eel River flows in part through a TNC preserve. Past and continuing logging, both upstream and downstream from the preserve, has created conditions conducive to exceptionally large floods, which have altered channel characteristics. These changes have contributed in a major way to declines of salmon and steelhead runs, making much of the channel unsuitable for rearing juvenile fish. Nonmigrating fishes thrive, especially threespine stickleback (*Gasterosteus aculeatus*) and Sacramento sucker (*Catostomus occidentalis*), as does the introduced California roach (*Lavinia symmetricus*). However, invasion from downstream by predatory Sacramento squawfish (*Ptychocheilus grandis*) is under way, and this may alter the populations of resident species. If logging and other abuses of the drainage were better regulated, the channel could return to a semblance of its original condition, but invasion of the introduced roach and squawfish is probably irreversible. Unfortunately, the protection TNC offers its section of river cannot compensate for watershed abuses outside the preserve or invasion of alien species. Whether or not this stream remains class II or becomes class III depends on how the native fauna adjusts to invaders and how well upstream terrestrial habitats are protected.

Class III Waters

Class III waters are natural in appearance but have been modified by human activity so that native biotic communities have been severely altered. Many native species are usually absent, and introduced taxa are common. Class III waters are unlikely ever to be restored to a natural state. For the most part, they are fortuitous refuges with characteristics favoring some, but not all, of their original aquatic inhabitants. Typical examples are stream sections between major dams and reservoirs that support a few native species.

Class III waters may nonetheless be very important for protecting remnant populations of native species, although requirements of the species must be recognized so that aspects of the environment that favor them will not be inadvertently altered. Because of their vulnerability to change, class III waters should not be relied upon for long-term preservation of natives, but rather as habitats that contain supplemental populations and gene pools. They can serve as sources of individuals to restock places being restored as class I or II. Management requires the same activities and knowledge as class II waters, except emphasis is on individual species or habitat types rather than on ecosystems. For example, introduced species are likely to be integral parts of the biotic community, so their elimination may be an unrealistic goal. However, management policies, such as manipulation of angling, could be oriented to keep introduced populations as low as possible. Examples of class III waters in California include the lower McCloud River, Britton Reservoir on the Pit River, and Suisun Marsh in the Sacramento–San Joaquin estuary.

The lower McCloud River in Shasta County consists of about 32 km of cold river sandwiched between McCloud Dam on its upper end and Shasta Reservoir downstream (see Williams, *this volume*, chap. 11). The river flows through a canyon to which access is difficult. The first 6 km are within USFS land featuring a public campground; the next 10 km flow through McCloud River Preserve of TNC, which features old-growth stands of Douglas fir; and the lower 16 km pass through land owned by two private fishing clubs. The surroundings, including much of the drainages of major tributaries, is managed for multiple use by the USFS and has been (and continues to be) heavily logged. McCloud Reservoir diverts water and traps sediment so the lower reach fluctuates less than formerly and is clearer and slightly warmer. McCloud River was once famous for its bull trout (*Salvelinus confluentus*) and was the only stream in California in which the species was found. It also supported large annual migrations of winter-run chinook salmon and steelhead, which ceased abruptly when Shasta Dam was closed in 1946. The bull trout became extinct within the last fifteen years, apparently because it was unable to adjust to changes in the river, and was replaced ecologically by introduced brown trout (*Salmo trutta*). Native rainbow trout, distinctive riffle sculpins (*Cottus gulosus*), and a well-developed amphibian fauna persist (Sturgess and Moyle 1978; Berg 1987). Thus, despite the presence of a TNC preserve focusing on the river, and general recognition of the high aesthetic qualities of the stream and its trout population, its fauna is highly modified. Extinction of bull trout and chinook salmon and the presence of two large dams indicate that restoration to class I or II would be nearly impossible.

Britton Reservoir, in the middle reach of the Pit River drainage, Shasta County, is the largest in a series of impoundments on a stream completely harnessed for hydroelectric power production. Despite the highly modified nature of the entire system, it is still dominated by native nongame fishes. The reservoir supports a fishery for various alien centrarchids, but native species still make up most of the biomass (Vondracek et al. 1989). Particularly noteworthy are large populations of tule perch (*Hysterocarpus traski*) and rough sculpin (*Cottus asperrimus*), the latter listed by the state of California as threatened. The tule perch population is at its upstream distributional limit, whereas the sculpins represent a downstream colonization from their native

haunts in Hat Creek. A native crayfish (*Pascifasticus fortis*), however, has been replaced by introduced species and is now considered endangered by both state and federal authorities. The reservoir has been a fortuitous refuge for native fishes because it is used solely for power production, resulting in rapid water turnover that mimics conditions of a giant riverine pool. Dams will prevent the system from returning to a higher class, but the system will probably continue to support native fishes because present water management is designed in part to favor the largest nesting population of bald eagles (*Haliaeetus leucocephalus*) in California (Vondracek et al. 1989). The eagles are a fully protected, endangered species that feed largely on native fishes.

Suisun Marsh is a large (34,000 ha) tidal marshland located in the highly modified Sacramento–San Joaquin estuary. These marshlands are intensively managed for duck hunting. One of the principal means of promoting duck populations is by encouraging growth of tules (*Scirpus* spp.) that require seasonally fluctuating salinities (from 0 to 15 g l^{-1}), a regime that also favors native estuarine fishes, although introduced species flourish as well (Moyle et al. 1985). Native fishes include two declining species now confined to the estuary—splittail (*Pogonichthys macrolepidotus*) and delta smelt (*Hypomesus transpacificus*)—as well as about twenty other kinds that vary widely in abundance. Two other native species are either extinct (thicktail chub, *Gila crassicauda*) or locally extirpated (Sacramento perch, *Archoplites interruptus*).

All the native species favor the most "natural" habitats remaining in the marsh—the dead-end sloughs—and show a high degree of ecological segregation, unlike the more widespread introduced species (Herbold 1987). The result is a highly unstable community to which new, alien fishes are constantly being added and native species are being eliminated. Recognition of the marsh as a preserve for native fishes could result in management measures to stabilize natural fish populations.

Class IV Waters

Class IV waters are natural-area refuges created entirely for the purpose of protecting selected species, usually fishes that are likely to become extinct if left in degraded natural environments. They often have a natural appearance, especially as aquatic and riparian vegetation develop, but because they are in reality unnatural, they generally require continuous monitoring and maintenance. When such refuges are created, a typical goal is to produce an environment as similar to the original as possible in its biological, chemical, and physical characteristics.

Class IV waters should be considered temporary solutions to the problem of imminent extinction, and as homes for backup populations of species with limited populations. Unfortunately, they may necessarily become permanent habitat for species whose natural habitats have been eliminated and which have little or no hope of recovery (for examples, see Williams, *this volume*, chap. 11). Successful natural-area refuges usually require considerable effort and money to establish, as well as insight into the biology of the species for which they are created. It is particularly important to reduce the effects of small populations on genetic variability. This can be done by keeping a number of subpopulations at separate localities, each representing carefully controlled intermixing from different founder populations, if available. It is also important to recognize that refuges with characteristics quite different from native habitats of the species may quickly select for individuals adapted for that particular set of environmental conditions, and possibly create a stock poorly adapted for reintroduction into restored natural habitats.

An example of a class IV water is a series of three pools constructed in 1978 along the outflow of Chimney Hot Springs in central Nevada as a refuge for Railroad Valley springfish (*Crenichthys nevadae*; C. D. Williams and Williams 1989). Previously, the outflow supported no fish because of the high temperature (63°C) of the spring water; however, water diverted into the pools cools sufficiently to support springfish in the lowermost pool, and seasonally in the other two. The populations require almost continuous monitoring, however, in that the fish have been extirpated twice, once when the spring outflow declined for unknown reasons and the pools dried, and once when damage by vandals increased flows, making the pools too hot for the springfish to survive.

Class V Waters

Class V waters are artificial refuges in which no attempt is made to re-create natural conditions. These "habitats" are important as temporary holding facilities for imperiled native species. They can vary from aquaria or concrete troughs in a fish hatchery, to ornamental ponds, to dirt-bottomed fish ponds. They are valuable and necessary facilities but should exist only until better refuges or preserves can be established. Sometimes life cycles of the fishes are completed through means such as artificial spawning techniques, artificial diets, and so on. Because fishes in such facilities survive under a narrower range of conditions than they would encounter naturally, the problem of maintaining genetic diversity and stocks that can survive in the wild is even more severe than in class IV waters.

Perhaps the best example of a class V water is Dexter National Fish Hatchery, New Mexico (Johnson and Jensen, *this volume*, chap. 13), where critically imperiled native western fishes such as Colorado squawfish (*Ptychocheilus lucius*), razorback sucker (*Xyrauchen texanus*), and bonytail (*Gila elegans*), among others, are being successfully reared and maintained. Others are listed in chapter 11 in this volume.

Class VI Waters

Artificial waters are the only places where a number of species exist. Typically, the container is glass and the water is mixed with alcohol or formaldehyde. Occasionally, the container is a freezer and the environment is ice or liquid nitrogen. It sits in the dark catacombs of a museum. These species' only hope is that someday we will develop the technology to resurrect them using preserved DNA.

A Strategy for Protecting Aquatic Biotas

There is a growing realization that preserving biotic diversity worldwide means protecting not only rare and endangered species but all native species of a region, even those that are now present in great abundance (N. Myers 1979a, b; E. O. Wilson 1988). As Scott et al. (1987) pointed out, there are numerous examples of species that went from great abundance to extinction in short periods of time. The best way to protect biotic diversity is to establish within geographic regions a protection system that contains a few large preserves with high diversity, combined with many smaller preserves that protect special habitats (e.g., R. I. Miller et al. 1987). Scott et al. (1987) suggested that this may be done most efficiently with Geographic Information Systems (GIS) technology, which uses computerized mapping of species' distributions. The prime areas for preserves are those which have the most species' distributional overlaps and the "right" conditions of physiography, land ownership, and so on. GIS and similar methods have been used mainly for mapping terrestrial systems but presumably could be applied to aquatic preserves if special problems of protecting aquatic habitats were recognized. We suggest a system of aquatic preserves that could be identified in the following steps:

1. Identify geographic regions for which an aquatic preserve system is desirable. In the West these will typically be regions of endemism (isolated drainage basins), especially of fishes, as they are the best-studied taxonomic group. This information is readily available (see Moyle 1976, and other state and regional works).

2. Within each region, identify waters with the highest percentage of native fishes or other taxa. If extensive distributional information is available, the GIS analysis might be appropriate here. Once potential preserves are identified (i.e., waters with a high percentage of native fishes), determine if this procedure has resulted in the omission of any native species. If so, determine where these fishes occur and include those waters as well in the list of candidate preserves.

3. Develop a priority list for acquisition and management. This list should be based on: (a) class of water; (b) presence of intact, native biotic communities; (c) amount of drainage included and other indicators of size; (d) protection against external, edge, and boundary effects; (e) ability to support minimum viable populations of large or otherwise important species; (f) redundancy as a positive feature; (g) difficulty of management; (h) presence of rare or endangered species; and (i) economic considerations.

Once a preliminary priority list is established, it could be refined by using the following criteria:

1. Each native species or assemblage/community should be represented in at least three class I or class II waters, if permitted by its past distributional pattern.

2. If a species or assemblage is less well represented than suggested in 1 above, then transplantation or restoration sites should

be selected to provide the minimum redundancy. For rare or endangered species, use of class IV or class V waters may be necessary, at least initially.

3. Preserves that already have some degree of protection will usually have a higher priority, even if they are slightly inferior in quality. Thus waters in national forests can become preserves with a change in management of their watersheds. New management objectives in California, for example, would include declaring them Aquatic Diversity Management Areas (ADMAs) to help protect aquatic communities.

Conclusions

High-quality waters dominated by native fishes and other natural biota are declining rapidly in numbers, especially in the American West. If natural diversity of regional aquatic biotas is to be perpetuated, preserve systems must be established now. This should be done as systematically as possible to ensure that the biota is protected against extinction. We are convinced that establishing preserve systems in each major western drainage is a realistic goal. Many potential preserves are already on public land, although formally designating them as such may require drastic changes in how the waters and their surroundings are used. A successful preserve system will require not only variety and redundancy but also a commitment to management of individual preserves according to criteria such as those set forth in this paper.

Summary

The decline of native fish faunas of western North America points out the need for a systematic regional approach to conservation that focuses not just on species but on the communities of organisms of which they are part. The preservation of aquatic biotas depends on protecting not only the aquatic habitats but surrounding terrestrial habitats as well. In the design of preserves, there is a need to consider (1) community-level biotic and abiotic interactions; (2) the strong relationships between habitat diversity and species diversity; (3) the importance of disturbance in maintaining species diversity; (4) minimum population sizes needed to guard against the stochastic, demographic, and genetic effects of isolation; and (5) the necessity of redundancy to guard against extinctions caused by catastrophic events.

Criteria for preserve design have primarily considered terrestrial systems, and thus can be applied to aquatic systems only if special constraints are added. Thus, aquatic preserves must be designed considering (1) protection of upstream areas (or better, protection of entire drainages), (2) barriers that prevent invasion of undesirable species from downstream but still permit migration of anadromous fishes, (3) wide buffer zones to protect aquatic habitats from changes occurring in surrounding terrestrial habitats, and (4) recolonization of streams by organisms washed out by natural floods.

In the long term, an aquatic preserve is likely to be effective only if it is part of a system that provides redundancy, habitat diversity, and protection of special habitats and rare species within each region of endemism. Potential aquatic preserves should be classed according to their ability to protect diversity. The system presented here divides a continuum of quality into six types of waters, ranging from class I (best) to class VI (worst). Most preserves should be class I or II, but, realistically, waters of all classes are likely to be part of any program of biological conservation.

Chapter 11

Preserves and Refuges for Native Western Fishes: History and Management

Jack E. Williams

Introduction

Preserves and refuges in the American West have historically focused on the protection of unique or spectacular landforms and geologic features like those prominent in Yellowstone, Grand Canyon, and Yosemite national parks. The national wildlife refuge (NWR) system has traditionally protected migratory waterfowl or large terrestrial vertebrates, and establishment of preserves and refuges for more obscure and less understood aquatic species is a recent facet of conservation. Consequently, few attempts have been made to examine and comment on the effectiveness of aquatic preserves.

The aquatic fauna of the western United States is characterized by its uniqueness (R. R. Miller 1959; Minckley et al. 1986). Unfortunately, development by a burgeoning human population has been devastating to the region's naturally sparse water resources. From large spring systems of west Texas (Brune 1975) to the once-mighty Colorado River (R. R. Miller 1961), western waters flow at a fraction of their historic levels. As a result, many native aquatic species are now extinct (R. R. Miller et al. 1989), and an increasing proportion of the remaining fauna has been classified as endangered or threatened (R. R. Miller 1961, 1972a; Minckley and Deacon 1968; J. E. Williams and Sada 1985a; J. E. Williams et al. 1985, 1989; J. E. Johnson

1987a). Between 1983 and 1987, twenty-two fishes from arid regions of the western United States were listed as threatened or endangered (U.S. Fish and Wildlife Service [USFWS] 1989c). The establishment of successful preserves and refuges is vital to protect regional biodiversity.

In this chapter the term *preserve* refers to areas where the biotic communities are largely natural and are managed mostly to protect their natural features. Populations of rare fishes typically are naturally occurring rather than introduced. *Refuges* are areas managed for one or several species rather than for an entire biota; the habitat may be natural or artificial. Populations in refuges may begin from a transplant and may or may not be within the native range of the species concerned. One of the largest artificial refuges for imperiled fishes is Dexter National Fish Hatchery, New Mexico, the subject of chapter 13 in this volume. The Dexter facility is dedicated toward preservation, production, and research toward recovery of diverse species at a single site. Figure 11-1 shows the locations of the preserves and refuges mentioned in this chapter.

My purposes in this chapter are to trace the historical development of preserves and refuges and to examine their effectiveness in protecting fishes in the West. By reviewing existing programs, management of presently protected areas can be improved and formation and management of new preserves and refuges can be optimized.

Historical Perspective

Development of major springs and their out-flows for recreation in arid parts of the western United States provided fortuitous protection for some native fishes. This development was a common practice in Texas (Brune 1975), and a number of rare species persist in such places, or at least did so until ground-water pumping caused some of the springs to fail. Some examples are Leon Springs pupfish (*Cyprinodon bovinus*) and Pecos gambusia (*Gambusia nobilis*) in Leon Springs (now dry) and Clear Creek gambusia (*G. heterochir*) in Clear Creek. Major populations of Comanche Springs pupfish (*C. elegans*) and Pecos gambusia were lost when Comanche Springs dried because its aquifer was pumped for agricultural irrigation, but stocks persisted in San Solomon Springs, Balmorhea State Park, Texas, where a formal refuge now exists (Johnson and Hubbs 1989; Echelle, *this volume*, chap. 9). Contreras Balderas (*this volume*, chap. 12) describes a number of similar developments in Mexico that protect native fishes. Use of surface outflows for water supply (as opposed to pumping from aquifers) also undoubtedly protects other habitats. Sometime prior to 1949, Hank and Yank's Spring in Arizona was impounded into a concrete "spring box" that was stocked for unknown reasons with Sonoran chub (*Gila ditaenia*). The habitat since has received supplemental stockings (Minckley and Brooks 1985) because it provides one of the few perennial habitats for this species in the United States.

The first active management of a now-imperiled western fish was undertaken in 1923, when the New Mexico Department of Game and Fish (NMDGF) built Jenk's Cabin Hatchery in the Gila Wilderness Area (WA) for propagating Gila trout (*Oncorhynchus gilae*; R. R. Miller 1950). This also probably was the first effort by a state government to protect a rare species of native fish. The NMDGF further protected Gila trout in 1958 by closing Main Diamond Creek to angling in what is now the Aldo Leopold WA (R. R. Miller 1950). Even earlier, in March 1955, the White Mountain Apache tribe in Arizona took action to protect the rare and then-undescribed Apache trout (*O. apache*) by closing streams on Mount Baldy to fishing (R. R. Miller 1972b). In another early effort the California Department of Fish and Game (CADFG) successfully transplanted the Paiute cutthroat trout (*O. clarki seleniris*) into North Fork Cottonwood Creek in 1946, following unsuccessful efforts in 1937 to establish it in Upper and Lower Leland lakes (USFWS 1985d).

Early designations of federal lands as wilderness areas, national parks, and national monuments also fortuitously provided valuable natural preserves for native fishes. The nation's first official wilderness, the Gila WA in New Mexico, was set aside in June 1924 by the U.S. Forest Service (USFS), although protection other than in name did not occur until passage of the Wilderness Act of 1964. The Gila WA protected genetically pure populations of Gila trout in three streams and provided an undisturbed watershed for reaches that now support some of the last populations of threatened spikedace (*Meda fulgida*) and loach minnow (*Tiaroga cobitis*) in New Mexico.

Death Valley National Monument (NM) was established in February 1933. Three pupfishes were protected by this action: *Cyprinodon n. nevadensis* in Saratoga and Valley springs, *C. s. salinus* in Salt Creek, and *C. s. milleri* isolated at Cottonball Marsh. The latter area is managed as a wilderness. In 1952 President Harry S Truman proclaimed 16.2 ha surrounding Devil's Hole as a disjunct part of Death Valley NM. This action protected the endangered Devils Hole pupfish (*C. diabolis*) (Deacon and Williams, *this volume*, chap. 5). In July 1938 Dinosaur NM was similarly ex-

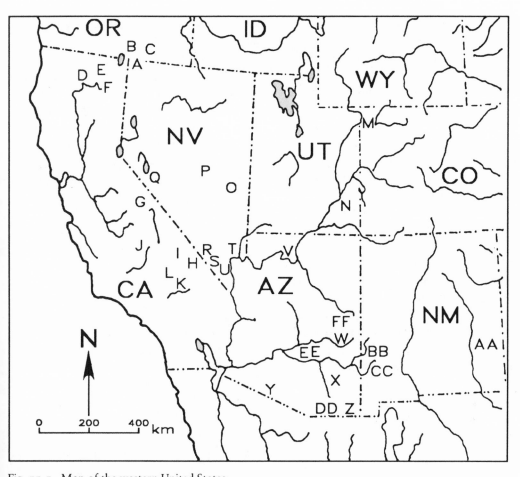

Fig. 11-1. Map of the western United States, showing locations of some refuges and preserves mentioned in text. Symbols (see Table 11-1 for details): A, Twelvemile Creek; B, Dace Spring; C, Borax Lake; D, McCloud River Reserve; E, Johnson Creek; F, Turner Creek; G, Owens Valley Native Fishes Sanctuary; H, Amargosa Canyon; I, Death Valley NM; J, Golden Trout WA; K, Desert Research Station; L, Lark Seep Lagoon; M, Dinosaur NM; N, Canyonlands NP; O, Hot Creek Reserve; P, Chimney Hot Springs; Q, Blue Link Refugium; R, Ash Meadows NWR and Purgatory Spring; S, Amargosa Pupfish Station; T, Moapa NWR; U, Hoover Dam Refugium; V, Grand Canyon NP; W, Ord Creek; X, Aravaipa Canyon; Y, Organ Pipe Cactus NM; Z, San Bernardino NWR; AA, Bitter Lake NWR; BB, Aldo Leopold WA; CC, Gila WA; DD, San Pedro National Riparian Conservation Area; EE, Boyce Thompson Arboretum; FF, Christmas Tree Lake.

panded to include canyons of the Green and Yampa rivers, which now act as important preserves for big-river fishes of the Colorado River system.

Official establishment of preserves and refuges for native fishes is a surprisingly recent phenomenon. Before 1970 there were few efforts to establish protected populations of native fishes or to set aside their habitats. In 1966 the White Mountain Apache tribe constructed Christmas Tree Lake specifically as a refuge for Apache trout and to provide a sport fishery for the species (Rinne et al. 1981). In 1967 the Nevada Fish and Game Commission established Hot Creek Preserve on Sunnyside Wildlife Management Area (WMA; now Wayne E. Kirch WMA) by constructing a small fish barrier to separate Moorman springfish (*Crenichthys baileyi thermophilus*) from introduced fishes in nearby reservoirs. This may be the first example of a state government protecting both a nongame species and its habitat. California followed close behind when it established the Owens Valley Native Fishes Sanctuary in April 1968 (Miller and Pister 1971; Pister, *this volume*, chap. 4).

Much smaller refuges such as aquaria have also been important in conservation efforts. The emergency establishment of an artificial refuge by the University of Texas in 1956 saved three individual Big Bend gambusia (*Gambusia gaigei*) that ultimately gave rise to the populations that exist today in Big Bend National Park (NP; J. E. Johnson and Hubbs 1989). Similar action in August 1968 temporarily saved Amistad gambusia (*G. amistadensis*) from extinction. This fish were moved to artificial pools at the University of Texas Brackenridge Field Laboratory when their habitat was destroyed by the rising waters of Amistad Reservoir. Unfortunately, the stocks were later contaminated by mosquitofish (*G. affinis*) and the Amistad gambusia became extinct (C. Hubbs and Jensen 1984). Other examples of emergency provisions that aided

in fish conservation efforts are given by J. E. Johnson and Hubbs (1989) and Minckley et al. (*this volume*, chap. 15).

The Nature Conservancy (TNC), a private conservation organization, first became involved in protecting the dwindling habitats of native fishes in Arizona when they established the Sonoita Creek Preserve in 1966. Three years later they made down payment on the Wood's Ranch, which contained part of Aravaipa Creek (Fig. 11-2), one of the least-disturbed streams in the Sonoran Desert; it supports seven of the thirteen fish species known from the Gila River watershed. Aravaipa Creek provides one of the best preserves for two threatened fishes, loach minnow and spikedace (J. E. Williams et al. 1985; USFWS 1988c, d). The Defenders of Wildlife took over and managed Aravaipa Preserve for a number of years, then TNC again acquired the property, which has been expanded to more than 17,000 ha upstream and downstream from a central wilderness area managed by the U.S. Bureau of Land Management (USBLM). Water rights established in 1887 have been acquired by TNC, ensuring perpetuation of the system. Also in 1969, TNC purchased 100 ha containing the Canelo Hills Ciénega in Arizona, thereby protecting habitat for state-listed Gila chub (*Gila intermedia*) and three more common fish species. Other TNC holdings, either temporary or permanent, that contributed to fish and aquatic habitat conservation in Arizona include the San Bernardino Ranch and Leslie Creek (now combined as the San Bernardino NWR, see below), Redfield Canyon–Muleshoe Ranch, Bingham Ciénega, Arivaca Ciénega, and the Hassayampa River Preserve.

Since 1970, three federal refuges or preserves have been established specifically for fishes. The Moapa NWR, Nevada, created in 1979 with the purchase of 4.9 ha of Moapa River headsprings, was first established by the USFWS to protect the endangered Moapa dace

Fig. 11-2. Aravaipa Creek, Arizona, within the
Aravaipa Canyon Wilderness Area. Photograph
by Tad Nichols.

(*Moapa coriacea*). Since then, San Bernardino
NWR in Arizona and Ash Meadows NWR in
Nevada were formed to protect vanishing na-
tive fish populations.

Dedication of Large Preserves on Public Lands

The establishment of large wilderness areas
and units of the national park system directly
protected many aquatic habitats and rare
fishes. The Gila, Aldo Leopold, and Golden
Trout wilderness areas all harbor populations
of imperiled trouts. Dinosaur NM and Canyonlands NP protect some of the best remaining big-river habitat in the Colorado River
drainage, occupied by endangered Colorado

squawfish (*Ptychocheilus lucius*), bonytail
(*Gila elegans*), humpback chub (*G. cypha*),
plus the increasingly rare razorback sucker
(*Xyrauchen texanus*). A large population of
humpback chubs also lives in the Little Colorado River in Grand Canyon NP and adjacent
Navajo Indian lands in Arizona. Death Valley
NM includes the entire native ranges of Cottonball Marsh, Salt Creek, Saratoga Springs, and
Devils Hole pupfishes. A subspecies of another
endangered species, Quitobaquito pupfish
(*Cyprinodon macularius eremus*), occupies a
single spring-fed pond in Organ Pipe Cactus
NM. A recent addition to this list is the San
Pedro Riparian National Conservation Area
in southeastern Arizona, where a 49.7-km
reach of stream has been set aside by the

USBLM to preserve riparian, wildlife, and fisheries values (Jackson et al. 1988).

With the exception of the San Pedro conservation area, preservation of fishes was not a concern or rationale for management when protection was granted to these areas. Yet when the Gila WA was formed in 1924, genetically pure populations of Gila trout were automatically protected while other stocks were not. Gila trout in Iron, Spruce, and McKenna creeks contain three of the five remaining natural, genetically uncontaminated populations, and the other two pure populations persist in the Aldo Leopold WA. All natural populations outside wilderness areas were contaminated by introductions of rainbow (*Oncorhynchus mykiss*) or cutthroat trout (*O. clarki*; USFWS 1979b). In addition to providing secure habitats, wilderness preserves support stocks that serve as sources for introductions into restored or artificial habitats. Main Diamond Creek Gila trout (Aldo Leopold WA) have been the source of various transplants into restored streams (USFWS 1979b).

Despite their wild and pristine appearances, most large preserves have major problems with introduced fishes. In 1986 surveys of Cataract Canyon in Canyonlands NP, only seven of twenty-three species collected were native (Valdez and Williams 1986). In terms of numbers, the difference between native and introduced fishes was even more disparate; about 83% of individual fishes caught were non-native. The effect of large numbers of introduced fishes on big-river species has thus been disastrous. Endangered Colorado squawfish comprised only 4.2% of the catch, and endangered humpback chub only 0.3%; bonytail and razorback sucker were not found. Introduced species thrive to the apparent detriment of the native fauna, despite seemingly pristine conditions.

There are two main conditions that allow introduced species to flourish in large pre-serves like Cataract Canyon. First, such preserves do not encompass entire drainage systems. Impounded reaches of the Colorado River both upstream and downstream provide habitat for non-native fishes intentionally or accidentally stocked. Lake Powell begins a mere 56 km downstream from Cataract Canyon, and in 1988, for example, striped bass (*Morone saxatilis*) from Lake Powell were first observed invading Canyonlands NP (R. Valdez, BIO/WEST, Inc., pers. comm.). Second, reservoirs above such preserves not only provide sources of invading species but also alter downstream habitats. Summer water temperatures are lower than natural because of release of hypolimnetic water for power generation. Turbidities, sedimentation patterns, and downstream movement of nutrients are altered due to silt capture in reservoirs. Recruitment of spawning gravels ceases, flows are decreased and stabilized, and flood flows are often eliminated.

The Green River in Dinosaur NM is adversely affected by Flaming Gorge Dam upstream near the Utah-Wyoming border. Abnormal flows and consistently low water temperatures preclude successful spawning by native warm-water fishes for a considerable reach (Vanicek et al. 1970; Tyus et al. 1987; Holden, *this volume*, chap. 3). Most reproduce at the mouth of the Yampa River within the monument (H. Tyus, USFWS, pers. comm.), from where they disperse into the Green River. Of two primary spawning areas remaining for Colorado squawfish in the Green River basin, one is on the Yampa River in Dinosaur NM (Tyus, *this volume*, chap. 19). Shallow margins along the Green River provide nurseries for young produced in the lower Yampa. Yampa River flows are not yet regulated, but a preliminary license has been issued for the Juniper–Cross Mountain Dam above Dinosaur NM. Unfortunately, the U.S. National Park Service (USNPS) did not secure water rights in the

Yampa River, which will be adversely affected by future water development.

Problems with non-native species argue strongly for large preserves that include entire drainage areas (Moyle and Sato, *this volume*, chap. 10). Cottonball Marsh is an excellent example of an area that contains all waters of an aquatic system (except, perhaps, during large floods). The entire marsh is included in a wilderness area, and through sheer remoteness, if nothing else, the Cottonball Marsh pupfish has never had to face an introduced species (except for the dogged efforts of a small cadre of scientists).

Habitats located near the edge of preserves are also vulnerable to outside threats. R. R. Miller and Fuiman (1987) described the precarious status of Quitobaquito pupfish despite the location of Quitobaquito Spring in Organ Pipe Cactus NM and a Man and the Biosphere Reserve. The water supply is threatened by groundwater mining and contamination by airborne pesticide drift from across the international boundary with Mexico. The San Pedro River is also potentially subject to groundwater mining, in both the United States and Mexico, and from pollution by mine wastes from extensive open-pit operations and smelting in the headwaters in Mexico (Minckley 1987; Jackson et al. 1988).

The proximity of human activities and population centers to many large preserves, on the other hand, offers excellent opportunities for nature study and public education. The extensive use of Quitobaquito Spring by birdwatchers could be capitalized on to exhibit the importance of preserving spring systems in the desert. In Death Valley NM, the USNPS has developed an interpretive tour through much of the Salt Creek area. An unobtrusive boardwalk was built to minimize physical impacts of high visitor use along the creek, and the pupfish population has remained robust.

Single-Species Refuges

Most recovery programs for endangered and threatened fishes include provisions for establishing new populations of rare species (J. E. Williams et al. 1988). Stocks of some imperiled species are collected from natural populations or bred artificially and released into historically fishless springs, reservoirs, or other artificial habitats created either especially for that purpose or for some other reason. Simons (1987), for example, documented more than a hundred releases of Gila topminnow (*Poeciliopsis o. occidentalis*) into isolated Arizona waters, many of which were developed for livestock watering.

Because recovery programs are usually oriented toward single species, few attempts have been made at establishing multiple-species refuges. Even rarer is the introduction of representatives from an entire community, including plants and invertebrates. The relatively simple community structure of desert hot springs may offer the best opportunity for such a project. When the Hoover Dam Refugium (Fig. 11-3) was established for Devils Hole pupfish, an endemic hydrobiid snail, riffle beetle (Elmidae), other invertebrates, algae, and substrate from Devil's Hole were stocked as well (Sharpe et al. 1973). Within a few years, however, most of the Devil's Hole biota had been replaced by colonizers, and only the pupfish and snail remained (J. E. Williams 1977).

Measuring the success of artificial refuges may be difficult, and it depends, of course, on their intended purpose. They can be appropriate and highly successful for public education; producing stocks for research, subsequent propagation, and transplanting; as temporary genetic reserves; or for recreation. Many rare fishes are adaptable to maintenance in aquaria for public education. Pupfishes, topminnows, and gambusias make fascinating displays. The

Fig. 11-3. Hoover Dam Refugium for the Devils Hole pupfish in Nevada. The refugium is 5.8 m long and a maximum of 3.0 m deep. Water is piped into the deep end and exits via an outflow box (left). Photograph provided by the U.S. Bureau of Reclamation.

Desert Research Station population of Mohave tui chubs (*Gila bicolor mohavensis*) has provided invaluable opportunities for students of California's Barstow School District, which owns the site. A Boyce Thompson Arboretum pond in Arizona was used as a source population for reintroductions of desert pupfish (*Cyprinodon m. macularius*) and Gila topminnow (Minckley and Brooks 1985). The White Mountain Apache tribe established a successful sport fishery for Apache trout in Christmas Tree Lake by periodically stocking it with the species (Rinne et al. 1979).

Artificial refuges that require a high level of intervention and maintenance present special management concerns. One of the most persistent single-species refuge populations has been the Devils Hole pupfish in the Hoover Dam Refugium. The population size varied substantially yet lasted fourteen years without augmentation (the sole remaining pupfish was removed in 1986; Baugh and Deacon 1988). The Amargosa Pupfish Station, a habitat of almost identical construction, was also built for Devils Hole pupfish, 24 of which were stocked in July 1980. Although the population decreased to only 7 fish in August 1984, the stock increased to 121 individuals in October 1987 and has maintained a comparable size since then (Baugh and Deacon 1988).

Both refugia provided insurance against extinction. They were needed, but the effort and costs were remarkably high. Water temperature at the Hoover Dam Refugium was regulated by controlling rates of flow from natural hot springs through pipes. As ambient conditions changed, it was necessary to alter flow manually to maintain the desired temperature of 33°C (Sharpe et al. 1973; J. E. Williams 1977). If the water cooled, reproduction ceased; whereas a slight rise in temperature could exceed the species' critical thermal maximum. Pipes supplying water to the refuge were destroyed several times by flash flooding. A new stock of thirty fish from Devil's Hole was introduced in 1988 (D. Buck, Nevada Department of Wildlife [NDOW], pers. comm.), but they died when a pipe broke in 1989 (D. Langhorst, NDOW, pers. comm.). The 1984 population crash at the Amargosa Pupfish Station resulted from a power failure that interrupted flow of pumped water. Clearly, high levels of maintenance for dams, wells, and an infrastructure of pipes and valves are disadvantages that are best avoided.

Unnatural selection pressures in such artificial refuges also may alter genetic composition or life-history strategies. Changes in genetics have been documented for stocks of salmonids reared in hatcheries (e.g., Cross and King 1983), but introduced populations of nongame fishes have seldom been examined for such effects (see Echelle, *this volume*, chap. 9). Both artificial refuge populations of Devils Hole pupfish experienced genetic bottlenecks as a result of equipment failure, and some startling examples of rapid phenotypic changes have been observed. J. E. Williams (1977) found significant changes in body proportions of Devils Hole pupfish within five years of their introduction into the refuge at Hoover Dam. Liu and Soltz (1983) described Devils Hole

pupfish introduced into Purgatory Spring, Nevada, as "definitely larger than the maximum natural size for this species and many were mis-shapen." They recommended the population be destroyed. Changes in life-history strategies have been documented when fishes are placed in novel habitats or subjected to new predators (Reznick and Bryga 1987).

Single-species refuges that depend on natural or seminatural habitats have been more successful. These refuges persist without artificial structures and water supplies. However, when reliable water is available from human development, artificial systems may also work. For example, Mohave tui chubs were introduced into Lark Seep Lagoon on the China Lake Naval Weapons Testing Station, California, in 1971, and now constitute the largest and probably safest population of this taxon anywhere. The water in this lagoon comes from leakage from an expanding water-treatment facility and irrigation overflow from a golf course. Thus far the supply has been constant or increasing. This success may also be a function of a large seminatural habitat (more than 4 ha), periodic deepening to maintain open water, and protection from vandalism and introductions of undesirable species (the U.S. Navy property is closed to public access; Feldmeth et al. 1985).

Occasionally even a natural water supply can fail. For uncertain reasons, but perhaps related to drought or an increase in groundwater pumping, flows of Chimney Hot Springs, Nevada (Fig. 11-4), decreased sufficiently for a transplanted population of Railroad Valley springfish (*Crenichthys nevadae*) to disappear (C. D. Williams and Williams 1989). A second stock was introduced when spring flow returned, and no decrease in spring discharge has been observed since. It is preferable, therefore, to examine potential refuge sites over a series of years to ensure a reliable water supply.

Lack of maintenance nearly caused a second extirpation of springfish from Chimney Hot Springs in 1988. In order to cool the 63°C water that originally issues from the spring, three pools were created along the outflow. If small dams are not properly maintained, water flows through the system too rapidly, and the critical thermal maximum of the springfish is exceeded. Water temperature exceeded 38°C in June, forcing springfish downstream where many became isolated and died as available habitat shrank during the hot summer.

A stable water supply appears to be present at Dace Spring on USBLM land in Oregon. The spring was fenced to exclude cattle and its out-

flow was deepened prior to introducing Foskett speckled dace (*Rhinichthys osculus* ssp.) in November 1979, and a population has become established. One of the primary advantages of Dace Spring is its location near the native habitat; both the water supply and its chemical characteristics are similar to conditions at Foskett Spring. Blue Link Refugium, established for Hiko springfish (*Crenichthys baileyi grandis*) at a remote locale far outside the native range, differs substantially from the native habitat (Sevon and Delany 1987). Vandalism, a problem at some refuges, is not likely because of the site's remoteness, but monitor-

Fig. 11-4. Chimney Hot Springs Refuge for Railroad Valley springfish in Nevada. Pools were created along the spring's outflow to allow sufficient cooling for the introduced springfish. The refuge is fenced to exclude livestock. In its natural habitats the fish is threatened by introduced species, water diversions, and livestock grazing. Photograph by J. E. Williams, 1982.

ing and maintenance will be difficult.

If a new population is to be established for recovery, the introduced stock must successfully adjust to a novel environment. Success may only be claimed after a self-reproducing population with a stable, integrated gene pool has been established and has persisted through a number of generations (Altukhov and Salmenkova 1987). Therefore, from the standpoint of genetic integrity alone, it may take a number of years to assess the viability of a stocking. Resistance to stochastic events such as floods and droughts may require even greater time to ascertain.

Protection of Natural Habitats

Protection of natural aquatic habitats and communities has been a major focus of efforts to conserve western fishes since the mid-1960s. Such protection may be achieved through acquisition, leases, or easements of vanishing habitats, or through additional protection for lands already in public ownership. Unlike establishing refuges, protection of fishes in their native habitats allows natural processes to continue and often protects a variety of other rare organisms as well (J. E. Williams et al. 1985).

Habitats on public lands may be protected in a number of ways, including designations as Areas of Critical Environmental Concern (ACEC) on USBLM holdings and as Research Natural Areas on USFS lands. Designating the habitat of an imperiled species as critical is often sufficient for the area to receive special protection. The secretary of the interior may also designate critical habitat for listed species pursuant to the Endangered Species Act (ESA) of 1973. In California, Turner and Johnson creeks are within designated critical habitat for the endangered Modoc sucker (*Catostomus microps*). Similarly, Twelvemile Creek in Oregon was designated critical habitat for the Warner sucker (*C. warnerensis*). On the Hig-

gins' Flat drainage of Johnson Creek, Modoc National Forest administrators reduced timber harvests and initiated reclamation on roads and previously logged areas in response to critical habitat designation. The USBLM reduced livestock grazing along Twelvemile Creek and acquired part of the stream. Legal obligations to protect critical habitat from destruction or adverse modification apply only to federal agencies, but local governments and private landowners often give special zoning and protection to such lands.

Combined efforts of public and private agencies protected Aravaipa and Amargosa canyons. As mentioned earlier, parts of the former were acquired by TNC and the Defenders of Wildlife while the central reach, managed by the USBLM, was designated wilderness. Of a total of seven native species protected in the stream, spikedace and loach minnow are federally listed as threatened, and roundtail chub (*Gila robusta*) is state listed. Other species include longfin dace (*Agosia chrysogaster*), speckled dace (*Rhinichthys o. osculus*), desert sucker (*Pantosteus clarki*), and Sonoran sucker (*Catostomus insignis*). Non-native fishes are rare and transitory, mostly because severe flash floods periodically decimate Aravaipa Canyon populations (Minckley and Meffe 1987). Federal lands in Amargosa Canyon were designated an ACEC in 1980, and the majority of private land was acquired by TNC in 1987. The habitat harbors Amargosa pupfish (*Cyprinodon nevadensis amargosae*) and an undescribed speckled dace (*Rhinichthys osculus* ssp.). Unfortunately, the preserve is frequently invaded from upstream by mosquitofish.

Conservation of rare species cannot, however, be assured simply through regulation and protective ownership. For example, TNC leases Borax Lake, Oregon, to protect Borax Lake chub (*Gila boraxobius*), and the lake is designated critical habitat. Abuse by off-road vehicles and unauthorized water diversions nevertheless continue to intermittently damage the habitat and population (USFWS 1987d). The underground aquifer supplying water for Borax Lake is threatened by geothermal energy wells being drilled on nearby public land.

Relatively small natural preserves may have even more concerns with invasion by introduced species than do large preserves on public lands. In-stream barriers have thus been constructed at several (Turner, Johnson, and Hot creeks, for example) to prevent populations of unwanted fishes from moving upstream. A barrier is being considered to protect native fishes of Aravaipa Canyon. As of 1988, introduced red shiners (*Cyprinella lutrensis*) were 1.5 km downstream from the confluence of Aravaipa and the San Pedro River, and by 1990 they had penetrated the lower two-thirds of the Aravaipa channel. Actions in addition to barrier construction to reduce the threat of this invading species have not yet been decided upon (W. Minckley, Arizona State University, pers. comm.). The Ord Creek preserve in Arizona was blocked by barriers and chemically treated twice to eliminate introduced trout that hybridized and competed with native Apache trout (Rinne et al. 1982). Similarly, Turner Creek in California was poisoned to remove Sacramento suckers (*Catostomus occidentalis*), which hybridized with endangered Modoc suckers. Many other such examples are known (Rinne and Turner, *this volume*, chap. 14). Such treatment is costly and often needs repeating as undesired fish circumvent barriers or are introduced by misguided or unknowing persons intent on "improving" sport fishing opportunities.

The protection of riverine fishes clearly poses special difficulties if parts of the stream flow outside preserve boundaries. One TNC preserve in California includes a pristine-appearing reach of the McCloud River (Fig. 11-5) that once provided spawning habitat for the largest runs of chinook salmon (*Oncorhynchus tshawytscha*) in California (L. Stone 1876, 1883). Chinook salmon no longer spawn there; they were stopped by Shasta and Keswick dams downstream on the Sacramento River. Loss of the massive influx of nutrients provided by dying salmon forever altered the character of this stream, and probably, along with introduced brown trout (*Salmo trutta*), led to extirpation of bull trout (*Salvelinus confluentus*) from California. Movement of nutrients and gravels into the preserve has been further reduced by an upstream reservoir. The bull trout maintained its only California population in the McCloud River and was listed as endangered by the state of California prior to extirpation.

If introduced species are absent, conservation of native fishes seems compatible with many multiple-use practices, as long as they are moderated to maintain, or at least not substantially reduce, habitat quality. Populations of Modoc suckers persist in Turner and Johnson creeks despite the presence of some private land, timbering, and livestock grazing. Better habitat conditions are, however, clearly found in those areas with minimal or no grazing pressure.

In another example, the Pecos gambusia (*Gambusia nobilis*) occupies nine gypsum sinkholes and two springs on Bitter Lakes NWR, New Mexico, despite high levels of hunting, sport fishing, and nature observation (USFWS 1983d). Pressures to expand sport fishing could, however, result in stocking of channel catfish (*Ictalurus punctatus*) into their habitat. Introduced fishes have already eliminated one transplanted population of Pecos gambusia from the refuge. Conflicts between sport fishing and rare fishes also occur on Ash Meadows NWR, Nevada. Largemouth bass (*Micropterus salmoides*) eliminated Ash Meadows pupfish (*Cyprinodon nevadensis mionectes*) from the main pool of Crystal Spring, but pupfish persisted in the outflow and reoccupied the spring when bass were eradicated (J. E. Williams and Deacon 1986).

Rehabilitation of Refuge Habitats

Scientists and managers are becoming increasingly aware that preservation of habitat is not in itself an adequate strategy for resource conservation. Restoration of damaged ecosystems can play a major role in preserving biological diversity (Cairns 1986, 1988; W. R. Jordan 1988), and such restoration is especially crucial in arid regions, where high water demand, storage, diversion, and consumption disturb aquatic systems. Native fish habitats in Ash Meadows NWR, Moapa NWR, Owens Valley Native Fishes Sanctuary, and San Ber-

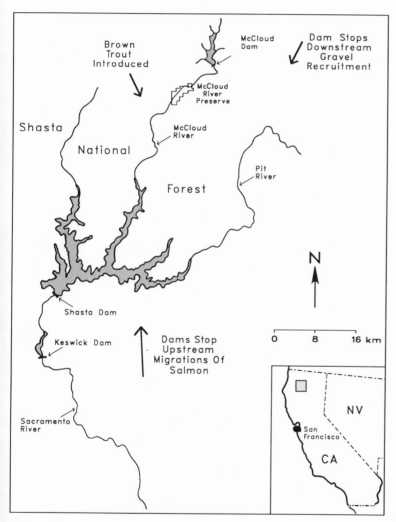

Fig. 11-5. Map of the upper Sacramento River basin, California, showing position of McCloud River relative to upstream and downstream dams, and other factors influencing the reach.

nardino NWR are subjects of ongoing restoration attempts.

Ash Meadows, along the Nevada-California border, is renowned for its diversity of fishes, aquatic invertebrates, and plants (S. F. Cook and Williams 1982). Aquatic habitats there were subjected to agricultural development and, more recently, modifications of springs for anticipated commercial and residential development (Deacon and Williams, *this volume*, chap. 5). Fortunately, most of the private lands were acquired by TNC in 1984; they were later transferred to the USFWS to establish the refuge, but not before some desert marshes had been drained, springs were pumped dry, reservoirs were created, and exotic species were introduced (Soltz and Naiman 1978; Deacon and Williams, *this volume*, chap. 5). Once the refuge was established, attention shifted to restoring these damaged ecosystems (Sada 1987a). Unfortunately, ecological requirements for many native aquatic species, especially the recently described hydrobiid snails (Hershler and Sada 1987), are poorly known. Initial restoration will focus on determining precise historical conditions, understanding requirements of endemic species, establishing baseline conditions, and eliminating immediate threats to endemic taxa (Pavlik 1987; Sada 1987b; USFWS 1990b).

The Owens Valley Native Fishes Sanctuary was the first refuge to be set aside and restored to benefit an entire native ichthyofauna. The cornerstone was the Owens pupfish (*Cyprinodon radiosus*), thought to be extinct when described in 1948 (R. R. Miller 1948). Rediscovery of Owens pupfish in a remote corner of Fish Slough in 1964 kindled efforts to secure the area as a refuge for all four species native to the valley (R. R. Miller and Pister 1971); also included are Owens tui chub (*Gila bicolor snyderi*), sucker (*Catostomus fumeiventris*), and speckled dace (*Rhinichthys osculus* ssp.).

Initial modifications involved constructing small dams to increase available habitat, control water levels, and exclude non-native largemouth bass and mosquitofish. Numerous transplants of native fishes have been made into the refuge, but reinvasion by bass necessitates periodic chemical treatment. One such incident occurred in July 1986, when earthquakes of 5.5 and 6.2 magnitude damaged upstream barriers, allowing access by largemouth bass. Most invasions, however, appear to be deliberate attempts by local anglers to establish new fishing areas. Beginning in 1989, a USBLM ranger regularly patrols the refuge in an effort to reduce this problem (E. P. Pister, CADFG, pers. comm.). The sanctuary's success depends on diligent efforts of local CADFG employees to maintain structures and eliminate undesired species as they appear.

Moapa NWR was established for Moapa dace (USFWS 1983a). Spawning habitat seemed limiting, so an artificial channel was constructed to direct the outflow from thermal springs. The effort was successful, and the refuge also protects a local form of springfish (*Crenichthys baileyi moapae*) and a number of endemic invertebrates. An endemic speckled dace (*R. o. moapae*) and roundtail chub (*G. robusta* ssp.) do not occur on the refuge but are nonetheless served by the protected water supply (Deacon and Bradley 1972; J. E. Williams 1978; J. E. Williams et al. 1985).

Most native fishes of the Río Yaquí drainage had been extirpated from the United States by the mid-1960s (Minckley 1973; McNatt 1974). Except for remnant populations in two small headwater streams (Rucker Canyon and Leslie Creek), their habitats were lost to livestock grazing, arroyo cutting, and groundwater pumping in excess of recharge (water mining; Hastings 1959; Hastings and Turner 1965; McNatt 1974). Artesian wells, ciénegas, ponds, and creeks on San Bernardino NWR in southeastern Arizona are now being restored to support these fishes, including endangered Yaqui chub (*Gila purpurea*) and Yaqui topminnow (*Poeciliopsis occidentalis sonoriensis*) and threatened Yaqui beautiful shiner (*Cyprinella formosus*) and Yaqui catfish (*Ictalurus pricei*). Leslie Creek in the Swisshelm Mountains north of the main refuge was recently added to the system. Other, nonlisted, native fishes (longfin dace, Mexican stoneroller [*Campostoma ornatum*], roundtail chub, and Yaqui sucker [*Catostomus bernardini*]) also are included in the management plan (USFWS 1987f). When the NWR is restored and restocked, nearly a fourth of the native ichthyofauna of Arizona will be locally secure.

Conclusions and Perspectives for the Future

The characteristics of some of the preserves and refuges mentioned in this chapter are summarized in Table 11-1, which further includes analyses of habitat types, isolation, physical modifications, effects of introduced species, and major factors regulating local populations of imperiled native fishes. Preserves and refuges are scored for their success, or lack thereof, in long-term perpetuation of native fishes.

Most national parks and wilderness areas have been relatively successful in long-term protection of native fishes. Preserves for species inhabiting springs, marshes, and headwater creeks have been most successful. Lower success in larger habitats is usually associated with the presence of nonnative species. Large headwater preserves, such as the Gila WA, provide sufficient isolation to preclude invasions by unwanted species. In some headwaters, such as Golden Trout WA, managers have successfully maintained fish populations by construction of in-stream barriers coupled with chemical treatments to remove non-native fishes. Turner, Johnson, and Ord creeks are examples of smaller preserves where invasion by undesirable species can only occur from downstream.

Larger streams and rivers face special problems, especially from invasions from outside. Numerous non-native fishes may invade preserves on major streams from both upstream and downstream. Preserves for big-river fishes, such as those in the Colorado River system, have achieved marginal success. Modifications to riverine areas outside the preserves must be controlled for continued maintenance of species within preserves, and this is a difficult task (Wydoski and Hamill, *this volume*, chap. 8).

Determining the success of preserves and refuges for large riverine fishes is further complicated by the fishes' long life expectancy. Lake Mohave along the Nevada-Arizona border harbors the largest known population of razorback suckers remaining in the Colorado River basin (Minckley 1983). Samples from Lake Mohave between 1981 and 1983 revealed that the suckers were twenty-four to forty-four years old, with no evidence of recruitment (McCarthy and Minckley 1987). Large numbers of these old fish gave the false impression of a large and viable population, which may have caused delays in their management toward recovery (Minckley et al., *this volume*, chap. 17). The same situation occurs in a number of other western suckers and minnows (Scoppettone and Vinyard, *this volume*, chap. 18).

Single-species refuges have shown mixed success (Table 11-1). Lark Seep Lagoon, Dace Spring, and Chimney Hot Springs have been more successful than most, although the latter required more than one release of fish to initiate and maintain a population. More successful single-species refuges are typically in natural areas or areas involving only slight modifications of existing habitats. Introductions of undesirable species have not occurred

Table 11-1. Characteristics of some preserves and refuges discussed in this chapter.[1]

Preserve/refuge	State	County	Ownership	Date estab.
Large Preserves on Public Lands				
Dinosaur NM	CO/UT	Moffat/Uintah	USNPS	1915
Grand Canyon NP	AZ	Coconino	USNPS	1919
Gila WA	NM	Catron	USFS	1924
Death Valley NM	CA/NV	Inyo/Nye	USNPS	1933
Organ Pipe Cactus NM	AZ	Pima	USNPS	1937
Canyonlands NP	UT	San Juan	USNPS	1964
Golden Trout WA	CA	Tulare	USFS	1978
Aldo Leopold WA	NM	Sierra	USFS	1984
San Pedro Riparian Natl. Conserv. Area	AZ	Cochise	USBLM	1986
Single-Species Refuges				
Christmas Tree Lake	AZ	Apache	Apache	1967
Boyce Thompson Arboretum	AZ	Pinal	state	1970
Hoover Dam Refugium	NV	Clark	USBR	1972
Lark Seep Lagoon	CA	San Bernardino	U.S. Navy	1972
Purgatory Spring	NV	Nye	USBLM	1972
Desert Research Station	CA	San Bernardino	Barstow	1978
Chimney Hot Springs	NV	Nye	USBLM	1978
Dace Spring	OR	Lake	USBLM	1979
Amargosa Pupfish Station	NV	Nye	USFWS	1980
Blue Link Refugium	NV	Mineral	USBLM	1985
Protected Natural Habitats				
Bitter Lake NWR	NM	Chaves	USFWS	1937
Ord Creek	AZ	Apache	Apache	1964
Hot Creek Reserve	NV	Nye	state	1967
Aravaipa Canyon	AZ	Graham/Pinal	USBLM, TNC	1969
McCloud River Reserve	CA	Shasta	TNC	1973
Amargosa Canyon	CA	San Bernardino/Inyo	USBLM, TNC	1980
Borax Lake	OR	Harney	USBLM, TNC	1982
Johnson Creek	CA	Modoc	USFS, pvt.	1985
Turner Creek	CA	Modoc	USFS, pvt.	1985
Twelvemile Creek	OR	Lake	USBLM, pvt.	1985
Rehabilitated Refuges				
Owens Valley Native Fishes Sanctuary	CA	Mono	LA, USBLM	1968
Moapa NWR	NV	Clark	USFWS	1979
San Bernardino NWR	AZ	Cochise	USFWS	1980
Ash Meadows NWR	NV	Nye	USFWS, pvt.	1984

[1]Key to codes, abbreviations, and acronyms:
Ownership (other than acronyms already given in text): Apache, White Mountain Apache tribe; Barstow, Barstow, California, School District; LA, City of Los Angeles; pvt., private.

No. rare fishes[2]	Habitats	Isolation	Modifications	Intro. spp.	Population regulation	Success
11/3	R	3	2	5	3	2
8/1	R	3	3	4	4	2
7/3	C	2	1	1	1	1
4/1	S	1	1	2	1	1
2/1	S	1	3	1	2	2
8/4	R	3	2	5	3	2
4/1	C	2	2	4	3	3
7/2	C	2	1	1	1	1
11/0	R	3	2	3	2	2
1/1	L	2	4	4	5	3
0/2	L	1	5	2	5	3
0/1	S	1	5	1	5	4
0/1	S	1	4	1	4	1
0/1	S	1	2	1	2	5
1/1	L	1	5	1	3	4
0/1	S	1	5	1	4	3
0/1	S	2	2	1	2	1
0/1	S	1	5	1	5	?
0/1	S	1	4	1	2	?
14/1	SL	2	2	3	2	1
1/1	C	2	2	5	2	3
1/0	SL	3	2	4	2	2
7/2	C	2	1	2	1	1
5/0	R	3	3	4	4	4
2/0	C	3	1	3	1	1
1/1	SL	2	2	1	1	1
4/1	C	2	3	4	2	2
4/1	C	2	3	3	3	?
3/1	C	2	3	2	3	?
4/2	SL	2	3-4	4	4	3
2/1	SC	2	4	3	4	3
8/4	SLC	3	3-5	3	4	?
4/4	SC	1	3	4	2	2, 3

(*Continued on page 188*)

Table 11-1. Continued

Habitats: S, isolated springs, marshes, swamplands; L, lakes, ponds, reservoirs; C, first- and second-order creeks; R, large creeks and rivers.
Isolation: 1, habitat well isolated within preserve/refuge, no outflow beyond boundary; 2, mostly within preserve/refuge, but parts may connect with areas outside boundary; 3, habitat with significant portion outside preserve/refuge boundary.
Modifications (amount of physical habitat alteration): 1, pristine; through 5, highly degraded.
Introduced species: 1, no effects, no introduced species; 2, effects minor or indiscernible, introduced species present; 3, moderately important, but other factors more so; 4, important; 5, very important, major artificial factor affecting rare species.
Population regulation: 1, naturally regulated; 2, population regulated by natural factors, but artificial factors (grazing, logging, channelization, etc.) also play a role; 3, natural and unnatural factors about equally important; 4, regulated mainly by artificial factors, but natural factors (flood, drought) also play a role; 5, populations regulated almost entirely by artificial factors.
Success: 1, population(s) persisted at historic levels for more than ten years; 2, persisted for more than ten years but at reduced levels; 3, population persists, but only with additional releases of rare fish; 4, population failed in five or more years, or with extirpation of one or more native fishes; 5, population failed in less than five years; ?, unknown or refuge/preserve recently established.
[2]Number of native taxa (species, subspecies, or undescribed forms) originally recorded/number of federally listed taxa present.

in those refuges where viable populations of rare species are established.

Many single-species refuges, including a number not discussed here, have not been in existence long enough to test their effectiveness. Most recovery plans consider three to five years' survival of a refuge population adequate to define it as successful. The recovery plan for Owens pupfish, for example, provides for establishment of five separate populations as the goal to allow removal of the species from the list of endangered and threatened wildlife (USFWS 1984h). According to that plan, refuge populations will be considered successful when a minimum of five hundred individuals has persisted for five years. Yet periodic invasions by exotic species have demonstrated a high vulnerability of these areas to disturbance. Also, five years is insufficient time to assess effects of stochastic events such as floods or drought, which may occur only as frequently as once in ten, twenty, fifty, or more years. Goals of recovery plans must be specifically designed to meet the needs of each species.

In models of extinction resulting from environmental stochasticity, the probability of survival of a species increases with increasing numbers of populations. Population size and habitat size are less important in determining survival. Therefore, a population broken down into a number of preserves or refuges would be expected to persist longer than a single population of equal initial size (Quinn and Hastings 1987). For many fishes, especially those in arid zones that are particularly vulnerable to losses by environmental stochasticity, refuges should be broadly scattered in order to lessen chances of a localized, large-magnitude catastrophe. Establishing more than one refuge also lessens the probability of extinction caused by invasion of non-native species.

Recent efforts to protect remaining natural habitats have been encouraging. Combining protection of public lands with acquisition or lease of sensitive private lands succeeded at Aravaipa Canyon, Amargosa Canyon, and Borax Lake (Table 11-1). Despite problems of vandalism, one of the best strategies for preservation of a single-locale endemic is the one used for the Borax Lake chub. It exists only in spring-fed waters of the Borax Lake system, the source of which is located on a small parcel of private land surrounded by USBLM property. TNC leased the lake to protect the chub, and 259 ha of private and public lands were designated critical habitat under the ESA. Outflow is to nearby Lower Borax Lake on USBLM land. The chub typically also occurs there, in natural channels between the two lakes, and in adjacent marshes and pools. If the main population were lost, these satellites would maintain the species until Borax Lake could be restored.

The proportion of native fishes in the American West listed by the federal government as endangered or threatened is staggering (Minckley and Douglas, *this volume*, chap. 1). The numbers are especially alarming when the depauperate nature of the native ichthyofauna of the West is compared with the diversity found in other regions. Of fifty-four endangered or threatened fishes listed in 1985, forty (74%) were from desert areas of the West (J. E. Williams and Sada 1985a). Recent surveys have shown that many more native fishes have declined in numbers and range and need formal protection (J. E. Williams et al. 1985, 1989; J. E. Johnson 1987a; W. F. Sigler and Sigler 1987; Moyle et al. 1989).

Faunal diversity can be conserved by identifying remaining habitats with high native species richness and setting them aside. A method has been proposed using Geographic Information Systems to identify such areas in terrestrial communities (Scott et al. 1987); a similar system should be applied to aquatic habitats. But an effective strategy for conservation of native fishes in the West cannot be restricted to a single approach such as protecting existing habitat, restoring damaged areas, or creating new, single-species refuges. Clearly, any and all methods must be implemented to increase chances of success. A new system for preserving diversity is needed. Preference for establishing new preserves should be given to remaining areas of high diversity, even if such communities do not at this time appear threatened. Declining species numbers indicate the need to identify and reverse causes of the decline.

Protecting the increasing numbers of imperiled fishes will be difficult, and our goal should be to ensure protection of fishes and aquatic communities before there is any need to invoke the ESA. Scott et al. (1988) argued that protecting biological diversity by focusing all efforts on endangered or threatened species is futile and often results in considerable expense and crisis management for individual species for which there is little hope of salvation. There is little argument against saving Colorado squawfish or California condor (*Gymnogyps californianus*), but the problems that result from focusing a majority of our restoration efforts and resources on single critically endangered species are obvious.

Chapter 12

Conservation of Mexican Freshwater Fishes: Some Protected Sites and Species, and Recent Federal Legislation

Salvador Contreras Balderas

Little has been published on the conservation status of threatened Mexican fishes. Deacon et al. (1979) reported 59 fishes endangered, threatened, or of special concern in Mexico; Contreras (1987) recorded 114; the International Union for the Conservation of Nature and Natural Resources (IUCN 1988) included 67 taxa; and J. E. Williams et al. (1989) listed 126. These differences in numbers are due to dissimilar criteria used to estimate endangerment, different inspection times, and inclusion or exclusion of subspecies or binational forms. Several lists (Contreras Balderas 1975, 1978a, b, 1987, in press) indicate ever-increasing numbers of imperiled species in Mexico, reflecting an increasing volume of available information as well as changes in ecological conditions in the country. This paper considers two positive aspects of this problem: current protection of some habitats and fishes, and new legislation that authorizes and enhances such protection.

I gratefully acknowledge the support of many people, impractical to list, who assisted in collections. Special thanks go to the anonymous reviewers for critical review of the manuscript from which this paper developed; my mistakes should not be credited to them; however, they made the paper better. Finally, thanks to Mrs. Cristina Franco for typing the manuscript.

Protection of Habitats and Fishes

Mexican freshwater habitats and fishes are protected mostly as fortuitous by-products of the protection of water, local scenery, or other resources. As such, protection is related to places rather than to habitats or species. For this reason, protected areas are described in alphabetical order by state rather than on a taxonomic basis, and each is referred to its nearest town as the point of geographic reference.

State of Chiapas

Lagunas de Colon (last visited 1982).

This little-known place spans the Mexico-Guatemala border 75 km southwest of Comitán de Dominguez. There are a few small and primitive pyramids, scarcely visible and unexplored. Several lagoons comprising the headwaters of the Río Grijalva support abundant aquatic birds and fishes. At least one endemic cichlid and a poeciliid (*Poeciliopsis* sp.) are apparently undescribed. Local campesinos care for the area and check on activities of visitors, and, besides agricultural development, there is little apparent human impact on the habitat. The waters are used for irrigation, recreation, and as sources of fish and game for food.

Lagunas de Montebello (last visited 1982).

A place long famous for its woodlands and lakes, Lagunas de Montebello on the Mexico-Guatemala border east-southeast of Comitán harbors numerous interesting plants and animals, which recently prompted authorities of the Mexican Secretaria de Desarrollo Urbano y Ecologia (SEDUE) to take steps to protect it. The area has been actively managed for a number of years as a recreational park. It is not known if direct consideration for the aquatic biota was responsible for its original designation. Two undescribed fishes are known from this area: a topminnow (*Poeciliopsis* sp.) and a cichlid (*Cichlasoma* sp.); but the fauna has scarcely been studied. Trout (likely rainbow trout, *Oncorhynchus mykiss*) have been stocked in some lakes, but their impacts are unknown. Fishing is permitted, while removal of other organisms is not, except by campesinos if it is their livelihood or source of food.

State of Chihuahua

Ojo de Galeana (= Ojo de Arrey; last visited 1986).

This large spring complex in the Río Santa María drainage is located within a hacienda a few kilometers south of Galeana. A series of springs, partially protected through conversion to aquatic gardens and swimming pools, also serve as a refugium for an undescribed cachorrito (pupfish; *Cyprinodon* sp.) and a rare cyprinid, sardinita de Santa María (*Cyprinella santamariae* [Evermann and Goldsborough]). Associated marshlands further support an endemic vole (*Microtus pennsylvanicus chihuahua* Bradley and Cockrum [1968]), a relict, low-elevation representative of a regionally montane mammal. The owners of this complex should be contacted to ensure continued protection of indigenous species, which could be accomplished with little inter-ference to maintenance of the facilities and grounds.

Ojo de Hacienda Dolores (last visited 1985).

Ojo de Hacienda Dolores is centrally located in a municipal park 11.2 km south of Jimenez along the Río Florido (Río Conchos basin); permanent caretakers are in residence. The large thermal spring harbors two endemic fishes: a cyprinodontid, the cachorrito de Dolores (*Cyprinodon macrolepis* Miller) and a poeciliid, guayacón de Dolores (*Gambusia hurtadoi* Hubbs and Springer). There are picnicking areas, and the spring pool, although relatively unmodified, is used for swimming. There are no major changes apparent in comparing my observations with those provided by C. Hubbs and Springer (1957) and R. R. Miller (1976a). The area is well maintained and clean of debris, although it is affected by intense recreational use and sometimes influenced by the use of soap for bathing.

Ojo de Julimes (last visited 1982).

This large thermal spring, near the mainstream Río Conchos, has also been converted to a swimming pool surrounded by a recreational park. It is inhabited by an undescribed cachorrito (*Cyprinodon* sp.) that may be a close relative of *C. pachycephalus* (Minckley and Minckley 1986; see below). Another special inhabitant is an undescribed crustacean (*Thermosphaeroma* sp.), a member of a group of hot-water isopods of scattered distribution from New Mexico, USA, to Aguascalientes, Mexico (Cole and Bane 1978; Bowman 1981).

Ojo de San Diego (last visited 1982).

Ojo de San Diego consists of a complex of hot springs and their outflows modified into swimming pools and baths. It is on a low hilltop that may represent an extensive, travertine spring mound adjacent to the Río Chuviscar, 57 km east of Ciudad Chihuahua. M. L. Smith and Chernoff (1981) and Minckley and

Minckley (1986) provided additional descriptive data for this place, which is the type locality and only known site of occurrence of the cachorrito cabezón (*C. pachycephalus* Minckley and Minckley), an endemic isopod (*Thermosphaeroma smithi* Bowman), the unnamed guayacón de San Diego (*Gambusia* sp.; assigned to the *G. senilis* complex by Miller 1976b), and one or more undescribed hydrobiid molluscs (J. J. Landye, Arizona Game and Fish Department, pers. comm.). The *Gambusia* was uncommon at my last visit. The thermal baths are a commercial venture, which dictates cleanliness of the grounds and waters, and the area was well kept and relatively undisturbed.

Ojo de Villa López (last visited 1985).

This is another large, highly modified spring adjacent to the Río Florido (Río Conchos system) within the town of Villa López. The outflow is dammed less than 100 m from the river with a structure scarcely a meter high to form a lake about 200 m long. Two apparently undescribed species are present: a cachorrito (*Cyprinodon* sp.; resembling *C. eximius* Girard) and the guayacón de Villa López (*Gambusia* sp.; similar to *G. hurtadoi*). Although management of dam gates and the extent of flood influence from the adjacent river are unknown, it seems likely that this habitat could be invaded at any time by congeneric forms, with consequent danger for the endemics. Both species are currently under study.

State of Coahuila

Bolsón de Cuatro Ciénegas (last visited 1989).

This extensive area of diversified aquatic and terrestrial habitats has received wide publicity in scientific and popular media as a center of endemism (Contreras Balderas 1969, 1977, in press; Minckley 1969c, 1978, 1984; Almada and Contreras Balderas 1984; J. E. Williams et al. 1985). The Cuatro Ciénegas basin has received limited federal protection as a recreation area since 1987. A recent symposium on the biota of Cuatro Ciénegas (Marsh 1984) included papers listing at least eight endemic fishes and perhaps another hundred endemic plants and animals from the valley. It is a partially closed drainage system, with waters remaining internal to evaporate or seep into the basin floor, as well as draining (both naturally and through man-made canals) to the Río Bravo del Norte (Rio Grande).

This oasis in the Chihuahuan Desert is an alltime favorite fishing and swimming place for local and regional peoples, and Indians depended on its abundant water, plants, wildlife, and fishes in the distant past. It is a hydrologically complex area, with a few of its major springs modified by development for recreation, and others canalized to transport water for use in agriculture and industry, mostly outside the basin.

Since its announcement to science in the late 1960s, people of many countries have been attracted to the area and may be found in the Cuatro Ciénegas valley at any time. Many groups are simply there to photograph and observe, but others, some authorized and some not, collect or otherwise interfere with the biota, potentially damaging some of the limited endemic populations. Such pressure has increased to a point where specific government regulation may be necessary.

These impacts on natural aquatic systems and the countryside, coupled with population increases in the town of Cuatro Ciénegas and larger numbers of people traveling over a major highway through the basin, may clearly be seen in the accumulation of glass, plastics, paper, and other debris of civilization (Contreras Balderas 1984). Some parts of the basin are more or less protected by private owners, especially when used for swimming, fishing, camping, or picnicking, but both public and private lands are becoming severely degraded. Canals interconnecting aquatic habitats and consolidating their outflows, local development of recreational facilities, and other physical damages to the system, although continuing, have yet to have severe impacts on the fauna. African cichlids (tilapias; *Oreochromis* sp.) were introduced in 1986, but no evidence of their spread or continued presence was found in 1989. Introduced water hyacinth (*Eichornia crassipes*) has, however, spread to choke some springheads and their outflows in the southeastern part of the basin.

La Alberca y El Socavon (last visited 1985).

This large spring and its outflow, explored thoroughly in 1978 and 1982, supported a park and swimming pool in the town of Múzquiz. The spring is failing, and after 1982 only enough water was available to serve the picnic area, with a small stream suitable for use by children. Relatively good protection was effected in the recreation area, but not to the supply of water, and the endemic platy de Múzquiz (*Xiphophorus meyeri* Schartl and Schroeder) has now lost more than 80% of its population and habitat and must be considered highly vulnerable and threatened. This species, the most northern and isolated of its genus, was described twice, by Schartl and Schroeder (1988) and by Obregon and Contreras Balderas (1988), with the former having nomenclatural priority.

State of Durango

Ojo de la Concha (last visited 1989).

The huge thermal springs known as Ojo de la Concha are located in the Río Nazas basin, 9 km west of Peñón Blanco, again in a municipal park. Several swimming pools with stairlike margins were built below the thermal outflows, forming excellent places for underwater observation of fishes. Unidentified tilapias have been introduced with little success, although several species are rare to common. This locality supports the most abundant

known population of the cachorrito de Nazas (*Cyprinodon nazas* Miller), a species comprised of three or four allopatric forms, and is the only place where it may be considered secure. I consider this species threatened in the Río Aguanaval and diminishing in abundance in the Río Nazas and Laguna de Santiaguillo basins. An undescribed endemic sardinita (shiner; *Cyprinella*, cf. *rutilus*) is soon to be named from La Concha (Contreras Balderas and Lozeno, unpub. data).

State of Nuevo León

Ojo de Apodaca (last visited 1989).

Ojo de Apodaca is a spring 1 km southwest of the town of Apodaca in the Río San Juan basin. In 1961 an old wall surrounded 75% of the spring's periphery. About 1964 the area was chosen to irrigate an experimental ranch. Since it was being used for recreation, the presidente municipal of the town of Apodaca ordered the wall completed and the area closed to ingress. Around 1975 several swimming pools and a park were constructed, and after 1980 the main spring was encircled by a second protective fence.

This last action corresponded with a time of severe regional drought when needs for water were high. However, since only authorized persons from Servicios de Agua y Drenaje de Monterrey (SADM) were allowed there, extraction of water was not exhaustive, and the aquatic biota (which includes one of the last wild populations of platy Apodaca, *Xiphophorus* cf. *couchianus* [Gordon]) survived unscathed. In May 1988 a contract was signed between the SADM and the Laboratorio de Ictiologia (Facultad de Ciencias Biológicas, Universidad Autónoma de Nuevo León [UANL]) to provide protection for the platy Apodaca and other local biota, to revegetate both aquatic and riparian habitats, and to keep the water clear and flowing.

This cooperative program is encouraging. It

seems to be the first official attempt in Mexico to protect a noncommercial species and its critical environment through limited, controlled, and managed use of habitat, centered on fish as well as water (not solely on water). This was done through the fine cooperation and understanding of Ingenieros Frederico Villarreal and José Luis Bueno of SADM, and Licenciado Francisco J. Elizondo Sepulveda, presidente municipal, Apodaca, Nuevo León, all of whom are to be commended.

Ojo del Potosí (Ejido Catarino Rodriquez; last visited 1989).

The original conditions at Ojo del Potosí are unknown. Several springs emerge from the base of a cliff. Sometime in the 1950s the spring outflow was enlarged and a wall was added to form a shallow marshland and reservoir that provided extensive habitat for three endemic species: the monotypic cyprinodontid genus *Megupsilon*, represented by cachorrito enano de Potosí (*M. aporus* Miller and Walters); cachorrito de Potosí (*Cyprinodon alvarezi* Miller); and a crayfish (*Cambarellus alvarezi* Villalobos). The latter two were first collected in 1948, and the former was not discovered until 1961.

Predatory largemouth bass (*Micropterus salmoides*) were introduced in 1974, to the detriment of the native biota (Contreras Balderas 1978b; M. L. Smith 1980), but the endemics persisted nonetheless. The habitat, described by R. R. Miller and Walters (1972), remained relatively stable at least from 1968 through 1984. Then, in 1985, the pool was reduced precipitously to only 15% of its former size as a result of intensified groundwater pumping. In 1986, 10% of the original habitat remained; by 1987 it was reduced to only 5%; and in 1989 only a shallow irrigation ditch remained.

The endemic species were clearly in critical danger of extinction, and stocks were removed to be maintained at the Laboratorio de Aqui-

cultura, UANL, with varying degrees of success. *Cyprinodon alvarezi* was readily propagated, but the other two have proven difficult to culture. A plan for protection of the area has been proposed by a group that constitutes what is essentially the first Mexican fish recovery team, formed by UANL personnel and with no legal status in the government. To date, the proposal has received sympathetic review by several authorities.

Bolsón de Sandia (last visited 1989).

This constitutes a hitherto unexplored basin 80 km southeast of Ojo del Potosí that was first detected from the air in 1983. It contained an undetermined number of springs and marshes, and was not reached by land until 1985, when two groups collected at two springs, Charco Azul (= El Barreño) and La Trinidad, and in a subsequent trip in that year in a pool named Charco Palma. All three localities yielded undescribed forms of cachorritos apparently related to *Cyprinodon alvarezi*. The first two also contained undescribed crayfish of the genus *Cambarellus*. On the next visit in 1987, La Trinidad was dry and its populations of fish and crayfish were gone. In 1988 Maria de Lourdes Lozano Vilano and party found ten other springs, already desiccated, and we will never know of their inhabitants. They also located a fourth flowing spring, La Presa, near Charco Azul, and a fourth form of cachorrito. La Presa and Charco Azul are populated by the most divergent pair of forms of *Cyprinodon* known in the basin.

A proposal has been submitted to protect this complex in situ, and an agreement should be signed and placed in effect in 1990. Stocks of fishes and crayfishes (except those from La Trinidad) are currently maintained in the Laboratorio de Aquicultura, UANL, in the charge of Arcadio Valdes. Their protection is part of the duties of the recovery team mentioned earlier.

State of San Luis Potosí

Venado-Moctezuma (last visited 1984).

Two large springs, one each in the towns of Venado and Moctezuma, are homes of the endemic solo goodeido (Goodeidae: *Xenophorus exsul* Hubbs and Turner). Each is protected as part of a municipal park, wherein the springs serve as swimming pools. No immediate dangers, excepting regional drought, are seen for the water supplies of these habitats; 3-m dams separate areas occupied by the original native aquatic communities from lower reaches of each system. Largemouth bass have been established downstream, but they had not yet attained the headsprings in 1984 when I, Diana Evans (IUCN), and students visited the area.

State of Tabasco

Baños del Azufre, Teapa (last visited 1980).

This is a large sulphurous spring, 6 km south of Teapa; its waters are considered medicinal. It acts as a critical refuge for two endemic poeciliids: the molly de Teapa (*Poecilia sulphuraria* Alvarez) and the guayacón bocón (*Gambusia eurystoma* Miller). The surroundings support a commercial hotel, trailer park, and resort, maintenance and upkeep of which are good. The spring has been converted into a well-kept swimming pool. Both fishes were common in 1980.

Protective Legislation

The recent promulgation of the Mexican Ley Federal del Equilibrio Ecologico y la Proteccion al Ambiente (SEDUE, 1988) has been announced in the United States and was briefly discussed by J. E. Williams et al. (1989). The objectives of this legislation are to protect the environment and ecological equilibrium, but the law also mentions (although does not define) species that are endemic, endangered, or threatened. It further establishes basic needs for environmental impact assessments for all public or private development plans, and provides levels of protection, either formally on public lands or as part of the environmental assessment process on private lands. At the same time, the law pursues a goal of increased production for the welfare of the people. It is new and must be followed by state enabling legislation as well as state and municipal regulations to be effective.

In part as a result of this legislation, January–March 1989 was a period of strong, widespread, and gratifying consultation among biologists, environmentalists, and politicians at different governmental levels, as well as with the general public and interested parties. Hopes are that the new laws, renewed interest, stronger conscience, and public awareness will provide a better setting for conservation efforts in Mexico, despite the growing and important demands for food production, development, recreation, and other human needs.

Conclusion

Large springs and their surroundings hold fascination for humans, especially in arid lands, and this, coupled with the not infrequent occurrence of endemic or relict species in such special habitats, resulted in the inadvertent protection of native fishes in Mexico. Of the fifteen such places described here that support rare fishes, thirteen have been modified and maintained at least in part for their recreational and therapeutic values. Seven of the fifteen areas are recognized or under consideration for conservation by private landowners or municipal and federal authorities, in part because of their aquatic habitats and biota.

Other such refuges exist in Mexico, and the present account must be considered preliminary. Rinne and Turner (*this volume*, chap. 14) mention a rare endemic trucha (trout; *Oncorhynchus* sp.) partially protected within the

Parque Nacional de Basasaechic, in the Río Mayo drainage of Chihuahua. The charalito Saltillo (*Gila modesta* [Garman]) has survived precariously for years in a small travertine-forming stream in a roadside park near Saltillo, Coahuila. The blind, subterranean bagre de Múzquiz (Ictaluridae: *Priatella phreatophila* Carranza) lives in a cave spring that is protected by the town of Múzquiz, Coahuila, as part of its potable water supply. It has been seen there recently once each in 1982 and 1984, and once from a new (as yet undisclosed) locality in 1989. A number of endemic species were apparently extirpated when large limestone springs near Parras de la Fuente, Coahuila, were modified near the turn of the century (R. R. Miller 1961, 1964c; R. R. Miller et al. 1989), yet a few persisted to 1975 except where subjected to pressures from introduced exotics (Contreras Balderas 1975); now they all seem extirpated (Contreras Balderas and Maeda 1985). Even in Parras, protection of spring sources through early development was preferable from the perspective of fish survival to pumping aquifers, which quickly destroys the surface waters of such areas (in this volume, Minckley and Douglas, chap. 1; Minckley et al., chap. 15).

Although the situation for aquatic habitats and freshwater fishes in Mexico is not promising, most government agencies are becoming more interested and responsive to public insistence for conservation. It should nonetheless be recognized that an appalling number of endemic species and affected aquatic habitats will almost certainly disappear before an adequate and rational system of species conservation and protected areas can be attained. Let us work toward success so the damage is minimized. The future is less than bright, but we have hope and good intentions.

Chapter 13

Hatcheries for Endangered Freshwater Fishes

James E. Johnson and Buddy Lee Jensen

Introduction

The desire to protect organisms by placing them into controlled environments may date to the first domestications of wild plants and animals. Ehrenfeld (1976) suggested Noah as the first practical conservationist because he protected animals during periods of environmental perturbations, then released them when conditions were more suitable for survival. Today, zoological, botanical, and other organizations continue to bring wild organisms into captivity for educational, economic, and ecological purposes. Protecting species in danger of extinction by placing them in refugia is a logical and beneficial outgrowth of this long-term, self-serving trend.

The environmental community was recently split by a controversy over whether to capture the last remaining California condors (*Gymnogyps californianus*) and attempt to breed them in captivity, or to allow them to remain free and take their chances with nature. The essence of that controversy covers the major points of protecting any endangered species in captivity. Arguments against capturing the condors included the following:

1. Human meddling has nearly destroyed the species; additional contact will only hasten its demise.
2. If all birds are removed from the wild, there will be less effort to protect their remaining habitat.
3. Birds reared in captivity will change. Reduced genetic variability or altered behavior will prevent them from readapting to life in the wild.

On the other hand:

1. There is presently no successful condor recruitment in the wild. Until problems limiting wild production are determined and corrected, the birds should be bred in captivity.
2. If the gene pool is lost, no amount of habitat protection will help the species.
3. The species is almost extinct. Only "hands-on" research will help us to learn why the species is continuing to decline and how to counteract the problems it faces.

The final decision to bring all surviving California condors into captivity was made on 23 December 1985 by the U.S. Fish and Wildlife Service (USFWS 1985c), and the last wild bird was captured 19 April 1987. There are presently twenty-seven adults in captivity, and the first chick was produced in 1988; its hatching made national news. Only time will tell if this was indeed the correct path of action for the species.

Captive propagation is nonetheless being used at present as a recovery tool for numerous endangered species. All remaining ($N = 18$) black-footed ferrets (*Mustela nigripes*) from the only known wild population near

Meteese, Wyoming, were taken into captivity between 1985 and 1987 in order to protect them from canine distemper (USFWS 1988b). Mexican wolves (*Canis lupus baileyi*) have been extirpated from the United States since 1970, but a few individuals were still alive in Mexico through the early 1980s (USFWS 1982a). Four were captured in the Mexican states of Durango and Sonora between 1979 and 1981 and brought into the United States to initiate a captive breeding program; they now number thirty. Red wolves (*C. rufus*) disappeared from the southeastern United States in 1976 when the USFWS captured the last individuals in Texas and shipped them to a captive propagation facility near Tacoma, Washington (Carley, n.d.; USFWS 1984d). Peregrine falcons (*Falco peregrinus*) have been reared in captivity by the Peregrine Fund since 1973 (USFWS 1984e), and the USFWS has been rearing whooping cranes (*Grus americana*) at Patuxent Wildlife Research Center since 1967 (USFWS 1986b).

Captive propagation is not an end in itself, but rather a means of perpetuating species until suitable habitat can be found, rebuilt, or created. Some captive breeding efforts have already begun this logical second step in recovery by reintroducing endangered species back into their natural habitats. For instance, peregrine falcons have been reintroduced into the Rocky Mountains for more than a decade (USFWS 1984e), and twelve red wolves were recently released in South Carolina by the USFWS (W. Parker, USFWS, pers. comm.).

While these efforts have made the national news and are rather well known to the public, another organized effort to protect endangered species in captivity has quietly proceeded for several decades. The purpose of this paper is to discuss ongoing efforts to protect some endangered fishes of western North America by placing representative samples in controlled environments until recovery actions reduce or eliminate threats to their survival.

Declines of many western fishes are documented elsewhere in this volume and will not be reviewed here. Early conservation efforts centered on status and distributional studies that, by noting declines of species and aquatic habitats, drew attention to the plight of these unique animals (R. R. Miller 1946a, 1961; Deacon et al. 1964, 1979; Minckley 1965, 1969b; Minckley and Deacon 1968; Deacon 1968b, 1979). As recently as twenty-five years ago, many native fishes were considered expendable in the interest of enhancing sport fishing. Reclamation efforts resulting in the use of piscicides on the Green (Holden, *this volume*, chap. 3), San Juan (VTN 1978; Platania and Bestgen 1988), and Gila rivers (Rinne and Turner, *this volume*, chap. 14) were extreme examples of that common and widespread practice. The formation of the Desert Fishes Council in 1968 (Pister 1985a, b, *this volume*, chap. 4) was one response to these growing problems; captive maintenance and propagation was another.

Fishes have many biological attributes that lend them to captive maintenance and production, including the following: (1) large numbers can be captured from the wild and transported with relative ease; (2) they can be bred and reared in captivity more readily than many other vertebrates; (3) fishes produce large numbers of offspring that can be stocked soon after hatching; (4) most fishes grow rapidly and mature quickly, minimizing generation time and holding costs; and (5) there is a long history of rearing sport fishes in captivity and stocking them into the wild, a custom and tradition that lends acceptance to captive rearing of endangered and threatened species.

There were a few early attempts to propagate disappearing fishes. Gila trout (*Oncorhynchus gilae*) were reared by the state of New Mexico between 1923 and 1939 at Jenk's Cabin State Fish Hatchery in what is now the Gila Wilderness Area (R. R. Miller 1950; USFWS 1984a), and the Arizona Game and Fish Department (AZGFD) propagated threatened Apache trout (*O. apache*) at Sterling Springs State Fish Hatchery between 1960 and 1974 (R. R. Miller 1972b; Minckley 1973; USFWS 1983b). Both efforts produced some fish but were terminated because of difficulties in rearing the native species compared with rainbow trout (*O. mykiss*). W. L. Minckley (Arizona State University [ASU], pers. comm.) attempted to rear razorback suckers (*Xyrauchen texanus*) at the Phoenix Zoo in Arizona in 1968 but was able to keep fish alive for less than a year due to water distribution problems. Several pupfishes (genus *Cyprinodon*) and live-bearing fishes (family Poeciliidae) were produced at ASU starting in the 1960s, and at least one pupfish (*Cyprinodon m. macularius*) has been successfully maintained there since 1976 (Minckley, pers. comm.). A. Peden and C. Hubbs attempted to save the now-extinct Amistad gambusia (*Gambusia amistadensis*) at Brackenridge Field Laboratory of the University of Texas in Austin and in Victoria, Canada (Peden 1973; Hubbs and Jensen 1984). The Steinhart Aquarium in San Francisco, California, maintained a population of Devils Hole pupfish (*Cyprinodon diabolis*) for several years before it died out (Castro 1971). Two populations of Mohave tui chub (*Gila bicolor mohavensis*) have been maintained in artificial ponds at the Desert Research Station (= Zzyzx Springs or Fort Soda), San Bernardino County, California, for more than twenty years (Moyle 1976; Taylor and McGriff 1985). There have been other attempts that are not noted here, but, for one reason or another, few succeeded. Starting in the 1970s, three long-term projects designed specifically to protect imperiled western fishes in artificial environments were initiated.

In Nevada, the U.S. Bureau of Reclamation established a Devils Hole Pupfish Refugium in Clark County, below Hoover Dam (Baugh and Deacon 1988). Twenty-seven pupfish

were stocked into the new habitat in October 1972 and were monitored carefully along with water flow, water chemistry, and aquatic organisms (Sharpe et al. 1973; J. E. Williams 1977; Sharpe 1983). Pupfish numbers varied in the new refugium, as did their morphology (J. E. Williams 1977), but a population survived until 1988.

In 1974 the Texas Department of Parks and Wildlife constructed the Comanche Springs Pupfish Canal at Balmorhea State Recreation Area, Texas, in association with San Solomon Springs (USFWS 1981b). That facility today supports an abundance of Comanche Springs pupfish (*Cyprinodon elegans*) as well as several endemic freshwater snails, amphipods, and the endangered Pecos gambusia (*Gambusia nobilis*), although questions have recently been raised about nutrition and average size of the captive fish (A. A. Echelle, Dexter National Fish Hatchery [memorandum to file], 3 August 1988).

The third facility, Dexter National Fish Hatchery (NFH), Dexter, New Mexico, operated by the USFWS, began working with threatened and endangered fishes in 1974. Since that time, propagation of imperiled fishes in artificial environments has centered on a continually expanding endangered species program at that hatchery (Stewart and Johnson 1981; J. E. Johnson and Rinne 1982; Maitland and Evans 1986; Rinne et al. 1986), which forms the major subject of the present report.

Dexter National Fish Hatchery

Initiation and Growth of Facility Commitment

Dexter NFH was constructed in the early 1930s by the USFWS to rear warm-water sport fishes, mostly largemouth bass (*Micropterus salmoides*) and channel catfish (*Ictalurus punctatus*). The hatchery is located in the

Pecos River valley of southeastern New Mexico at an elevation of 1067 m. Water is pumped from a shallow aquifer at a constant temperature of 18°C. Culture facilities are available in a holding house with sixteen 1360-l concrete tanks, a hatching battery and laboratory, and outdoor installations that have recently been improved to include four 1.9-m by 12.2-m raceways and forty-eight earthen ponds varying from 0.04 to 0.7 ha in surface area.

New laws, responsibilities, and improved methods of fish transport brought changes to the mission of Dexter NFH in the early 1970s. The idea of using it as an endangered fish refuge was tested in 1974 when C. Hubbs and A. A. Echelle provided stocks of Big Bend (*Gambusia gaigei*), Amistad, and Pecos gambusias, Comanche Springs and Leon Springs pupfishes (*Cyprinodon bovinus*), and fountain darter (*Etheostoma fonticola*). All six species survived and reproduced in the hatchery's ponds (Stewart and Johnson 1981), even though *E. fonticola* and *C. elegans* typically occupy only flowing waters in nature. The reason these stream fishes survived and reproduced seems to involve patterns of water circulation in the Dexter NFH ponds. Inflowing water constantly spills in near the mouth of 2-m by 3-m concrete "kettles," which also control water outflow, creating a zone of continuous water movement. Stream fishes apparently find sufficient habitat in such places to meet their requirements.

Following success with all six species, the USFWS began to expand its work on endangered species at the Dexter station. No guidelines or policies existed for such an installation, and no recovery plans for endangered or threatened fishes had yet been completed. By 1979, however, recovery plans had been developed for eight species (J. E. Johnson 1987a), and those plans, along with assistance from the scientific community (J. E. Johnson 1979), helped establish goals as follows: (1) to establish a refuge to protect genetic stocks of imperiled fishes in case wild populations were lost, (2) to initiate research efforts to determine and alleviate threats to survival of imperiled fishes, and (3) to propagate sufficient quantities of selected species of a quality that would allow reintroductions into suitable habitats within their likely historic ranges.

During the early 1980s Dexter NFH continued to expand its role in endangered species maintenance and recovery. Additional fishes were cultivated, some more successfully than others, and a few were even removed from the facility when their status in the wild was better understood, in order to make space for higher-priority species (Rinne et al. 1986). Twenty-four taxa of native fishes have been held at Dexter NFH since 1974 (Table 13-1); twelve are housed there at present (J. E. Johnson and Hubbs 1989; unpub. data). The present staff consists of a manager, three fishery biologists, a fisheries researcher, and three support personnel. Two academic research biologists (W. L. Minckley, ASU, and Anthony A. Echelle, Oklahoma State University) have each spent a year at Dexter NFH under Intergovernmental Personnel Act agreements between the USFWS and their respective institutions.

Special Concerns

An endangered fishes rearing and holding facility has several unique concerns and requirements not often or always considered in other kinds of hatcheries. For instance, fishes brought from the wild may harbor unknown parasites or disease organisms that, if allowed to spread, could contaminate other rare species. Strict quarantine measures preclude most of these problems, but a fungus (*Ichthyophonus* sp.) contaminated stocks of Comanche Springs pupfish and Yaqui topminnow (*Poeciliopsis occidentalis sonoriensis*), and several species have contracted Asian tapeworm (*Bothriocephalus acheilognathi*) while at Dexter NFH, including Colorado squawfish (*Pty-*

Table 13-1. Fishes held at Dexter National Fish Hatchery, Dexter, New Mexico, 1974–1989. Those marked with an asterisk are currently being maintained.

Scientific name	Common name	Natural distribution[1]
Gila cypha Miller	humpback chub	AZ, CO, NV, UT, WY
G. ditaenia Miller	Sonoran chub	AZ, SON
G. elegans Baird and Girard	*bonytail	AZ, CA, CO, NM, NV, UT, WY, BCN, SON
G. nigrescens (Girard)	*Chihuahua chub	NM, CHI
G. pandora (Cope)	Rio Grande chub	NM, TX[2]
G. purpurea (Girard)	Yaqui chub	AZ, SON
G. r. robusta Baird and Girard	roundtail chub	AZ, CA, CO, NM, NV, UT, WY, SON, CHI, SIN
G. r. jordani Tanner	*Pahranagat roundtail chub	NV
G. r. seminuda Cope and Yarrow	*Virgin roundtail chub	AZ, NV, UT
Cyprinella f. formosa (Girard)	*beautiful shiner	NM, CHI
C. f. mearnsi (Snyder)	Yaqui beautiful shiner	AZ, CHI, SON
Plagopterus argentissimus Cope	*woundfin	AZ, CA, NV, UT, SON(?), BCN(?)
Ptychocheilus lucius Girard	*Colorado squawfish	AZ, CA, CO, NM, NV, UT, WY, SON, BCN
Catostomus bernardini Girard	Yaqui sucker	AZ, SON, CHI
Xyrauchen texanus (Abbott)	*razorback sucker	AZ, CA, CO, NM, NV, UT, WY, SON, BCN
Cyprinodon bovinus Baird and Girard	*Leon Springs pupfish	TX
C. elegans Baird and Girard	Comanche Springs pupfish	TX
C. m. macularius Baird and Girard	*desert pupfish	AZ, CA, SON, BCN
C. pecosensis Echelle and Echelle	Pecos pupfish	NM, TX
Gambusia amistadensis Peden	Amistad gambusia	TX
G. gaigei Hubbs	*Big Bend gambusia	TX
Poeciliopsis o. occidentalis (Baird and Girard)	*Gila topminnow	AZ, CA(?), NM, SON
P. o. sonoriensis (Girard)	Yaqui topminnow	AZ, SON
Etheostoma fonticola (Jordan and Gilbert)	fountain darter	TX

[1] Abbreviations for Mexican states are as follows: BCN, Baja California del Norte; CHI, Chihuahua; SIN, Sinaloa; and SON, Sonora.
[2] M. L. Smith and Miller (1986) tentatively recorded *Gila pandora* from Mexico, but this has yet to be thoroughly documented.

chocheilus lucius), beautiful shiner (*Cyprinella formosa*), and woundfin (*Plagopterus argentissimus*). An external parasitic copepod (*Lernaea* sp.) is common on wild *P. lucius*, *P. argentissimus*, and chubs (*Gila* spp.). While treatment for *Lernaea* is relatively simple by hand picking or chemical means, additional handling of new fish may further weaken them and result in mortalities.

A second problem unique to rearing rare fishes is that few have ever been propagated in captivity, and techniques developed for production of warm-water sport fishes are often inappropriate. Fishes from springs and small streams (e.g., poeciliids, cyprinodontids, some cyprinids) needed little encouragement to survive and reproduce in ponds at Dexter NFH. However, several species failed to reproduce naturally in captivity, while others produced only limited numbers of offspring, necessitating development of new culture techniques (Toney 1974; Hamman 1981, et seq.; Inslee 1982a, b; Rinne et al. 1986). Spawning techniques for selected species are discussed later.

A rare species may someday be reestablished in nature entirely from captive stock, and in fact such attempts are presently under way for red wolves and Socorro isopods (*Thermosphaeroma thermophilum*; USFWS 1982b, 1984d). If a species is to be reestablished entirely from such a source, maintenance of genetic diversity is vital. Vrijenhoek et al. (1985), Meffe (1986, 1987), Meffe and Vrijenhoek (1988), and J. E. Johnson and Hubbs (1989) have discussed genetics of the various fish stocks at Dexter NFH that are being used in ongoing reintroduction efforts. Periodic checks are made on the genetic health of fishes at the hatchery (A. A. Echelle et al. 1983; Buth et al. 1987; Ammerman 1988; Ammerman and Morizot 1989; Minckley et al. 1989; Echelle, *this volume*, chap. 9) and will be continued to ensure that captive populations adequately represent the taxa being protected. Recognition of the importance of genetics in

game fish production—reviewed by Kincaid and Berry (1986), Kapuscinski and Jacobson (1987), and Ryman and Utter (1987), among others—has added even greater impetus to the endangered species conservation effort.

Another unusual aspect of an endangered fishes rearing center is the unique value of each individual removed from nature. A rare fish species may be represented in the wild by only a single population, or only a few hundred (or fewer) individuals. Removal of any of these rare organisms may appreciably reduce stability, breeding success, or genetic viability of the population. Before individuals are taken, an assessment of the impact of that action must demonstrate that it does not jeopardize the species. Thus, individuals of a rare species in captivity are also valued differently than sport fishes in similar facilities. Disease outbreaks, pump failures, handling errors, and myriad other incidents are encountered in sport fish hatcheries on a regular basis, and the loss of a few individuals, a pond, or an entire stock, while serious, is not fatal to the species. Similar problems may also be expected at an endangered species hatchery, but because they could be so devastating, special efforts must be made to minimize such events and reduce their impacts.

Water distribution and disposal are other problems unique to endangered species hatcheries. Recirculation (reuse) of water through several ponds or raceways is a standard practice in many hatcheries. However, this practice increases the chances of spreading disease, as well as the possibility of mixing species. Therefore water at Dexter NFH is passed only once through a given pond or other system. Most taxa maintained at Dexter NFH belong to three families (Cyprinidae, Cyprinodontidae, Poeciliidae), and hybridization between any two species of the same family may be possible. Hybrids of related species are often fertile (A. A. Echelle et al. 1989; Echelle, *this volume*, chap. 9) and could easily be overlooked, resulting in genetic contamination of a refuge stock. This in turn could result in alterations of wild stocks if such fish were used for reintroduction.

Protective measures that prevent spread of disease also limit accidental mixing of fishes. Nets, protective clothing, and other equipment used on-site are never moved from pond to pond without thorough cleaning to reduce chances of transfer of disease, fish eggs, or fishes themselves. Personnel are trained to avoid accidental mixing of stocks or pathogen transfers and to maintain constant surveillance for such events. Public access to the hatchery grounds is monitored to prevent vandalism or inadvertent problems. The possibilities of fishes (or diseases) being transferred physically by birds, mammals, or amphibians that frequent the ponds nonetheless remains a threat.

Yet another special management aspect of an endangered fishes facility is the control of wastewater to prevent fish from escaping. Loss of rare individuals into adjacent surface waters would not only reduce viability of the captive population but also would pollute local habitats with non-native species, which might reduce the viability of local species. Dexter NFH solved the problem of escape by draining wastewater into a series of small, enclosed sumps; evaporation and percolation prevent water, and therefore organisms, from ever reaching the nearby Pecos River.

A final facet of an endangered fishes rearing center, and perhaps its most controversial, is the priority assigned to various recovery actions. Finite space, water, and money limit the amount of hatchery consideration for recovery efforts for a given fish. Some actions take more space and time than others, and moving a species from a simple holding regimen into active propagation for reintroduction, or bringing new species into the facility, may result in a need for space occupied by another species. Such decisions are always difficult

when they deal with organisms that, by definition, are all on the verge of extinction. Priorities are placed on the recovery aspects of endangered species as dictated by the Endangered Species Act (ESA; USFWS 1983e). The species most threatened in the wild have higher priorities, except when no wild habitat remains within the likely historic range; in those cases species are assigned a lower priority because there is little chance for recovery in nature (J. E. Johnson and Rinne 1982; Rinne et al. 1986). If pond space is needed for a high-priority species, a lower-priority fish must be removed from the hatchery. A species extirpated in the wild with no chance of ever being reintroduced would be removed even though loss of the hatchery population could mean extinction. While this may seem to contradict the ESA, it must be remembered that the purpose of the act is "to provide a means whereby the ecosystems upon which endangered species and threatened species depend may be conserved" (USFWS 1983e). If no suitable habitat remains in nature and there is little chance to create it, recovery is impossible; a low priority, although difficult to accept, is thus correct. With luck, an aquarium, a zoo, or some similar private, local, or agency facility will maintain the organism.

The only fish species to have faced this "triage" policy decision was the Amistad gambusia, which had no historic habitat remaining in nature. Unfortunately, the species was lost in captivity due to contamination with mosquitofish (*Gambusia affinis*) before the decision was made to remove it from Dexter NFH (see below). Some other rare western fishes in need of captive propagation or maintenance under refuge conditions, but excluded from Dexter NFH because of a lack of space or low priority, include, among others, the humpback chub (*Gila cypha*), Clear Creek gambusia (*Gambusia heterochir*), Pecos pupfish (*Cyprinodon pecosensis*), Tularosa pupfish (*C. tularosae*), and Devils Hole pupfish.

Fig. 13-1. Colorado squawfish, F_1 of the 1974 year class, about 50 cm total length. Photographed at Dexter National Fish Hatchery, New Mexico, by J. E. Johnson, May 1981.

Species Accounts

The following accounts summarize work on selected fishes held at Dexter NFH. The species that has received the most space and effort, the razorback sucker, is thoroughly discussed in chapter 17 of this volume and is not covered here.

Ptychocheilus lucius (Colorado Squawfish; Fig. 13-1)

This largest native minnow in North America, and the only major predatory fish of the Colorado River, once reached 1.8 m in total length (TL) and 36 kg in weight (Minckley 1973). Historically, Colorado squawfish ranged throughout the larger rivers of the Colorado River system, sometimes migrating into smaller tributaries, from Colorado and Wyoming to Mexico. Its present distribution is limited to the upper basin of the Colorado River in Colorado, Utah, and New Mexico, including the Colorado, Green, Yampa, and San Juan rivers and their major tributaries (Tyus et al. 1982a; Tyus, *this volume*, chap. 19; USFWS 1978b, 1989a). Colorado squawfish were extirpated

from the lower Colorado River basin by about 1970 (Minckley 1973), but reintroduction efforts initiated in 1985 in the Salt and Verde rivers in Arizona have begun to reestablish them (J. E. Johnson 1987b; USFWS 1989a).

The first Colorado squawfish were brought into captive propagation at Willow Beach NFH in Arizona from the Green, Yampa, and Colorado rivers in Utah and Colorado (Table 13-2), and later were transferred to Dexter NFH. A total of 122 Colorado squawfish has been taken from the wild for brood stock since 1973, but 60 of these were juveniles that failed to survive transfer to the hatchery, and not all the remaining wild adults have contributed gametes to the captive gene pool. Frankel and Soulé (1981) recommended that at least 50 individuals should contribute genetic material to a captive population in order to retain genetic heterogeneity for even a few generations. While it is not likely that 50 wild Colorado squawfish contributed to the Dexter NFH gene pool, recent electrophoretic monitoring indicates that juveniles continue to adequately represent wild populations of the Green and Colorado rivers (Ammerman and Morizot

1989). The present (1989) brood stock consists of 14 wild fish, plus 185 F_1 individuals from a 1974 year class and 243 F_1 fish from 1981. Wild fish continue to be used as brood stock whenever possible, but F_1 fish are depended on to supplement production when wild fish are unavailable.

It is standard procedure in sport fish hatcheries to select for certain traits (high fecundity, high food conversion, docile temperament, disease resistance) in brood fish in order to produce more and larger fish (Kincaid and Berry 1986). In a put-and-take sport fish stocking program these attributes are consid-

ered beneficial, but for captive stocks that may be used to reestablish a wild population or a species, such characteristics could reduce fitness of the stock. For instance, several long-lived imperiled fish species produce a variety of growth rates within a single cohort (Minckley 1983; unpub. data), with some individuals growing rapidly and maturing early while others grow more slowly and mature at a greater age. This variation may help a species meet changing reproductive or survival conditions over an individual's life span of thirty to fifty years (Tyus 1986). If only the fastest growing fish were selected for propagation

Table 13-2. Colorado squawfish brought into Willow Beach and Dexter National Fish Hatcheries as brood stock for captive propagation, 1973–1988.

Dates	Numbers	Capture localities
Originally Transferred to Willow Beach NFH (all wild adults)		
July 1973	8	Yampa River, CO
November 1975	14	Colorado River, Grand Junction, CO
May 1976	8	Green River, UT
April 1978	11	Colorado River, Grand Junction, CO
May 1978	11	Green River, Jensen, UT
October 1979	3	Green River, Jensen, UT
November 1979	4	Colorado River, Moab, UT
Originally Transferred to Dexter NFH		
March 1980	217 F_1 (1974 year class)	Willow Beach NFH, AZ
August 1981	16,000 F_1 and F_2 (1981 year class)	Willow Beach NFH, AZ
September 1981	98 F_1 (1974 year class)	Willow Beach NFH, AZ
September 1981	13 wild adults[1]	Willow Beach NFH, AZ
November 1981	73 F_1 (1974 year class)	Hotchkiss NFH, CO[2]
October 1981	5,300 F_1 and F_2 (1981 year class)	Willow Beach NFH, AZ
April 1987	2 wild adults	Hotchkiss NFH, CO[3]
August 1987	60 wild juveniles	Green River, UT[4]
June 1988	1 wild adult	Colorado River, Grand Junction, CO

[1] Fish brought to Willow Beach NFH from the wild (see above) and transferred to Dexter NFH.
[2] Fish spawned at Willow Beach NFH, transferred to Hotchkiss NFH, then transferred a second time to Dexter NFH.
[3] The original capture site was Colorado River at Grand Junction, CO.
[4] Fish were captured as larvae and small fry, reared at Vernal, UT, and transferred to Dexter NFH. All died; none contributed to the captive gene pool.

and reintroduction, survival of new populations could be significantly reduced in habitats experiencing periodic drought or other conditions that might eliminate one or more year classes.

The only factor used to select wild Colorado squawfish for captive production at Dexter NFH was their original susceptibility to capture. No additional selection takes place, but the protected habitat of a hatchery places different selective pressures on individuals than would wild habitats, and these pressures may influence the genetic makeup of hatchery-reared F_1 populations.

Spawning and rearing techniques for Colorado squawfish were discussed briefly by Toney (1974) and more thoroughly by Hamman (1981, 1986, 1989). Brood fish spawned in holding ponds at Dexter NFH, and even in concrete raceways in Willow Beach NFH when gravel substrate was provided (Toney 1974), but they produce only a few young in this way. It is not known how many Colorado squawfish are hatched in holding ponds, because the ponds are managed for adults, and cannibalism on small individuals is high. A temperature of about 20°C is needed for natural spawning, along with fluctuating water levels (Holden and Wick 1982; Valdez and Williams 1987), but forced (hormone-induced) maturation can occur at a slightly lower temperature (17°C; Hamman 1981). Injection of common carp (*Cyprinus carpio*) pituitary (4.0 mg kg^{-1} body weight) and hand stripping of gametes increases production by several orders of magnitude and facilitates control of progeny through use of mass culture techniques. Adults survive this annual manipulation with little obvious trauma.

In 1987 Dexter NFH had a reintroduction commitment for 100,000 yearling Colorado squawfish of approximately 7 cm TL. Thirty-one females were injected with carp pituitary to induce ovulation; twenty-five responded to produce 2,112,681 eggs. Eight thousand eggs

were shipped to other facilities for research, and an estimated 617,770 hatched and grew to "swim-up fry," a 29.3% survival rate. About half those fry (337,620) were stocked in ponds for growth, and the remainder were shipped to other facilities for testing, growth, and stocking in other parts of the Colorado River basin (Table 13-3). In October 1987, 103,110 fish averaging 7.6 cm TL were harvested from Dexter NFH ponds and stocked into the Salt and Verde rivers in Arizona. During this same year, an unknown number of adult Colorado squawfish spawned naturally in their holding ponds, but only 12 young-of-year fish were recovered in October.

Colorado squawfish are easily grown under hatchery conditions and readily accept pelleted food. Feed conversion from fry to 5 cm TL at Dexter NFH in 1987 was 1 kg of fish for every 8 kg of food, and the ratio was 1:4.9 in fish from 5.0 to 7.6 cm TL. Adult squawfish will take pelleted foods but also feed on forage fish (including goldfish [*Carassius auratus*] and young trouts [*Oncorhynchus* sp.]), if available. In the wild, squawfish become piscivorous at about 10 cm TL (Vanicek and Kramer 1969), and cannibalism is extreme in ponds if

Table 13-3. Distribution of Colorado squawfish from Dexter National Fish Hatchery, 1987.

Destination	Number
Page Springs State Fish Hatchery, AZ (fry)	100,000
Willow Beach National Fish Hatchery, AZ (fry)	125,000
USFWS, Grand Junction, CO (fry)	50,000
USFWS, Vernal, UT (fry)	5,150
Stocked in AZ waters (Salt and Verde rivers) (75 mm TL)	103,110
Total	383,260

Fig. 13-2. Wild-caught bonytail, about 50 cm total length, from Lake Mohave, Arizona-Nevada. Photograph by W. L. Minckley, March 1981.

fish are crowded or improperly fed. In captivity, feeding and growth seem to cease when water temperatures drop below 13°C.

Reintroduction of Colorado squawfish began in 1985 on the Salt and Verde rivers in Arizona (J. E. Johnson 1987b) as one part of the recovery program for this endangered species (USFWS 1978b, 1989a). Success will be evaluated after the effort is completed in 1994 by determining if natural reproduction occurs. Several restocked fish have been recaptured in both streams (W. Kepner and J. Brooks, USFWS, pers. comm.), but high numbers of non-native predatory fishes, especially channel and flathead (*Pylodictis olivaris*) catfishes, severely limit all native fishes in these waters and may thwart the reintroduction attempt (Marsh and Brooks 1989; USFWS 1989a). A second Colorado squawfish stocking program is planned for the mainstream lower Colorado River, Arizona-California, in part to determine if a sport fishery may be developed for the species (USFWS 1987b).

Gila elegans (Bonytail; Fig. 13-2)

This minnow is the rarest of the endemic big-river Colorado River fishes. Bonytail were once found throughout the mainstream and major tributaries of the Colorado River system (Tyus et al. 1982a). A remnant population persists in Lake Mohave, an artificial impoundment on the Colorado River immediately downstream from Hoover Dam. The few wild fish captured from this reservoir over the past ten years varied from an estimated thirty-four to forty-nine years old (Minckley et al. 1989), indicating little or no successful recruitment since impoundment in 1954. A few bonytail also exist in Lake Havasu, Arizona-California, and in upper Colorado River basin waters. Taxonomic problems within the genus *Gila* confuse their status in the latter area (J. E. Johnson 1976; USFWS 1989b).

Eighteen adult bonytail were collected from Lake Mohave between 1976 and 1981 as brood stock for a propagation program at

Dexter NFH (Minckley et al. 1989). Not all fish contributed to the gene pool, however, as some failed to survive capture and transportation trauma and others did not respond to hormone application by producing gametes. Six female and five male wild-caught bonytail parented the 1780 existing F_1 brood fish that were produced in 1981 (Hamman 1982c); all the wild fish are now dead.

Culture methods for bonytail (Hamman 1982c, 1985b) include intraperitoneal injection of carp pituitary at 4.0 mg kg^{-1} of body weight at 24-hour intervals in 20°C water and hand stripping of gametes. At 18°–20°C, eggs begin to hatch at 99 hours and finish by 174 hours; swim-up occurs 48–120 hours later. In 1983, twenty-four females produced between 1015 and 10,348 eggs per fish (mean = 4990), and survival to swim-up fry varied from 17% to 38% (Hamman 1985b). Eight bonytail artificially spawned in 1988 to produce fry for the state of California averaged 0.62 kg in weight and produced 15,380 eggs per kilogram. Numbers of eggs per female varied from 3672 to 16,795, and averaged 9507.

Bonytail brood fish held in ponds at Dexter NFH annually produce 10,000–20,000 young through natural spawning. Approximately 110,000 of these have been stocked back into Lake Mohave between 1981 and 1987. Even though bonytail apparently do not recruit in Lake Mohave, individuals appear to be able to survive there and thus may serve as a source of genetic material well into the twenty-first century (Rinne et al. 1986; Minckley et al. 1989). Reintroduction goals have not been set (USFWS 1989b), and the only goal at Dexter NFH has thus been to maintain bonytail and perfect methods for its captive production.

Reintroduction of Colorado squawfish has proceeded under the experimental, nonessential regulation of the ESA. Continued disagreement about this designation for bonytail, mostly because of its extreme rarity, has delayed the use of reintroduction as part of a recovery effort. Experimental, nonessential designation greatly reduces the protection that reintroduced populations receive under the act (USFWS 1984f), and such reintroductions are thus more acceptable to land and water owners and managers than the full protection afforded to most endangered species (J. E. Johnson 1987b). The easing of restrictions is opposed by those who wish to protect the little remaining wild riverine habitat for this most endangered of Colorado River fishes, and because experimental designation may confuse recovery goals (USFWS 1989a). There are valid points to both arguments, but the longer that recovery efforts are stalled and survival of the species depends solely on the captive population at Dexter NFH, the greater are the chances that an operational error will result in extinction.

To preclude this possibility, additional stocks of bonytail are now established in urban lakes in Tempe, on the Buenos Aires National Wildlife Refuge (NWR), and at The Nature Conservancy's Hassayampa Reserve, all in Arizona. This controversy may have run its course, as both the USFWS and the AZGFD have agreed to proceed with the experimental regulations for stocking bonytail into Arizona streams, and a *Federal Register* proposal has been prepared for publication.

Plagopterus argentissimus (Woundfin; Fig. 13-3)

Woundfin, a small, silvery minnow characterized by modified rays that form a stiff spine in its dorsal fin, were once distributed throughout the lower Gila River downstream from Tempe, Arizona, and the Colorado River from Yuma, Arizona, up to and including the Virgin River basin in Utah (R. R. Miller and Hubbs 1960; USFWS 1985a). The species has disappeared from all its historic range except for a short reach of the Virgin River, where it is now threatened by invasion of red shiners (*Cyprinella lutrensis*), Asian tapeworm, and

Fig. 13-3. Woundfin, about 100 mm total length, from the Virgin River, Utah. Photograph by J. E. Johnson, October 1986.

extensive water developments in one of the fastest-growing parts of Utah (Greger and Deacon 1982; USFWS 1985a, 1986c; Deacon 1988).

Approximately 1000 woundfin were brought into Dexter NFH between 1978 and 1987, but many were lost to parasites and disease. Of special note is the impact of the protozoan ichthyophthiriasis, or "ich" (*Ichthyophthirius multifilis*), which, in conjunction with stress from hauling and treatment for fungus (250 mg l^{-1} formalin for an hour every other day for a two-week period), continues to limit both young and adult woundfin. In 1987, for instance, the USFWS brought 235 fish to Dexter NFH, but only 170 survived transportation and treatment. An additional 2015 fish of all age classes were obtained in 1988, before reclamation of the Utah portion of the Virgin River to reduce or eliminate red shiners. As of January 1989, only about half of these survived; again, mortality was mostly the result of ich infestations (D. Hales, USFWS, pers. comm.). Heckmann et al. (1987) noted ich on wild woundfin in the Virgin River, but it has not been identified as a threat to that population (USFWS 1985a).

Both artificial and natural production methods for woundfin have been attempted at Dexter NFH (Rinne et al. 1986), with better success realized from natural efforts like providing various water flows and substrates in ponds, circular tanks, and raceways. Woundfin have spawned under most treatments, but no method produced more than a few hundred offspring. In 1987, 464 fry were produced from 157 adults. In 1988, 135 adults produced 1703 fry. Fecundity of captive woundfin is about 200 eggs per female; most fish spawn the second spring after hatching, and it appears that most survive two reproductive seasons (D. Hales, pers. comm.). On 12 May 1987 natural woundfin spawning was observed in two raceways, and fry were first observed on 19 May. This is similar to earlier observations at the University of Nevada, Las Vegas, where spawning in an artificial stream was first noted on 3 May 1980, and young were first observed on 14 May; water temperatures varied between 19° and 26°C (Greger and Deacon 1982).

Production methods for woundfin will continue to be tested at Dexter NFH until sufficient numbers of fish can be produced to initiate a

Fig. 13-4. Chihuahua chub, about 25 cm total length, from the Mimbres River, New Mexico, brood stock at Dexter National Fish Hatchery. Photograph by B. L. Jensen, November 1981.

reintroduction effort. In 1985, five experimental, nonessential woundfin reintroduction sites were identified in central Arizona (Hassayampa, Verde, Salt, and Gila rivers, and Tonto Creek), but no fish have been stocked (J. E. Johnson 1987b). Asian tapeworm infestations in the Virgin River prevent direct stocking of woundfin into these sites for fear of spreading the parasite. At Dexter NFH, adult woundfin are treated for tapeworms, but only their offspring will be reintroduced, and then only after they have been certified not to harbor the parasite.

Gila nigrescens (Chihuahua Chub; Fig. 13-4)

An endemic minnow of the Guzman Basin, mostly in Chihuahua, Mexico, the Chihuahua chub reaches its northern distributional limit in the Mimbres River of southwestern New Mexico (R. R. Miller and Chernoff 1980). Koster (1957) and R. R. Miller (1961) believed the chub to be extirpated from the Mimbres River (and the United States) due to loss of

required pool habitats, but in 1975 the species was rediscovered in a small spring-fed tributary (Rogers 1975), 124 years after it had last been collected in the system; the population numbered fewer than a hundred individuals. Following extensive flooding in December 1978, all the remaining wild fish that could be found (ten) were moved to Dexter NFH. Not all were removed, however, nor was the habitat completely destroyed, because a population numbering perhaps a few hundred fish reestablished itself during a later period of less violent flooding and greater stream stability (USFWS 1986a).

At Dexter NFH, captive Chihuahua chubs in ponds have been slow to increase in numbers through natural recruitment, but as no reintroduction effort has been authorized, no induced spawning has been attempted. By 1987, numbers had increased to 373 fish, likely more than survive in the Mimbres River, especially after a series of severe floods in June 1988 (unpub. data). An attempt in 1987 to increase genetic heterogeneity of the captive population was unsuccessful when 21 new fish brought from the Mimbres River population into the hatchery failed to survive.

Several Chihuahua chubs from the Río Piedras Verdes, Chihuahua, Mexico, were transferred into Dexter NFH in 1978 as backups to Mimbres River fish. Because of possible differences between the two stocks (R. R. Miller and Chernoff 1980), and because Mimbres River fish did well, the stocks were not mixed and the Mexican fish were eliminated from the hatchery in 1985.

Recently, an initiative was begun to introduce Chihuahua chubs into Galinas Creek, a tributary of the Mimbres River on Gila National Forest lands (USFWS 1986a). Several other sites in New Mexico have been considered and rejected by the USFWS, U.S. Forest Service, and New Mexico Department of Game and Fish. No stocking has been considered for the mainstream Mimbres River because of the population that persists there. It has been a policy of the USFWS not to supplement an established wild population of an endangered fish species from captive stocks because of the chance of introducing disease, parasitic organisms, or genetically inferior fish into the population. The only exception to this policy has been the stocking of bonytail in Lake Mohave (see above), where little or no natural recruitment seems possible.

Cyprinodon spp. (Pupfishes)

Several species of pupfishes have been maintained at Dexter NFH over the past sixteen years, including Comanche Springs, Leon Springs, and desert pupfishes (*Cyprinodon m. macularius*). Comanche and Leon springs were two of the largest springs in west Texas before they were lost to water-table declines from groundwater pumping (Brune 1975). The two pupfish endemic to those springs also occupied adjacent surface waters, from which they were transferred to Dexter NFH in 1974 to maintain captive gene pools in case the wild populations were lost (USFWS 1981b, 1985b).

In 1974 a non-native pupfish (sheepshead minnow, *C. variegatus*) was illegally introduced into Leon Creek, the only remaining natural habitat of the Leon Springs pupfish, and it began to hybridize with the native species (C. Hubbs et al. 1978; C. Hubbs 1980). Rapid and drastic action was initiated, and rotenone was applied to the entire 8 km of permanent water in Leon Creek on 13 February 1976 in an attempt to remove all pupfish that possessed *C. variegatus* characteristics (C. Hubbs et al. 1978). Pupfish that appeared to be pure *C. bovinus*, as well as endangered Pecos gambusia and other fishes and invertebrate species native to the creek, were held for reintroduction in nearby temporary refuges until the stream detoxified. The rotenone application failed to remove all the *C. variegatus*–like individuals, so on four separate occasions between November 1976 and August 1978, determined seining efforts were fielded during which all suspect pupfishes were removed; only those individuals with *C. bovinus* phenotypes were returned to the water. After August 1978, either as a result of artificial (seine/rotenone) or natural selection (C. Hubbs 1980), only *C. bovinus* phenotypes could be found, and this situation persists today. In this instance recovery efforts appear to have protected the wild population, and they were far less hazardous to the endangered species because of backup stocks at Dexter NFH.

Cyprinodon elegans is endemic to two isolated spring systems in southwest Texas (USFWS 1981b). The type locality, Comanche Springs, Pecos County, ceased to flow in 1955. The species is presently restricted to several springs (principally San Solomon, East Sandia, Phantom Lake, and Giffin springs) and their collective outflows in and around Balmorhea State Park, Reeves County. In 1974 the Texas Department of Parks and Wildlife

Fig. 13-5. Desert pupfish, about 40 mm total length, progeny of brood stock from Santa Clara Slough, Sonora, Mexico. Photograph by R. Clarkson, August 1980.

constructed an artificial pupfish canal at Balmorhea State Park specifically for *C. elegans* (A. A. Echelle and Hubbs 1978). This on-site refuge, which uses water from San Solomon Spring, combined with the presence of the species in nearby Phantom Lake and Giffin springs, appeared to have reduced the threat to the species sufficiently that Comanche Springs pupfish were removed from Dexter NFH in 1986.

Prior to 1968, sheepshead minnows were illegally introduced into Lake Balmorhea, an irrigation reservoir that collects water from several of the Reeves County springs. Comanche Springs pupfish that were swept down the main irrigation canal into Lake Balmorhea readily hybridized with the non-native species (Stevenson and Buchanan 1973), but neither *C. variegatus* nor the hybrids were able to ascend the rapid waters of the canal to contaminate the springhead populations. In 1988, however, *C. variegatus* was discovered in East Sandia Spring (A. A. Echelle, pers. comm.), probably the result of another illegal transfer, and it now threatens to contaminate the entire wild population of *C. elegans*. Efforts are under way to assess the extent of *C. variegatus* dispersal and determine methods to remove it from the spring systems, their feeder canals,

and Lake Balmorhea. A stock of *C. elegans* was moved back into Dexter NFH in 1989.

The desert pupfish (Fig. 13-5) is endemic to the Gila River basin, lower Colorado River, and Salton Sea/Laguna Salada basins in Arizona, California, and northern Mexico (*C. m. macularius*), Quitobaquito Spring, Arizona (*C. m. eremus*), and the neighboring Río Sonoyta in Mexico (possibly another, undescribed subspecies [R. R. Miller and Fuiman 1987; Schoenherr 1988; Hendrickson and Varela 1989]). *Cyprinodon m. macularius* had been extirpated from the United States since 1950 (Minckley 1973), until specimens were obtained from springs and lagoons on the northeast side of the Colorado River delta in Sonora, Mexico, and brought to ASU and Dexter NFH for propagation.

In 1987 Hendrickson and Varela (1989) found desert pupfish to be far more common and widespread than expected on the Colorado River delta in Mexico. They attributed the increased abundance to the effects of uncontrolled flows of the Colorado River in 1983 and 1984. As the water receded, vast areas of marginal habitat were created in which the pupfish does well. Hendrickson and Varela (1989) predicted a rapid contraction in desert pupfish range and abundance as

Fig. 13-6. Pecos gambusia from Balmorhea State Park, Texas, male (above) and female (below). The female is about 35 mm total length. Photograph by J. E. Johnson, July 1985.

water levels declined. The Colorado River is again under control of upstream dams, and essentially all its water is diverted for irrigation and domestic water supplies; deltaic habitats will disappear rapidly under this regimen.

Desert pupfish have done well at Dexter NFH with no special handling, and they have been reintroduced in Arizona into eight natural or seminatural habitats within their historic range and two outside it (in the Bill Williams River system). In 1987 pupfish remained at two of the former and one of the latter sites (J. E. Williams et al. 1985; Hendrickson and Varela 1989). Schoenherr (1988) listed six refuge populations of Salton Sea stocks of desert pupfish that were established as an independent effort by the state of California.

Poeciliid Fishes (Live-bearers)

Most poeciliids do well at Dexter NFH and reach population maxima during the late summer months. Cold winters decimate most species, however, and may reduce populations to less than a tenth of their summer maxima. It is not unusual to begin the spring reproductive period with only a few hundred fish of each species, but populations quickly rebuild. Such fluctuations appear common in wild poeciliid populations as well (C. Hubbs and

Broderick 1963; J. E. Johnson and Hubbs 1989).

Six taxa of live-bearing fishes have been housed at Dexter NFH, but now only two species are present. Three founding species brought to the facility were Pecos gambusia, Big Bend gambusia, and Amistad gambusia, all from Texas. The last was lost from the wild in 1968 when its only known habitat, Goodenough Spring, Val Verde County, was inundated by the filling of Amistad Reservoir on the Rio Grande (Peden 1973). A stock of the then-undescribed poeciliid was obtained as Goodenough Spring was flooded; the fish were moved to the University of Texas, Austin, and then in 1974 to Dexter NFH (J. E. Johnson and Hubbs 1989). As no natural habitat remained for the species within its historic range, no reintroduction efforts were initiated. Sometime prior to 1979, both the Austin and Dexter NFH stocks became contaminated by mosquitofish, and the species was forced to extinction (C. Hubbs and Jensen 1984).

Pecos gambusia (Fig. 13-6) occupy four isolated, spring-fed areas adjacent to the Pecos River in New Mexico and Texas (J. E. Williams et al. 1985). The species was originally brought to Dexter NFH because little information was available on its status in nature. As more data were accumulated (A. A. Echelle and Echelle 1980; USFWS 1983d), the USFWS decided that the species was secure enough to remove it from the hatchery (J. E. Johnson and Hubbs 1989). At about the same time, A. A. Echelle et al. (1983) identified three groups of Pecos gambusia in the wild, with the most divergent population approaching the species level of differentiation.

The example of the Pecos gambusia brings up an interesting question. How should a genetically diverse species be protected at a refuge like Dexter NFH, which has only a limited amount of available pond space? Pecos gambusia from Phantom Lake Spring are the most heterozygous of the four populations, and certainly are the most endangered as that spring continues to fail (Brune 1975; A. F. Echelle et al. 1989). Other stocks are less heterozygous, but each has unique features. Should populations of each of the four spring areas be held to ensure protection of genetic diversity? Should populations be mixed to create a "superguppy" (Vrijenhoek et al. 1985; Meffe and Vrijenhoek 1988)? Or should one population be selected to represent all the others? Also, if the Phantom Lake Spring gambusia is saved but the spring is lost, how can the surviving individuals be used to effect recovery? No answers are available to these questions, nor are any efforts under way to move any stock of Pecos gambusia back into Dexter NFH.

Of the three founding *Gambusia* species at Dexter NFH, only *G. gaigei* still faces a severe threat to its survival in nature. Big Bend gambusia have proven to be one of the most difficult species to maintain at Dexter NFH. The species is endemic to a series of thermal springs that rise along the Rio Grande in Big Bend National Park (C. Hubbs et al. 1977; J. E. Williams et al. 1985). Ambient water temperatures in these springs vary from 20° to 50°C, with the gambusia inhabiting those between 20° and 34°C (J.E. Johnson and Hubbs

1989). During some winters, water temperatures at Dexter NFH fall substantially below the lower tolerance limit for *G. gaigei*, even though the kettle portions of hatchery ponds remain near 18°C. Since Big Bend gambusia were brought to Dexter NFH in 1974 it has been necessary to obtain additional wild stocks five times. A recently established procedure moves several hundred fish indoors into heated aquaria in winter. This has maintained the species since 1986, but selection of surviving individuals by USFWS personnel rather than through natural means may not be a good management procedure. The species continues to survive principally because of its ability to live in artificially ponded spring waters at Big Bend National Park (C. Hubbs et al. 1977; USFWS 1984b; J. E. Johnson and Hubbs 1989).

Two subspecies of Sonoran topminnow—*Poeciliopsis o. occidentalis* (Gila topminnow) and *P. o. sonoriensis* (Yaqui topminnow)—have also been held at Dexter NFH (USFWS 1984c; J. E. Johnson and Hubbs 1989). Gila topminnows are endemic to the Gila, Concepción, and Sonora rivers of southern Arizona and northern Mexico, and Yaqui topminnows to the Río Yaquí of Mexico and extreme southeast Arizona (Vrijenhoek et al. 1985).

Yaqui topminnows were held at Dexter NFH until their natural habitat in the United States (San Bernardino Creek and surrounding springs) was protected by USFWS land acquisition (San Bernardino NWR; J. E. Johnson 1980a; J. E. Williams et al. 1985). Gila topminnows were brought to Dexter NFH as part of a major reintroduction effort on national forest lands in Arizona (Brooks 1985; Simons 1987; J. E. Johnson and Hubbs 1989; Simons et al. 1989). Starting in 1981, one hundred habitats were stocked with Gila topminnows from Dexter NFH or the Boyce Thompson Arboretum (the latter a topminnow brood stock maintained by AZGFD). Fish at approximately a third of the reintroduction sites survived for more than five years, prompting the USFWS to initiate steps to downlist both subspecies to threatened status (see, however, Simons et al. 1989). More recently, the U.S. Bureau of Land Management has also become interested in reintroducing Gila topminnows and has initiated a pilot program that should expand to more than twenty sites within two years.

All the original Dexter NFH stock of Gila topminnows came from the Boyce Thompson Arboretum stock, which in turn came from Monkey Springs, Santa Cruz County, Arizona (Vrijenhoek et al. 1985), and perhaps from Cocio Wash, Pinal County, Arizona (J. E. Johnson, unpub. data). Vrijenhoek et al. (1985) found no genetic variability in the Monkey Springs fish and suggested that a more heterozygous population should be used for reintroduction attempts. Acting on this recommendation, the USFWS eliminated the existing captive stock of Gila topminnows in 1985 and replaced it with fish from Sharp Spring, Santa Cruz County, Arizona, that demonstrated greater genetic diversity (J. E. Johnson and Hubbs 1989). Survival of reintroduced Gila topminnows from this more heterozygous stock can now be compared with the success of the earlier stockings to test the importance of genetic diversity in establishing and perpetuating populations.

Summary and Conclusions

This paper reviews the short history of attempts to protect native southwestern fishes by bringing selected species into captivity. Dexter National Fish Hatchery, Dexter, New Mexico, has been the center of that effort since 1974, serving both as a refuge and a production center for reintroducing imperiled native fishes back into their natural habitats. While not all recovery efforts have used the facilities at Dexter, all the recovery plans for these species call for their protection away from wild habitats, and almost all recommend captive propagation for eventual reintroduction. In addition, many recovery efforts for these species, including habitat renovation and reclamation of historic waters, would not have been attempted without Dexter NFH underwriting the survival of species through a protected gene pool. This has allowed biologists to concentrate on recovery instead of the more urgent demands of simply ensuring survival. Fish from Dexter NFH have also been made available for research, eliminating impacts of removing specimens from wild populations.

The successful program at Dexter NFH has engendered requests for help to establish similar programs elsewhere. Personnel from Dexter NFH have assisted in development of a razorback sucker–rearing program at Ouray NWR, Utah, and have worked with the Pyramid Lake Paiute Indian tribe in Nevada to culture endangered cui-ui (*Chasmistes cujus*). They also helped set up the hatchery facility and culture program for the Klamath Indian tribe's Lost River sucker (*Deltistes luxatus*) and shortnose sucker (*Chasmistes brevirostris*) recovery efforts in Oregon, and advised personnel from the Universidad Autónoma de Nuevo León and the state of Nuevo León, Mexico, regarding refugia for their imperiled fishes.

Recovery efforts for endangered fishes have little past knowledge to draw upon, and many attempts may not succeed. As knowledge of recovery successes and failures continues to grow, a higher proportion of successes may be anticipated, but the scientific community and the public in general must not be discouraged by failures, even when they may lead to an occasional extinction. When species that are facing extinction are grist for the recovery mill, some failures must be anticipated. Captive propagation will help to reduce the failure rate while providing additional data on the needs of species and safeguarding their gene pools.

Chapter 14

Reclamation and Alteration as Management Techniques, and a Review of Methodology in Stream Renovation

John N. Rinne and Paul R. Turner

Introduction

The native fish fauna of western North America is depauperate in numbers of species but rich in endemism (R. R. Miller 1959; Minckley et al. 1986). Despite adaptations to some of the harshest aquatic environments in the world (R. R. Miller 1946b; Deacon and Minckley 1974; Naiman and Soltz 1981), western fishes have been unable to withstand the inroads of human activities. Many are imperiled; some are already extinct (J. E. Johnson and Rinne 1982; J. E. Williams et al. 1985, 1989; R. R. Miller et al. 1989). This problem was explicitly recognized in the Endangered Species Preservation Act of 1966, and subsequent legislation continues to address the ever-increasing declines of animal and plant species and their habitats (Williams and Deacon, *this volume*, chap. 7).

Three primary management activities resulting from the Endangered Species Act (ESA) of 1973 are listing, protection, and recovery (J. E. Johnson and Rinne 1982). The first two officially initiate and monitor the conservation process for rare species. The last, recovery, is the key to effective reversal of their declines. As a result of the ESA and its amendments, documents termed "recovery plans" were drafted by multiagency "recovery teams." Such plans began appearing for fishes in 1975

(*Devils Hole Pupfish Recovery Plan*, U.S. Fish and Wildlife Service [USFWS] 1977), and currently, forty-seven (61%) of seventy-seven endangered and threatened fishes listed for the United States have recovery plans in place (U.S. Department of the Interior 1989).

The objectives of most recovery plans include habitat concerns (Table 14-1). Thus, recovery efforts for imperiled taxa have followed a philosophy that "as goes the habitat, so goes the species." Natural aquatic ecosystems in the West are rapidly disappearing (J. E. Williams et al. 1985). This paper reviews activities in habitat improvement, alteration, and renovation for preservation and conservation of selected native western fishes. We also review the state of the art in chemical renovation techniques. Finally, procedures are delineated for future endeavors, to facilitate more effective management of native fishes and prevent further extinctions of this valuable, already limited, resource.

Management Strategies

Salmonid fishes have received the most attention of any group in the American West because they dominate as game and food fishes. Recent conservation efforts for salmonids also arise from the recognition of their remarkable diversity. About fifty nominal species of trout

Table 14-1. Habitat and renovation considerations included in recovery plans for threatened and endangered species of the western United States (compiled from recovery plans current in 1988).

Species	Habitat proposals			
	Maintenance	Enhancement	Alteration	Restoration
Apache trout	X	X	X	X
Gila trout	X	X	X	X
Greenback cutthroat trout	X	X	X	X
Lahontan cutthroat trout	X	X	X	X
Little Kern River golden trout	X	X	X	X
Paiute cutthroat trout	X	X	X	X
Borax Lake chub	X	X	X	X
Chihuahua chub	X	X	X	X
Colorado squawfish	X	X	X	X
Humpback chub	X	X	X	X
Pahranagat roundtail chub	X	X	X	X
Mohave tui chub	X	X	X	X
Moapa dace	X	X	X	X
Woundfin[2]	X	X		X
Cui-ui	X	X	X	X
Comanche Springs pupfish[3]	X	X		
Devils Hole pupfish	X	X		X
Leon Springs pupfish[3]	X	X		
Owens pupfish	X	X	X	X
Warm Springs pupfish	X	X	X	
Pahrump poolfish	X	X	X	
Clear Creek gambusia	X	X		X
Pecos gambusia	X	X		
Sonoran topminnow	X	X	X	X
Unarmored threespined stickleback	X	X	X	X

[1]1, congeners, hybridizing with native form; 2, predatory fishes; 3, competitive species of fishes; and 4, other organisms (frogs, crayfishes) suspected of interacting to the detriment of the native form.
[2]Woundfin recovery plans were altered in 1988 to include rotenone application to the Virgin River for eradication of red shiner; see text for discussion and further explanation.
[3]Both species have been (Leon Creek, Texas) or may soon be (Balmorhea area, Texas) adversely affected by introduction of sheepshead minnow (*Cyprinodon variegatus*), which has (or may) necessitated use of renovation and restocking to preserve the species (Echelle, *this volume*, chap. 9, and memorandum to file [3 August 1988], Dexter National Fish Hatchery, New Mexico).

| Poisoning planned | | Target |
Yes	No	organisms[1]
X		1, 2, 3
X		1, 2, 3
X		1, 4
X		1, 2, 3
X		1, 2, 3
X		1, 4
	X	
	X	
	X	
	X	
X		2, 3, 4
X		1, 2
X		2, 3, 4
	X	
	X	
	X	
	X	
X		2, 3, 4
X		2, 3, 4
X		2, 3, 4
	X	
X		2, 3
X		2, 3, 4
X		2, 3, 4

have been described from the region. Various interpretations of this array of forms (which is due to a general lack of reproductive isolation) have resulted in contradictory classifications that have hampered effective management. Behnke and Zarn (1976) and Behnke (1979) emphasized this genetic diversity and its implications for management. Both Behnke (1979) and Rinne (1988b) suggested that these especially adapted forms represent unique evolutionary units. Accordingly, rare species, subspecies, and even races are now being designated for perpetuation in a variety of habitats (USFWS 1979a, b). It is clear that other, nonsalmonid, nongame species of fishes should be managed with the same philosophy (Meffe et al. 1983; Vrijenhoek et al. 1985; Minckley and Douglas, *this volume*, chap. 1). From physical alteration (commonly termed "stream improvement") to augmentation as extreme as creation of artificial spawning channels or hatcheries, habitat management is a fundamental consideration in recovery for imperiled native fishes.

Physical Habitat Alterations

Stream improvement—changing physical features to increase or enhance habitat for selected fishes—has been extensive in the western United States (Needham 1936; Duff et al. 1988). Most efforts have been, and will likely continue, on a comparatively small scale (Platts and Rinne 1985), each performed as a local "solution" to a perceived problem. Habitat enhancement for rare species (Table 14-1) may either be focused and limited (designed to reduce an immediate threat of degradation and eventual species loss) or relatively diffuse but major in terms of time and money for longer-term scenarios. In order to estimate the extent of the use of stream improvements and reclamation for past and present management of western fishes, we contacted game and fish departments in thirteen western states; their responses form a partial basis for this report.

Improvement structures

Extensive stream improvement on federal lands in the West was instituted in the early 1930s by the Civilian Conservation Corps (CCC) (H. S. Davis 1936), mostly through installation of single-log and multiple-log structures (Fig. 14-1), rock gabions, trash catchers, and concrete-block and rock-boulder structures in ways that increased numbers, sizes, and depths of pools (Jester and McKirdy 1966; Duff 1980). At the onset, H. S. Davis (1936) questioned the value of such habitat improvements in terms of both cost-benefit and their often myopic, Band-Aid approach. Mullan (1962) and Maughan and Nelson (1982) more recently echoed these comments.

McKirdy (1964) examined 1600 structures installed by the CCC, and reported 470 (29%) functioning with no maintenance after thirty years. Tarzwell (1938) and Jester and McKirdy (1966) reported positive benefits to stream habitat, fishes, and invertebrate fish foods. However, Carufel (1964) evaluated more than two hundred structures on nine Arizona streams and could substantiate claims of neither increased fish populations nor of greater angling pressures or successes. Rinne (1981) reported increased habitat availability and greater numbers of Gila trout (*Oncorhynchus gilae*) in areas with structures but alluded to cost of installation as an increasing problem in their use. Attrition rate of structures on several streams varied from 25% to 100% in five years, so maintenance also became prohibitive. The costs of CCC structures are unknown, but were likely $25 or less per unit. Structures cost about $100 each in the 1950s and 1960s. Rinne (1981) reported costs of $600–$700 each in the Mimbres River system in New Mexico. Today, cost per unit conceivably approaches $1000 unless volunteer labor is available.

Little ongoing installation of structures was reported by agencies contacted by questionnaire, and we do not specifically present those data. However, we suspect that such projects are done (as informally reported by Wyoming) on a day-to-day manner in many areas. Most data in agency files are not generally available; for example, hundreds of projects completed in California (E. Gerstung, California Department of Fish and Game [CADFG], pers. comm.) are undocumented other than for general locale and accounting of funds expended. Most habitat improvement in Washington has been directed toward Pacific salmons, a group we excluded from consideration.

The only report of a structure built to assist a nongame fish was from Wyoming, for the endangered Kendall Warm Springs speckled dace (*Rhinichthys osculus thermalis*; Binns 1978). Yet, with funds provided by the USFWS through the University of Nevada, Las Vegas, the state of Nevada built a dam creating a refuge for White River springfish (*Crenichthys baileyi*) as early as 1967 at Hot Creek, and many other such projects are known (and reported in the literature) for California, Arizona, Oregon, and elsewhere (Williams, *this volume*, chap. 11).

Barriers

Isolation above waterfalls or other physical barriers, and even the simple presence of long, dry reaches downstream from headwaters (Fig. 14-2), have proven a boon for native fishes, especially those which suffer detrimental interactions with non-native species. Genetically uncontaminated (unhybridized) stocks of native trout are often found only above such barriers (Fig. 14-3; Mello and Turner 1980; Rinne and Minckley 1985), and relict populations of a number of nongame species are similarly restricted to such refugia. It is therefore not surprising that barriers are a part of recovery programs.

Stone-concrete barriers superimposed on existing bedrock features were used in McKnight, Iron, and Little creeks for Gila

A

B

Fig. 14-1. Single (A) and multiple (B) log stream-
improvement structures in McKnight Creek, New
Mexico. Photograph by J. N. Rinne, August 1977.

Fig. 14-2. Dry reaches of streams such as this in McKnight Creek, New Mexico, serve as partial to highly effective barriers to upstream movement of undesired salmonid species. Photograph by J. N. Rinne, July 1978.

trout in New Mexico (USFWS 1979a). Other streams supporting this species have long, generally dry reaches (Main and South Diamond and McKnight [Fig. 14-2] creeks) or natural waterfalls (McKenna, Spruce, and Big Dry creeks) that prevent upstream movement of non-native salmonids (Mello and Turner 1980).

In Ord Creek, Arizona, an Apache trout (*Oncorhynchus apache*) population was assumed to be protected by a 3-m log barrier constructed in 1964. By the mid-1970s, how-

ever, brook and brown trout (*Salvelinus fontinalis, Salmo trutta*) invading (or moved by anglers) from downstream had increased to constitute 90% of the fish population above the barrier (Rinne et al. 1982). A secondary 2-m rock-filled gabion was built prior to reclamation of the stream. A cinder-block barrier was constructed in 1979 on Lee Valley Creek, and gabions have been built between 1980 and 1987 on Centerfire, Snake, Fish, Home, and Hay creeks (Apache-Sitgreaves National Forest, Arizona) to guard more than 80 km of Apache trout habitat from intrusion by nonnatives.

Artificial barriers vary from complex and costly to simple and relatively inexpensive. Schaeffer and Templeton barriers to protect golden trout (*O. aguabonita*) on the South Fork of Kern River, California, originally cost $100,000, and nearly another $40,000 was later required to update and maintain them (R. Giffen, U.S. Forest Service, pers. comm.). The cost of artificial barriers in Arizona and New Mexico varied from $1000 (Lee Valley Creek) to $25,000 (Iron Creek). Almost 30 m^3 of material blasted from a steep bedrock run in Canones Creek, New Mexico, protected a native population of Rio Grande cutthroat trout (*O. clarki virginalis*) at the nominal cost of explosives (Rinne and Stefferud 1982). Artificial barriers have also been applied in recovery efforts for other subspecies of cutthroat trout in Wyoming and Colorado, but no estimates of costs or effectiveness are available.

Barriers have also been built to assist endangered Sonoran topminnows (*Poeciliopsis o. occidentalis, P. o. sonoriensis*) in Arizona, with varying success (see Minckley et al., *this volume*, chap. 15). On the San Bernardino National Wildlife Refuge, Arizona, barriers for attempted management of *P. o. sonoriensis* consisted of vertical and inclined water drops varying from half a meter to ten or more meters high to isolate springs, artesian wells,

Fig. 14-3. (A) Natural falls on the East Fork White River, Arizona, have succeeded in preserving a genetically pure population of Apache trout at its type locality. (B) Falls on Chitty Creek, Arizona, protected a native population, possibly of Gila trout, for years, but were finally bypassed by introduced rainbow trout, and the indigenous form was genetically swamped. We suspect anglers transferred fish upstream on Chitty Creek. Photographs by J. N. Rinne, August 1978 and 1979, respectively.

and ponds from mosquitofish invasions, which jeopardized topminnows. Habitats were renovated, restocked with topminnows, and reinvaded by mosquitofish with discouraging consistency over a period of seven years; all the presumed barriers failed to be effective. In this case, unauthorized transfers of fish by the visiting public may have been a factor, but dispersal of mosquitofish during heavy rainfall through overland sheet flow or up substantial inclines (greater than a meter high) was also observed (B. Robertson, USFWS, pers. comm.). Their presence upstream from the refuge, and transport downstream during flood, should also be investigated. Mosquitofish share remarkably high capabilities for dispersal, colonization, and survival with common carp (*Cyprinus carpio*), bullhead catfishes (*Ameiurus* spp.), and green sunfish (*Lepomis cyanellus*), all of which have proven to be undesirable "weeds" introduced to western North America.

Major Installations: Hatcheries and Fishways

Numerous salmonid fishes, from regional runs of economically important Pacific salmon to local and isolated subspecies of cutthroat and other trouts, are maintained by capturing adults that enter now-degraded spawning streams, stripping and fertilizing their gametes, rearing young in hatcheries, and restocking as fingerlings. Such is practiced for Eagle Lake rainbow trout (*Oncorhynchus mykiss aquilarum*) at Eagle Lake, California; Bonneville cutthroat (*O. clarki utah*) of Bear Lake, Idaho-Utah; and others (W. F. Sigler and Sigler 1987; Moyle and Sato, *this volume*, chap. 10). Stocks of Colorado squawfish (*Ptychocheilus lucius*), bonytail (*Gila elegans*), razorback sucker (*Xyrauchen texanus*), and other species are now being propagated in hatcheries to be reintroduced into the wild (Rinne et al. 1986; Johnson and Jensen, *this volume*, chap. 13). Installation of a fishway

for native Colorado squawfish is being considered as part of water development in the Colorado River basin (Tyus, *this volume*, chap. 19). One of the most significant alterations of stream habitat for conservation and management of a native nonsalmonid fish is described by Scoppettone and Vinyard (*this volume*, chap. 18) for cui-ui (*Chasmistes cujus*), an endangered, stream-spawning sucker of Pyramid Lake, Nevada.

Biological Manipulation: Reclamation (Poisoning) of Systems

A second major activity that enhances the prospects for survival of native fishes is removal of competing or predatory, typically non-native, fishes (R. R. Miller and Pister 1971; Rinne et al. 1982; Meffe et al. 1983; Meffe 1983b). Histories of some of these efforts are reviewed here as an aid in planning stream renovation projects in native fish management. Despite extensive use of fish toxicants in the American West (Tables 14-2, 14-3), the literature is woefully lacking in documentation of extent, results, or even the techniques used in such operations. In fact, most data are available only in the "gray literature," a formidable volume of processed reports in state game and fish department and federal agency files, or in the field notes or memories of persons involved.

Removal of undesirable species of fishes through use of piscicide was first used as a management tool in the United States in 1934 in Michigan (Krumholz 1948), where, under the supervision of Milton B. Trautman, two small private ponds were treated with powdered derris (0.04–0.1 mg l^{-1}) to remove common carp and goldfish (*Carassius auratus*). Since that first attempt, innumerable projects have been designed and executed to eliminate native "trash" or "rough" fishes (many now imperiled) and establish or enhance game fish.

The first project to bring such practices into the public eye was initiated in 1962 on the upper Green River, Colorado-Utah-Wyoming (Holden, *this volume*, chap. 3). "The Green River incident" fueled a controversy that contributed to a growing concern for native animals. R. R. Miller (1963, 1964a) reviewed the Green River operation from the perspective of native species and issued a plea for judicious use of poisons. C. Hubbs (1963) also questioned the use of rotenone to improve sport fishing, noting that it may have serious ecological impacts on native faunas. Others defended the Green River operation as necessary, expedient, and justified (Holden, *this volume*, chap. 3). Passage of federal endangered species legislation in the 1960s prompted various state laws that collectively mandate the protection of native faunas from human activities, including toxicants. Reclamation projects designed to enhance both nongame and game fishes in the West nonetheless continue to be enthusiastically pursued.

Based on responses to our questionnaire, Arizona, California, and Colorado have not only been active in renovations but have also maintained in-house records of results (Tables 14-2, 14-3). Many streams (more than two hundred) have been treated, and if one assumes an average of 12 km per renovation (calculated for sixteen Apache trout reclamation projects in Arizona), more than 2400 stream km have been poisoned in these three states alone. Additional activities in a stream in Montana (13 km), the Green River project (715 km intended, almost 800 km actually treated at some level), the Gila River in Arizona (at least 48 km), and 160 km of the San Juan River in New Mexico (Olson 1962a), increase the total to nearly 4000 km, a distance equivalent to that from Boston to Phoenix by highway. By comparison, Lopinot (1975) reported treatment of about 11,000 km of streams in the midwestern United States between 1963 and 1972.

Extent and results of piscicide use

We spent substantial time in a "paper chase" for data on stream renovations in Arizona to document the extent of poisoning as well as to evaluate actual and potential impacts. Arizona was selected not because of its special importance but because of our longer and more extensive experience with its fishes, published and unpublished sources of information, and personnel who performed the renovations.

Use of piscicides in Arizona (Fig. 14-4) began early in the history of the technique and was mostly directed toward trout management. By the 1950s Hemphill (1953, 1954) had used toxaphene to eradicate fish in Lyman and Becker lakes (both in the Little Colorado River basin) and San Carlos Reservoir on the Gila River. The first treatment was disrupted by an unexpected runoff dilution of the 0.1 mg l^{-1} application to about 0.05 mg l^{-1}. Lyman Lake was then drained to assess the effectiveness of the operation. Results downstream, in the Little Colorado River, were unrecorded. Targets of the treatments in all three habitats were introduced species—carp and yellow perch (*Perca flavescens*) in Lyman and Becker lakes, and stunted crappie (*Pomoxis* sp.) and yellow bass (*Morone mississippiensis*) in San Carlos Reservoir. The only native species noted was roundtail chub (*Gila robusta*; reported as "bonytail chub" by Hemphill [1954]; see Minckley [1973]), which was numerous in Lyman Lake. Hemphill (1953) also noted that attempts to use toxaphene in flowing waters "have been markedly unsuccessful due to its slow action," testimony that applications in streams were being practiced at the time.

Between 1958 and 1962, stream renovations in Arizona were designed to eliminate native nongame fishes thought to deter the establishment and production of trout. The earliest involved 18.5 km of the Black River, from

Table 14-2. Summary of responses from some western states on poisoning of stream habitats, 1950–1988.

State	Streams, years, km treated	Target species	Species of concern
Arizona[1]	Black R., 1958, 18.5 km	native suckers	rainbow trout
	Gila R., 1960, 48 km	"rough" fish, common carp	warmwater sport fishes
	Little Colorado R., 1961, 23 km	common carp	rainbow trout
	Wet Beaver Cr., 1962, 16 km	smallmouth bass, native fishes	rainbow and brown trouts
	Chevelon Cr., Woods Can. Lk., 1968, 30 km	golden shiner	rainbow trout
California	Misc. streams, pre-1972	"rough" fishes	"non-native" trouts
	Misc. streams, 1970–present	non-native and hybrid trouts	Lahontan, Paiute, "cutthroat" trouts
	Little Kern R., 1970–present	rainbow and hybrid trouts	Kern R. golden trout
	S Fk. Kern R., 1970–present	rainbow trout	Kern R. golden trout
	McCloud R., 1970–present	rainbow trout, non-native fishes	redband trout, Modoc sucker
	Owens R. basin, 1969–present	non-native fishes	Owens pupfish
Colorado	53 streams, pre-1949	—	sport fishes
	45 streams, 1950–1959	—	sport fishes
	23 streams, 1960–1969	—	sport fishes
	30 streams, 1970–1979	—	sport fishes
	76 streams, 1980–1988	—	"native trout"
	nine streams, 1980–1988	brook, brown, rainbow, hybrid trouts	greenback, Colorado River, Rio Grande cutthroat trouts
Montana	lakes only, 1972	—	westslope cutthroat
	Elkhorn Cr., 1980, 13 km	—	"cutthroat" trout
	four lakes, no date	hybrid trout	westslope cutthroat
New Mexico	23 streams, 1948–1962, 539 km	"rough" fishes	sport fishes
	Rio Grande, Caballo Res., 1960	"rough" fishes	sport fishes
	San Juan R. and tribs., 1961, 160 km	"rough" fishes	sport fishes
	Bonita and Negrito cr., 1961	"rough" fishes	sport fishes
	Ute Lk. watershed, 1962, 104 km	"rough" fishes	sport fishes
	McKnight Cr., 1980, 8 km	Rio Grande sucker	Gila trout
	six streams, 1981–1988, 42 km	brown, rainbow trouts	Gila trout, Rio Grande cutthroat trout
Utah	Green R., 1962, ca. 100 km	"rough" fish, common carp	rainbow trout
	Virgin R., 1988, 75 km	red shiner	woundfin, Virgin roundtail
Wyoming	Green R. watershed, 1962, 600 km	"rough" fish, common carp	rainbow trout
	Arnica Cr., 1985, 1986, 25 km	brook trout	cutthroat trout

[1]Projects geared toward recovery of Apache trout are presented separately in Table 14-3.

Comments

unsuccessful
unsuccessful
unsuccessful
unsuccessful

unsuccessful

—

—

—

—

—

see Minckley et al., *this
volume*, chap. 15

—

—

—

—

—

—

—

—

—

unsuccessful
unsuccessful

—

—

successful
variable success

see Wyoming, below
see text

short-term success

—

Crosby Crossing to Diamond Rock Lodge, in June 1958. The objective was to remove "mountain suckers" (desert sucker, *Pantosteus clarki*) to reduce perceived competition with a planned stocking of rainbow trout. Approximately 76 l of rotenone was dripped into the river at three stations. The estimated kill was more than 1134 kg of rough fish and 100 kg of trout. Numbers of fry killed (presumably including small species as well as juveniles of larger fishes) were not estimated; they "numbered in the thousands." Two days later, "catchable-sized" rainbow trout were stocked at the rate of 155 per kilometer. Based on "last reports," the treatment was "quite successful and no appreciable numbers of rough fish had been observed" (Foster 1958). Surveys in the 1960s and later indicated that native suckers were again the dominant fish species (W. L. Minckley, Arizona State University, and J. Novy, Arizona Game and Fish Department [AZGFD], pers. comm.).

The next recorded project took place in January 1960. The objective was again to remove native rough fish and carp from San Carlos Reservoir and the inflowing Gila River to enhance warm-water game fishes (Gruenwald 1960). A helicopter dispensed 340 l of 10% emulsified toxaphene to achieve concentrations of 5.0 mg l^{-1} over the reservoir and 48 km upstream on the Gila River main stem. Heavy rainfall increased stream flow, and ultimately the lake volume, from 95 to more than 635 ha-m in a two-week period, and dilution rendered the effort unsuccessful, although concentrations must have been sufficient to kill fishes for a time.

In June 1961, 16–23 km of the Little Colorado River from Lyman Lake to Springerville was treated with 60% toxaphene diluted to 0.5 mg l^{-1} (J. Bruce 1961). The kill consisted largely of carp, Little Colorado spinedace (*Lepidomeda vittata*; now listed as threatened), green sunfish, and small numbers of brown trout. No roundtail chubs were noted,

Table 14-3. Poisoning of Arizona streams for recovery and enhancement of native Apache trout, 1962–1988. Compiled from AZGFD files, Rinne (1985b), and Minckley and Brooks (1985).

Stream, location, and year	Km treated	Target Species	Origin and year of introduction
Mineral Cr., Apache Co., 1962	unknown	brook trout	Ord Cr., Apache Co., 1962, 1967, 1968[1]
Grant Cr., Apache Co., 1963	4.7	rainbow trout	as above
K.P. Cr., Greenlee Co., 1963	18.0	rainbow trout	as above
North Canyon Cr., Coconino Co., 1963[2]	unknown	rainbow trout	Bonito Cr., Apache Co., 1963[3]
North Canyon Cr., 1967[2]	unknown	Apache trout	Ord Cr., Apache Co., 1968, 1978[1]
Grant Cr., Graham Co., 1965[2]	11.0	brook, brown, rainbow trout	Ord Cr., Apache Co., 1968, 1969, 1971
Ash Cr., Graham Co., 1965[2]	11.0	brook, brown, rainbow trout	Ord Cr., Apache Co., 1965, 1968
Marijilda Cr., Graham Co., 1968[2]	14.0	brook and rainbow trout	as above
Ord Cr., Apache Co., 1977	16.0	brown and brook trout	Ord Cr., Apache Co., 1977
Ord Cr., Apache Co., 1980	16.0	brook trout	Paradise Cr., Apache Co., 1980[3]
Hurricane Cr., Apache Co., 1982	8.5	rainbow trout	East Fork White R., Apache Co.[4]
Bear Wallow Cr., Apache Co., 1982	15.0	rainbow and hybrid trout	Soldier Cr., Apache Co.[5]
Bear Wallow Cr., 1987	15.0	rainbow and hybrid trout	as above
Lee Valley Cr., Apache Co., 1982	4.8	brook trout	as above
Lee Valley Cr., 1987	4.8	brook trout	East Fork White R., Apache Co.[4]
Home Cr., Apache Co., 1987	18.0	rainbow trout	as above
Wildcat Cr., Apache Co., 1989	6.5	rainbow trout	as above

[1]Stocks propagated at AZGFD Sterling Springs Hatchery.
[2]Habitat outside the native range of Apache trout.
[3]Now repopulated by brook trout (D. Parker, USFWS, pers. comm.).
[4]Presumably stocked directly from the indicated habitat.
[5]Stocks propagated at USFWS Williams Creek National Fish Hatchery.

Comments
successful, present today
failed, now inhabited by hybrid trout
failed, now inhabited by hybrid trout
successful, removed and replaced (see below)
successful, present today
successful, present today
failed, now inhabited by hybrid trout
failed, now inhabited by brook and hybrid trout
failed
failed
successful
failed, re-treated (see below)
successful, present today
failed, re-treated (see below)
successful, present today
successful, present today
to be stocked, 1989

although they were in Lyman Lake in 1953 (Hemphill 1954).

In May and June 1962, 16 km of Wet Beaver Creek, Yavapai County, were treated to remove smallmouth bass (*Micropterus dolomieui*), "western white sucker" (actually Sonora sucker, *C. insignis*), "Gila mountain sucker" (desert sucker), speckled dace (*Rhinichthys osculus*), and rountail chub (Bassett 1962). Trout plantings had not succeeded the previous year, purportedly due to interactions with smallmouth bass and roundtail chub. The creek is in rugged terrain (canyon walls to 366 m high) and remote, and a helicopter, pack animals, and hiking were required to dispense 170 l of emulsifiable and 9 kg of powdered rotenone. Only "one large intermittent pool above the spring was left untreated; it contained "only speckled dace . . . and mountain suckers," a "possible food source for brown trout at a later date." The project was deemed successful. All but "young western white suckers" were removed (Bassett 1962). More than eleven thousand fingerling brown and rainbow trout were either dropped into the stream from helicopters or brought in by off-road vehicles following renovation. Native fishes, along with trout, smallmouth bass, other introduced centrarchids, and ictalurids were there when Minckley (pers. comm.) sampled in 1966.

In the early 1960s an unauthorized introduction of golden shiners (*Notemigonus crysoleucas*), presumably as bait, followed by their spread to fishing lakes and natural streams of the Mogollon Rim region (Minckley 1973), stimulated renovation in 1968 of Wood's Canyon Lake and the Chevalon Creek system downstream more than 30 km to the new Chevalon Reservoir. Emulsifiable rotenone killed introduced salmonids, golden shiners, and native suckers and minnows, including the last recorded specimen of roundtail chub from the Little Colorado basin (Minckley, pers. comm.). Within two years, golden shiners again occupied essentially the entire drainage and had spread to adjacent drainages, where they continue to be a management problem. Whether they were reintroduced or survived treatment is unknown.

Stream reclamation for maintenance of native Apache trout began in Arizona in 1962 (AZGFD files). That species suffers detrimental interaction with non-native salmonids—hybridization with rainbow and cutthroat trout, predation from brown trout, and apparent competition from brook trout—and soon disappears when those species become abundant (USFWS 1979a). Of fourteen renovations for which we found specific information (Table 14-3), most failures were due to apparent genetic recontamination of stocked native trout by rainbow trout (Rinne 1985b; Rinne and Minckley 1985), either from survivors of treatment or as a result of unauthorized stocking. Successful renovations were in relatively simple (nonheterogeneous) habitats, in remote areas that precluded restocking of alien forms, and in places where natural barriers (normally dry channels) prevented reinvasion by non-native fishes.

In one instance, renovation to establish a threatened trout species outside its natural range destroyed native fishes that may have been unknown to science. "Suckers and minnows" were recorded during rotenone treatment of Grant Creek, on the Pinaleño Mountains; voucher specimens were not preserved. No suckers and only two minnows—longfin dace (*Agosia chrysogaster*) and Yaqui chub (*Gila purpurea*)—had ever been recorded from the endorheic Willcox Playa drainage into which Grant Creek drains. That basin was prehistorically part of the Río Yaqui basin of Mexico rather than the Gila River basin that presently surrounds it on three sides (Hendrickson et al. 1981). A speckled dace, presumably originating with stocked Apache trout, was later captured from Grant Creek

(Minckley, unpub. data). Such an incident is an unfortunate distraction from an otherwise commendable program as well as a loss of irretrievable scientific information of unknown importance.

Record keeping as well as application of more detailed methodology improved after the *Apache Trout Recovery Plan* (USFWS 1979a) called for reestablishment of a diversity of populations within the native range as a recovery goal. For example, Bear Wallow Creek was treated with antimycin by the AZGFD, who used bioassay to determine the frequency of drip stations required to sustain concentrations of 10 μg l^{-1}. A 4.5-l container applied toxicant for about an hour, and drip stations were refilled to achieve at least two hours of application between stations. The piscicide was detoxified with an oxidant (potassium permanganate [KMnO$_4$]) at concentrations of 1.0 mg l^{-1} below an artificial rock-masonry fish barrier.

Use of poisons in lower-elevation, warm-water habitats of Arizona has been mostly limited to eradication of introduced carp, centrarchids, and catfishes in ponds and lakes. We made no effort to review the use of piscicides in such habitats at any elevation. Warm-water habitats discussed above include the San Carlos Reservoir, the Gila River, and the Little Colorado River basin (Lyman Lake and the stream below). Application of antimycin in attempted management of native topminnows (Marsh and Minckley 1990) is discussed in chapter 15. In September and October 1988, rotenone was used in the Virgin River, Utah, to attempt removal of red shiners (*Cyprinella lutrensis*) from 35 km of habitat occupied by endangered woundfins (*Plagopterus argentissimus*) and Virgin River chubs (*Gila robusta seminuda*). Failure to establish a detoxification station on schedule, complicated by other miscalculations, resulted in decimation of fish populations at least 75 km down-

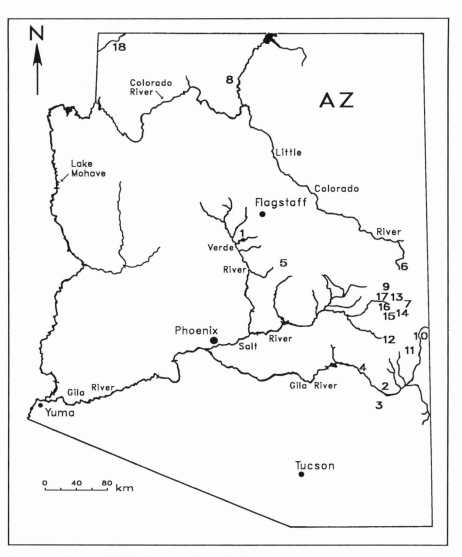

Fig. 14-4. Geographic distribution of recorded stream renovation projects in Arizona, 1950–1989; details are in text and in Tables 14-2 and 14-3. Site designations are: 1, Wet Beaver Creek; 2, Gila River–San Carlos Reservoir; 3, Ash, Grant, and Marijilda creeks on Mount Graham; 4, Bylas Springs; 5, Chevelon Creek–Woods Canyon Lake; 6, Lyman Lake–Little Colorado River; 7, East and North forks, Black River; 8, North Canyon Creek; 9, Mineral Creek; 10, Grant Creek; 11, KP Creek; 12, Bear Wallow Creek; 13, Lee Valley Creek/Lake; 14, Home Creek; 15, Wildcat Creek; 16, Reservation Creek/ Lake; 17, Ord Creek; and 18, Virgin River.

stream in Arizona and Nevada (Minckley 1989a). This incident has yet to be detailed in agency reports or the open literature but is generally discussed below.

Hindsight in Action: Evaluation of Habitat Improvements and Renovation

Platts and Rinne (1985) concluded that both riparian and stream "improvements" in the Rocky Mountain region could be beneficial for fishes but suggested that not all systems should or could be enhanced. Some streams (Mullan 1962) or stream reaches (Moyle et al. 1983) are unsuited for fishes because of precipitous channels or highly fluctuating flow regimes, and no amount or kind of alteration will create acceptable habitat.

Another reason that anticipated benefits have not been realized is the failure to recognize that watershed degradation is a primary cause of habitat deterioration and the decline or ultimate loss of fishes (Maughan and Nelson 1982; Rinne 1988a). Much stream improvement has been directed toward treatment of symptoms rather than causes of habitat deterioration (Heede and Rinne 1989). Even in headwaters, "stream improvement" may fail to enhance declining or already reduced fish populations that result from cumulative, continuing watershed problems (Rinne 1985a, 1988a; Szaro and Rinne 1989). Accelerated recovery activities for several large-river nongame fishes through reintroduction (Minckley 1983; J. E. Johnson 1985; Minckley et al., *this volume*, chap. 17) also might fail because of deterioration of aquatic habitats as a consequence of watershed damage. Furthermore, non-native predatory species substantially reduce chances for success (Marsh and Brooks 1989), and their removal or control may be nearly impossible. Large river drainages are complex, economically and politically as well as ecologically, and it is difficult to effect changes that could lead to habitat improvement.

Stream modifications to benefit native species can have the opposite effect. For example, in-stream log structures benefited populations of endangered Gila trout (Rinne 1981) by enhancing survival during drought and in winter in Main Diamond Creek (Regan 1966; Hanson 1971). But too many installations or too large a structure, combined with little harvest in small headwater streams, may cause overpopulation (Nankervis 1988) and reduced mobility (Rinne 1982). Introduced rainbow or brown trout may also be enhanced differentially, causing "improvement" to speed the demise of a native form.

Some stream habitat improvements have apparently been done for no apparent reason, as is strongly indicated by the fact that only a few have been monitored or evaluated. Even when they are evaluated, few projects have been followed for as long as ten years. Long-term evaluation of projects is required to justify their continued use (Platts and Rinne 1985).

Barriers are more simply discussed and evaluated. They either function to exclude an undesired species, or they do not. Further, there are ample natural examples that may be used to develop specifications for artificial structures. Permanency is clearly a major consideration, and installations that supplement natural barriers seem most effective. Public access is a major problem associated with this technique. Unknowing children or adults, as well as vandals, may move fishes and nullify the benefits of a barrier; public exclusion (or intense education) may be necessary. Another problem, especially in complex drainages of multiple ownership, is the undetected presence of an undesired species upstream from a barrier and its downstream dispersal into a controlled area.

Permanent installations—fishways, hatchery alternatives to natural spawning areas, and so on—are major investments requiring thorough evaluation. Care must be taken to ensure that the species will use the facility and that new selective forces will not foster excessive change in the population. All life-history stages must be accommodated. The literature on Pacific salmon is filled with such questions and answers pertaining to problems of stock maintenance. Expensive and complex installations and programs to save native nongame species will likely be few, but they may be necessary.

The effects of stream renovation on native fishes also need careful evaluation. Some streams (Moyle et al. 1983) should never have been poisoned. In the absence of sustained stocking, repeated renovation, and maintenance of fish barriers, trout frequently disappear and native fishes reestablish. Examples in Arizona include the Black River, Wet Beaver Creek, and the Little Colorado River. Several streams in the White Mountains and elsewhere in Arizona, classed as "brown trout fisheries" in management plans (Stephenson 1985), are dominated in numbers and biomass by native and non-native cyprinids and catostomids (Rinne, unpub. data) and will remain so unless they are managed in a put-and-take manner for salmonids. Attempted renovations for native trout most frequently failed because of incomplete kill of the target species or unauthorized reintroductions. Benefits to one native species also may result in loss or change in populations of another. We have no idea what genetic or other information was lost with the annihilation of "suckers and minnows" from Grant Creek in the isolated Willcox drainage of Arizona, of a population of Rio Grande sucker (*Pantosteus plebeius*) in McKnight Creek, New Mexico (see below), of roundtail chub from the Little Colorado River, or of innumerable stocks of speckled dace, mountain suckers, and other species destroyed by poisoning in other streams in the West.

Rolston (*this volume*, chap. 6) emphasizes the ultimate goal in the battle against extinc-tion of western fishes as preservation of the evolutionary process, not just its products. We therefore should not be attempting to conserve individuals of a species, but rather the dynamic genetic units they represent. If this goal is to be attained, genetic variation must be represented in recovered populations so that the species can continue to evolve in a natural manner. This is why habitat is so important. Preservation of species under controlled conditions of an artificial refuge, such as a hatchery or zoo, is to be avoided if possible.

Thus, when viewing post-renovation reestablishment of a native species or fauna, the question of its genetic makeup is important. Resilience, resistance, or chance may result in differential survival of individuals with a specific genetic makeup distinctly different from the original. Spring or headwater taxa have a greater chance of surviving poisoning than those living downstream. Accordingly, a genetically different population, reduced in heterogeneity and survivability under subsequent natural or human-induced disturbance (e.g., drought, harsh winters, flooding, or wildfire), may disperse throughout a system.

Piscicide application is probably responsible for the absence of native fishes in some areas of the West. Razorback sucker, Colorado squawfish, humpback chub (*Gila cypha*), and bonytail have not been recorded in Wyoming since the Green River renovation of 1962 (Anonymous 1977). Even though habitat alterations resulting from Flaming Gorge Dam may have had more impact on native fishes than the rotenone application (Holden, *this volume*, chap. 3), all four endangered or proposed endangered species were apparently extirpated by rotenone in that part of their geographic ranges. Razorback sucker and Colorado squawfish have recently been rare or absent in the San Juan River of New Mexico, along with roundtail chub (J. Brooks, USFWS, pers. comm.). The San Juan was poisoned concurrent with the closure of

Navajo Dam (Table 14-2; Olson 1962a, b). The sucker and squawfish almost certainly were present in the past, and the chub was abundant. Roundtail chubs are apparently gone from the Little Colorado River basin (Minckley 1985b), which suffered repeated applications of piscicides.

The Gila River in eastern Arizona is almost devoid of native fishes (Minckley and Clarkson 1979), although five species were present thirty to fifty years ago and suitable habitat seems present today. This area, as with the San Juan and Little Colorado rivers, was poisoned (Table 14-2). Fish populations in tributaries include the same species that were originally in the mainstream. Why have they not repopulated? Minckley (1985b) speculated that some stocks of native fishes may have been genetically adapted to certain river reaches, and once extirpated cannot be replaced.

Many of these ideas are subject to testing in the Virgin River, Arizona-Nevada-Utah, where two events in 1988 decimated a mainstream fauna. First, rotenone treatment of a short segment for eradication of red shiners accidentally removed most native fishes from a major portion of the river. Only woundfin and roundtail chub were held aside for reintroduction. Repopulation was anticipated to occur from a short untreated segment upstream. In January 1989 additional catastrophe struck when Quail Creek Reservoir, an off-channel impoundment of about 4000 ha-m capacity, abruptly failed, scouring the stream with a discharge greater than $566 \text{ m}^3 \text{ sec}^{-1}$.

Natural recolonization may occur by (1) in situ reproduction by survivors, (2) dispersal into the mainstream from tributaries, or (3) a combination of these. If the first prevails, the original fauna will presumably reestablish, with the possibility that some species had been forced to population levels so low as to result in genetic bottlenecks (which could affect survivorship). The second alternative may result in a major faunal change. The Virgin River mainstream, as in many western systems, has (or had) a distinctive fauna of streamlined, large-finned, current-adapted forms, including the two imperiled species plus speckled dace and desert and flannel-mouth (*Catostomus latipinnis*) suckers. The tributary fauna includes Virgin spinedace (*Lepidomeda m. mollispinis*), which scarcely occurred in the mainstream, plus forms of speckled dace, desert sucker, and flannel-mouth sucker characterized by short, stubby bodies and small fins, quite distinct in appearance from their relatives in the mainstream (Minckley 1989a). Morphological differences between mainstream and tributary populations are distinct enough to allow quantification of the situation, and genetic bases of these differences are under study. The third alternative, mixing of options 1 and 2, will result in a situation too complex for us to speculate on.

A fourth possibility is that non-native species will dominate a new community composed of red shiners, carp, bullheads, and green sunfish. Studies of the Virgin River system clearly provide remarkable opportunities to obtain data of potentially wide application.

Renovation Procedures: Review and Recommendations

Use of poisons to remove unwanted fishes has become an established management technique. Eschmeyer (1975) reviewed materials and methods of stream reclamation. Lennon et al. (1971) outlined application procedures, summarized case histories for most known fish toxicants in aquatic habitats in the United States, and reviewed two state manuals on stream reclamation methods. Gilderhaus et al. (1969) analyzed results of field trials of antimycin on twenty ponds and lakes and five streams. Slifer (1970) recommended improved techniques for calculating the amounts of rotenone and antimycin needed for stream reclamation, and methods of neutralizing both poisons. Lopinot (1975) and Cumming (1975) briefly summarized the use and history of fish toxicants in the United States. Since then, little new information has appeared on the protocols of piscicide use.

Recent practical experience with western streams and their faunas reveals some distinctive problems associated with the use of toxicants in management of native fishes. Here we attempt to evaluate environmental impacts and reasons for success or failure, and make recommendations for future activities.

Piscicides

Three major piscicides have been used in fisheries management in the United States. The first of these, and the longest used, is rotenone ($C_{23}H_{10}O_6$), a natural plant derivative commonly known as derris or derris root (Krumholz 1948), which has been used by man for centuries to kill fishes for food (Leonard 1939). Rotenone is produced by some species of Leguminosae, a family of plants essentially worldwide in distribution. Powdered derris root contains about 4% rotenone. It inhibits cellular respiration in fishes, and death is caused by suffocation. Rotenone (or its carrier, typically a petroleum product; see below) repels fishes (Lennon et al. 1971), which may allow them to escape by seeking springs or other low-concentration or untreated areas. It persists in the environment for only a few days or weeks.

The second compound is toxaphene ($C_{10}H_{10}Cl_{18}$), a chlorinated hydrocarbon, liquid-emulsion insecticide that is highly toxic to fishes (Neghberbon 1959). Tanner and Hayes (1955) noted toxicity to fishes seven months after toxaphene treatment of a Colorado reservoir. It was briefly tested and used in the 1950s but proved too slow acting for general use, and this, along with residue persistence in the environment, resulted in its discontinuance.

Antimycin ($C_{28}H_{40}N_2O_9$), an antibiotic produced in cultures of streptomyces (Dunshee et al. 1969; C. R. Walker et al. 1964; Lennon 1966), was introduced to fisheries management in the 1960s (Lennon 1966) and has been widely used since the early 1970s; Strong (1956) described it chemically. It persists briefly in the environment, typically disappearing in minutes or hours. Antimycin inhibits electron transport to cellular oxidative pathways (Strong 1956; C. R. Walker et al. 1964). As such, it acts quickly and irreversibly within tissues of the gills, resulting in suffocation.

Of the three, antimycin has become preferred. It is apparently not detected by fishes at concentrations typically applied, and its toxic action appears irreversible (Gilderhaus et al. 1969). Many experienced biologists consider the apparent inability of fishes to detect the formulation—and thus avoid it—one of its major advantages (B. Rosenlund, USFWS, pers. comm.). A rapid natural degradation rate (only minutes in highly oxidative streams) and the comparatively low impact of antimycin on aquatic invertebrates (if applied at the usually recommended 10 μg l^{-1}) are other major advantages.

Probabilities of Total Fish Removal

New Mexico case histories

Reclamation of streams has been a major part of Gila trout recovery in New Mexico. The first renovation in that state to benefit a native species involved rotenone treatment in 1970 to ensure the absence of non-native trout from McKnight Creek, Mimbres River drainage. That stream, though outside the native range, was selected as a refugium for Gila trout transplanted from the type locality (Main Diamond Creek, Gila River drainage). The transplant was successful, although a native population of Rio Grande suckers was eliminated in the process. The *Gila Trout Recovery Plan* (USFWS 1979b) called for treatment of four additional streams within the natural range in 1981–1988 to eliminate non-native and hybrid trout. One population each of four pure stocks of Gila trout are now (or soon will be) established (Loudenslager et al. 1986).

Antimycin treatment to remove brown trout from 3 km of upper Iron Creek, Gila Wilderness Area, illustrates some typical problems in reclamation of remote streams with extensive marshy areas and innumerable small springs. The entire reach was electrofished prior to poisoning in order to move Gila trout to temporary refugia (P. R. Turner and McHenry 1985). Treatment was conducted during low flow in early July 1981 by a team with renovation experience on Apache trout streams in Arizona. All side canyons, springs, and marshes were marked in advance, and 3.8-l containers designed to release antimycin and maintain concentrations of 10 μg l^{-1} for an hour were placed at 50- to 100-m intervals. Sand coated with antimycin (see below) was applied by hand to standing waters. The entire system above an artificial rock-concrete fish barrier was treated in two phases on successive days. Some upstream reaches (treated the first day) were re-treated the second day while the lower reach was being poisoned. Potassium permanganate was used for detoxification at the artificial barrier, and brown trout were killed for only 200–300 m downstream.

In 1982, spot sampling captured a single age-1 brown trout. Supplemental transplants in 1984 and 1985 from uncontaminated upstream reaches helped establish a reproducing population of Gila trout that was monitored annually. No additional brown trout were taken until August 1986, when a mature, age-5 female was removed. Four age-1 brown trout were removed in 1987, and twelve age-2 individuals were taken in 1988. Thus, the goal of eliminating an introduced predatory species was not realized, and managers must contend with progeny of brown trout that survived the treatment.

Only one of the four reclamation projects for Gila trout was successful in eliminating non-native trout with a single antimycin application. In that instance, the 6-km reach contained few side channels, no intermittent sections, and no tributaries that could be used by fishes during low flow (early July) when treatment occurred. Even there, a native population of speckled dace survived, presumably along stream margins in areas of groundwater inflow. Complete removal of fishes is unlikely with a single treatment (Tables 14-2, 14-3). This is especially true when dealing with complex streams with associated marshlands and inflowing springs (Rinne et al. 1982).

Piscicide Selection and Concentrations

Most western biologists now use antimycin almost exclusively in stream renovation. Both rotenone and antimycin are used in Colorado and California. Supply problems with antimycin resulted in the CADFG using rotenone almost exclusively in 1987–1988, except in critical projects involving golden trout (Gerstung, pers. comm.).

Antimycin is available as liquid concentrate and as a coating on sand. The first may be diluted as appropriate for applying through drip stations or by spraying. Sand is typically spread by hand or by a broadcast applicator. Although target application rates are usually between 2 and 50 μg l^{-1} (Mullan 1973, 1975), actual concentrations often vary because of natural uncontrollable factors or human error.

The AZGFD commonly uses 10 μg l^{-1} antimycin. In Ord Creek, a pilot study demonstrated that caged brook trout lost equilibrium in about 45 minutes at that concentration (Rinne et al. 1982). In agreement with laboratory and field studies by Berger et al. (1969) and Gilderhaus et al. (1969), all caged fish were moribund in 110 minutes. In West Creek, Rocky Mountain National Park, five hours were required to kill introduced brook trout at antimycin concentrations of 7.0 µg l^{-1}. Ultimately, 18 µg l^{-1} was applied. Fluorescein dye was used to monitor movement of the toxicant in the stream visually (Rosenlund and Stevens 1988). In California, concentrations of 10–20 µg l^{-1} removed brown trout from golden trout habitat in South Fork of Kern River (Meyers 1977); antimycin sanding was applied in marshy areas.

Lennon and Berger (1970) reported a 65% kill of brook trout in small ponds with antimycin formulation on sand at an estimated concentration of 3.5 µg l^{-1}. The fish began to react in three hours and died in forty-eight. By comparison, a concentration of 10 µg l^{-1} produced a total kill in eight hours. Berger et al. (1969) conducted tests with antimycin in ponds and laboratory containers, and reported fish to be most susceptible at warmer water temperatures. Hard and high-pH waters reduced toxicity to trout, and high alkalinities induced more rapid degradation.

Initially (1939–1950), rotenone was usually applied at concentrations of 0.5 mg l^{-1}. However, frequent failures tended to stimulate use of greater concentrations, calculated most frequently on the basis of past experiences of the applicator. This often translated into a calculated amount plus a "certain excess for a margin of safety." Inconsistent results and unstable commercial products were early problems. As a result, a number of emulsified formulations (mostly 2.5% to 5% rotenone in a petroleum-based carrier) were developed. Nine were tested by Shannon (1969). Liquid formulations making up eight of the products tested were malodorous and repelled fish. Accordingly, escape had to be denied target species to ensure success. A review of techniques and equipment for renovation with rotenone was provided by Hooper (1955). W. R. Turner (1959) suggested that concentrations of 1.0 mg l^{-1} were sufficient for removal of fish from ponds, but cautioned workers to carefully consider local environmental conditions.

Effects on Other Organisms

Houf and Campbell (1977) reported no effects of either rotenone or antimycin on macrobenthos in Missouri ponds. Meyers (1977) similarly reported "no evident effects" on benthic invertebrates of antimycin applied at 20 µg l^{-1} in a stream. In contrast, Jacobi and Degan (1977) estimated 50% reduction in biomass of aquatic macroinvertebrates two days after antimycin was applied at 17–44 µg l^{-1} in Seas Branch Creek, Wisconsin. Although short-term reductions in biomass of benthos varied from 0% (crayfishes) to 100% (*Baetis* spp. and *Gammarus* spp.), all common taxa were present and benthic biomass had recovered to pretreatment levels after a year. For this same treatment, Avery (1978) estimated that the maximum concentration of antimycin had, in fact, been 61 µg l^{-1} for 7.5 hours. Mean density of benthic organisms recovered to pretreatment levels in seven months, but recovery took from one and a half months for the coleopteran *Optioservus* and chironomids, to twenty-five months for trichopterans of the genus *Brachycentrus*. Avery (1978) concluded that no invertebrate taxon was eliminated by antimycin treatment in Seas Branch Creek. He also stated, however, that when threatened or endangered taxa were present, or when in-stream cover was lacking, it would be more appropriate to use habitat improvement than piscicide application to enhance sport fisheries. Minckley and Mihalick (1982) reported almost complete decimation of stream invertebrates after application of antimycin at an estimated 10 µg l^{-1} in Ord Creek, Arizona. We suspect that actual concentrations of antimycin in Ord Creek were 20–30 µg l^{-1}.

Pre- and posttreatment (less than a week before and three months following) samples of benthic invertebrates were taken in Big Dry Creek, New Mexico, to evaluate impacts of two antimycin treatments of 10–20 µg l^{-1}.

Mangum (1984, 1985) found that sediment dwellers (chironomid dipterans and oligochaetes), a mayfly (*Epeorus* sp.), and a stonefly (*Isoperla* sp.) were adversely affected by the treatment, but the overall effects on the macroinvertebrate community were minimal. He concluded that antimycin would probably have less impact on macroinvertebrates than rotenone (Mangum 1984).

Detoxicants

Potassium permanganate ($KMnO_4$) is the most common chemical used to augment oxidation and breakdown of both antimycin and rotenone, but chlorine is also used. Concentrations of either compound sufficient to detoxify rotenone, however, may be toxic to fishes and require treatment with sodium thiosulfate after sufficient contact time with the piscicide (Dawson 1975). Lower effective concentrations of antimycin make it feasible to apply sublethal concentrations of $KMnO_4$ (1.0 mg l^{-1}) or chlorine (0.5 mg l^{-1}) for detoxification. Deactivation half-life for antimycin in soft water at 12°C was, however, about eighty minutes for chlorine versus only seven minutes for $KMnO_4$. Use of chlorine to deactivate antimycin in most streams would thus be too slow to prevent fish kills for a considerable distance downstream. $KMnO_4$ will also kill fish (Marking and Bills 1975), especially in low-temperature waters or those with high pH and hardness.

Many of the early treatments in Arizona made no mention of detoxification, and we assume that none was performed. Therefore, far more stream kilometers must have been treated than were reported. In Ord Creek (Rinne et al. 1982), antimycin was successfully detoxified with sodium hypochlorite at concentrations of 1.0 mg l^{-1}; caged fish survived 500 m downstream from the detoxification site. Meffe (1983b) did not detoxify because of a lack of connection between treated spring runs and the adjacent Gila River; water

was ponded downstream, then disappeared through seepage or evaporation. Meyers (1977) reported no use of detoxicant in South Fork of the Kern River, California, since the downstream reach was to be treated the following year and extensive falls provided natural oxidation. Trout and Sacramento suckers (*Catostomus occidentalis*) were nonetheless killed 1 km downstream from the falls (J. Stefferud, U. S. Forest Service, pers. comm.). Most renovation projects for native Apache trout in Arizona (Table 14-3) and all recent treatments for Gila trout in New Mexico included detoxification of antimycin with $KMnO_4$, the success of which was unrecorded.

Considerations for Renovation Projects

Temporal-spatial concerns

Late summer–early autumn treatments are designed to remove autumn-spawning brook and brown trout prior to reproduction, and age-0 and older spring-spawning cutthroat and rainbow trout. Treatment in summer also reduces the probability of eggs or larvae surviving in the substrate. Late summer drought and low flow conditions, as well as warmer water temperatures in many parts of the West, may also be advantageous.

Antimycin treatments in August of two successive years were used to eliminate brook trout in a Wyoming stream (Gresswell 1991). In Colorado, 68% of all renovations were conducted in August and September (Rosenlund, pers. comm.). Most projects for Gila and Apache trout in Arizona and New Mexico have been conducted prior to the summer monsoon characteristic of that region. Treatment during drought avoids the possibility of dilution that results from unexpected spates and takes advantage of lesser volumes of water as well as lowered habitat heterogeneity. Continuous stream flow nonetheless appears to enhance effectiveness of piscides through mixing and through minimizing the filtering or chemical effects of subsurface percolation of water through substrates separating intermittent pools.

Pretreatment Surveys

All reaches of the stream to be treated should be surveyed once or more in the year prior to treatment. Ideally, this survey should define the distributions of both native and non-native fishes within the channel, its tributaries, and off-channel springs and marshes. Presence of age-0 and spent or gravid individuals of the target species should be noted. The occurrence of target fish as larvae in either springheads or within the substrate (e.g., salmonids in redds) will help dictate the timing of treatment. Such information is especially important in cold streams and in relatively lengthy reaches where times of spawning and rate of larval development are likely to be prolonged or variable because of different water temperatures in differing sections (e.g., spring-fed tributaries versus open channels).

A major effort should be made to assess the presence and potential impacts of treatment on rare, sensitive, or imperiled nontarget organisms. Pretreatment sampling must include invertebrates, which should be sent to taxonomic experts, especially if the fauna has not been thoroughly inventoried. When localized populations are found, it is necessary to maintain sufficient numbers of each taxon for reintroduction. This is especially required if all upstream reaches of a system (including spring-fed sources) are to be treated. If possible, natural refugia such as spring sources should be left untreated to ensure survival of endemic forms.

A resurvey should be performed immediately before treatment to determine changes that may have occurred due to flooding or other events, and to reconfirm the presence of potential problems. Just prior to poisoning, stream discharge should be determined for each tributary and at various points along the channel by using either a current meter (U.S. Geological Survey methodology), a float (Robins-Crawford method; Orth 1983), or conductivity meter (salt dilution method; Engstrom-Heg 1971b). Likewise, transit time for a prism of toxicant through a series of test reaches should be determined by use of either fluorescein (or other) dye or salt (Slifer 1970). If possible, specimens of the target species (and other species of concern) should be assayed for susceptibility and response to the piscicide of choice under conditions of water quality, temperatures, and flow expected in the treatment area. Timing of poison contact, required effort, and fewer surprises during the operation can all be derived from such bioassays.

We recommend detailed measurement and flagging of the selected reach(es) prior to treatment. Different colors should be used to mark the channel at 100-m intervals, the uppermost point of tributaries to be treated, and potential problem areas (springheads, waterfalls, and marshy areas) that will require special attention. Determination of necessary spacing between application containers will be simplified by accurate definition of distances as well as by the bioassay recommended above. Detailed ground surveys will locate waterfalls and intermittent reaches to be depended on as permanent or temporary fish barriers, and will also pinpoint such problems as impassable reaches or the need for trail maintenance. Careful measurements and notes may also be translated into a field map that shows relative distances and locations of landmarks, providing data to estimate numbers of personnel, application points, and requirements for auxiliary equipment such as backpack sprayers.

Piscicide application

We recommend multiple treatments in all projects. When access is restricted (in remote wilderness areas), back-to-back (one- to three-day) treatment will ultimately save in time and travel costs. In California, projects to enhance golden trout commonly involve an initial anti-mycin treatment at 10 μg l^{-1} followed by an application of rotenone at 1.0 mg l^{-1} before a crew departs the area. Because rotenone dis-tresses fishes, causing them to swim erratically from treated areas, it is possible to locate places where they survived the initial treat-ment (S. Stevens, CADFG, pers. comm.). Rote-none is commonly used alone in high-gradient California streams where toxic concentrations of antimycin cannot be maintained for more than 100 m because of rapid degradation in-duced by turbulence in rapids and cascades.

Our experience in headwater streams in New Mexico suggests that treatment during intermittent flow reduces mixing, and there-fore effectiveness, of piscicides. In addition, some target fishes (especially small individu-als) may survive treatment in places with cool groundwater inflow (or underflow), beneath undercut banks, and in springheads. Treat-ment at night when some fish are more active in open water may be a viable alternative. It may be desirable to treat streams with exten-sive marshes and backwaters when low flow withdraws water from such places and enables better access by treatment crews (Gerstung, pers. comm.).

We recommend application by drip contain-ers (Engstrom-Heg 1971a). These should meter the toxicant at a constant rate for four to ten hours, depending on the species and target concentration. Gresswell (1991) obtained good results using containers constructed by drilling a small hole in the pan of animal waterers. To reduce the likelihood of particles blocking the units and altering flow rate, stream water used to dilute the piscicide should be filtered. Refilling drip stations may be necessary, depending on flow rate and con-tainer size. Because of the longitudinal exten-sion of a piscicide prism, operating the upper-most drip station for a period and then starting

downstream booster stations in a stepwise fashion as the prism moves downflow in-creases contact time.

Because of antimycin's rapid natural degra-dation in high-gradient (greater than 3%) western streams, booster stations must be placed at 50- to 200-m intervals to ensure maintenance of toxic concentrations. In con-trast, only twenty-eight stations were required in a 25-km reach of the low-gradient (1.8%) Arnica Creek drainage, tributary to Yellow-stone Lake, where antimycin at concentrations of 8.0 μg l^{-1} eliminated brook trout (Gress-well, in press). Although the actual amount of antimycin remaining in a stream after a given distance of travel cannot yet be quantified, it is probably satisfactory to decrease the concen-tration applied at booster stations if they are less than 100 m apart. For example, workers in soft-water mountain streams in Colorado apply antimycin at 8.0 μg l^{-1} at the first drip station, then use 4.0 μg l^{-1} at boosters placed at 61- to 76-m intervals (Rosenlund, pers. comm.). To ensure maintenance of toxicity, many workers recommend placing target fish in live cages just upstream from each booster station. This is especially appropriate if on-site pretreatment bioassay data are unavailable.

In most projects at least some intermittent sections, backwaters, seeps, or marshes must be treated separately because of poor mixing with the channel. In most cases these may be sprayed with portable backpack spray units designed for applying liquid pesticides. Gress-well (1991) found a knapsack sprayer more efficient than either galvanized or collapsible-bag fire-suppression pumps. In any case, avail-able units should be field tested in advance, and backup units should be available. Dis-pensing toxicants into problem areas with hand-held spray bottles offers a simple, light-weight alternative (Stevens, pers. comm.). In all cases safety precautions as noted on pesti-cide labels must be observed.

Detoxification

We recommend use of $KMnO_4$ at 1.0–2.0 mg l^{-1} for detoxifying both antimycin and rotenone. We prefer metering the chemical using 19-l constant-flow Mariotte bottle dispensers. The elongation of piscicide prisms makes it necessary to detoxify far longer than might be expected. Under normal conditions, detoxification should probably continue for one and a half to two times as long as the time of piscicide application. Without exception, live cages containing the target fish species should be placed at about 100-m intervals downstream from a detoxification station and monitored periodically by permanently assigned personnel. Duplicate drip containers and premeasured $KMnO_4$ should be available, and if fish in live cages become stressed or begin dying more than 200 m below the original detoxification station, a backup should be started immediately. The potential for a disastrous fish kill below a target reach is greater for rotenone than for antimycin, and if the former is used, a concentration of $KMnO_4$ greater than 2.0 mg l^{-1}, detoxified in turn with sodium thiosulfate, may be necessary.

Conclusion: Stream Improvement and Renovation in the Future

Much of the early (pre-1970) stream improvement and reclamation was directed toward enhancement of game fish populations. Stream habitat improvement is no longer conducted at yesterday's scale. The decrease in such activities reflects increased costs and increasing awareness that such projects are often a short-term approach to the larger problem of watershed deterioration (Platts and Rinne 1985; Heede and Rinne 1989). Stream reclamation projects have also become more specific and limited in scope than previously, although they still deal with attempted enhancement of individual species.

Becker (1975) established that little or no consideration was given to rare, threatened, or endangered species in reclamation projects in Wisconsin. The philosophy that renovation was done for the "good of mankind" was used there to justify wholesale slaughter of tons of native nongame fishes. A few short years ago this also seemed to be the case in the American West. Calling these operations "nongame fish control" instead of "nongame fish eradication" made them more acceptable. In Arizona, renovations were done under federal Dingell-Johnson programs designed to enhance game fish and were called "manipulation of environmental conditions pertaining to minor jobs of a developmental nature." Entire systems, such as the upper Green River in Wyoming and the upper San Juan River in New Mexico, were treated, and populations of native nongame species were eliminated or drastically reduced. Most game fish populations are improved only temporarily, for three to five years.

The Endangered Species Preservation Act of 1966, Endangered Species Conservation and National Environmental Policy acts of 1969, and finally the Endangered Species Act of 1973 changed the course and goals of stream renovation activities. This is reflected in the transition in the 1970s from stream renovation to remove rough fish to projects mostly for the removal of non-native introduced species in the 1980s. Introduced populations of brown, brook, and rainbow trout began to be targets of poisoning to benefit native cutthroat, golden, Apache, Gila, and other trouts.

Reclamation activities to enhance native, warm-water, nongame species have not occurred as frequently as for indigenous salmonids, but they are increasingly being considered, proposed, and implemented. Piscicides have been used in small systems to remove predatory mosquitofish in an attempt to save endangered Sonoran topminnows. Minckley et al. (*this volume*, chap. 17) describe the success of predator removal in enhancing recruitment of razorback suckers in an isolated habitat in Lake Mohave, Arizona-Nevada. The Virgin River mainstream was treated with rotenone to remove red shiners, which are detrimental to endangered woundfin populations. We anticipate that the use of fish toxicants in endangered species management will continue and intensify in the future.

It is clear that if enough effort is expended to remove undesired fishes, we will ultimately succeed. All fishes can be removed from a system with multiple treatment by the poisons available to modern fishery managers. However, we plead for a judicious approach to their application. Almost half a century ago stream reclamation with poisons was considered a boon for management of sport fish populations. In retrospect, many of these same projects are now considered disasters for native fishes. We must take care not to consider these same techniques an unqualified boon to native fish management in the future. The use of toxicants must be critically evaluated alongside assessments of the capability of a habitat to support the target species and its associated community, and the possibilities for other types of habitat improvement. Indeed, we must value and be good students of the history of habitat reclamation and improvement, or we surely will be condemned to repeat our mistakes. A myopic focus on single-species management that destroys other species of fishes and other organisms is unlikely to withstand the ethical test of promoting "the integrity, stability and beauty of the biotic community" (A. Leopold 1949).

SECTION V

No Time to Lose:
Management for Short-lived Fishes

Many small fishes live only briefly. Some, like male pupfishes, which invest tremendous amounts of energy not only into actual reproduction but also into ancillary activities such as territorial defense, may live only a few weeks after achieving sexual maturity. Other species, especially the smaller minnows, mature and reproduce in their second summer of life, and perhaps a third.

Such short-lived species find themselves in severe trouble when their habitats are modified. If a year class fails, natural mortality may be so high that whole populations, or an isolated species, may be jeopardized. Loss of a second year class may result in extinction. Fortunately, and likely adaptively, they tend to be locally abundant, widespread in distribution, remarkably resistant to extirpation, or all of the above. It is difficult to kill them, modify all their habitat, or find and eliminate the last few pairs of individuals from a system. Females of some live-bearing species even store sperm so that one mating can serve to fertilize eggs for a number of broods (or breeding periods) when males are absent. Nonetheless, at least twenty-five of the forty North American taxa of fishes that have become extinct since the mid-1800s were short-lived.

Chapter 15 reviews examples from the cyprinodontoid fishes. Some have declined to extinction, some have become rare, and a few may be recovering following a major decline. The pupfishes and allied families of the western United States and northern Mexico have much in common in addition to small sizes and short life cycles. Most live in places that are marginal, at best, for any kind of fish life because of severe environmental conditions. Despite their tolerance of extremes of salinities, temperatures, and other physical and chemical factors, a major proportion of the seventy-six taxa known from the desert West are either extinct (9.2%), threatened or endangered (38.2%), or classed as rare or vulnerable (30.3%). The minority that remains (seventeen taxa, 22.3%) is either of unknown status or deemed secure.

Recovery efforts for cyprinodontoid fishes often include reintroductions into new or renovated habitats, either after propagation in a refugium or directly from one place to another. Chapter 16 reviews successes and failures of such activities in the West and analyzes patterns in a data set involving records for almost five hundred stockings of forty taxa. Luckily, most short-lived fishes are relatively easy to propagate. They quickly build large populations under artificial or seminatural conditions, and tend to do the same when introduced into suitable habitat. But problems exist in the selection of introduction sites, or perhaps in the definition of a "suitable habitat" for reestablishment. Perhaps most important is a consideration of the conflicts between political expediency (and decision-making processes) and reintroduction and reestablishment of listed species. Discussion of the question of what constitutes recovery is especially germane to the battle against extinction.

Artist's depiction of pupfish interactions, used as a commemorative logo for the 1988 symposium giving rise to this volume. Artwork prepared and contributed by Barbara Terkanian, Arizona State University.

Chapter 15

Conservation and Management of Short-lived Fishes: The Cyprinodontoids

W. L. Minckley, Gary K. Meffe, and David L. Soltz

Our treatment of representatives of four western American families of fishes—Cyprinodontidae (pupfishes), Goodeidae (Empetrichthyinae; poolfishes and springfishes), Fundulidae (killifishes), and Poeciliidae (live-bearers)—in a single chapter has a natural basis. All four families are in the order Cyprinodontiformes (Parenti 1981), and for shorthand purposes we collectively refer to them as cyprinodontoids. All are ecologically and physiologically similar, and all present comparable problems and opportunities with respect to conservation and management. Most live a relatively short time, tend to be ecological generalists, and have wide physiological tolerances. Finally, they occupy similar habitats and play comparable ecological roles. Unfortunately, a major proportion of western American cyprinodontoids are endangered or of special concern. Our purpose is to provide a general review of their biology and management, enumerate the group members and their status, and discuss specific examples of declines and recoveries. We also comment on more general issues in endangered species conservation.

Ecological Settings

Natural Habitats

Cyprinodontoid fishes typically live in environments that are marginal for other fishes, including isolated springs, riverine marshlands (ciénegas), seeps, and inlets and shorelines of rivers and lakes. They all frequent shallow water, often in or near dense beds of aquatic vegetation, and their environments are marked by physical and chemical harshness. Even crystal-clear springs that vary little in discharge or other apparent features may have low pH or dissolved oxygen, dangerously elevated temperatures, high concentrations of toxic salts, or any combination of these (Sumner and Sargent 1940; Sumner and Lanham 1942; Deacon and Minckley 1974; J. E. Williams and Wilde 1981).

As we might expect, such environments are typically species-poor, and it is not unusual for a single kind of cyprinodontoid to occur alone or with only one or a few other fishes. Additionally, the American West had a naturally depauperate fish fauna, with piscine predators conspicuous in their absence. Consequently, western cyprinodontoids are evolutionarily inexperienced with other species, and especially so with aggressive predators.

Recent Habitat Changes

Aquatic habitats began to undergo modifications as soon as western North America was colonized by Europeans. All major drainages were affected, especially in the past hundred years, with many suffering dramatic alterations in habitat structure and species composition. Virtually all these changes are a result of human demands for water, and the future portends no changes in this pattern.

Documented, visible, and large-scale habitat deterioration began in earnest with a cycle of arroyo cutting in the late 1800s, during which mature, physically buffered watercourses were severely eroded. The causes are debated, but overgrazing by domestic livestock and climatic change have both been implicated (Hastings 1959; Hastings and Turner 1965; Cooke and Reeves 1976; Hendrickson and Minckley 1985). Incision of stream channels resulted in lowered water tables, which desiccated watersheds and replaced continuously flowing systems with dry, occasionally flooding arroyos. Flood-straightened channels carried water swiftly with increased erosive power. Marshes dried, springs failed, and streamside backwaters and inlets disappeared as watercourses were increasingly constrained by steep arroyo walls.

Shortly thereafter, damming of streams became widespread in response to changes in channel form and further human population increases. Dams reduced natural variation in discharge, stabilizing wildly fluctuating rivers and replacing long reaches of streams with placid lakes. Reservoirs designed for storage and delivery of water for irrigation and flood control drowned lotic habitats, blocked former fish migration routes, and generally disrupted ecological relations both up- and downstream. Groundwater pumping also increased, destroying aquatic habitats through further lowering of water tables and drying even the largest springs (Brune 1975). Springs are critical habitats for many cyprinodontoids (J. E. Johnson and Hubbs 1989; Meffe 1989), and groundwater mining is a particularly destructive form of water use.

Compounding these alterations, and perhaps as catastrophic, was the introduction and establishment of non-native species. At present, exotic forms outnumber, outcompete, and prey upon cyprinodontoid and other native fishes that evolved under biologically de-

pauperate conditions, more often than not resulting in their demise. Introduced species from biologically rich areas such as the Mississippi River basin, southern Mexico, or tropical Africa flourish under the prevailing artificially stabilized conditions.

Thus the western American cyprinodontoids, formerly abundant if inconspicuous, find themselves trapped in disappearing and altered habitats, and interacting in new, complex, and invasive communities. Habitat loss and deterioration have allowed many to survive only in remote enclaves such as springheads that have not yet been affected by groundwater declines or exotic species. These vestiges of a formerly widespread and abundant fauna constitute the raw materials available for management and recovery.

Life History Background

Most cyprinodontoids live no more than a year, although individuals may persist for two or more years, especially in captivity. This short life span has both positive and negative features for management. Most species can be bred quickly and easily under artificial conditions, and large populations build rapidly. Likewise, a few colonists in a new habitat can rapidly result in large, viable populations in nature. Short generation times also mean rapid turnover of individuals, however, and their status can change abruptly from abundant to rare, or extirpated. Thus, almost continual monitoring is necessary.

Cyprinodontoids are largely food generalists. Most often, however, cyprinodontoids and goodeids tend to be herbivores, preferring algae and detrital materials, and fundulids and poeciliids are carnivores, eating mostly invertebrates.

Physiologically, species of all four families are broadly tolerant of environmental extremes. Most can withstand wide thermal ranges, and killifishes and pupfishes in particular occupy habitats in a wide array of salinities from fresh to as high as three times seawater (Deacon and Minckley 1974; Soltz and Naiman 1978). Some also live with other chemical variations, such as low concentrations of dissolved gasses. Some springfishes in Nevada live in springs with dissolved oxygen levels consistently less than 1.0 mg l^{-1} (J. E. Williams and Wilde 1981), concentrations lethal to other fishes. Physiological tolerances are rarely a major obstacle in cyprinodontoid conservation and management.

The greatest life-history difference among these families regards reproduction. Poeciliid females are internally fertilized, gestation occurs entirely within the female's body (except for one South American species), and offspring are born precocious and fully formed (Constantz 1989). There are no special requirements for reproduction other than seasonally warm water temperatures. Most other fishes in the United States are egg layers; their offspring go through embryonic development and larval stages in the open environment.

Little is known of the reproductive biology of the egg-laying goodeids (goodeids [Goodeinae] in Mexico, where most species occur, bear their young alive) or western fundulids, but spawning by cyprinodontids (*Cyprinodon* spp.) requires a suitable substrate (which may vary by species) and involves complex behavioral interactions leading to mating. Males of all species of pupfishes except one (*Cyprinodon diabolis* in Devil's Hole, Nevada) maintain breeding territories. Thus, pupfish reproduction may be more affected than others by environmental change; exotic fishes, in particular, may disrupt mating behavior.

Species and Their Status

We list western American cyprinodontoids of interest to the geographical scope of this book

in Table 15-1. For each we have compiled information on natural range, habitat, current status, and appropriate references. The official status is listed where available (rare, threatened, endangered, and so on, from federal or state compilations or as presented by J. E. Williams et al. [1985, 1989] or J. E. Johnson [1987a]), or was estimated by us based on current range and likelihood of survival.

Cyprinodontids are relatively numerous, with thirty-six taxa in three genera (*Cualac*, *Cyprinodon*, and *Megupsilon*), twenty-nine named species and subspecies, and seven recognized but as yet unnamed forms. There are five taxa of fundulids in three genera (*Fundulus*, *Lucania*, and *Plancterus*). Goodeids (Empetrichthyinae) are represented by two genera (*Crenichthys* and *Empetrichthys*) with ten named species and subspecies. The majority of these egg-laying fishes are now restricted to springs or spring-influenced habitats, although a few are truly riverine forms. Twenty-five poeciliid species in three genera are represented in the American West. *Gambusia*, which dominates in number of species, occurs primarily in springs and spring-fed habitats, whereas *Poeciliopsis* and *Xiphophorus* inhabit both springs and streams.

A summary of the status of these fishes (Table 15-2) indicates that cyprinodontids and goodeids are in a more precarious situation than fundulids or poeciliids. Six taxa of goodeids and pupfishes are already extinct, and only two of forty-six are considered relatively secure. Poeciliids, fundulids, and goodeids have about the same proportion of their species classed as threatened or endangered (20%–30%). One of five fundulids is probably endangered, and the others are considered rare or secure. None of the goodeids is estimated to be secure (Courtenay et al. 1985). In summary, only eight (10.5%) of the western cyprinodontoids are thought to be secure;

Table 15-1. Ranges, habitats, status, and pertinent references for cyprinodontoid fishes of the southwestern United States and northern Mexico.[1]

Species	Native range	Habitat
Cyprinodontidae		
Cualac tessellatus	La Media Luna, SLP	never common, but in various habitats in springs and marshes
Cyprinodon alvarezi	Ojo del Potosí, NLE	deeper waters, spring-fed pond; 18°–23°C
C. atrorus	Cuatro Ciénegas basin, COA	ephemeral pools and saline lakes, avoids headsprings; highly euryhaline and eurythermal
C. bifasciatus	Cuatro Ciénegas basin, COA	thermal headsprings and outflows; relatively stenohaline and stenothermal
C. bovinus	Leon Creek, Pecos County, TX	springs, outlet streams, and marshes; eurythermal
C. diabolis	Devil's Hole, Nye County, NV	restricted to thermal, spring-fed, limestone cavern
C. elegans	Comanche, Phantom Cave, San Solomon springs, Pecos and Reeves counties, TX	spring-marsh complex and irrigation outflows; often in swift currents
C. eximius	Río Conchos, CHI; tributaries of Rio Grande, TX	shallow river edges, marshes, creek mouths, and springs
C. fontinalis	Ojo de Carbonaria, CHI	springs, outflows, and marshes
C. latifasciatus	Parras, COA	unknown, presumably springs and outflows
C. macrolepis	Ojo de Hacienda Dolores, CHI	spring pool and outlet; 24°–33°C; intolerant of cold
C. macularius		
C. m. eremus[2]	Río Sonoyta, Quitobaquito Spring, AZ, SON	spring-fed pond; river pools
C. m. macularius	lower Colorado River basin, AZ, CA, SON, BCN	stream edges, marshes, backwaters; springs and outflow marshes; ephemeral lakes and shore pools
C. meeki	upper Río Mezquital, DGO	nonthermal springs, stream edges, marshes; near aquatic vegetation

Status	References
endangered due to restricted distribution, development, and exotic species	R. R. Miller 1956; J. E. Williams et al. 1985
endangered due to restricted occurrence, pumping for irrigation, exotic species	R. R. Miller 1976a, 1981; Contreras Balderas, *this volume*, chap. 12
rare, variable in abundance year to year; declining water levels may force syntopy and hybridization with *C. bifasciatus*	R. R. Miller 1968c, 1981; Minckley 1969c, 1978, 1984; E. T. Arnold 1972
vulnerable due to limited range; hybridizes with *C. atrorus*	R. R. Miller 1968c, 1981; Minckley 1969c, 1978, 1984; E. T. Arnold 1972
endangered due to restricted range, exotic species, and potential habitat alterations	R. R. Miller 1961; A. A. Echelle and Miller 1974; C. Hubbs et al. 1978; S. E. Kennedy 1977; C. Hubbs 1980
endangered due to restricted distribution, small natural population size, and potential for water developments	R. R. Miller 1948; Soltz and Naiman 1978; Deacon 1979; Baugh and Deacon 1988
endangered due to overpumped groundwater, habitat alteration, exotic species	J. R. Davis 1979; USFWS 1981b; Ono et al. 1983; A. A. Echelle et al. 1987
vulnerable due to habitat degradation and loss	R. R. Miller 1976a, 1981; J. R. Davis 1980; Garrett 1980a
threatened due to restricted range and potential water developments	M. L. Smith and Miller 1980
extinct prior to 1953, presumably due to habitat alterations	R. R. Miller 1964c, 1976a, 1981; R. R. Miller et al. 1989
rare due to restricted distribution; single large spring modified for swimming	R. R. Miller 1976a, 1981; Contreras Balderas, *this volume*, chap. 12
endangered due to habitat loss, exotic species, groundwater pumping, and pesticide blowover	R. R. Miller 1943a; R. R. Miller and Fuiman 1987; Schoenherr 1988; Hendrickson and Varela 1989
endangered due to habitat loss and exotic species	Minckley 1973; Black 1980; R. R. Miller and Fuiman 1987; Schoenherr 1988; Hendrickson and Varela 1989
endangered due to spring dewatering, industrial pollution, channel incision	R. R. Miller 1976a, 1981; Contreras Balderas, *this volume*, chap. 12

Table 15-1. Continued

Species	Native range	Habitat
C. nazas	Laguna Santiaguillo, Ríos Nazas-Aguanaval, COA, DGO, ZAC	river edges, marshes, backwaters; springs
C. nevadensis		
C. n. amargosae	Amargosa River, Inyo County, CA	river edges, marshes; eurythermal (4°–40°C)
C. n. calidae	Springs near Tecopa, Inyo County, CA	warm springs (to 42°C) above Amargosa River
C. n. mionectes	large, low-elevation springs, Ash Meadows, Nye County, NV	springs, spring runs, and marshes
C. n. nevadensis	Saratoga Springs and marshes, Inyo County, CA	springs and marshes
C. n. pectoralis	small, higher-elevation springs, Ash Meadows, Nye County, NV	low-discharge, warm (30°–31°C), constant springs
C. n. shoshone	Shoshone Springs, near Shoshone, Inyo County, CA	spring, its outlet, associated marshes
C. pachycephalus	Baños de San Diego, CHI	spring runs and pools, to 43.8°C
C. pecosensis	Pecos River, NM, TX	river channel and edges, backwaters, springs, sinkholes
C. radiosus	Owens River, Inyo County, CA	river edges, marshes, sloughs, and springs; eurythermal
C. rubrofluviatilis	Red and Brazos rivers, OK, TX	river edges, channels, backwaters, over sand bottoms; euryhaline and eurythermal
C. salinus		
C. s. milleri	Cottonball Marsh, Inyo County, CA	pools and channels surrounded by salt crusts; salinities to 100 g l^{-1}
C. s. salinus	Salt Creek, Inyo County, CA	pools, edges, and channels of creek; euryhaline and eurythermal
C. tularosa	White Sands area, Otero County, NM	constant temperature, warm springheads, outlets, and marshes
Cyprinodon sp.[3] (Guzman pupfish)	Guzman complex and Laguna de Bustillos, CHI	river channels, edges, marshes, and springs and outflows

Status	References
vulnerable due to water developments	R. R. Miller 1976a, 1981; Contreras Balderas, *this volume*, chap. 12
vulnerable due to restricted distribution, exotic species	R. R. Miller 1948; Soltz and Naiman 1978
extinct; springs developed as baths; possibly hybridized with *C. n. amargosae*	R. R. Miller 1948; Pister 1974; Soltz and Naiman 1978; R. R. Miller et al. 1989
endangered due to agricultural development and exotic species	R. R. Miller 1948; Soltz and Naiman 1978; Deacon and Williams 1984; J. E. Williams and Sada 1985b
rare due to limited distribution	R. R. Miller 1948; Deacon 1967, 1968a; Soltz and Naiman 1978
endangered due to ditching, impoundment and reduced flow, and restricted range	R. R. Miller 1948, 1981; R. R. Miller and Deacon 1973; Soltz and Naiman 1978
endangered, considered extinct until 1986	R. R. Miller 1948, 1967; Pister 1974; Taylor et al. 1988
endangered due to restricted range, development of springs, and hybridization with *C. eximius*	M. L. Smith and Chernoff 1981; Minckley and Minckley 1986
endangered due to hybridization with *C. variegatus*, habitat degradation and loss	A. A. Echelle and Echelle 1978; A. A. Echelle et. al. 1987; A. A. Echelle and Conner 1989
endangered due to water development and exotic species	R. R. Miller 1948; R. R. Miller and Pister 1971; Naiman and Soltz 1978
secure and widespread, although groundwater use increasing	A. A. Echelle et al. 1972
rare due to limited distribution	LaBounty and Deacon 1972; Naiman et al. 1973; R. R. Miller 1981
rare due to limited distribution	R. R. Miller 1943b; Soltz and Naiman 1978
rare due to limited distribution	R. R. Miller and Echelle 1974; R. R. Miller 1981
unknown, several populations scattered over vast area	R. R. Miller 1981; Minckley, unpub. data

Table 15-1. Continued

Species	Native range	Habitat
Cyprinodon sp. (Julimes bighead pupfish)	Thermal spring near Julimes, CHI	open sandy spring runs (to 44°C)
Cyprinodon sp. (Monkey Spring pupfish)	Monkey Spring, Santa Cruz County, AZ	marshes and pond
Cyprinodon sp. (Palomas pupfish)	Palomas Basin, CHI	springs, marshes, ephemeral playa lake
Cyprinodon sp.[4] (Cachorrito de Sandia)	Sandia Basin, NLE	springs, outflows, and marshes
Cyprinodon sp. (Cachorrito de Villa López)	Villa López, CHI	large, impounded spring pond
Cyprinodon sp.[5] (Whitefin pupfish)	Río Yaquí and Río del Carmen, CHI	river edges, marshes, backwaters, springs
Megupsilon aporus	Ojo de Potosí, NLE	densely vegetated, spring-fed marsh
Fundulidae		
Fundulus limi	San Ignacio, BCN	springs and marshes
F. parvipinnis	Coastal CA, BCN, and BCS	estuaries, creek mouths; coastal
Lucania interioris	Cuatro Ciénegas basin, COA	shallow, highly variable, and marginal marshes and springs
L. parva	Pecos River, TX, NM; introduced in UT and CA	river edges, backwaters; springs, marshes
Plancterus z. zebrinus	Pecos River, NM, TX; lower Rio Grande, TX; introduced in AZ, UT	shallow, sandy, river edges, channels, and backwaters
Goodeidae (Empetrichthyinae)		
Crenichthys baileyi		
C. b. albivallis	Upper White River, White Pine County, NV	vegetated warm springs and their outflows and marshes
C. b. baileyi	Ash Spring, Lincoln County, NV	as above
C. b. grandis	Crystal and Hiko springs, Lincoln County, NV	as above
C. b. moapae	headwater springs of Moapa River, Clark County, NV	as above
C. b. thermophilus	Moorman and Moon River springs and Hot Creek, Nye County, NV	as above
C. nevadae	Railroad Valley, Nye County, NV	spring pools and warm outflows (30°–32°C)

Status	References
endangered due to restricted range, diversion, and bathing	Soltz, unpub. data
extinct due to exotic species	Minckley 1973; this chap.
unknown; population at Palomas, CHI, extirpated 1970s through groundwater pumping in NM	R. R. Miller 1981; Minckley, unpub. data
endangered due to limited range and water development	Contreras Balderas, *this volume*, chap. 12
vulnerable due to limited range and water development	Contreras Balderas, *this volume*, chap. 12
likely secure due to wide range	Hendrickson et al. 1981; R. R. Miller 1981
endangered due to water developments and exotic species	R. R. Miller and Walters 1972; J. E. Williams et al. 1985; Contreras Balderas, *this volume*, chap. 12
unknown; verbal reports of habitat loss and aquaculture at type locality	G. S. Myers 1927; R. R. Miller 1943c; Minckley, unpub. data
presumably secure due to wide distribution	C. L. Hubbs et al. 1979
rare, difficult to collect; presumably secure in extensive habitats	C. L. Hubbs and Miller 1965; Minckley 1969c, 1978, 1984
locally abundant	A. A. Echelle, pers. comm.
widespread and locally common	Shute and Allan 1980
vulnerable due to water development, herbicide use, and exotic species	J. E. Williams and Wilde 1981; Baugh et al. 1985; Courtenay et al. 1985
endangered due to development, water manipulations, and exotic species	as above
as above	as above
vulnerable due to water development and exotic species	as above
rare due to development and exotic species	as above
rare due to exotic species and irrigation development	C. L. Hubbs 1932; LaRivers 1962; C. D. Williams and Williams 1981

Table 15-1. Continued

Species	Native range	Habitat
Empetrichthys latos		
E. l. concavus	Raycraft Ranch Spring, Nye County, NV	spring pools, outflows, and marshes
E. l. latos	Manse Ranch Spring, Nye County, NV	as above
E. l. pahrump	Pahrump Springs, Nye County, NV	as above
E. merriami	Springs in Ash Meadows, Nye County, NV	deep parts of large spring pools
Poeciliidae		
Gambusia affinis	Rio Grande, Pecos rivers, NM, TX; widely introduced throughout region	river channels, margins, backwaters; springs, marshes, and artificial habitats of all kinds
G. alvarezi	Ojo del San Gregorio, CHI	spring and outflow; heavily vegetated, flowing water
G. amistadensis	Goodenough Spring, Val Verde County, TX	spring and outflow; vegetated, flowing water
G. gaigei	Boquillas and Graham Ranch springs, Brewster County, TX	clear, warm-water springs, outflows, and marshes
G. geiseri	Comal and San Marcos springs, central TX; widely introduced in west TX springs	cool, clear, high-discharge springs; often in swift water
G. georgei	San Marcos Spring and River, TX	shallow margins over mud bottoms with little vegetation
G. heterochir	Clear Creek, Menard County, TX	springhead, heavily vegetated, thermally variable
G. hurtadoi	Ojo de Hacienda Dolores, near Jimenez, CHI	thermally variable, heavily vegetated spring; clear, rich in carbonates
G. krumholzi	Río de Nava, near Piedras Negras, COA	unknown
G. longispinis	Cuatro Ciénegas basin, COA	weedy, ephemeral marshes along spring runs and streams
G. marshi	Cuatro Ciénegas and upper Río Salado basins, COA	headsprings, streams, marshes, canals, ponds, and lakes
G. nobilis	springs in Pecos River basin, TX, NM	clear, spring-fed streams and marshes; high carbonates and heavy vegetation
G. senilis	Río Conchos basin, CHI, DGO, and Devil's River, TX	stream channels, edges, backwaters; springs, outflows, marshes; eurythermal

Status	References
extinct due to groundwater pumping and springhead filling	R. R. Miller 1948; Minckley and Deacon 1968; this chap.
endangered, occurs only in refugia outside original range	as above; Soltz and Naiman 1978
extinct due to groundwater pumping	R. R. Miller 1948; Minckley and Deacon 1968; this chap.
extinct, apparently due to exotic species	R. R. Miller 1948; Minckley and Deacon 1968; Soltz and Naiman 1978; this chap.
possibly the single most abundant freshwater fish in the world	Lee and Burgess 1980; Courtenay and Meffe 1989
vulnerable due to restricted range; hybridizes with *G. affinis*	C. Hubbs and Springer 1957; A. A. Echelle et al. 1989; Minckley, unpub. data
extinct, habitat flooded by Amistad Reservoir	Peden 1970, 1973; C. Hubbs and Jensen 1984
endangered; all extant fish descended from three individuals	C. Hubbs and Springer 1957; C. Hubbs and Broderick 1963; C. Hubbs et al. 1986
likely secure, based on numbers and wide distribution	C. Hubbs and Springer 1957; Harrell 1980; Milstead 1980
possibly extinct; hybridizes with *G. affinis*	C. Hubbs and Peden 1969; J. E. Johnson and Hubbs 1989
endangered due to restricted range; hybridizes with *G. affinis*	C. Hubbs 1957, 1959, 1971; J. E. Johnson and Hubbs 1989
rare; known only from a single locality	C. Hubbs and Springer 1957
unknown; likely vulnerable due to limited range	Minckley 1963
vulnerable due to limited range; difficult to collect and assess	Minckley 1962, 1969c, 1978, 1984
probably secure; widespread and common within range	as above; Meffe 1985a
endangered due to limited range	C. Hubbs and Springer 1957; A. A. Echelle and Echelle 1980; A. F. Echelle and Echelle 1986; A. F. Echelle et al. 1989
unknown, but likely secure due to wide distribution	C. Hubbs and Springer 1957; Guillory 1980

Table 15-1. Continued

Species	Native range	Habitat
G. speciosa	undefined; south TX and northwest Mexico	springs, outflows, marshes; stream margins
Gambusia sp. (guayacón de San Diego)	Baños de San Diego, CHI	thermal springs and outflows; up to 43.8°C
Gambusia sp. (guayacón de Villa López)	Villa López, CHI	springs, outflows, and marshes
Poeciliopsis lucida[6]	Ríos Fuerte, Sinaloa, and Mocorito, SON, SIN	slow-moving to rapid streams and deep channels
P. monacha	Ríos Mayo, Fuerte, and Sinaloa, SON, SIN	harsh headwater arroyos, carved from bedrock; highly variable temperature, discharge, light and food
P. occidentalis		
P. o. occidentalis	Gila River, AZ, NM; Ríos de la Concepción and Sonora, SON	springs, marshes, river margins, backwaters; slow to moderate current, near cover
P. o. sonoriensis	Río Yaqui, AZ, SON, and lower Río Mayo, SON	as above
P. occidentalis ssp.	upper Río Mayo, SON	as above
P. prolifica	Río Yaqui to Río Grande de Santiago, SON, SIN, and NAY	streams, along sides of deep, slow pools
Xiphophorus couchianus	Río San Juan basin, NLE	spring pools and slow streams, with dense vegetation
X. gordoni	Cuatro Ciénegas basin, COA	as above
X. meyeri	La Alberca y El Socavon (Múzquiz), COA	as above

[1]Abbreviations for Mexican states are as follows: BCN, Baja California del Norte; BCS, Baja California del Sur; CHI, Chihuahua; COA, Coahuila; DGO, Durango; NAY, Nayarit; NLE, Nuevo León; SLP, San Luis Potosí; SIN, Sinaloa; SON, Sonora; and ZAC, Zacatecas.

[2]The population in the mainstream Río Sonoyta, SON, may represent an as-yet-undescribed subspecies (R. R. Miller and Fuiman 1987).

[3]Status of this widespread form(s) is uncertain; a number of distinct taxa could be involved.

[4]Diversity in pupfishes in this small, intermontane basin is remarkable (Contreras Balderas, pers. comm.), and they may represent a species flock.

[5]The two pupfishes (Río Yaqui and Río del Carmen) are superficially similar, but no critical comparisons of their morphology or other features have yet appeared.

[6]A number of unisexual forms of *Poeciliopsis*, not considered here, also occur in northwestern Mexico (Schultz 1989).

Status	References
unknown, only recently defined as a full species	Rauchenberger 1989
vulnerable due to limited range	M. L. Smith and Chernoff 1981; Minckley and Minckley 1986; Contreras Balderas, *this volume*, chap. 12
endangered due to water development	Contreras Balderas, *this volume*, chap. 12
unknown, presumably secure based on wide distribution	Thibault and Schultz 1978
unknown, but likely tenuous due to limited range and peripheral habitat	as above
endangered in USA due to habitat loss and predation by *G. affinis*; likely secure in Mexico due to extensive range	Minckley 1973; Minckley et al. 1977; Meffe et al. 1983; Vrijenhoek et al. 1985; Simons et al. 1989
as above	as above; Galat and Robertson, in press
unknown, likely secure due to wide and remote range	Vrijenhoek et al. 1985
unknown, but likely secure due to wide distribution	Thibault and Schultz 1978
endangered due to habitat degradation and loss	Rosen 1960; J. E. Williams et al. 1985
rare due to limited distribution	R. R. Miller and Minckley 1963; Minckley 1969c, 1978, 1984
endangered due to water development and limited range	Schartl and Schroeder 1988; Obregon and Contreras Balderas 1988; Contreras Balderas, *this volume*, chap. 12

about the same number (11.8%) are poorly known, and their status is not estimated.

Research and Management: General Perspectives

Management of threatened and endangered fishes must include a strong research component if it is to succeed. Intelligent management decisions must be based on at least a minimum amount of natural history information. More preferred, of course, are extensive data on the ecology and genetics of species of concern, including life-history data, accurate historical and contemporary distributions, knowledge of hierarchical population structure and gene flow, interactions with syntopic species, physiological responses to abiotic conditions, foods, reproductive and feeding behaviors, and so forth.

Two conceptual contexts must be included in management decisions. First is an appreciation of the historic and evolutionary aspects of a species' biology. The fishes we are attempting to conserve today have existed for millennia (G. R. Smith 1981a; Minckley et al. 1986) and have survived under natural conditions that must have included long sieges of remarkable adversity. Yet, our technological interference during the most recent instant in geologic time has led to rampant problems for these organisms, and ultimately to their rapid demise. Thus, perhaps the best and most effective management is to promote "naturalness" by protecting wild habitats. In practice this may be impossible when severe water depletions have already occurred, exotic species have become widespread and abundant, or humans demand the development of a water system. Nonetheless, short of complete habitat protection, management plans should recognize and strive to perpetuate as closely as possible the conditions under which a species evolved.

The second concept is recognition that no organism lives in isolation; each is one part of a complex and dynamic community that contributes to its own evolutionary milieu, just as it contributes to that of other species. Western fishes should not be managed as single, isolated entities, but as integral parts of the ecosystems they originally occupied. Entire watersheds or aquifers should thus be set aside for conservation efforts (Moyle and Sato, *this volume*, chap. 10) rather than isolated habitat patches for a species of special concern. Nonetheless, in the real world of conservation biology (Soulé 1986), economic, physical, and

Table 15-2. Status summary (numbers of taxa in each indicated category) for cyprinodontoid fishes of the southwestern United States and northern Mexico.

Families	Extinct	Threatened/ endangered	Rare/ vulnerable	Presumably secure	Status unknown	Total taxa
Cyprinodontidae	3	18	11	2	2	36
Fundulidae	0	1	1	3	0	5
Goodeidae	3	3	4	0	0	10
Poeciliidae	1	7	7	3	7	25
Totals	7	29	23	8	9	76
(%)	(9.2)	(38.2)	(30.3)	(10.5)	(11.8)	(100.0)

other constraints come into play, and species management in unnatural habitats is necessary. A lack of community perspective may result in successful manipulations for one species but decline and loss of others, which is a situation that must be avoided.

Research and Management: Case Studies

Case studies were selected from among examples in Table 15-1 to illustrate the marked contrasts and problems in cyprinodontoid conservation and management. Dealing with restricted species endemic to small habitats may seem straightforward and relatively simple, assuming, for example, that a spring(s) supporting an isolated species can be set aside, that its aquifer is secure, and that one can act rapidly and effectively enough to prevent extinction. In some cases, however, these requirements cannot be met. Consequently, the Monkey Spring pupfish (*Cyprinodon* sp.), one species and two subspecies of poolfish (*Empetrichthys merriami*, *E. latos pahrump*, and *E. l. concavus*), and the Amistad gambusia (*Gambusia amistadensis*) were lost (among others), despite attempts to save some of them, and we relate circumstances of their extinctions.

Another poolfish (*E. l. latos*) and the Leon Springs and Owens pupfishes (*C. bovinus*, *C. radiosus*), all of limited distributions, have thus far been spared extinction, but these successes have been far from meeting the goal of conserving whole communities. The poolfish persists in unnatural communities in alien surroundings. The Owens pupfish occupies managed habitats within its native range but in a variably complete natural community. Only the fortunate Leon Springs pupfish survives with most of its original ecosystem members in a natural remnant of its native range. Similar accounts, illustrating additional varia-

tions, exist for the Clear Creek and Big Bend gambusias (*G. heterochir*, *G. gaigei*) in Texas (J. E. Johnson and Hubbs 1989) but are not described here. We further refer the reader to one of the most important success stories in native fish salvation and recovery, that of the famous Devils Hole pupfish, whose history is described by Baugh and Deacon (1988) and Deacon and Williams (*this volume*, chap. 5).

Widely ranging species present other problems, varying from the complicated arena of international politics through decisions as to which of a number of populations are to be conserved and how to make (and who makes) that decision. The most extensive and well-documented attempt at management toward recovery of a formerly wide-ranging, short-lived fish is that for the Sonoran topminnow (*Poeciliopsis occidentalis*). We summarize those data as pertinent to later discussion and refer the reader to other papers for details (Meffe et al. 1983; Vrijenhoek et al. 1985; J. E. Johnson and Hubbs 1989; Quattro and Vrijenhoek 1989; Simons et al. 1989; Hendrickson and Brooks, *this volume*, chap. 16). The desert pupfish (*C. macularius*) and the Pecos gambusia (*G. nobilis*) are two other species we selected as examples. The first may be nearing extinction, and only rapid and decisive action will save it. The Pecos gambusia seems relatively secure, at least in the short term, because of a fortuitous occurrence of populations on already protected lands and co-occurrence with other endangered forms already under protection.

Extinctions: Obituaries for Five Taxa

Monkey Spring pupfish

This still unnamed species, once restricted to a single, small, isolated spring system in southeastern Arizona, disappeared in 1971 due to predation by an introduced fish and human error. Had certain events been anticipated and contingency plans made, the species would have been perpetuated with far less effort than was expended in failure.

The first known reference to Monkey Spring was by Pumpelly (1870), who traveled "to see some springs which were forming a heavy deposit of calcareous tufa." Photographs in 1889 (Hastings and Turner 1965, pls. 11a, 12a), a time of drought accompanied by severe overstocking of cattle, showed Monkey Spring and its environs largely stripped of vegetation and trampled by livestock; a small, natural ciénega was fed by the spring. By 1895 the ciénega had been obliterated by construction of a dike impounding Monkey Lake (Hastings and Turner 1965, pl. 9a) to create part of an irrigation system that exists today in modified form. Fishes were first sampled by F. M. Chamberlain (1904), a U.S. Fish Commission biologist (Jennings 1987) who described the area as follows:

> This spring . . . flows about 50 [miner's] inches of water at [a temperature of] 80°[F], strongly impregnated with lime. [Although varying in definition, a miner's inch equalled about 1.5 ft^3 (45 l) min^{-1}.] The original wasteways show limy incrustations resembling lava streams. The water now enters a pond about 150 yd. long, 30 to 50 yd. wide and about 6 ft. deep at deepest point. This is an artificial pond supported by a dike upon which willows a foot in diameter are growing. There may have been a small pool naturally formed before the dam was built which furnished habitation for the small native fish. The present pond is almost filled with a dense growth of Chara and another plant not identified. It is bordered by a tall growth of sedges.
>
> A second reservoir about 40 yds. square lies on the hillside a quarter of a mile farther down and perhaps at 75 ft. lower level, and the same height above the valley. The original spring flow was quite steep into the canon below, which has a width of 30 to 100 ft. with rocky walls at intervals. The water is now conducted by a ditch around the hill.

Monkey Spring originally flowed over an arroyo terrace, upon which travertine deposits

from its carbonate-rich water formed a natural dam. Travertine is a freshwater limestone formed when water with high levels of dissolved carbon dioxide carries supersaturations of calcium carbonate to the surface. Pressure release, temperature change, and agitation drive off the gas, altering pH to promote precipitation of insoluble carbonate. Photosynthesis by algae and higher plants also induces travertine deposition because it also removes carbon dioxide from water. The point at which travertine begins to form varies with concentrations of gasses and solutes, water volumes, and so on.

The paleoecology of this system was studied in 1964 by mapping travertine deposits and core sampling sediments (Minckley 1973; G. Cole, G. Batchelder, and Minckley, unpub. data). Travertine began to form and accumulate about 100 m from the source as hard, rough-textured veneers, tortuous channels, and enclosed tubes on the terrace surface. Included were casts of cattail leaves (*Typha* sp.), sedges, mosses, charophytes, twigs, and logs, as well as scattered shells of a minute endemic apulmonate snail (Hydrobiidae), which lives today in the spring source. Where water spilled into the arroyo 200–300 m downflow, "veils" of carbonates were deposited (Hendrickson and Minckley 1985, fig. 20). Diversion of water for irrigation beheaded the system, and the veils and other features are now weathering away in the absence of continuing deposition.

The old Monkey Lake bed was a white, powdery marl 20–50 cm thick, including calcified charophyte remains (stems and frustules), sedge stems and nutlets, shells of pulmonate snails (mostly *Physa* sp.) and sphaeriid clams, and a scattering of catfish (see below) and turtle (Sonoran mud turtle, *Kinosternon sonoriense*) bones. Beneath this were layers of rich black peat bearing rare charophytes, abundant sedge stems, roots, and nutlets, and hydrobiid shells; other molluscan remains were uncommon. The peat represented old ciénega deposits. Coring encountered rock at 1.5 m beneath the surface on the east side; the lower ends of samples included bits of travertine. On the west, angular gravels washed from the adjacent slope were interbedded with peat, and bottoms of cores were pure gravel, indicating that the maximum depth of the depression was about 4 m.

Isolation of fishes must have originated with canyon incision, which left precursors of Monkey Spring and its biota on the alluvial terrace. A nearby rhyolite-andesite intrusion (E. D. Wilson et al. 1960) probably forces water to the surface to form the spring (Hendrickson and Minckley 1985) and resists erosion more than other local rocks to form an abrupt barrier to further cutting. The outflow rises just upstream from this barrier, which also may have isolated and protected it. Over a period of time, terrace deposits were overlain, armored, and augmented by travertine. Humans' enhancement of the natural dam to form Monkey Lake and diversion of flow around a hill to a reservoir increased both its isolation and the amount of habitat.

Few quantitative data are available for the fish population. Chamberlain (1904) clearly selected only a few specimens, depositing fifty-four adult pupfish, forty-one topminnows, and three chubs at the U.S. National Museum (R. R. Miller, University of Michigan Museum of Zoology [UMMZ], pers. comm.). Chamberlain noted:

> The upper pond is abundantly inhabited by Poecilia [Sonoran topminnow] Cyprinodon, and Leuciscus [Gila chub, *Gila intermedia*]. In addition, a catfish has been introduced [later referred to Yaqui catfish, *Ictalurus pricei*, by Miller and Lowe (1964)]. A hardshell turtle was seen. The longest Leuciscus seen was about 8 inches, only small ones were taken in the seine. In second reservoir only the Poecilia and a few Cyprinodons were seined. A few frogs.

Carl L. Hubbs sampled the area in September 1938. His unpublished field notes

(field no. M38-206, UMMZ) estimated Monkey Lake to measure 60 by 150 m. The spring and pond were surrounded by "grassy slopes [that were], partly wooded." Pupfish were abundant. About two hundred were preserved (R. R. Miller 1943a), along with series of chubs and topminnows; two juvenile catfish were caught.

Robert R. Miller (field no. M50-60, UMMZ) visited the site in April 1950. The spring was rimmed by a "sparse willow border," and cattle had trampled edges of the outflow. Monkey Lake had been drained two years earlier and his collection was in the spring, its outflow, an earthen canal leading to a "large earthen reservoir," which he was told was built about 1943 (perhaps it was reconstructed from that reported by Chamberlain), and in the reservoir. Pupfish were rare in the outflow and uncommon in the reservoir, topminnows swarmed everywhere, chubs were common, and the introduced catfish was reported by the rancher but none was collected. Results of ten seine hauls in the reservoir (estimating numbers of topminnows) were as follows: 67 (3.9%) Gila chubs; 62 (3.6%) pupfish; and 1591 (92.5%) Sonoran topminnows. The rancher was interested in stocking largemouth bass (*Micropterus salmoides*) in the pond, but Miller "told him it would wipe out the *Cyprinodon* and *Gila* and urged him to construct a small pond near the spring source for the *Cyprinodon*. He seemed impressed."

According to Hastings and Turner (1965), Monkey Lake was drained when "a former owner bulldozed the bottom, inadvertently removing the seal that retained the water, and opening an intricate system of natural piping in the travertine underneath." When Minckley (unpub. data) began work there in 1964, Monkey Lake remained dry, the headspring was fenced from livestock, and an earthen canal diverted water for domestic use and into the same irrigation pond seen by Miller. Topminnows were abundant from headspring to pond, chubs were rare in the spring and out-flow and abundant in the reservoir, and pupfish swarmed along margins of the pond but were not taken elsewhere. The catfish was represented only by skeletal remains on the dry floor of Monkey Lake.

Stocks of pupfish were moved for various purposes to artificial ponds and aquaria at Arizona State University (ASU) and the Deer Valley office of the Arizona Game and Fish Department (AZGFD), and attempts were also made to establish them in the headspring and its outflow (Minckley 1973; Schoenherr 1974, 1988). Adults remained in the headspring for more than two years and grew far larger than in the irrigation pond (to almost 60 mm total length versus perhaps 40 mm) but apparently did not reproduce. Elsewhere, they became abundant in artificial ponds, overwintering to reproduce in their second and third summers.

Events in 1968 spelled doom for the species. First, six adult largemouth bass were reportedly stocked into the irrigation pond, and gill nets were used to remove that same number. Next, the pond was scheduled to be drained and repaired. Chubs and topminnows were living in the spring and outflow (as were pupfish, introduced to the headspring), so all were expected to reappear naturally. More than five hundred pupfish were nonetheless moved from the pond to AZGFD holding facilities for restocking if necessary. All pond fishes then were destroyed by drying; no catfish or bass were found.

Only topminnows had immigrated to the refilled pond by autumn 1968, so approximately four hundred of the wild-caught pupfish were reintroduced on two occasions in winter–early spring 1969. Nine adult chubs from a downstream irrigation structure were also restocked. Pupfish had bred by April 1970, and their perpetuation seemed assured. We were unaware, however, of another unauthorized stocking of an unknown number of adult largemouth bass. They reproduced prolifically, and the shorelines swarmed with juveniles when the pond was next visited in late June. The only pupfish found were a few in the headspring, where they persisted with no recruitment and finally disappeared in 1971. The local form of chub, perhaps equally as distinct from its relatives as the pupfish (Rinne 1976; DeMarais 1986), also had disappeared. Only topminnows persisted.

By the time the catastrophe was detected, the pupfish at AZGFD had all been used for reintroduction or had expired. Research stocks at ASU were depleted and never attained production sufficient to save the species. Fish kept in aquaria were "far more secretive and 'nervous' than desert pupfish, remaining that way throughout life, and demonstrated low reproductive success" (Minckley 1973). Populations in outdoor ponds had also waned, in part a result of inattention, and the species was lost. The outflow canal from Monkey Spring was lined with concrete in the early 1970s (Gerking and Plantz 1980). Topminnows live today in the headspring and also are present in dense, marginal vegetation of the irrigation pond, where they are heavily preyed upon by bass and bluegill (*Lepomis macrochirus*), and with their populations maintained in part by immigration.

Three empetrichthyine poolfishes

Three of four modern taxa of poolfishes, genus *Empetrichthys*, all from Nye County, Nevada, disappeared between 1948 and 1962 (Minckley and Deacon 1968; Deacon 1979). The fourth, *E. l. latos*, was saved from extinction and is discussed in a later section.

As is often the case in these situations, the disappearance of poolfish (*E. merriami*) from Ash Meadows is shrouded in mystery. The species never was common. Collectors on the Death Valley Expedition of 1891 preserved only six specimens (Gilbert 1893). Three more were captured in 1930, twenty-two in the period 1936–1942 (R. R. Miller 1948), and the "last specimen was seen in 1948" (Soltz and Naiman 1978). All collection sites were in large springs (described by Miller [1948] as varying to maxima of 15 m across and 9 m deep), wherein the fish lived near the bottom. LaRivers (1962) attributed its rarity to competition from a great abundance of pupfishes. Minckley and Deacon (1968) speculated that its demise reflected the added competition or other interactions resulting from establishment of exotic species (Deacon and Williams, *this volume*, chap. 5).

The reasons are clear for the disappearance of three other poolfish taxa from the adjacent Pahrump Valley. Their springs simply stopped flowing and desiccated as a result of water mining (Minckley and Deacon 1968), when water volumes pumped from deep beneath the valley floor exceeded recharge rates of the aquifer. Springs on the floors of desert basins reflect underground water tables forced upward by artesian pressure of montane aquifers. Water for these vast reservoirs is often from distant recharge areas, and its underground movement may be remarkably slow. For example, water issuing from some springs in Ash Meadows moves 100 km or more in a complex underground journey that takes ten thousand to thirty thousand years (Winograd and Pearson 1976). Numerous similar situations are known in the West (Riggs 1984). Once such a store of water is depleted, it is gone for a long time.

The first deep wells in Pahrump Valley were drilled in 1910, and fifteen pumps were removing 525 ha-m per year by 1916 (Minckley and Deacon 1968). Original depth to water is unknown, but by 1951 it lay 11.3 m below the surface as thirty-nine wells removed 3100 ha-m of water. The water table had fallen to between 21 and 26 m in 1961, when sixty-four wells pumped about 4500 ha-m. Manse Spring, the only system for which long-term discharge data are available, declined from an estimated 10.2 m^3 min^{-1} in 1875, to 5.4 m^3 in 1916, to 4.4 in 1951, and to 2.5 m^3 min^{-1} in 1966. Its outlet ceased to flow in 1971 or 1972, and the pool dried in 1975. Manse Spring poolfish were transferred to other places in 1971 and saved, although they remain listed as endangered (U.S. Fish and Wildlife Service [USFWS] 1980b, 1989c). Other systems dried earlier. The two Pahrump springs, both originally used for irrigation, failed in 1957 (between 1955 and 1957 according to Soltz and Naiman [1978]). Raycraft Spring was bulldozed full of soil in an attempt at mosquito control in 1957 (Minckley and Deacon 1968; 1955 according to Soltz and Naiman 1978), probably after its decline to a stagnant pool created habitat suitable for the pest. No permanent natural surface waters now remain in Pahrump Valley.

These desert springs, although isolated in space, were not small. They typically consisted of one or more major pools, an outflow channel, and associated marshes. The springhead at Raycraft Ranch (supporting *E. l. concavus*) was 1.6–8.2 m wide, 13.1 m long, and to 0.5 m deep, with a slight current in the pool and a swiftly flowing outlet channel (R. R. Miller 1948). Discharge was estimated by Waring (1920) as 38 l min^{-1}. Manse Ranch Spring pool (*E. l. latos*) was 3.0–16.4 m wide, 19.6 m long, 0.3–1.8 m deep, and its current was "swift in the outlet"; a second smaller, fishless spring trickled to the main pool from 45 m away. Miller (1948) gave no dimensions for the two Pahrump Ranch springs (inhabited by *E. l. pahrump*) but noted that a northern pool was dredged in 1941 to the detriment of native fishes and occupied by common carp (*Cyprinus carpio*) in 1942. Poolfish were rare in a southern spring in 1942, but a population was present "in a marshy area about 200 yards from the source."

Amistad gambusia

Reasons for extinction of this Texas species are also clear-cut: destruction of habitat followed by human error (J. E. Johnson and Hubbs 1989). The species was discovered in 1968 in Goodenough Spring, Val Verde County, after Amistad Reservoir had been completed and began to fill on the adjacent Rio Grande (Peden 1970, 1973). The first specimens were captured in April; the reservoir level rose to permanently inundate the habitat in July; and a few individuals were collected in August in flooded brush and cacti near the spring, which was by then covered by 7 m of water. In April 1969 a diver could not detect clear water issuing from the spring, which then lay 23.2 m beneath the turbid reservoir surface; no *G. amistadensis* could be found.

Goodenough Spring was among the three largest warm-water springs in west Texas. It emerged from beneath a limestone cliff, discharged 1.9–18.4 m^3 sec^{-1} of water, and flowed 1.3 km to the Rio Grande. A momentary extreme of 101 m^3 sec^{-1} reflected temporary flooding from the normally dry arroyo above it. Floods in the Rio Grande historically inundated the source to a depth of 12 m. Water temperatures varied from 20.6° to 29.4°C, with 95.6% of 159 readings over twenty-eight months falling between 26.7° and 29.4°C (Peden 1973).

Gambusia amistadensis occupied the headspring, creek, and creek–Rio Grande confluence. It co-occurred with at least fifteen other fishes, many piscivorous, and including western mosquitofish (*G. affinis*), which interact to the detriment of most spring-inhabiting gambusias and other poeciliids. Peden (1973) could not find it in other local springs. Another survey in 1979 also failed to find the species, and C. Hubbs and Jensen (1984) concluded that it was unlikely to exist in nature.

After 1968, a stock in an artificial pool at the University of Texas (UT) in Austin was its only known buffer against extinction (C. Hubbs and Jensen 1984). Fish taken to Vic-

toria, Canada, proved difficult to maintain in aquaria and died. At the University of Texas the fish suffered mortality from low winter temperatures and also were intolerant of summer water temperatures greater than 35°C. Individuals were maintained indoors to ensure against winterkill, and a second population was established in a concrete pond adjacent to an identical pond holding Big Bend gambusia.

Both ponds were fed by well water from a single T-fitted outlet, which, unfortunately, had no valve; water passed from the T to the edge of each pond and was left flowing on winter nights to prevent cold mortality. Someone had incorrectly replaced a hose, cutting off water flow, and cold mortality occurred in 1972. When the mistake was discovered, both hoses were repositioned incorrectly beneath the surfaces of both ponds. Fish swam from pool to pool through the hose, and both stocks may have been contaminated (C. Hubbs and Williams 1979).

In 1974 a stock of *G. amistadensis* had been transferred from the concrete pool to Dexter National Fish Hatchery (NFH), New Mexico. When the potential contamination at UT was detected, Dexter fish were replaced from original stock at Austin. As soon as the hatchery stock began to flourish, routine monitoring at Austin was discontinued.

Fish from the indoor stock at Austin were then determined to be mosquitofish rather than Amistad gambusia. The fish at Dexter were also examined and only mosquitofish were found, which also proved to be the case in the one remaining pond at UT; the other refuge pond had dried. Thus, despite natural co-occurrence of Amistad gambusia and mosquitofish, prolonged contact under artificial conditions must have resulted in extirpation of the springhead form. The USFWS (1987g) removed *G. amistadensis* from the official list of endangered species due to its presumed extinction.

Successes with Species of Restricted Geographic Range

Pahrump poolfish

This taxon was restricted to Manse Spring, which failed due to overpumping in Pahrump Valley, Nevada. It was abundant in 1936–1942 (R. R. Miller 1948), and remained so into the early 1960s. However, establishment of exotic goldfish (*Carassius auratus*) resulted in population depression after 1962 or 1963 (Deacon et al. 1964), emphasizing its precarious state.

This last event was accompanied by progressive lowering of the spring, and in 1971 twenty-nine fish were moved to Corn Creek Spring on the USFWS Desert Game Range near Las Vegas, where the species persists today. In 1972 sixteen individuals from Corn Creek or Manse Ranch Spring were introduced into a refuge complex named Shoshone Ponds, which had been constructed and set aside by the U.S. Bureau of Land Management (USBLM 1987) for the species in Spring Valley, White Pine County. These were followed by fifty more in 1976, and these fish founded a population that also still survives. Both Corn Creek Spring and Shoshone Ponds are fed by thermal water, natural in the former and from a drilled artesian well in the latter.

A third population originated from an unrecorded number of poolfish transferred to a reservoir at Spring Mountain Ranch State Park, Clark County, in 1983. This habitat, unlike the others, is fed by surface runoff and fluctuates seasonally in temperature. The reservoir stocking was nonetheless successful, although poolfish remain torpid during winter (Baugh et al. 1988). Two other places were apparently stocked with Pahrump poolfish, but these failed to maintain themselves. Soltz and Naiman (1978) noted "relatively large reproducing populations . . . in ponds at Corn Creek, near Las Vegas, and in an isolated canyon above the Colorado River. A third population was established in an artificial refugium in Ash Meadows, but it died out in 1977." Other artificial and transitory stocks were also kept for short periods at ASU and the University of Nevada, Las Vegas (J. E. Deacon, UNLV, pers. comm.).

Problems in management at Corn Creek have included appearance of mosquitofish on two occasions, both resulting in depression of poolfish populations that necessitated removal of the endangered form and chemical eradication of the pest. Potentially predatory exotic bullfrogs (*Rana catesbeiana*) were present, but fifty-one frogs examined by Withers (memorandum to file [27 November 1985], Nevada Department of Wildlife, Las Vegas) had eaten no poolfishes. Other persistent problems include encroaching cattails (*Typha* sp.) at Corn Creek, maintenance of artesian wells at Shoshone Ponds, and needs for intensive monitoring in attempts to interdict unauthorized introductions of non-native fishes at all three refugia.

Owens pupfish

This pupfish is endemic to the Owens River drainage in east-central California. It was originally abundant in marshes, springs, sloughs, and irrigated wet pastures from Owens Lake near Lone Pine, Inyo County, north to Fish Slough, Mono County (D. H. Kennedy 1916). Details of its distribution in springs and streams around Owens lake were never documented. Owens pupfish were thus distributed over a substantial geographic area, although isolated in a single intermontane basin.

Early in the century these shallow marshes were dramatically reduced by channelization and export of much of the surface water from Owens Valley through the Los Angeles Aqueduct. At about the same time, predatory game fishes (largemouth bass, brown trout [*Salmo trutta*), and rainbow trout [*Oncorhynchus mykiss*]), became established, followed by dramatic declines of the pupfish. It was, in fact, considered extinct when named by Miller (1948). The species was rediscovered in Fish Slough in 1956 by California Department of Fish and Game (CADFG) personnel, and again in 1964 by C. L. Hubbs, R. R. Miller, and E. P. Pister. The population numbered no more than a few hundred fish. In August 1969 their water supply failed, and about eight hundred fish were held in a live-cage in another channel to ultimately serve as founders of the managed populations of today. The story of its apparent extinction, rediscovery, management, and recovery was detailed by R. R. Miller and Pister (1971) and Pister (*this volume*, chap. 4).

The Owens pupfish is one of the most extensively and successfully managed cyprinodontoids. B. J. Turner (1974) found little genetic differentiation among any of the pupfishes of Nevada and California, including *C. radiosus*. A comparative study of the evolution of thermal tolerance revealed Owens pupfish to be just as resistant to variation as other pupfishes, and more so to low temperatures, presumably because of its high-elevation (to 1200 m) habitats (J. H. Brown and Feldmeth 1971). Generalizations about pupfish breeding systems and basic life-history information were successfully applied in its management. There is also an extensive "gray literature" that we chose not to cite, consisting of CADFG reports, several environmental impact reports, and an uncompleted doctoral dissertation. Much of this is cited in recovery plans for this species (USFWS 1984h) and for the Owens tui chub (*Gila bicolor snyderi*; USFWS 1989d).

Basically, conservation of the Owens pupfish may be attributed to the vision and dedication of Edwin Philip Pister, longtime CADFG fisheries manager for Inyo and Mono counties, California, and a founder of the Desert Fishes Council. He combined basic fisheries management methods and knowledge of the species's biology to quickly and successfully build and stock three refugia—two in Fish Slough and one at Warm Springs about 40 km away (R. R. Miller and Pister 1971).

Creation of refugia involved damming headsprings to erect fish barriers, thus simultaneously providing habitat and barriers to access by exotic predators, particularly largemouth bass. Exotic species were eradicated with piscicides before pupfish were reintroduced. All of the first three refugia were successful, although little shallow water, preferred by most pupfishes, was (or is) present. All sites have deep water created as a consequence of barrier dams, which provide habitat for other native fishes (Owens tui chub, Owens sucker [*Catostomus fumeiventris*], and speckled dace (*Rhinichthys osculus* ssp.]) as an additional benefit of the program.

However, deep, open water also creates a number of problems in management. It attracts vandalism in the form of unauthorized introductions of game fish as well as providing preferred habitat for these introduced species. Several illegal stockings of largemouth bass have been made, necessitating piscicide treatment after removal for restocking of as many native fishes as possible. A second problem is mosquitofish, which repeatedly colonized one spring in Fish Slough and either colonized or was illegally introduced into two others. Extensive operations to remove them have been undertaken several times, with varying success. On the other hand, extensive development of emergent vegetation, mostly cattails, quickly reduces shallow-water habitat. Cattails must be removed on a regular basis at considerable expense. Pupfish in one refuge slowly declined to disappearance by 1984, presumably due to siltation and overgrowth of vegetation. Efforts are under way to expand the overall population by stocking other refugia in Owens Valley. The CADFG has stocked six other potential sites since 1986, and the species has persisted in at least two of these.

Leon Springs pupfish

Cyprinodon bovinus was described by Baird and Girard (1853) from Leon Springs, Texas, which had ceased to flow by 1958 as a result of groundwater pumping (Brune 1975). The pupfish population (as well as a stock of Pecos gambusia, see below) was extirpated early in development of the system for irrigation use. Carl L. Hubbs was unable to find Leon Springs pupfish in 1938, and Hubbs (1957), Miller (1961), and the U.S. Bureau of Sport Fisheries and Wildlife (1966) listed the species as extinct.

Springs at the type locality were developed for irrigation before 1908, and a Leon Springs Irrigation Company existed in 1911. The springs were impounded in 1918, and discharge could not be measured in 1920 because water from Lake Leon backed over the outflows. The lake was apparently filled by spring flow at least until 1932. Local irrigation by artesian flow was in seasonally low supply from 1939 to 1946, alternated between creek, lake, and spring or artesian sources from 1947 to 1951, and after 1951 was from pumped groundwater. Records of water use (C. Hubbs 1980) also suggest a decline in supply: more than 617 ha-m yr^{-1} until 1931; more than 493 until 1944; more than 370 before 1960; and as little as 123 ha-m yr^{-1} in 1971. In addition, rotenone was applied to the lake to remove common carp as a management action favoring sport fishes (Knapp 1953). By 1978 lake levels were being maintained by pumped groundwater to serve local gravity-flow irrigation, and plans existed to develop the area into a residential subdivision (C. Hubbs 1980).

Rediscovery of *C. bovinus* was largely by chance. In summer 1964, while returning to Arizona from Mexico, automotive failure briefly marooned one of us (WLM) in Fort Stockton, Texas. While awaiting repair, he inquired about possibilities of local springs and pupfishes, using preserved specimens from Mexico as examples. He was assured by mechanics repairing his vehicle that they regularly seined the same kind of fish for bait from a spring along Leon Creek. Directions and name of the owner were obtained, the area was visited in December 1965 (Minckley and Arnold 1969), and pupfish were collected that were later identified and redescribed as *C. bovinus* by A. A. Echelle and Miller (1974).

Permanent water exists as two semi-isolated reaches in Leon Creek, which originates in seeps and flows 1 km to join another 1-km-long outflow from Diamond-Y Spring. A combined, permanent flow then passes another kilometer or so and percolates into the ground. The channel then becomes ill-defined and dry for about 2 km, then water reenters from seeps and springs to form a second 2.7-km reach of perennial flow that ends in two livestock-watering tanks. The reach upstream from the Diamond-Y inflow sometimes dries, and the downstream extent of both the upper and lower segments varies with climatic conditions (C. Hubbs 1980). The water is saline, with conductance varying between 13,000 and 17,000 μmhos cm^{-1}; salt encrustations are common along the banks, which are mostly vegetated by sedges and other low marshland plants (S. E. Kennedy 1977).

The pupfish population and its habitat remained relatively stable from 1965 to 1974. Then, sheepshead minnows (*Cyprinodon variegatus*), a species from Atlantic and Gulf coastal habitats, appeared, presumably as a result of live-bait operations. By November 1975 an evident hybrid swarm had formed, and genetic swamping of the native species had become a clear possibility. Management of the Leon Springs pupfish has since involved its protection at Dexter NFH (Johnson and Jensen, *this volume*, chap. 13) and efforts to remove the *C. variegatus* genome from the pupfish's habitat.

Well before the danger of genetic swamping (and even today), the entire range of this pupfish population lay within an oil and gas field, and its major water source, Diamond-Y Spring, was less than 800 m downhill from an operating gas-cracking plant (S. E. Kennedy 1977). The oil field was developed in the 1940s, and the pupfish and a number of other species had already survived well drilling as well as more than three decades of pumping operations before their discovery! The field operators have proven cooperative, and an earthen dike now protects the spring from potential pollution.

Efforts involved in eradicating the *C. variegatus* genome from Leon Creek were described by C. Hubbs et al. (1978) and C. Hubbs (1980). Hybrids were present in the entire lower reach in January 1976, and it was treated with rotenone in February. Prior to treatment, six 100-m segments were seined. All macroinvertebrates and fishes (excluding pupfishes) were set aside for reintroduction, and a stock of *C. bovinus* was moved to Dexter NFH as insurance against a possible disaster. Rotenone was applied by overall spraying and by drip stations placed at spring inflows. Areas of observed pupfish breeding activity were further treated with antimycin-A to kill eggs that might be present. After the piscicides had dissipated, fish and macroinvertebrates were reintroduced, and *C. bovinus* and associated fishes and invertebrates were also transferred downstream from the upper section.

In August 1976 the lower reach of Leon Creek supported all the reintroduced fishes except *Gambusia geiseri* (which likely succumbed to habitat extremes and was probably introduced into the system anyway), even including a natural hybrid swarm between *G. nobilis* and *G. affinis*, which later resumed pretreatment conditions. However, phenotypic traces of genetic contamination by *C. variegatus* persisted near inflowing springs and in the terminal ponds. On the next visit, in November 1976, all pupfish with an apparently introgressed phenotype were removed by seining. Fish appearing to be *C. bovinus* were inhabiting springs, while those with apparent hybrid influence were in shallower, colder water. In March 1977 hybrids seemed absent from parts of the lower segment, and the livestock ponds had dried, but suspected hybrids were present elsewhere; introgressed individuals were again seined and destroyed.

In November 1977 hybrids had again become common in much of the lower reach, and a need for additional chemical treatment was indicated. The lower reach was scheduled to be poisoned twice in April 1978, two weeks apart, the first time with antimycin-A and the second with rotenone. This plan was disapproved by the USFWS because it would necessitate killing *Gambusia nobilis*, which was by then listed as endangered. Meanwhile, pupfish with hybrid phenotypes had appeared in the upper reach, and it was deemed too hazardous to treat both reaches.

Volunteers were diverted (and their numbers expanded) to a selective seining operation as an immediate alternative to chemical treatment. Three or four fine-meshed seines were in contant use during a three-day effort. All parts of the system were seined repeatedly and all suspect pupfishes were removed. In August 1978 no obvious hybrid phenotypes were discernible, but perhaps 2% of all pupfishes examined varied toward the *C. variegatus* phenotype and were removed anyway. Samples since that date have (amazingly) included only *C. bovinus*, and subsequent allozyme analyses of 176 specimens from throughout the Leon Creek system detected no evidence of introgression of genes from *C. variegatus* (A. A. Echelle et al. 1987).

Problems with Space: Conserving Formerly Widespread Species

Desert pupfish

The desert pupfish originally occupied much of the lower Colorado River drainage, including the Salton Sea, Gila River, and Colorado River delta, as well as the independent Río Sonoyta (R. R. Miller and Fuiman 1987). It was sometimes remarkably abundant in riverine sloughs, marshes, and shoreline pools of desert lakes, as well as springs (Miller 1943a; Barlow 1958a, b, 1961; Minckley 1973). Most populations in the United States are now extirpated, with the remainder restricted to isolated refugia (Schoenherr 1988). Desert pupfish remained abundant at a few localities on the Colorado River delta in Mexico in 1986, but these may have been an artifact of recent discharge events (Hendrickson and Varela 1989); observations in that same area between 1976 and 1984 indicated that they were rare (Minckley, unpub. data). After almost a decade of effort by members of the Desert Fishes Council, the desert pupfish was listed as endangered in March 1986 (USFWS 1986d); no recovery plan has yet appeared.

Much of the research on desert pupfish has focused on its taxonomic status. It was described from the San Pedro River, Arizona, by Baird and Girard (1853). R. R. Miller (1943a) documented occurrence of one morphological form (*C. m. macularius*) throughout the Gila and lower Colorado rivers, and another, recently named *C. m. eremus* by Miller and Fuiman (1987), in Quitobaquito Spring, Organ Pipe Cactus National Monument, Arizona (Río Sonoyta basin). The form occupying the mainstream Río Sonoyta may comprise another distinctive and yet unnamed subspecies (McMahon and Miller 1985). Pupfish in the Salton Sea, which may have originated from desert springs inundated when the Colorado River entered to flood that basin in

1904–1907 (Miller 1943a), from fish carried by those floodwaters (B. W. Walker 1961), or both, were recognized as *C. m. californiensis* by C. L. Hubbs et al. (1979) and others (Loiselle 1980, 1982; Deacon et al. 1979). However, there is little electrophoretic or morphological indication of differentiation of Salton Sea pupfish from those of the Colorado River (Miller 1943a; B. J. Turner 1983). In addition, the Salton Sea (and its Mexican counterpart, Laguna Salada) has been filled by the meandering river and then desiccated as many as seven times between 1840 and 1907 (Carpelan 1961), and pupfish populations must have been repeatedly mixed and isolated as well. Salton Sea pupfish were referred to *C. m. macularius* by R. R. Miller and Fuiman (1987).

Other works on desert pupfish have included numerous behavioral analyses as well as investigations of life history and physiological tolerances. Behavior studies focused on breeding of different populations (Cowles 1934; Barlow 1961; Kynard and Garrett 1979) and comparative work on pupfish breeding systems in general (Liu 1969; Kodric-Brown 1977, 1978, 1981). The desert pupfish more often than not lives in harsh and variable habitats that strongly influence its life history (Kinne 1960, 1965; Kinne and Kinne 1962a, b; Sweet and Kinne 1964; Crear and Haydock 1970; Soltz and Hirshfield 1981). It is able to adjust to environmental temperatures from near freezing to above 38°C, salinities greater than 100 g l^{-1}, and dissolved oxygen less than 1.0 mg l^{-1} (Lowe et al. 1967; Lowe and Heath 1969; Hillyard 1981). Other specific aspects of its life history and population biology were described by Cox (1972a, b), Naiman (1979), Walters and Legner (1980), and McMahon and Tash (1988).

Despite what has been learned, desert pupfish underwent a well-documented and drastic decline following 1950, particularly after 1970 in the Salton Sea region (Schoenherr

1979, et seq.; G. F. Black 1980). Biologists were acutely aware of its disappearance by the mid-1970s, almost certainly due to remarkable increases in populations of newly introduced, non-native sailfin and shortfin mollies (*Poecilia latipinna*, *P. mexicana*) in the 1960s, followed by population explosions of African tilapias (*Oreochromis* spp., *Tilapia zilli*) in the 1970s and 1980s, and their progressive invasion of irrigation drains, shore pools, and other places that had previously acted as pupfish refugia (Schoenherr 1979, 1985, 1988).

Diffuse management efforts commenced in the late 1970s, including monitoring of transplanted and natural populations, rearing under artificial and seminatural conditions, introductions to establish additional refuge stocks, and manipulations to exclude or remove exotic fishes to enhance existing and often declining stocks (Schoenherr 1988; Hendrickson and Varela 1989). State and federal agencies and groups of individuals focused on their local populations, and no plan for recovery throughout the original geographic range has yet been developed.

Efforts to maintain desert pupfish in refugia have nonetheless been extensive. Only three natural populations exist in California, but Schoenherr (1988) listed six managed refugia and plans for two more for Salton Sea fish. R. R. Miller and Fuiman (1987) noted another California refuge not listed by Schoenherr. In Arizona, no native *C. m. macularius* populations remain, but Schoenherr (1988) reported refuge stocks of fish derived from the east side of the Colorado River delta in at least six managed refugia in Arizona, varying from natural springs, through ponds and springs in state parks, to stocks perpetuated artificially at universities and elsewhere. Hendrickson and Varela (1989) recorded introduction of the same pupfish stock into ten natural or seminatural habitats in Arizona since 1982 (including those recorded by Schoenherr

1988), eight within its native range and two outside. The fish persisted in 1987 at only two of the former and one of the latter. In addition, at least four populations of *C. m. macularius* were successfully maintained in small artificial ponds in public and private parks or educational institutions within their native range in the United States, and another in Mexico.

The Quitobaquito form of desert pupfish was maintained in at least two refugia on Organ Pipe Cactus National Monument in addition to its natural habitat until just prior to its listing as endangered, when it was removed from the refugia by the U.S. National Park Service (USNPS). The USNPS was concerned that listing might include critical habitat or other provisions that would necessitate substantial efforts to maintain the populations, which had been established in habitats fed by artificial, less-than-permanent water supplies (Minckley, unpub. data). Additional Quitobaquito pupfish exist at educational institutions (three stocks; Miller and Fuiman 1987), and three populations live in habitats outside their native ranges in Arizona (within the original range of *C. m. macularius*)—one in headwaters of the Santa Cruz River and two others in the San Pedro River drainage. One of the latter is almost certainly a mixture of *C. m. eremus* and *C. m. macularius*, and there is concern that this or other stocks may spread to contaminate habitats useful for recovery of the native form.

Sonoran topminnow

Poeciliopsis occidentalis exists as two subspecies in the United States: *P. o. occidentalis* (Gila topminnow) in the Gila River basin, and *P. o. sonoriensis* (Yaqui topminnow) in the uppermost Río Yaquí. We restrict our review to the Gila River form, which was "one of the commonest fishes in the southern part of the Colorado River basin" prior to 1940 (C. L.

Hubbs and Miller 1941) but declined so dramatically in the next two decades (R. R. Miller 1961; Minckley 1973; Minckley et al. 1977; Meffe et al. 1983) that it was placed on the federal endangered species list in 1967. Its extirpation was attributed to habitat degradation and elimination through predation by introduced mosquitofish.

Early research on topminnows included that by Schoenherr (1974, 1977, 1981) and Constantz (1974, 1975, 1976, 1979, 1980) on reproduction, habitat use, behavior, and interactions with mosquitofish. Meffe et al. (1983) and Meffe (1983a) expanded on this work to include additional analyses of impacts of mosquitofish and document further declines of natural populations. Meffe (1984) and Minckley and Meffe (1987) argued that moderate flooding might benefit this and other native fishes by removing exotics. Vrijenhoek et al. (1985) studied genetics of natural populations in the United States and Mexico, and Quattro and Vrijenhoek (1989) correlated genetic and fitness differences among populations. Concurrently, there were several reports of new topminnow populations and losses of others (J. E. Johnson and Kobetich 1969; McNatt 1979; Rinne et al. 1980; Collins et al. 1981; Meffe et al. 1982). Management of the Sonoran topminnow consists of a combination of monitoring natural stocks, manipulations to recover declining native populations, stocking and monitoring new populations within the natural range, maintenance of stocks at universities, museums, and hatcheries, and an ongoing program of basic research.

Topminnows had already largely been forced to springs when Minckley (1973) began an intensive study of Arizona fishes in 1963. He found only five natural populations in as many years. Efforts in the 1970s added eight localities, so thirteen natural populations were known in 1979. No more populations were subsequently found, and the thirteen had declined to ten in 1986; six of these were threatened by mosquitofish (Simons et al. 1989).

Problems with management of natural populations are illustrated by attempts to recover stocks at Bylas Springs, Arizona (Meffe 1983b; Marsh and Minckley 1990). Topminnows occupied three small, otherwise fishless, springs (S-I, S-II, and S-III) flowing from near a stony escarpment to desiccate, percolate into unconsolidated alluvium, or trickle over a substantial cut bank into the Gila River. S-I and S-III were discovered in 1968 (J. E. Johnson and Kobetich 1969), and S-II was found in 1981. Unusual winter floods in 1977–1978 allowed mosquitofish to enter S-I, and mosquitofish and red shiner (*Cyprinella lutrensis*, recorded only once) to enter S-III; S-II remained isolated. Topminnows immediately declined in the two contaminated systems, surviving only at the headsprings.

More than 150 topminnows, along with large but unestimated numbers of endemic and indigenous invertebrates, were removed from S-I in March 1982 and the habitat was poisoned with antimycin-A. No live fishes were evident three weeks later, and the native animals were restocked. By July, both topminnows and invertebrates were common, but mosquitofish had reappeared. Since the Gila River had not flooded again, the alien must have survived the poisoning (Meffe 1983a).

Concrete V-notch weirs about 0.6 m high were constructed on all three springs in winter 1983–1984 as barriers to reinvasion by mosquitofish. The sources of S-I, S-II, and the areas around all three barriers were fenced from livestock. Stocks of topminnows and invertebrates were again removed, and S-I was treated a second time with antimycin-A in April 1984. More than two hundred topminnows were restocked a month after renovation, and they reproduced prolifically. But mosquitofish again survived in marshes below the barrier and reinvaded when the creek bypassed its weir after heavy rains in summer 1984. Mosquitofish comprised 24% of all fish taken in December 1985, 69% in September 1986, and 98% by July 1987 (Simons 1987). Only nine topminnows were observed in the headspring in 1989, none was seen elsewhere, and mosquitofish swarmed in the system (Marsh and Minckley 1990). Topminnows were extirpated from S-III by mosquitofish by 1984. The spring was poisoned in April 1984 and remained fishless in December 1985 (Brooks 1986a) and summer 1986, when thirty to forty topminnows were introduced from S-II. An uncontaminated topminnow population persisted in 1989. However, the channel of S-III had by 1989 bypassed its barrier, and mosquitofish, if they reinvade, have ready access to the upper parts of the system.

S-II has retained a small, intact population of topminnows since its discovery in 1981, although only a few fish were there in 1989 as a result of overgrowth by cattails after fencing. The fish were gone in 1990. Although unsightly, and seemingly damaging, cattle grazing and trampling clearly preclude growth of emergent vegetation in these systems, and open water, albeit highly disturbed, is maintained. Cattails also invaded and filled fenced pools above all three weirs, forming distinct mounds that completely displaced open water and forced flow around the barriers. Thus, even where flooding initially passed around a weir, cattails would have ultimately caused it to bypass anyway. Where fences failed, cattle ate and trampled the vegetation and open water persisted (Marsh and Minckley 1990).

Thus, a decade of recovery and maintenance efforts did not curtail declines in topminnow populations at Bylas Springs. Although we have no doubt that two of the three populations would already be gone if mosquitofish had not been partially controlled, the facts remain that (1) the topminnow stock at S-I has been removed and replaced twice, and is again near extinction through depredations by mosquitofish that resisted two attempts at eradication; (2) a native stock at S-II was lost as a result of encroachment of vegetation after fencing designed to protect its spring from livestock; and (3) one population (S-III) was lost to mosquitofish, necessitating restocking from S-II after the non-native was removed. This record is not encouraging.

On a broader scale, the first attempts at recovery of topminnows through reintroduction began in 1964 (Minckley 1969b), when fish from Monkey Spring, Arizona, were moved to a variety of locales. These fish were all eliminated by flooding, pesticides, or mosquitofish. Through the early 1980s (Minckley and Brooks 1985), ninety-two documented sites were stocked, six of which still persist (Simons 1987; Simons et al. 1989).

More systematic recovery attempts began in 1981, when a memorandum of understanding (MOU) between the AZGFD, the U.S. Forest Service (USFS), and the U.S. Fish and Wildlife Service consolidated plans for reintroduction of the topminnow on USFS lands. Criteria for monitoring and eventual downlisting or delisting were also established, to be formalized in modified form in a subsequent recovery plan (USFWS 1984c) as follows: (1) downlist when twenty populations have been successfully reestablished in the wild, within their historic range, and have survived for at least three years; and (2) delist (before 1987) when (a) at least 50% of existing, natural, reclaimed, or newly discovered populations have been secured through removal of or protection against invasion of mosquitofish and other predatory species, and through protection of the habitat by management plans, cooperative agreements, land acquisition, or other means, and (b) fifty populations have been successfully reestablished in the wild, within their historic range, and have survived for at least three years; or thirty populations have been successfully reestablished and have survived at least five years.

Reintroductions were expedited by the Boyce Thompson Arboretum, Arizona, and Dexter NFH. An arboretum pond was provided for topminnow (and desert pupfish) propagation, and the hatchery has been devoted to rearing endangered western fishes since 1974 (Johnson and Jensen, *this volume*, chap. 13). Topminnows from Monkey Spring were propagated in both places for AZGFD restocking efforts. Hendrickson and Brooks (*this volume*, chap. 16) further review this production and reintroduction program.

Ironically, Dexter NFH was the scene for an incident involving Sonoran topminnows that resulted in extirpation of one endangered population by another. Both Sonoran topminnows and Pecos gambusia had been held at the hatchery from 1976 through 1981. The gambusia were removed in 1981, but the topminnows were retained. A pure sample of 1200 topminnows was preserved in August 1981. However, in October 1982, a second sample of about 300 presumed topminnows was almost half Pecos gambusia; topminnows declined to virtual elimination by 1984 (Minckley and Jensen 1985). Fortunately, none of the contaminated stock was placed in natural habitats.

Reintroductions of topminnows under the MOU began in 1982. To date, about 28% of the ninety-nine reintroductions made under the formal recovery plan persist (Simons et al. 1989), and the goal of downlisting *P. o. occidentalis* is being pursued. Simons et al. (1989) reviewed the reintroduction effort and agreed, in principle, that downlisting was appropriate. Downlisting to threatened status reduces a species' legal protection very little while providing a tangible achievement for politicans and the public. These authors expressed surprise, however, at the elimination of protection for natural populations after 1987 (delisting criterion 2a, above), which

they attributed to "emotional and motivational [political] considerations quite apart from biology." The recovery plan otherwise directs natural populations to be protected and enhanced without time limitations (USFWS 1984c). They further questioned the likelihood of long-term viability of many "successful" topminnow reintroductions. Most were in artificial habitats such as ponds and pump-filled catchments that will not persist without human support, and no funds or plans exist for their maintenance. Even those in natural systems may be transitory, since mosquitofish are an ever-present danger and floods have already eliminated some well-established topminnow stocks (Collins et al. 1981; Meffe et al. 1983; Brooks 1985). Furthermore, regional rainfall averaged 51 cm (varying from 43 to 61 cm) during the effort (1982–1986), while averaging 36 cm (20–51 cm) over the preceding fifty-one years; a major drought would extirpate many reintroduced populations. Such climatic variation is unpredictable, and persistence of a population for three to five years cannot adequately predict long-term survival.

New data also entered the picture when Vrijenhoek et al. (1985) demonstrated little genetic variation within native topminnows in a given system. Most differentiation was among the remaining natural stocks. With two exceptions, all reintroduced populations have been derived from Monkey Spring, so that genome may be relatively secure, but all the other lineages remain in jeopardy. In addition, based on allozymes surveyed, Monkey Spring fish are homozygous and have lower survivorship, growth rates, fecundity, and developmental stability than more heterozygous stocks (Quattro and Vrijenhoek 1989). Because heterozygosity appears to impart greater adaptability to individuals, enhancing their persistence in new environments (Frankel 1983; Meffe 1986; A. A. Echelle et al. 1989), other stocks might be preferable for the recovery effort. Topminnows from a more heterozygous stock are now in culture at Dexter NFH for future use (Brooks 1986a).

Simons et al. (1989) ended their review by recommending revision of delisting criteria for topminnows to meet the challenge of preserving genetic diversity. Refugia in the Gila River basin should be established for as many unique lineages of topminnows as possible, which, based on current data, would involve three or four geographic areas. Additional MOUs will be required to expand efforts to other federal and state lands, and an experimental, nonessential classification (USFWS 1984f), reducing protection to reintroduced populations but also reducing political resistance to the program, is being sought (Hendrickson and Brooks, *this volume*, chap. 16).

Pecos gambusia

The Pecos gambusia is endemic to a variety of springs and spring-influenced habitats of the Pecos River drainage of southeastern New Mexico and west Texas. Although it must have once been widespread (J. E. Johnson and Hubbs 1989), the earliest records indicate a historical distribution not much greater than at present (A. A. Echelle and Echelle 1980; A. F. Echelle and Echelle 1986; A. F. Echelle et al. 1989); a number of old records are questionable and cannot be verified. C. Hubbs and Springer (1957) suggested that the Pecos gambusia was forced into spring-fed habitats through competitive interactions with mosquitofish, a species indigenous and widespread in nonspring habitats of the lower Pecos River basin. The Pecos gambusia has apparently existed for a long time isolated at relatively few sites in four disjunct regions.

This species is closely associated with low-elevation springheads of low to moderate conductivities and moderate temperatures (Bednarz 1979; A. F. Echelle and Echelle 1986). It is usually found in marshes and quiet side channels in association with dense aquatic vegetation and submerged debris. It tends to be most abundant near spring sources and is replaced downflow by mosquitofish, especially where conditions fluctuate or become ephemeral. Foods include a variety of animal matter, particularly surface and water-column invertebrates, along with some filamentous algae (Bednarz 1979).

The problems encountered by this species include groundwater pumping (Brune 1975), which eliminated populations when Comanche and Tunis springs, and perhaps others in west Texas, were dried. Predatory fishes have had some impacts, such as the apparent elimination of Pecos gambusia by green sunfish (*Lepomis cyanellus*) in Lake St. Francis and perhaps elsewhere on Bitter Lake National Wildlife Refuge (NWR; Bednarz 1979; Brooks and Wood 1988). Likewise, there is some hybridization with mosquitofish and large-spring gambusia (*Gambusia geiseri*; Peden 1970; A. A. Echelle and Echelle 1980; Milstead 1980; Rutherford 1980); some populations of the former and all of the latter were likely introduced (C. Hubbs and Springer 1957; A. A. Echelle et al. 1989).

Mosquitofish do not appear to affect Pecos gambusia as much as they do some other poeciliids, although initial effects of their establishment may have resulted in population reduction of the native species, with subsequent recovery to near original sizes. Hybridization between the two species likewise does not appear to have had significant impacts on Pecos gambusia; limited introgression has been detected (citations above), and a hybrid swarm has persisted in one spring run for more than twenty years (C. Hubbs et al. 1978; C. Hubbs 1980).

The large-spring gambusia, because it also inhabits springheads, may be a potentially more serious problem than mosquitofish. Several instances of possible reductions of Pecos gambusia by this form have been reported (A. A. Echelle and Echelle 1980), but hybridization is infrequent, and large-spring gambusias appear to do poorly in saline, thermally fluctuating waters such as lower Leon Creek. Interactions are thus tempered and do not appear to jeopardize the continued survival of Pecos gambusia.

On the other hand, both morphological (A. F. Echelle and Echelle 1986) and genetic (A. A. Echelle and Echelle 1980; A. F. Echelle et al. 1989) analyses indicate relatively strong differentiation among populations of Pecos gambusia. In both data sets the Balmorhea, Texas, populations are more differentiated than others, including the presence of a unique color morph wherein some males are bright golden yellow in contrast to the typical silvery coloration. Each existing population has unique features, and if genetic variability is to be maintained, each should be perpetuated. Decisions on if or how this should be done have been deferred by the U.S. Fish and Wildlife Service (J. E. Johnson and Hubbs 1989; in this volume, Johnson and Jensen, chap. 13; Hendrickson and Brooks, chap. 16).

Future and ongoing efforts to recover Pecos gambusia should concentrate in three areas. First, major efforts should be made to maintain natural populations. This possibility exists for some stocks that are being cared for as part of management schemes for other endangered species as well as themselves (e.g., along with Comanche Springs pupfish [*Cyprinodon elegans*] in Balmorhea State Park, and Leon Springs pupfish [*C. bovinus*] at Diamond-Y Spring, both in Texas), and on already secured federal lands (Bitter Lake NWR, New Mexico). Long-term projections that some major springs in Texas will fail in the next two decades (A. F. Echelle et al. 1989) underscore the need for both immediate and long-range planning. Second, annual or more frequent monitoring should be continued throughout its native range to ensure against dangerous population or habitat changes, increased hybridization with other gambusiines, or losses to exotic fishes presently or poten-

tially introduced. Third, a search should be continued and intensified for sites in which to establish new stocks. The latter may hold little promise, since previous attempts have resulted in discovery of only a few promising localities (A. A. Echelle and Echelle 1980; Hendrickson and Brooks, *this volume*, chap. 16). A new site for the divergent Balmorhea population is especially important.

Summary and Discussion

Conservation must always look ahead because hard-fought gains can be lost instantaneously and permanently by momentary lapses of vigilance. This is even more important in managing endangered species, where the stark finality of extinction of an irreplaceable genetic lineage is the harvest of failure.

Goals

In the short term, the first goal of endangered species management is to save the target species from extinction. This may be thought of as a form of "salvage biology" that parallels the long-established discipline of salvage archaeology. Genomes of a population of animals or plants are analogous to the information hidden in an archaeological site; the spiral toward population extinction is similar to wanton destruction of a prehistoric occupation site, burial ground, or hunting camp. It is interesting that humans will slow or stop major developments, or at least fund an archaeological "salvage dig," then savor the results and support with their dollars and interest the perpetuation of artifacts in safe, permanent repositories. Not until very recently has there been a tendency to provide comparable support for perpetuating populations of rare animals or plants, and it was far more recently that people considered it worthwhile to save an imperiled fish that they could (or would) not eat!

The next step is to ensure perpetuation

of local populations—to secure habitat and work to gain the wisdom to maintain endangered species until recovery can be effected. This may be done in several ways: through application of appropriate legislation, which is now available; through in situ protection of organism and habitat; by transferring individual representatives into a refuge in anticipation of better times; and by renovation or creation of habitat wherein it may be recovered. We must then determine as soon as possible what factor(s) limit the population, and remove or alleviate that influence.

The next stage—recovery—has, since endangered species legislation, come to be defined in two very different ways. One espoused by agencies, now charged by law to perform recovery, requires that a population level be judged to be somewhere above "endangered" (i.e., in imminent danger of extinction) or "threatened" (i.e., in danger of becoming endangered); this allows the removal of the taxon or population from an official list, at which time agency responsibility is reduced or dissolved. A second, biological, definition, more difficult to delineate and attain and less acceptable to most bureaucrats, says that recovery consists of reestablishment of a population size, dispersion, and structure that will define itself by allowing an organism to proceed along independent evolutionary pathways comparable to those it followed prior to disruption by human interference. After recovery under this second definition is achieved, the organism may be allowed to program its own future. Implicit in this definition is the need for the species to occupy a position in a community, since no species evolves in total isolation.

What Happens in Practice?

Although attempts to save taxa and populations often fail, in some cases they do succeed; incidents of both are described above. Before and during the early 1960s, without the support of philosophies and the legislation of the developing environmental movement, isolated poolfish like *Empetrichthys latos pahrump* and *E. l. concavus* were doomed when technological modernization and development (electric pumps) were applied in and near their habitats. Efforts by concerned individuals nonetheless saved *E. l. latos*, the Owens pupfish, the Leon Springs pupfish, and other species like Big Bend gambusia (C. Hubbs and Williams 1979; J. E. Johnson and Hubbs 1989). Other efforts, such as those for the Amistad gambusia and Monkey Spring pupfish, failed. Nonetheless, when we are dealing with species of highly restricted distributions, the direct acquisition and protection of natural communities is feasible and is being successfully applied in native fish management in the American West (Williams, *this volume*, chap. 11). Use of artificial and seminatural habitats is also being applied with some success.

Formerly widespread species like desert pupfish and Sonoran topminnows present totally different kinds of problems. Once abundant in diverse habitats, these fishes clearly declined to endangered status over a period of only a few decades (and far less for pupfish in the Salton Sea basin). The phenomenon was spatially analogous to that which occurred with natural desiccation in Death Valley. As interconnections dried up, native fishes retreated to remnant springs.

What happened to desert pupfish and topminnows, however, was not natural. Human development of water resources was rapid in ecological time, and especially so in a geological context. Many western fishes must have suffered comparable restrictions in range during dry cycles of the past. However, after those periods waters were reconnected and fish populations were reunited. Dams and other man-made barriers preclude that possibility now; they form blockages to redispersal and gene flow that will remain insurmountable for the foreseeable future.

During arid climatic cycles of the past, pupfish and topminnows also must have held on in isolated enclaves such as perennial springs and ciénegas, to redistribute in better times. These natural refugia are again serving such a purpose during the artificial "drought" created by humans. But this time many of the refugia have been destroyed; in many that remain, pressures from swarms of aggressive exotic species have created another level of abuse that native species apparently cannot tolerate. Introduced fishes present a problem in native fish management in the West that merits substantially greater investments of time and effort. As noted by Minckley and Douglas (*this volume*, chap. 1), ways to eradicate or control non-native fishes are often identified as the single major problem confronting managers of imperiled remnant stocks.

Widespread species suddenly reduced in abundance and isolated in remnant populations seem to confuse managers. The desert pupfish, for example, is a riverine species that once ranged throughout lower elevations of the *entire* Gila and lowermost Colorado River basins. It is *not* characteristic of springs, although it occurred in a few, or desert marshes, but rather of the margins of larger rivers. Nonetheless, management efforts have emphasized springs, pools, and marshlands as refugia, and reestablishment of a riverine population of the species is never mentioned. The historic perspective is lacking. More important, how does one reestablish such a species in a reasonably natural community when the rivers are dry?

Prospects for the Future

Meffe and Vrijenhoek (1988) presented two zoogeographic models of gene flow in western American fishes based on contrasting isolation or connectedness of habitats. They argued that fishes like the Sonoran topminnow and other stream-dwelling forms historically enjoyed high levels of gene exchange among populations in what they termed a "stream hierarchy model." In contrast, strongly isolated fishes such as some pupfishes and gambusias were highly insulated from gene flow (a "Death Valley model").

Populations of formerly widespread cyprinodontoids are now broken into isolates of variable permanence and security. How should we define recovery for such a species: as a historic whole or for the individual populations that remain available? Should the intent be to reestablish desert pupfish and Sonoran topminnows throughout their former ranges? This is unrealistic. Introduced poeciliids, cichlids, and other predators and competitors can scarcely be eradicated from the desert pupfish's native range in the Salton Sea basin, and Arizona rivers and marshlands now dry as a result of upstream diversion, damming, and overpumping of groundwater cannot be recreated for Sonoran topminnows or pupfish without unreasonably vast and expensive alterations of patterns of human use.

When managers decide which of these models applies to a given taxon or population, they may restore interpopulation gene flow if it is apparent that such exchange occurred in the recent past; or they can avoid such interchanges if the populations were historically isolated and it is deemed that such isolation should be maintained (Meffe and Vrijenhoek 1988). In practice, a continuum from isolation to panmixia obviously existed in the recent past, and each taxon or population must be treated individually. But the decision of how to treat a species must be made expeditiously, as soon as the facts are in. The geographic ranges and population status of Sonoran topminnow, Pecos gambusia, desert pupfish, and others continue to deteriorate, despite the availability of biological data that would permit their recovery.

The reactions and responses of managers and scientists alike to new information merits some comment, as it pertains to overall management problems for endangered fishes. As an example, the demonstration by Turner that the diverse pupfishes in Death Valley, as well as desert pupfishes in refugia of the Colorado River (B. J. Turner 1974, 1983, 1984), were electrophoretically similar resulted almost immediately in questions to Minckley (unpub. data) from agency personnel as to the "need" to maintain certain remnant stocks. Lack of differences at loci surveyed was immediately transferred to a conceptual lack of importance. It is notable that the same reactions were obtained from a number of individuals following suggested nomenclatural changes over the years, such as the suppression of subspecific names in the Mexican stoneroller (*Campostoma ornatum*) by Burr (1976) and a decision by Chernoff and Miller (1982) that beautiful shiners (*Cyprinella formosa mearnsi*) from the uppermost Río Yaquí were indistinct from adjacent populations. Holden (*this volume*, chap. 3) reviews similar taxonomic questions that have influenced management efforts for endangered big-river chubs (genus *Gila*) of the Colorado River basin for more than thirty years, and the controversy continues today (M. E. Douglas et al. 1989).

On the other hand, a study by Vrijenhoek et al. (1985) promoted needed consideration of population genetics in the management of Sonoran topminnows. It also had some interesting side effects. They demonstrated that more than 50% of the total genetic variation throughout its geographic range (Ríos Mayo, Matape, Sonora, and Concepción in Mexico, Río Yaquí and Gila River, United States and Mexico) was due to differences among three major geographic groups (subspecies). About 25% of the variation was due to gene frequency differences among localities (among demes) within river systems, and 21% was attributable to variation (heterozygosity) within samples at single localities. They further discovered that isolated remnant stocks at the northern periphery of the species' range in

Arizona had low variability (A. A. Echelle et al. 1989), and that fish from Monkey Spring, used for most reintroduction efforts, had no demonstrable genetic variation at the loci examined.

Based on the latter information, Vrijenhoek et al. (1985) recommended use of another, more variable, population from Sharp Spring, Arizona, as brood stock, on the assumption that heterozygosity and fitness correlate positively (evidence for this is accumulating for fishes [A. F. Echelle et al. 1989; Quattro and Vrijenhoek 1989]). The Monkey Spring stock was removed from Dexter NFH, and the Sharp Spring stock was obtained to be placed in production for future restocking.

Proposals now have been generated for testing relative fitness of individuals from Monkey and Sharp springs in different habitats (Hendrickson and Brooks, *this volume*, chap. 16). Problems that will likely arise in such comparisons include the following: (1) suitable habitats for additional reintroduced populations in Arizona are becoming scarce; (2) habitat-to-habitat variation is so great that paired comparisons may be almost impossible; (3) the original stock at Dexter was eliminated, so direct comparisons can only be made by using fish already stocked and established at diverse reintroduction sites; and (4) the original stock may have been a mixture of Monkey Spring fish and an extirpated stock from Cocio Wash, Arizona (J. E. Johnson and Hubbs 1989; Johnson and Jensen, *this volume*, chap. 13). Genetics of Cocio Wash fish were never assayed electrophoretically (nor were actual brood fish from AZGFD stock at Boyce Thompson Arboretum or Dexter NFH).

We do not forward these potential problems to criticize past or projected studies, but simply to illustrate some of the complexities to be anticipated in the future. There is no doubt that the Sonoran topminnow presents an unparalleled opportunity for advances in management of short-lived endangered species. More is known about it than perhaps any other native western fish; it remains locally abundant, is easily manipulated, and responds well to laboratory care and experimentation. Detailed population analyses based on knowledge of genetics, age structure, growth rates, relative reproductive success, and behavior, along with appropriate environmental data, should lead to further successes as well as yielding scientific information of general application.

A. F. Echelle et al. (1989) performed genetic studies like those on the Sonoran topminnow on the four remaining Pecos gambusia stocks. About 48% of the detected variation was within samples (heterozygosity) from a given system (more than twice that demonstrated by Vrijenhoek et al. [1985] for the Sonoran topminnow), leaving 52% of the variation to occur between systems. Thus, although population heterozygosity is relatively high, large differences among the populations remain; these apparently reflect a long history of restricted gene flow. The stock in the area of Balmorhea State Park is the most divergent, while the others are more similar among themselves.

A. F. Echelle et al. (1989) concluded that each population of Pecos gambusia should be perpetuated separately to ensure maintenance of the genetic variability that characterizes this species. The question is (J. E. Johnson and Hubbs 1989), how and where are these stocks to be maintained if springs of the region continue to fail? Available transplant localities are few (A. A. Echelle and Echelle 1980), and using these would involve moving Balmorhea stock into proximity with another distinctive population (Hendrickson and Brooks, *this volume*, chap. 16). This same type of plan is tentatively being applied to Sonoran topminnows in Arizona, where three geographic populations identified in the Gila River basin are used to restock habitats only within their former, presumed geographic ranges (D. Hen-

drickson, AZGFD, pers. comm.). The desert pupfish awaits further consideration of its past distribution, future assessment of genetic structure of the remnant populations, and synthesis of such data into a viable recovery plan.

Conclusions

If we wish to place research and management of cyprinodontoids into a broader perspective, it is instructive to briefly review the last twenty or thirty years and ask what progress has been made and where efforts might be going. Early studies, during times when little was known of most of these fishes, involved taxonomy, definition of distributions, and natural history descriptions. Conservation efforts were exploratory, trial-and-error attempts to save species or populations from extinction; almost all involved transfer of fishes, sometimes within their native ranges and habitats but often elsewhere (in part, Hendrickson and Brooks, *this volume*, chap. 16).

Interest in cyprinodontoids increased exponentially in the 1960s and 1970s, partly because their precarious survival states were recognized and publicized. Basic studies continued, as they must in the future, but other techniques and ideas became incorporated into the scenario, including testing of general ecological theory; that is, r- and K-selection, predator-prey relationships, and concepts of community ecology. The emergence and perfection of electrophoretic methods allowed the determination of genetic variation and relationships (A. A. Echelle et al. 1989; Echelle, *this volume*, chap. 9), and new methods in molecular genetics such as mitochondrial DNA and DNA fingerprinting are now being applied. Progress in study and application of new techniques has been rapid.

One of the major factors enabling this rapid transition was the USFWS's development of Dexter NFH as an endangered fishes rearing and experimental station (Johnson and Jensen, *this volume*, chap. 13). Solely because of Dexter NFH, large numbers of critically endangered fishes were reared for experimentation and other recovery efforts that would otherwise have been impossible. We consequently derived knowledge invaluable for recovery. The USFWS, especially Region 2 in Albuquerque, New Mexico, is to be commended for insightful, aggressive, and continuing support of the facility.

Application of this (and other) information, however, still remains a problem, and we must not become complacent. In the decade between the first and second American Fisheries Society lists of endangered, threatened, or special concern fishes (Deacon et al. 1979; J. E. Williams et al. 1989), 136 new taxa were added and only 25 were removed; this gives a 1989 total of 362 taxa of fishes in North America that warrant protection. The 25 taxa were removed from the list for reasons such as taxonomic changes, acquisition of new data on status, or extinction, but "not a single fish warranted removal from the list because of successful recovery efforts" (J. E. Williams et al. 1989). Of 48 fishes that were changed from one category to another but remained on the list, 7 had improved in status, 22 had declined, and 19 were reclassified because new data indicated they were either more common or rarer than we had earlier believed.

Conservation and management of native western fishes is at a critical point in time and development. For more than twenty years workers with this remarkable fauna have concentrated on the necessary task of documenting change, establishing institutional and political clout, and, perhaps most important, developing an infrastructure of competent and dedicated persons with the common goal of fighting the pervasive loss of biodiversity. With all this in place, and both old and power-

ful new tools with which to work, the challenge is to move to a new plateau and break new ground. No sensible person would argue that continued monitoring and surveys are unimportant; they are in fact critical to document changes in populational and species status. But if nothing else is accomplished, we will simply have a detailed record of extinction. What, then, should be done?

Recent developments in population genetics provide unifying principles around which conservation of western fishes may be organized. In the past decade, population genetics has come into focus in fishery management and conservation (Meffe 1986, 1987; Ryman and Utter 1987; and reviewed by Echelle, *this volume*, chap. 9) and has become an organizing and unifying principle of conservation biology. We now can recognize and document historical levels of isolation and gene flow (M. H. Smith et al. 1989), and we may thus apply an evolutionary perspective to recovery efforts.

If this perspective can be incorporated into the ecosystem and community approaches necessary to habitat management, a strong and viable recovery program will result. Formerly widespread species will be managed in large areas also inhabited by the plants and animals with which they were historically associated, and with some level of the habitat diversity in which they evolved. Refuge design, habitat preservation, and reconstitution of degraded ecosystems will require enlightened and innovative approaches, but these remain the keys to rational maintenance of natural ecosystem integrity. If people can be educated in time to recognize the value and necessity of biodiversity (E. O. Wilson 1988), the presently imperiled cyprinodontoids and other such native fishes will exist another twenty years, or twenty centuries, into the future.

Chapter 16

Transplanting Short-lived Fishes in North American Deserts: Review, Assessment, and Recommendations

Dean A. Hendrickson and James E. Brooks

Introduction

One of the oldest ways to establish new fisheries is to transplant individuals of a species from one place to another. As a result, once geographically restricted species like the common carp (*Cyprinus carpio*), mosquitofish (*Gambusia affinis*), and rainbow trout (*Oncorhynchus mykiss*) have become essentially cosmopolitan. While many transplants have attained the desired results, the environmental impacts of human-mediated movement of fishes, especially outside their native ranges, have been or are potentially severe (Courtenay and Stauffer 1984; Mooney and Drake 1986; Huenneke 1988). Many species are now endangered as a result of interactions with alien fishes. It may seem odd, therefore, that transplanting is a common management technique for threatened and endangered species.

The popularity of transplantation among endangered species managers, and, as a consequence, the public, is indicated by a proliferation of literature and popular attention given to recent transplants (Booth 1988; D. E. Brown 1988; Conway 1988; Griffith et al. 1989). For endangered fishes in particular, J. E. Williams et al. (1988) noted that thirty-two of thirty-nine completed recovery plans "call for one or more forms of introductions." They went on to outline important points that must be considered in introduction programs, from planning through stocking and follow-up.

Additional evidence of increased use of transplants in recovery is the 1982 amendment (section 10j) to the Endangered Species Act (ESA; U.S. Fish and Wildlife Service [USFWS] 1984f). This amendment facilitates transplantation programs by providing for establishment through reintroductions of "experimental" populations of endangered species for recovery and research purposes. Section 10j grew from a recognition of the utility of reintroductions to recovery efforts, as well as a need to alleviate the often-insurmountable political barriers confronting such programs. Management agencies pondering the intrusion of a new endangered species into their jurisdiction are understandably hesitant. As currently mandated by the ESA (e.g., USFWS 1986f), coordination and paperwork associated with fully protected species are not to be taken lightly. Section 10j facilitates implementation of programs by relieving agencies receiving stockings of listed species from most of the responsibility they would otherwise have for populations with full protection.

Given the current popularity of transplantation, our purpose here is to.document the extent of the technique's application in conservation of native short-lived fishes of the desert

West and to assess its utility. We provide a summary of data on transplants and overviews of selected programs. We analyze the data set to assess overall utility of transplants in achieving progress toward recovery. In hopes that revelation of common problems or factors in success might assist managers in implementing future programs, we describe results of our search for similarities and differences among the programs and stocking sites. Finally, drawing from reviews of these data and personal experience, we discuss the practical realities of transplantation efforts and make recommendations for the design and execution of future applications.

J. Deacon, F. Hoover, C. Hubbs, E. Lorentzen, W. Minckley, E. Pister, D. Sada, A. Schoenherr, C. Swift, and J. E. Williams provided unpublished files on transplants. Williams also provided unpublished data compiled by the Endangered Species Committee of the American Fisheries Society.

Role of Transplants in Endangered Fish Management

Endangered species managers seek means to buffer their target species against extinction. Often, the easiest and most certain way to accomplish this goal is simply to increase population size; however, sound programs also address maintenance of geographic dispersion (Soulé 1986; Soulé and Simberloff 1986; Quinn and Hastings 1987, 1988; Gilpin 1988) and genetic diversity of target taxa (Schonewald-Cox et al. 1983; Meffe 1986, 1987; Ryman and Utter 1987; Lacy 1988; Meffe and Vrijenhoek 1988). Transplants are a convenient way to achieve progress toward these goals. Simple creation of populations where none previously existed not only increases population size for the target species but may also increase geographic and genetic diversity of population structure (Lacy 1987;

Ehrlich and Murphy 1987). Such diversity may act as a buffer against the impacts stochastic processes exert on species and gene survival (Quinn and Hastings 1987).

Often, decreases in population size and geographic range may result in stochastic extinction of isolated demes. Populations once interconnected, albeit sporadically, may thus become permanently isolated. Again, reintroductions might be applicable in this case as a human-mediated replacement of natural interdemic gene flow.

If adequate habitat is no longer available, its restoration might seem an obvious precursor to any transplantation attempt, although such is not necessarily the case. While adequate habitat is certainly an ultimate requirement for recovery, ascertaining habitat suitability need not necessarily precede stocking for experimental purposes. Where knowledge of habitat needs is lacking, as is often the case, experimental transplants may serve as a valuable heuristic tool. Rare organisms are inherently difficult to study, but stocking can alleviate rarity, albeit artificially and perhaps only temporarily. If habitats prove adequate, the same effort may result in the establishment of new populations, speeding progress toward recovery and saving the expense of habitat evaluation.

Methods

This study is largely based on responses to our requests to natural resource managers and university and private biologists throughout the North American deserts for basic data, in any format, on transplants of native species. We looked for information on species, locations (numbers and descriptions), nature of stocking (origin, numbers of individuals, dates), and a measure of success based on whatever evaluation criterion was used. We consulted unpublished records, reports, draft and approved recovery plans, and published papers.

We restricted our analysis to short-lived fishes, defined as species that typically live three or fewer years. Although not a natural grouping, this agglomeration is useful because management of short-lived species is inherently different from that of longer-lived forms. Typically, small populations of small species are more prone to extinction than are similar-sized populations of large (long-lived) species. Conversely, in large populations, long-lived species are at greater risk of extinction (Pimm et al. 1989). Countering this biological reality, which complicates the conservation of short-lived forms, practical problems associated with their management are often smaller and simpler due to their more restricted ranges. Habitat protection and management for the Devils Hole pupfish (*Cyprinodon diabolis*; Deacon and Williams, *this volume*, chap. 5) presented fewer logistic difficulties than protecting habitat in the vast Colorado River basin for the Colorado squawfish (*Ptychocheilus lucius*), bonytail (*Gila elegans*), humpback chub (*G. cypha*), and razorback sucker (*Xyrauchen texanus*; see, in this volume, Minckley et al., chap. 17; Tyus, chap. 19).

In general, when faced with lack of data on age and growth, we assumed that small species are short-lived. Habitat size and quality may affect fecundity, growth rate, age at first reproduction, and maximum age. Although quantitative relations have not been determined, habitat size is positively correlated with body size in fishes of the intermountain West (G. R. Smith 1981b).

We analyzed data on stockings made as part of conservation efforts by professional biologists. Humans have transplanted fishes for millennia regardless of their qualifications, but anonymous individuals, often referred to as "Bait-Bucket Charlie," are rarely docu-mented. Knowledge of clandestine actions is thus based on a biased subsample of an unknown total. Despite the important impacts illicit stockings have had on receiving faunas, we chose not to deal with such activities.

Our analyses were restricted to data from stockings into "wild" habitats, to focus on managers' abilities to foster self-sustaining populations. We excluded movement of fishes to hatcheries, aquaria, and other places directly dependent on human maintenance. Due to the inherent complexities of evaluating the success of stocking programs, we used each program's independent evaluation of success without regard for criteria. If an introduction was not declared successful or unsuccessful, we used presence (= success) or absence (= failure) at the time of last survey as the criterion.

Some terminology requires definition. We use *transplant* for any movement by humans of fishes from one locale to another. We define two types of transplants: sites inside (= *reintroductions*) and sites outside (= *introductions*) the native range of a species. *Native range* is where a species historically and naturally occurred, plus the interconnected waters in which it would reasonably have occurred. If " historically" and "naturally" are dropped from this definition, it equals the "historic range" of J. E. Williams et al. (1988). We prefer this deviation for describing stockings because "historic range" may not always equate to true native range. One example involves the Yaqui catfish (*Ictalurus pricei*), which was established in the Gila River basin of Arizona at a site outside its native range (Minckley 1973). Although that stocking was an introduction by our definition, and the population is now gone, the Gila drainage (or the particular site, Monkey Spring), has become part of the historic range under the definition of Williams et al. (1988). Historic range may be changed through human actions, but native range remains unchanged except by natural dispersal.

Table 16-1. Native southwestern fishes transplanted for conservation purposes. List includes total numbers of sites stocked for each, total number of stockings, number of sites stocked more than once, and number and percentage of sites stocked at which a population was successfully established. Taxa are numbered in decreasing order of transplantation success. Due to the effects of the single very large program (taxon 27) on summary statistics, overall means for each column were calculated with and without it.

Species	No. sites	No. stockings	Sites with > one stocking	No. successful	Success rate	References[1]
1. Crenichthys nevadae	3	4	1	3	100.0	1
2. C. baileyi grandis	2	4	1	2	100.0	1
3. Etheostoma fonticola	1	2	1	1	100.0	1
4. Rhinichthys osculus nevadensis	2	2	0	2	100.0	1
5. Gasterosteus aculeatus ssp.	1	1	0	1	100.0	2, 3
6. Gila bicolor snyderi	1	2	1	1	100.0	1, 4
7. Cyprinodon bovinus	1	3	3	1	100.0	5
8. Gila purpurea	5	6	1	5	100.0	6, 7
9. Lepidomeda mollispinis pratensis	1	1	0	1	100.0	1
10. Poeciliopsis occidentalis sonoriensis	4	6	2	4	100.0	6, 7
11. Rhinichthys osculus ssp. (OR)	1	2	1	1	100.0	1
12. Eremichthys acros	1	1	0	1	100.0	1
13. Lepidomeda m. mollispinis	1	1	0	1	100.0	9
14. Cyprinodon nevadensis mionectes	3	3	0	3	100.0	1
15. Empetrichthys l. latos	9	10	1	5	55.6	1, 9
16. Rhinichthys o. osculus (AZ)	2	2	0	1	50.0	6, 10
17. Cyprinodon salinus	4	8	2	2	50.0	11
18. C. tularosa	2	2	0	1	50.0	12
19. Agosia chrysogaster	2	2	0	1	50.0	6
20. Cyprinodon m. macularius	22	32	2	8	36.4	7, 13, 14, 15
21. Moapa coriacea	3	3	0	1	33.3	1, 16
22. Rhinichthys osculus ssp. (CA)	3	3	0	1	33.3	11
23. Cyprinodon radiosus	21	30	5	6	28.6	1, 4, 17
24. Gila bicolor mohavensis	16	24	7	4	25.0	11, 18, 19, 20

Taxon					
25. *Cyprinodon macularius eremus*	9	9	0	2	22.2	7, 15
26. *Gasterosteus aculeatus williamsoni*	5	7	1	1	20.0	2, 3
27. *Poeciliopsis o. occidentalis*	208	230	15	38	18.3	7
28. *Gambusia gaigei*	11	14	1	2	18.2	21, 22
29. *Cyprinodon diabolis*	6	8	2	1	16.7	9
30. *Gambusia nobilis*	33	37	4	5	15.2	23
31. *Cyprinodon nevadensis amargosae*	7	10	3	1	14.3	11
32. *Cyprinodon* sp. (Monkey Spring, AZ)	2	4	2	0	0.0	6, 8
33. *Gila robusta jordani*	1	1	0	0	0.0	1
34. *Plagopterus argentissimus*	5	6	1	0	0.0	6, 7, 8
35. *Gila orcutti*	2	2	0	0	0.0	11
36. *Tiaroga cobitis*	2	2	0	0	0.0	8
37. *Gambusia geiseri*	1	1	0	0	0.0	24, 25
38. *Meda fulgida*	2	2	0	0	0.0	6, 8
39. *Cyprinodon nevadensis shoshone*	1	1	0	0	0.0	1, 11
40. *C. n. nevadensis*	1	2	1	0	0.0	11
Grand totals	407	490	58	107	26.3	
Grand totals excluding taxon 27	199	260	43	69	34.7	
Grand means	10.2	12.2	1.4	2.7	48.4	
Grand means excluding taxon 27	5.1	6.7	1.1	1.8	49.2	

Summaries by families

Cyprinidae (16 taxa)					
Totals	49	61	11	19	38.8
Means	3.1	3.8	0.7	1.2	
Cyprinodontidae (12 taxa)					
Totals	79	112	20	25	31.6
Means	6.6	9.3	1.7	2.1	
Goodeidae (4 taxa)					
Totals	15	19	3	11	73.3
Means	3.7	4.7	0.7	2.7	

Table 16-1. Continued

Species	No. sites	No. stockings	Sites with > one stocking	No. successful	Success rate	References[1]
Poeciliidae (5 taxa)						
Totals	257	288	22	49		
Means	51.4	57.6	4.4	9.8	30.3	
Poeciliidae excluding taxon 27 (4 taxa)						
Totals	49	58	7	11		
Means	12.2	14.5	1.7	2.7	22.4	
Gasterosteidae (2 taxa)						
Totals	6	8	1	2		
Means	3	4	0.5	1	60.0	
Percidae (1 taxon)						
Totals	1	2	1	1	100.0	

[1], American Fisheries Society, Endangered Species Committee, unpub. data; 2, Camm Swift, Los Angeles County Museum, pers. comm.; 3, USFWS (1985e); 4, USFWS (unpub. data) and E. P. Pister, CADFG, pers. comm.; 5, C. Hubbs (1980); 6, Minckley and Brooks (1985); 7, AZGFD files; 8, Minckley (1973); 9, J. E. Deacon, pers. comm.; 10, Johns (1963, 1964); 11, R. R. Miller (1968b); 12, Hendrickson and Brooks, unpub. data; 13, USFWS (1986d); 14, Schoenherr (1988); 15, Hendrickson and Varela (1989); 16, C. L. Hubbs et al. (1974); 17, Shumway (1987) and USFWS (1984h); 18, USFWS (1984g); 19, F. Hoover, CADFG, pers. comm.; 20, Hoover and St. Amant (1983); 21, C. Hubbs and Springer (1957); 22, C. Hubbs and Broderick (1963), USFWS (1984b), and C. Hubbs et al. (1986); 23, Brooks and Wood (1988); 24, C. Hubbs, Univ. Texas, pers. comm.; 25, C. Hubbs (1982).

Results

Stocking records for forty taxa in the families Cyprinidae, Cyprinodontidae, Goodeidae, Poeciliidae, Gasterosteidae, and Percidae are summarized in Table 16-1; 490 transplants of these fishes to 407 sites were made between 1936 and 1988. Most received a single stocking, but 58 sites received 2 to 4 separate transplants. As a result, 107 new populations were considered successful. The total site success rate (across all taxa) was about 26%, although taxon-specific success varied between 0% and 100% (mean = 48%). Excluding one exceptionally large program (Gila topminnow) from our calculations resulted in adjustment of overall success for all other taxa to about 35%.

These data may be reduced to three groups based on degree of success. Fourteen species successfully established populations at all sites stocked. Nine failed at all sites, and seventeen enjoyed intermediate levels of success (14–56% of sites stocked). All large programs fell into the intermediate group, while all totally successful and unsuccessful programs were small (one to five sites). An indication of inverse relationship between program size and rate of success is apparent (Fig. 16-1).

Selected Species Accounts

Gila topminnow (*Poeciliopsis o. occidentalis*)

An extensive recovery program involving reintroduction has been carried out with Gila topminnows in Arizona. This poeciliid, one of two subspecies in the state, was one of the most common fishes in the lower Gila River basin before 1940 (C. L. Hubbs and Miller 1941). A steady decline since then, due to habitat loss and introduction of exotic mosquitofish (Minckley et al., *this volume*, chap. 15), resulted in its listing as endangered in 1967 (USFWS 1983c). By 1977 only thirteen natural populations persisted in the United States. This decreased to eleven in 1981, and ten in 1988 (Simons et al. 1989). All populations were confined to isolated springs and streams, mostly on private land and comprising a minuscule proportion of the original native range.

Reintroductions at about ninety sites during the 1960s and 1970s resulted in only two populations surviving by 1980 (Minckley and Brooks 1985; Simons 1987). Reintroduction efforts continued, however, and by the time a recovery plan appeared (USFWS 1983c), topminnows were being stocked on U.S. Forest Service lands. Habitats varied from windmill-fed metal tanks to small earthen impoundments ("stock tanks"), natural springs, ciénegas (Hendrickson and Minckley 1985), and streams. An interagency agreement afforded reintroduced fish legal status similar to that later provided to "experimental" populations (USFWS 1984f).

The recovery plan identified reintroduction as the principal means for recovery. The downlisting criterion of twenty reintroduced populations surviving for at least three years was surpassed by 1985 (Brooks 1985, 1986a). In 1987 topminnows remained in 35 of 191 reintroduction sites (Simons 1987); thirty populations surviving in 1987 had persisted at least three years, and twenty-three had survived five years or more. All kinds of sites were considered "wild" habitats, and populations were counted equally toward recovery, regardless of the type or quality of habitat.

The recovery plan stipulated that 50% of natural populations be protected prior to delisting. However, this condition was a requirement only until 1988. If delisting had not occurred by 1988, the plan inexplicably waived protection of natural populations. Delisting could occur once fifty reintroduced stocks were established and had survived three or more years in "wild" habitats within the natural range, or thirty populations survived five years. Thus, if all three-year-old populations extant in 1987 survived until 1989, the Gila topminnow could be removed from the endangered species list.

Unfortunately, early planning did not address conservation of the total genetic and life-history diversity in this taxon. Geographically isolated populations varied in life-history characters (Constantz 1979; Meffe 1985b), and long after stocking was initiated the widely reintroduced genetic stock (Monkey Spring, Santa Cruz County) was found to be homozygous at all twenty-five loci electrophoretically surveyed by Vrijenhoek et al. (1985). In laboratory studies of three stocks, Monkey Spring fish had the lowest survival, growth, fecundity, and developmental stability. Higher values of these factors were correlated with heterozygosity in other popula-

tions (Quattro and Vrijenhoek 1989). Thus, a more heterozygous stock from Sharp Spring, Arizona, would have presumably been more fit for reintroduction than homozygous fish from Monkey Spring, and more genetic diversity would have been protected if Sharp Spring stock had been used.

Future plans call for replication of each of four geographically discrete topminnow gene pools at several sites within its hydrographic sub-basin. Use of the three other native Arizona stocks (Ciénega Creek, Sharp Spring, and Bylas Springs) began in earnest in 1987, although one successful stocking with Sharp Spring fish was made in 1981. Limited availability of habitat for continued reintroduction has permitted only sixteen sites so far to be stocked with other than Monkey Spring fish,

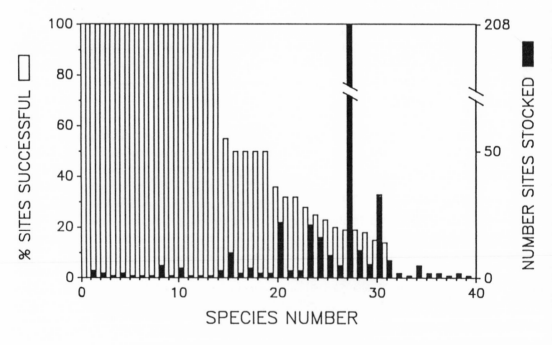

Fig. 16-1. Percentage success compared with absolute project size as measured by numbers of sites stocked for transplantation programs for southwestern short-lived native fishes. The X-axis is species number as given in Table 16-1.

and only seven of these may be considered wild. The success of these sixteen stockings is yet to be assessed.

Desert pupfish (*Cyprinodon macularius*)

Like Gila topminnow, the desert pupfish (*Cyprinodon m. macularius*) was once wide-ranging and abundant in southern Arizona (C. L. Hubbs and Miller 1941). Unlike the topminnow, however, habitat destruction and interactions with exotic fishes brought about extirpation of all natural populations in Arizona shortly after 1950 (Minckley 1973). Small, highly localized, natural populations persist in California near the Salton Sea (Schoenherr 1988) and at several localities along the lower Colorado River in Mexico (Hendrickson and Varela 1989).

Despite its early decline, the desert pupfish was not listed as endangered until 1986 (USFWS 1986d). Perhaps partly because of this, its recovery program has been far less extensive than the topminnow program. Although a recovery plan is still in early draft stage, reintroductions already have been carried out using stock from natural populations in Sonora, Mexico. That same stock is maintained as a large population at Dexter National Fish Hatchery (NFH), New Mexico (Johnson and Jensen, *this volume*, chap. 13). In southeastern California, remnant populations have provided gene pools for transplants at nine sites, only two considered wild, but all successful (Schoenherr 1988). Two of eight wild sites stocked in Arizona have been successful; however, their long-term persistence is uncertain: one was nearly extirpated from unknown causes when last surveyed in 1987, and the other was invaded by exotic fathead minnow (*Pimephales promelas*) in 1986 but apparently remained healthy (Hendrickson and Varela 1989). Additional reintroductions in Arizona have been hampered by political resistance related to its pending, and now final, status as endangered.

The only natural population of another subspecies, the Quitobaquito pupfish (*C. m. eremus*), is in Quitobaquito Spring on Organ Pipe Cactus National Monument, Arizona. One successful reintroduction of this form was accomplished following chemical renovation of the Quitobaquito system to remove introduced golden shiner (*Notemigonus crysoleucas*; Minckley 1973). Transplants outside its native range have been successful (Hendrickson and Varela 1989), and two established stocks now threaten to escape and contaminate reintroduction sites proposed for *C. m. macularius*.

Yaqui chub (*Gila purpurea*)

The Yaqui chub, restricted to the intermontane Willcox Playa basin of Arizona and upper Río Yaqui basin, Sonora and Arizona (DeMarais 1991), has been successfully reintroduced into five sites in Arizona. Three habitats have required repeated renovations to remove exotic species, and have thus received multiple stockings plus extensive manipulations directed toward restoration of historic conditions on the San Bernardino National Wildlife Refuge (NWR; J. E. Johnson 1985; USFWS 1987f). Two other sites received no substantive attention following a single stocking. Habitat destruction and interactions with exotic species resulted in the Yaqui chub's listing by the USFWS (1984i) as endangered.

Mohave tui chub (*Gila bicolor mohavensis*)

Mohave tui chubs became endangered following introduction of the exotic arroyo chub (*Gila orcutti*), which resulted in introgressive hybridization (C. L. Hubbs and Miller 1943). A single genetically pure stock at Soda (formerly Zzyzx) Springs, California, provided for nineteen individual transplants to fourteen sites within and outside its historic range (St. Amant and Sasaki 1971; Hoover and St. Amant 1983). Although some lasted as long as twenty years, populations persist today at only two wild habitats. Flooding, poor water quality, and inadequate water quantity caused other populations to fail (USFWS 1984g).

Unarmored three-spine stickleback (*Gasterosteus aculeatus williamsoni*)

The unarmored three-spine stickleback was listed as endangered in 1970 (USFWS 1985e) because of range reductions through extensive urbanization and genetic introgression with *G. a. microcephalus* (introduced coincidently with trout). One of five stockings into wild habitats succeeded in establishing a new population. The failure of others appears to have been caused by exotic fishes and marginal habitat conditions.

Pecos gambusia (*Gambusia nobilis*)

The Pecos gambusia has been reintroduced into several habitats (Bednarz 1979; USFWS 1983d; Bouma 1984). Endemic to springs and gypsum sinkholes of the Pecos River basin of eastern New Mexico and western Texas (C. Hubbs and Springer 1957), the species was negatively affected by mosquitofish through predation, competition, and hybridization. Exotic large-spring gambusia (*G. geiseri*) may similarly affect Pecos gambusia near Balmorhea, Texas.

Pecos gambusia were stocked into habitats at Bitter Lake NWR, New Mexico, during the early 1970s (Bednarz 1979) and in 1981 (Bouma 1984). Of thirty attempts, only two were successful (Brooks and Wood 1988). Most stockings were in gypsum sinkholes, where high salinities may have precluded success. The historic distribution in Bitter Lake NWR sinkholes is poorly documented (A. A. Echelle and Echelle 1980), which clouds the issue of which populations are natural. Two sinkholes (designated 7 and 27) are thought to harbor natural populations, but it is possible that they were stocked prior to ichthyofaunal surveys. Such may also be the case with

Pecos pupfish (*Cyprinodon pecosensis*) on Bitter Lake NWR.

Elsewhere, A. A. and A. F. Echelle's (unpub. data) 1982 transplant of nearly one hundred Pecos gambusia from Giffin Spring into a Comanche Springs pupfish refugium pond at Balmorhea State Park, Texas, was successful. The presence of and resulting impacts by *Gambusia affinis* and *G. geiseri* in several habitats in the Balmorhea area warrant further renovations and transplants.

A. F. Echelle and Echelle (1986) observed greater genetic variability in Pecos gambusia from the Balmorhea area than in other populations. Springs at Balmorhea are declining in flow, so they recommended conservation of the genome within its historic range but outside the Balmorhea area. Rattlesnake Spring in Carlsbad Caverns National Park was proposed as a potential reintroduction site. However, another population of Pecos gambusia already exists at Blue Spring near Carlsbad. Should the USFWS move to ensure preservation of the species within an area, and thus duplicate the Blue Spring population? Or should they maximize genetic variability of the species and transplant fish from Balmorhea? Which population to introduce remains a question.

Big Bend gambusia (*Gambusia gaigei*)

The Big Bend gambusia has been in almost continuous jeopardy of extinction since its discovery and description by C. L. Hubbs (1929). Known only from two springs within Big Bend National Park near Boquillas Crossing, Texas, it suffers not only from limited range but also from declining spring flows and the presence of mosquitofish. Conservation efforts from the 1950s onward have focused on improvement and development of habitat and preventing invasion by mosquitofish. Habitat renovations to remove mosquitofish, followed by transplants of *G. gaigei*, have taken place within and outside the native range (C. Hubbs and Broderick 1963; C. Hubbs and Williams 1979; J. E. Johnson and Hubbs 1989).

Mosquitofish first invaded *G. gaigei* habitats in the mid-1950s, extirpating the latter at its type locality (Boquillas Spring). This prompted renovation of Spring 4, the only remaining natural habitat occupied by the species, to remove the mosquitofish. A stock of *G. gaigei* was retained, and a series of reintroductions in 1956 all failed. All the Big Bend gambusia were lost except for four individuals at the University of Texas, Austin. After loss of one of these, the last three representatives of the species were stocked in the spring of 1957 into a refuge pond constructed at Rio Grande Village, Big Bend National Park. A population became established, but several instances of invasion by *G. affinis* and green sunfish (*Lepomis cyanellus*) necessitated repeated renovations and reintroductions. After more habitat constructions in the mid-1970s, a large population of Big Bend gambusia now lives at Rio Grande Village, in both a refuge pond and an overflow drainage ditch. More recent attempts (1989) resulted in establishment of the species in Spring 1 and its associated marsh. To date, a total of fourteen stockings, many representing repeated attempts at single sites, have resulted in establishing two populations. Unfortunately, this species's preferred warm-spring habitat is limited, and most already contain mosquitofish.

About fifty *G. gaigei* from a laboratory stock were transplanted to Dexter NFH in 1974. Once their survival indicated that the water quality was acceptable, the stock was replaced with about sixty-five fish from the refuge at Big Bend National Park. High winter mortality outdoors at Dexter was common and necessitated additional transplants to the hatchery. Presently, large numbers (more than a thousand) winter indoors and are released outdoors in warmer seasons.

Woundfin (*Plagopterus argentissimus*)

The critically endangered status of woundfin, and of the whole fauna of the Virgin River, to which the woundfin is now restricted, was documented by Deacon (1988). Four reintroductions of woundfin during the early 1970s into the Gila River basin of Arizona failed (Minckley and Brooks 1985). Additional experimental, nonessential reintroductions were approved in 1985 (USFWS 1985a), but technical problems have thwarted implementation of the ruling. Little is known of the suitability of the habitat at proposed reintroduction sites. The Virgin River is a medium-sized stream with discharge, morphometry, and fish community unlike those found at most of the proposed locales.

The status of the woundfin continues to deteriorate. Adequate stock for reintroductions may not be available either through hatchery propagation or from the Virgin River. Attempts to culture fish at Dexter NFH failed to produce numbers sufficient for reintroduction. Even if adequate numbers could be obtained from the Virgin River, the presence of Asian tapeworms (*Bothriocephalus acheilognathi*) in fish from that sole donor population, and their absence from proposed receiving habitats, preclude direct transfers. Topical treatment to rid woundfin of tapeworms can produce helminth-free fish for future stocking.

Additional problems arose when the Virgin River was subjected to management attempts to remove another threat to woundfin, the exotic red shiner (*Cyprinella lutrensis*). Rotenone treatment was attempted in a reach between artificial barriers, but it went awry, causing general fish mortalities through a much longer reach than intended (Arizona Game and Fish Department [AZGFD] files). Red shiners are a documented threat to woundfin (Deacon 1988), and the renovation was clearly necessary, but another unforeseen event complicated the issue before success could be determined or its impacts outside the target reach assessed. Failure of a dike at Quail Creek Reservoir and a resulting flood of nearly double the magnitude of any natural event in the hundred-year record imposed another anthropogenic insult.

The cumulative effects of these incidents remain to be evaluated, but the Virgin River and woundfin provide an exemplary case of the myriad difficulties associated with endangered fish management. It is almost certain that reintroduction will become a critical part of future recovery efforts for this species (USFWS 1985a).

Discussion

Limitations of Data and Recommendations for Future Studies

The quality of our data set is highly variable, and this influenced the analysis and made us hesitant to draw firm conclusions. The large sample size and restrictions originally placed on the use of data nonetheless allow some tentative generalizations that permit suggestions for improvements in study design and data collection to assist future investigators.

There are interesting indications, for example, of a direct relationship between numbers of fish stocked and the likelihood of success at a site. Success may also be improved by repeated stocking. Issues confounding our ability to draw firm conclusions include incomplete data (numbers of fish stocked and detailed records of supplemental stockings, for example, were not consistently available), lack of knowledge of habitat suitability, and a tendency across all reintroduction projects to fail to clearly define a protocol. We were left wondering, for example, if sites inherently destined to be successful under almost any stocking regime were more likely than others to receive multiple stockings because they "looked good" to project personnel; or were

they successful because of repeated stockings? Similarly, did "good-looking" single-stocked sites receive larger numbers of fish because they looked good, or were larger stockings responsible for producing higher rates of success? Stockings rarely had a rigorous a priori experimental design, but rather proceeded in "seat-of-the-pants" or "gut-feeling" directions. We strongly advise against continuing this apparent trend and recommend extensive prior planning and subsequent strict adherence to experimental protocols for future transplants.

Other problems involve the definition of "success." Simons et al. (1989) discussed the simple numerical goals set in the Gila topminnow recovery plan and the evidence that those subjectively determined and (at least in part) politically motivated criteria for success were inadequate. Aquatic (and other) habitats in desert are notoriously unpredictable. Persistence or viability of transplanted stocks should be tested against habitat perturbations (drought, flood, etc.) of reasonable severity to ensure meaningful longevity, and goals should reflect biological considerations free of politics. In most cases, far more than three to five years are required to test the ability of reintroduced populations to be self-sustaining in the face of natural environmental extremes, yet survival of a set number of reintroduced stocks of topminnow for those short periods were forwarded as criteria for downlisting and delisting.

We were further unable to extract reliable statistics on persistence times. Future reporting of this statistic by managers would assist in determining times required to accurately assess transplant success. In the present data set, populations were defined simply as successes or failures. It was not uncommon, however, for some stockings to produce "successful" populations for as much as five to ten years before failing. For short-lived species with tendencies to disperse, five to ten years of occupation of areas suitable in wet periods but marginal or dry in drought must have been commonplace. Perhaps such natural phenomena should be incorporated into recovery efforts.

The impacts of illicit transplants were not formally addressed, although many are noteworthy. Illicit transplants are generally detected only outside the native range. Movements of a taxon within its native range are unlikely to be detected. The ecology of invasions of exotic fishes after introduction nonetheless merits attention for the light such studies might shed on the ecology of reintroductions. Amazing rates of invasion have been documented for some small short-lived fishes, including sheepshead minnow (*Cyprinodon variegatus*; A. A. Echelle and Conner 1989), Arkansas River shiner (*Notropis girardi*; Bestgen et al. 1989), and Plains minnow (*Hybognathus placitus*; Hatch et al. 1985) in the Pecos River; and red shiner in the Colorado River basin (C. L. Hubbs 1954; Koehn 1965; Minckley and Deacon 1968; Minckley 1973; Deacon 1988).

Finally, data on transplants of threatened and endangered fishes are at present remarkably difficult to obtain. Our compilation, even though it is certainly not complete, clearly demonstrates that difficulties in obtaining data were not due to a paucity of stockings, but rather to a lack of publications reporting them. Nearly all the data came from unpublished files. We join Minckley and Brooks (1985) and J. E. Williams et al. (1988) in pleading for publication of basic information on transplants of all fishes, and especially for native fishes of special concern.

Philosophy, History, and Analysis of Transplantation Programs

J. E. Johnson (1980a) outlined two alternative philosophies for endangered species reintroductions. One may either "determine the reason for the original extirpation and correct it before attempting reintroduction, [or] produce many individuals and spread them abundantly through historic range (Johnny Applefish method)." He went on to propose solutions to political impediments to reintroductions of listed fishes, including "an EXPERIMENTAL [emphasis his] classification within the Endangered Species Act to allow special management regulations for reintroduced populations."

In the intervening years we have seen amendment of the ESA to incorporate "experimental" status and supplemental provisions such as "non-essential" that waive most of the protection otherwise given a species by the act. These new bureaucratic expediencies have been implemented in the desert West to allay the apprehensions of politicians and agency personnel, permitting, for example, stocking of Colorado squawfish in Arizona and approval to stock woundfin there, both with experimental, nonessential status.

Despite J. E. Johnson's (1980a) emphasis of an experimental status for reintroduced populations, now provided under section 10j (or similar predecessor agreements, such as for the Gila topminnow), practical implementation of reintroductions has proceeded in a far from rigorous scientific fashion. The "experimental" designation should be interpreted literally. It is only through such an approach that we will gain the biological, physical, and chemical habitat data required for management. As an example, despite the fact that more than two hundred sites have received topminnows, there was no underlying experimental design for subsequent data collection and analysis that might have resulted in improved knowledge of topminnow habitat requirements. The sites selected and stocked presented a complex array of conditions, and no attempts were made to erect control or treatment groups for testing specific hypotheses. Follow-up monitoring received limited funding and, again, there were few plans for hypothesis testing. Monitoring data compiled without a rigorous a priori design was not amenable for a posteriori statistical testing (Brooks 1986a); consequently, results were anecdotal and inconclusive.

We must recognize, however, that despite a lack of experimental design, significant accomplishments accrued from the "Applefish" approach. Such is the case with topminnows, a taxon for which there are certainly more individuals in more sites today than there would have been without the program. This approach offers undeniable advantages of low initial investment in research and habitat evaluation and short implementation time, concurrent with possibilities (albeit with unknown probability) of some margin of success. The political expediency of the "Applefish" approach should not be ignored. If habitats are politically accessible, then simple, logistically obtainable goals may be proposed and attained, and bureaucratic success of such a program can be ensured. The topminnow program provides an example of this, but attainment of the recovery plan's criteria for downlisting to threatened, or delisting, further illustrates major conflicts between political advantages that may be only short term, and long-term biological recovery (Simons et al. 1989).

Reintroduction programs using the "Johnny Applefish" approach will likely experience a high rate of failure of individual plantings, and political implications of high failure rates must be weighed against any perceived advantages. Reintroduction programs, especially those involving extensive hatchery production, are expensive, and taxpayers may object when they appear unsuccessful. Repercussions of the low success rate of topminnow reintroduction efforts are already being manifested by resource managers with jurisdiction over additional prospective reintroduction sites not only for topminnows but also for desert pupfish and other taxa. Since passage of section 10j, all transplants must be designated either "essential" or "nonessential." The former carries full protection of the ESA, and any management activities that might affect fully protected populations, either beneficially or adversely, must be preceded by consultations. Resource managers are hesitant to accept this additional burden, especially if it is probable that reintroduced populations will not survive.

Designation of populations as nonessential reduces costs for land-management agencies, but simply obtaining the designation can be a major undertaking, and the ramifications may be far-reaching. The required announcements of proposed and final rulings in the *Federal Register* can be expected to take a minimum of two years, and will certainly delay stockings. A proposal to stock topminnow and desert pupfish into about thirty-five sites on federal lands in Arizona was stalled in this process for more than six years (AZGFD files). Thus the bureaucracy surrounding the establishment of experimental populations has impeded rather than facilitated. The need to list and approve each proposed experimental reintroduction site individually through *Federal Register* publication has effectively brought studies of topminnows to a standstill. Furthermore, the geographic distribution of experimental, nonessential sites interspersed among sites with fully protected stocks has generated an administrative nightmare.

Most species reintroduced into the American deserts on a large scale have thus far proven relatively simple to culture. Both the short-lived topminnows, gambusias, and pupfishes and the longer-lived Colorado squawfish, razorback sucker, and bonytail are readily produced in hatcheries (Johnson and Jensen, *this volume*, chap. 13). Other short-lived stream fishes in critical need of captive production, or imminently faced with its likelihood, may pose greater challenges. Preliminary attempts to culture woundfin met with limited success. Although reintroductions have been approved, they are not possible until large-scale propagation can provide the fish. It is clear that attaining this goal with this stream-obligate taxon will require more intensive hatchery management than simple release into ponds, which is essentially all that topminnows and desert pupfish require. Other stream obligates, such as the loach minnow (*Tiaroga cobitis*) and spikedace (*Meda fulgida*) of the Gila River basin, may present new problems. Both have reintroduction proposed as part of their draft recovery plans.

Restriction of reintroductions to documented historic range is certainly advisable if adequate habitat is available or can be restored. Protection of occupied habitat and restoration of historic conditions at former, documented sites of occurrence should remain among the highest priorities. It is only in places where a species persists that we can be certain all its needs are met. It is also here that the species and its associated fauna evolved, and should continue evolving.

In many cases historic habitats have degraded beyond the point of restoration. Thus, for species with critical needs, managers may be forced to use locales within the native range where no historic collection records exist. Many habitats used for recovery in the desert West are, in fact, recent and man-made, and often distant from historic occurrences (Simons et al. 1989; Minckley et al., *this volume*, chap. 17). Their use may be the only option, but these areas should always have a lower priority than protection and restoration of native habitat.

Augmentation, the reintroduction of fish into habitats where natural, conspecific populations persist, is a technique that may be advisable under certain circumstances. We recommend augmentation only after it has been demonstrated as the sole solution to problems affecting a population. Such might be the case where lack of recruitment is occurring and it is not possible to increase recruitment through means other than stocking. We suspect that this situation occurs, and we present a hypothetical situation to illustrate it: a target species suffering near-total and consistent mortality as the result of heavy, size-limited predation on larvae, with excellent survivorship of postlarval size classes, and control of predators not possible. Stocking of postlarval fish would be an obvious means of increasing recruitment into the adult population. In such a case, we concur that augmentation should prove useful, but we caution that care should be taken in selection of genetic stock, and in genetic management of propagation.

Summary and Conclusions

Transplants of many types of short-lived fishes have been common in desert habitats of western North America. These have met with relatively low rates of success; it is clear that this approach to recovery is far from a panacea. In some cases the probability of survival of the target taxon has been increased, and in at least two instances (*Gambusia gaigei* and *Empetrichthys l. latos*) transplantation clearly saved the species from extinction. In almost all cases, however, although the species may appear more secure, we have yet to demonstrate the biological bases for that conclusion. Does increasing the number of homozygous populations increase the probability of survival of the Gila topminnow? More populations of a stock clearly provide a numerical buffer against extinction but do nothing toward perpetuating the total genome of a species.

Rigorous application of scientific methodology will improve any transplantation program. The aquatic communities we strive to perpetuate would profit from a priori attention to study design. The underlying biological basis of endangered species recovery must be subjected to more rigorous professional examination than in the past. Goals have sometimes been formulated more on the basis of politics than on realistic expectations. The ability to conduct comprehensive experiments with transplantation of endangered species has also become limited by politics, and means should be sought to alleviate bureaucratic constraints.

Even though the labels "experimental" and "nonessential" are unacceptably inconsistent with scientific and biological reality, they have proven a mixed blessing. By permitting reintroductions where they otherwise might not have occurred, the legislation has provided opportunities for research. However, they have not been without costs. In the case of Colorado squawfish and woundfin, and perhaps in the future with Gila topminnow, desert pupfish, and other species, we have or will have populations that occupy habitats that we have little or no ability to protect.

The Salt and Verde rivers of Arizona contain experimental, nonessential Colorado squawfish. Without the protection of the ESA, the fish are living in the last remaining reaches of unique, uncontrolled, hot-desert rivers, which in themselves are critically endangered as a North American habitat type. It is difficult to expect establishment and long-term persistence of self-sustaining populations of native imperiled fishes in the rapidly developing, water-hungry West unless the habitats upon which reestablished populations depend are protectable. The proposed experimental, nonessential stockings of woundfin in the upper Gila River, Arizona–New Mexico, and Colorado squawfish in the lowermost Colorado River, Arizona-California, also are questionable. The native fish fauna of western North America now consists of taxa that cannot afford to lose a single population or the opportunity to expand their ranges into any available habitat. We contend that woundfin, among others, is in that position, yet "nonessential" reintroductions are proposed.

SECTION VI

Problems of Time and Space: Recovery of Long-lived Species

Unlike small species, larger western fishes live a long time. For example, some razorback suckers from an Arizona-Nevada reservoir were forty-four years old when they were sacrificed for age determination in 1981–1983. From the human perspective, some of these fish hatched in the 1930s when the editors of this book had just mastered walking and talking, and at least twenty of the twenty-nine contributors were yet to be born!

The first chapter of this section provides documentation for the past and present status of the razorback sucker and its biology, past and current efforts toward recovery, and recommendations and predictions for its future. These large fish need large habitats such as major streams of the Colorado River basin, to which they are endemic. Because they live a long time, they need not reproduce each year, and an apparent strategy would be to forestall reproduction when there is low probability that offspring would survive, then emphasize "good" years. Therein lies a problem, assuming such a strategy exists, because no significant recruitment has been recorded anywhere in the Colorado River basin for at least the last twenty (and perhaps thirty) years. We might conclude that no "good" years have been detected; however, razorback suckers do, in fact, reproduce annually, but their larvae do not survive. Experiments and observations support a tentative conclusion that predation by non-native fishes precludes larval survival. Predation by introduced species is also a factor in low survival of reintroduced stocks, so management options may be more limited by the biological pollution of non-native species than by the vast physical and chemical habitat changes wrought by humans.

Lakesuckers, discussed in chapter 18, share great longevity with the razorback sucker. Many survive at least thirty years, and individuals more than forty years of age are not uncommon. Some of these unique western fishes, such as the June sucker of Utah Lake, also seem endangered by predation, although water development that limits access to riverine spawning grounds seems more important for others. Cui-ui have only one place to spawn, the lower Truckee River. Pyramid Lake itself is too alkaline for developing eggs to survive. A deltaic barrier formed by declining lake levels prevented access to the river, and recruitment failed for almost twenty years. Only their long life span allowed cui-ui to persist. This problem was circumvented by developing a fishway by-passing the delta and a management protocol that included hatchery production; new year classes have now appeared.

Another problem involves introgressive hybridization between lakesuckers and other native suckers, presumably due to some form of habitat disruption or a scarcity of one species and great abundance of the other. This process is already implicated in the disappearance of an original form of June sucker from Utah Lake and is evident in some other species of the group.

The Colorado squawfish forms the subject for chapter 19. Unlike the suckers, this largest native minnow in North America (achieving a length of 1.8 m [six feet]) persists in what appear to be viable, reproducing populations in part of its native range in the Green and Yampa rivers, Utah and Colorado, and perhaps in the upper Colorado main stem in Colorado. It is long gone from the vast lower Colorado River basin downstream from Grand Canyon, and rare where it once lived in many upstream tributaries.

A unique ontogeny of behavior characterizes this large obligate piscivore in the Green River. Adults are solitary and relatively sedentary along shorelines and in eddies during the nonreproductive season. Then they migrate to spawn, sometimes 200 km or more, to specific, canyon-bound river reaches. Young hatch to drift downstream and grow through their juvenile stages in lower flat-water parts of the system, then mature to establish their own residence areas from which they migrate to spawn. Human-induced changes, such as dams that interdicted necessary movement routes, and effects of potential non-native competitors and predators, are described and interpreted. Population sizes are estimated and compared with the relative degree of water exploitation in the Green (largest squawfish population and least developed by humans) and Colorado (far fewer fish and more development) rivers, and prospects for the future are reviewed. Timely and unimpeded flow of water through an extensive and diversified reach of river must be provided if this remarkable top carnivore and its ecosystem are to be preserved.

WANTED

FOR FUTURE GENERATIONS

Colorado Squawfish *(Ptychocheilus lucius)*

Bonytail Chub *(Gila elegans)*

Razorback Sucker *(Xyrauchen texanus)*

These protected species have been stocked as part of research on imperiled Colorado River fishes. If you catch one RETURN IT TO THE WATER alive and notify the Arizona State University Center for Environmental Studies at 965-2977 or the Arizona Game and Fish Department Nongame Branch at 942-3000.

Billboard notifying fishermen and others of the presence of state and federally listed fishes stocked in waters used extensively by humans. A combination of research and education on endangered fishes in urban lakes (Marsh 1990) has proven highly successful based on data obtained on growth rates and species interactions coupled with positive public interest, response, and cooperation.

Chapter 17

Management Toward Recovery of the Razorback Sucker

W. L. Minckley, Paul C. Marsh, James E. Brooks, James E. Johnson, and Buddy Lee Jensen

Introduction

Our efforts to prevent extinction are selective for large, colorful, and highly visible species. There is general support for recovery of the whooping crane, for example, and to protect the bald eagle. Other organisms receive less attention, especially when they are small, secretive, or difficult to observe. With few exceptions, imperiled freshwater fishes fall into this second category, even when a species is large, conspicuous, and originally of wide distribution.

The razorback sucker, *Xyrauchen texanus* (Abbott), is one of these. A large and uniquely shaped species of a monotypic genus (Fig. 17-1) endemic to the vast Colorado River basin of western North America, and once ranging from Wyoming to northwestern Mexico (Holden 1980a; Minckley et al. 1986), this fish has largely disappeared (Fig 17-2). Only a few isolated populations and scattered individuals exist today. Not only was the razorback sucker scarcely known except to specialists until a few years ago, it was one of a number of species that were regarded as inconsequential or even detrimental to management of sport fisheries. Until recently these species were acknowledged, if at all, as oddities or of scientific interest, and many remain essentially ignored.

Development of environmental awareness has begun to change this pattern. Public and professional concerns for imperiled native animals, including fishes, began to rise in the 1960s (Pister, *this volume*, chap. 4). However, the Colorado River basin had already been greatly altered by this time, and native fishes declined apace with degradation. Managed fishes were relatively well known, but such was not the case for these threatened or endangered native species, many of which were rare before they attracted attention. Long histories and traditions of research and accumulated knowledge on which to base management decisions were lacking. It was evident that major voids in information existed, and some species had become too rare to study.

This paper describes an effort to recover an imperiled species. We first review decline of the razorback sucker through a detailed assessment of the historic record. Emergence of concern for the species and development of data on its status and biology are summarized, along with political events that allowed formulation of a recovery strategy. The rationale for selecting propagation and reintroduction as recovery options is described, and development and implementation of a culture and stocking program are outlined. An interim evaluation of successes and failures is followed by a projection of needs for future actions. We end with a discussion of criteria and potentials for recovery of the razorback sucker and other long-lived freshwater fishes.

A

B

Fig. 17-1. Razorback suckers: a 46-cm male (A) and 63-cm female (B), trammel netted from Lake Mohave, Arizona-Nevada, February 1979 and March 1990, respectively. Photographs by W. L. Minckley.

Abbreviations and Sources of Information

Acronyms are applied to names frequently used in the text as follows: Arizona Game and Fish Commission and Department (AZGFD), Arizona State University (ASU), California Department of Fish and Game (CADFG), Nevada Department of Wildlife (NDOW), U.S. Bureau of Land Management (USBLM), U.S. Bureau of Reclamation (USBR, including when named the Water and Power Resources Service), U.S. Department of the Interior (USDI), U.S. Fish and Wildlife Service (USFWS, including, except for literature citations, when named the Bureau of Sport Fisheries and Wildlife [USBSFW]), national fish hatchery (NFH), and state fish hatchery (SFH). Finally, the Colorado River basin has been divided into upstream and downstream political units at Lee's Ferry, Arizona, 24 km below Glen Canyon Dam, by agreement among the basin states.

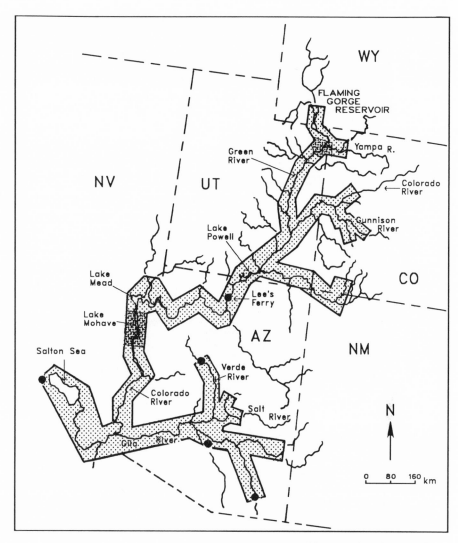

Fig. 17-2. Present and historic distribution of razorback suckers in the Colorado River basin of western North America. Dense stipple represents zones of present concentration, dots are for archaeological records, and the entire historic range is outlined.

The terms *upper* and *lower* basins are applied to these two regions, as is *basin* to the entire watershed; *Colorado River* is understood. Figure 17-3 provides a map of the basin and some place-names.

Comments on sources of information are important because of the volume and scattered nature of data on fishes of this vast region. Three basic problems impede work on imperiled fishes here and for that matter in most of western North America: (1) too few

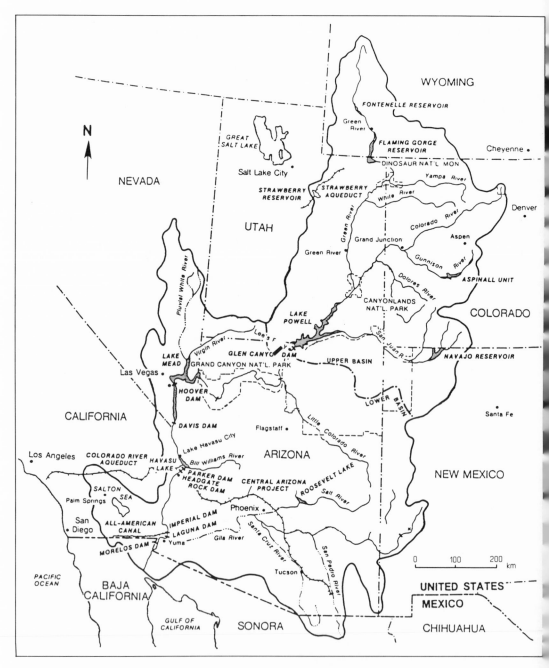

Fig. 17-3. Map of the Colorado River basin, showing some place-names mentioned in text.

researchers are involved; (2) publication outlets are limited, manuscript backlogs are large, and time to appearance of papers is long; and (3) most studies are funded by short-term contracts from state or federal agencies, which results in staggering numbers of processed reports, or "gray literature," that may be limited in availability and reliability.

The last item presents a multifaceted problem especially pertinent to the present work. Not only are agency reports often difficult to obtain, they may also be interim in nature. Time constraints sometimes necessitate less-than-final reduction and interpretation of data, and errors or premature conclusions appear. We attempted to avoid these pitfalls by personal contacts with investigators when questions arose, or if this was impractical, by our discussion in the text of possible errors or misinterpretations.

Editors of scientific journals resist the citation of gray literature, a policy we understand but condone only in part. Data in such reports have become critical to efforts toward perpetuation of native fishes, and they must be taken into account. The present paper could not have been written without reference to such materials, since about half of the more than 350 works we cite are of the gray category. We are also certain that there are other relevant reports that we failed to uncover. One basic rule we followed was the personal examination of each original work by one of us (which obviously should apply to any kind of literature). Those cited from other sources are clearly indicated as such and marked "not seen" in the literature cited.

We also depended on personal contacts to document both historic records and results of projects dealing with razorback suckers. The species has become so scarce that it rarely goes unnoticed, and occurrences recorded in memoranda, field notes for general studies, and records obtained during management operations were brought to our attention.

Research on razorback suckers has been supported by ASU, the USBR, and the USFWS. Gail C. Kobetich (USFWS) was instrumental in obtaining initial funding and support. More than two hundred students in classes from ASU gave of themselves in efforts on Lake Mohave, the lower Colorado River, and elsewhere, and we thank them all. The editorial efforts of P. H. Eschmeyer and S. P. Vives improved the manuscript. Finally, we gratefully acknowledge the efforts in our behalf of many colleagues in obtaining unpublished reports and other materials. Names and affiliations of individuals cited as providing data in personal communications (hereinafter pers. comm.) are listed in the appendix to this chapter.

The Trend toward Extinction

Abundance in Aboriginal and Early Historical Times

Few absolute data exist on the abundance of Colorado River fishes before and during exploration and settlement by Europeans. Historical statements must furthermore be interpreted with care (Forman and Russell 1983), since exaggerations must have existed then, as now. Early records were based on a few sampling sites at river crossings, near military installations, or at camps of explorers or survey parties. Inaccessibility and difficulty of sampling the turbulent, canyon-bound rivers precluded more than cursory evaluations in the upper basin, but many streams of the lower basin were studied, and their fish faunas were relatively well described. Razorback suckers were distinctive, large in size, and edible, and thus attracted attention.

In prehistoric times, razorback suckers were well enough known to lower-basin Indians to be specifically named: *tsa'xnap* by Yumans (Forde 1931) and *suxyex* by Cocopahs (Gifford 1933). Records for the species as food for Native Americans are frequent

(Ellis 1914; Rostlund 1952; K. M. Stewart 1957; LaRivers 1962), as are its remains in archaeological sites (R. R. Miller 1955, 1961; C. L. Hubbs 1960; Minckley and Alger 1968; Minckley 1976). It was also subject to a special fishing technique; V-shaped weirs constructed by Indians along shorelines of an ancient lake that once occupied the Salton Sea basin (Wilke 1978, 1980; Mehringer 1986) were placed at distances from shore and depths that closely correspond to those used by breeding aggregations in today's lacustrine habitats. Bones of razorback suckers and bonytail (*Gila elegans*) in excavated Indian camps attest to the success of this fishery.

One of the first published records for the razorback sucker was its original description as *Catostomus texanus* from the "Colorado and New Rivers," Arizona and Baja California del Norte, Mexico (Abbott 1861). Earlier, Bartlett (1854) noted "a fish resembling the buffalofish of Mississippi, between Yuma and Gila Bend" that could only have been a razorback sucker. He considered its flesh "soft and unpalatable." The species was redescribed as *Catostomus cypho* by Lockington (1881) from a specimen caught near the confluence of the Gila and Colorado rivers. In 1890 Gilbert and Scofield (1898) "found [razorback suckers] extremely abundant at Yuma and at all points below as far as the Horseshoe Bend, and in Hardee's Colorado [the Mexican Río Hardy]." Snyder (1915) reported on observations made in 1894 by E. A. Mearns on the Colorado River delta in Mexico and at Yuma, and at Gila City (the present Dome, Arizona), on the Gila River. He quoted Mearns's unpublished notes as follows: "The flesh of this species is excellent and of fine flavor; very large specimens, refusing to take a hook, were snared in deep holes among the rocks. A line with several hooks attached was allowed to sink to the bottom and when a school moved over it the line was brought out with a sharp swish of the pole, a sucker usually being

hooked." In 1904, F. M. Chamberlain caught twelve razorback suckers in one of three seine hauls at the junction of the Gila and Colorado rivers, along with larger numbers of bonytail and introduced common carp (*Cyprinus carpio*) and black bullheads (*Ameiurus melas*; R. R. Miller 1961). Grinnell (1914) caught the species from the Colorado mainstream near Mellen (Topock), Arizona, in 1910 but wrote nothing of its abundance.

Razorback suckers appeared in numbers in the Salton Sea after it was filled by Colorado River floods of 1904 and 1905 (Evermann 1916). Odens (1989) related an account by a resident, George Utley, who camped near the sea in 1909 and said that razorback suckers "came up the irrigation canals in search of food, splashing around." Utley reported securing 147 fish by hand axe and .22-caliber rifle in an hour. The fish persisted in the Salton Sea until after 1929 (G. A. Coleman 1929) but disappeared when salinities resulting from evaporative concentration presumably exceeded their tolerance (B. W. Walker et al. 1961).

The genus *Xyrauchen* was erected by Eigenmann and Kirsch (in Kirsch 1889) based on specimens from the Gila River near Fort Thomas, Arizona. Elsewhere in the Gila River basin, Gilbert and Scofield (1898) recorded it from the Salt River between Tempe and the mouth of the Verde River. Chamberlain (1904) seined one from the Salt River near Roosevelt in 1904 and obtained local testimony that the species also occupied the Gila River near the Arizona–New Mexico boundary (near Safford and Duncan). He reported it "rather common" in the San Pedro River in the 1800s, when it was sold in nearby Tombstone as "buffalo." The species remained abundant enough in the late 1800s and early 1900s in central Arizona to be harvested for human and animal food, and fertilizer (R. R. Miller 1961; Minckley 1973; Ellison 1980). A commercial fishery existed in one Salt River reservoir until 1949 (C. L. Hubbs and Miller 1953), and razorback suckers persisted in the Verde River and canals near Mesa into the 1950s (Wagner 1954; Minckley 1983).

There is no question that razorback suckers were originally widespread in the upper basin, but statements regarding their abundance are ambiguous. D. S. Jordan (1891) wrote that "suckers and round-tails [*Gila robusta*] become abundant" in the Gunnison River below Black Canyon (he collected at Delta, Colorado). In addition to razorback suckers (reported as *Xyrauchen cypho*), Jordan's party also collected one specimen of "what seems to be a new species of *Xyrauchen*" in the Uncompahgre River near Delta (*X. uncompahgre*, which later proved to be a hybrid flannelmouth [*Catostomus latipinnis*] × razorback sucker; C. L. Hubbs and Miller 1953). He noted that "the water is full of fishes of the species enumerated above as found in the Gunnison." Razorback suckers were recorded as "very abundant" at Blake City (Green River Station) on the Green River, Utah, but actual numbers collected were not given. D. S. Jordan (1891:5) noted that "species of *Catostomus* and *Xyrauchen* reach a considerable size, and are food-fishes of poor quality," yet later in the same report (p. 24) he wrote, "*Xyrauchen cypho* is very abundant, reaching a weight of 10 pounds [4.5 kg], and is a good food-fish." The specimens he caught were about 20 cm long (presumably standard length [SL]). In contrast, Evermann and Rutter (1895) reported sampling at Green River, Wyoming, where they collected seven species of fishes (numbers not indicated), including razorback suckers: "These represent the result of almost constant seining for the greater part of a day, and thus indicate the paucity of fishes in this stream."

D. S. Jordan and Evermann's (1896b) statement that razorback suckers were "very abundant where the water is not too cold" clearly referred to the entire range as based on literature of the time, and inferred they were scarcer upstream. Ellis (1914) wrote that the fish lived in larger streams throughout western Colorado but failed to address abundance. He recorded three specimens, 29–40 cm long (presumably total length [TL]), caught in 1912 from the Grand (now Colorado) River at Grand Junction. The species was "usually marketed with Flannel-mouth Suckers," mostly "because of its large size." Alternatively, razorback suckers may not have been abundant enough to be independently marketable. He also noted another lower basin occurrence as follows: "Professor Junius Henderson has told the writer that *X. texanus* is taken in numbers by the Mohave Indians from the Colorado River near Fort Mohave," which remains near the major center of abundance in that region, in what is now Lake Mohave.

Kidd (1977) quoted a resident of Delta, Colorado, who fished commercially in the Gunnison River between 1930 and 1950 and took fifty Colorado squawfish (*Ptychocheilus lucius*) and a large number of razorback suckers "in one of the better years." Wiltzius (1978) also reviewed occurrences of razorback suckers in the Gunnison, citing T. K. Chamberlain (1946) as reporting them common in the lower part of that stream. Wiltzius concluded that the species began its decline before 1912, assuming that D. S. Jordan's (1891) indicated abundance was correct.

Other, later works provided general overviews of the species. D. S. Jordan and Evermann (1923), in the popular *American Food and Game Fishes*, noted razorback suckers as abundant and of considerable value as food. Simon (1946, 1951) reported only a few specimens from the Green River in Wyoming. Beckman (1953, 1963) recorded the species in Colorado with no further comment. W. F. Sigler and Miller (1963) compiled widely scattered records for Utah from the Colorado River at Moab, White River near Ouray, and from the Green River at several places between the town of Green River, Utah, and Hideout Canyon near the Wyoming border.

After the 1950s, most general works cited or dealt specifically with razorback suckers' increasing rarity; for example, for Arizona see: R. R. Miller and Lowe (1964, 1967), Minckley (1973); for California: Shapovalov et al. (1959), Leach et al. (1974, 1976, 1978), Moyle (1976); for Colorado: Everhart and Seaman (1971), J. E. Johnson (1976), Langlois (1977), Wooding (1985); for Nevada: LaRivers (1962), Deacon and Williams (1984); for Wyoming: G. T. Baxter and Simon (1970); and throughout its range: R. R. Miller (1964b, 1968a, 1969b, 1972a), Deacon (1968b, 1979), Holden et al. (1974), Reiger (1977), Ono et al. (1983), and J. E. Williams et al. (1985).

Evidences and Patterns of Decline

Natural populations of razorback suckers have disappeared from the entire Gila River basin. The fish almost certainly was common throughout that system and in the lower Colorado mainstream eighty to one hundred years ago. One population of large adults persists in Lake Mohave on the mainstream Colorado River, but the remainder of the river downstream from Lake Powell (Figs. 17-2, 17-3) yields only isolated adults and a few juveniles. The species was widespread but apparently less abundant in the upper basin. Data for the past twenty-five years indicate little change in this pattern, except that the species is now excluded from the Green River upsteam from Dinosaur National Monument. Razorback suckers have therefore declined to endangered status in their former center of abundance in the lower basin, and concern exists that the smaller upper-basin populations are following in turn.

Lower Colorado River basin

Suspicions that razorback suckers were disappearing from the lower basin were expressed shortly after Lake Mead was impounded by the closure of Boulder (Hoover) Dam in 1935. Dill (1944), R. R. Miller (1946a), Wallis (1951), and Jonez and Sumner (1954) mentioned this trend, and R. R. Miller (1961) and Minckley and Deacon (1968) summarized earlier work and presented additional supporting data. Even then, essentially all records were of adults. The historic absence of young (small) fish from collections (C. L. Hubbs and Miller 1953) is striking.

There is little evidence that the species was ever common in Marble and Grand canyons upstream from Lake Mead, perhaps because no early collecting was done in the inaccessible gorges. Eight records are known: one fish taken by an angler at Bright Angel Creek in 1944; a second near Lee's Ferry in 1963; four adults taken or seen in the Paria River, one in 1978 and three in 1979; a single fish photographed and released above Bass Rapid in 1986; at least three observed in Bright Angel Creek in April 1987; and three taken each year in 1989 and 1990 from the mouth of the Little Colorado River (C. O. Minckley and Carothers 1980; Carothers and Minckley 1981; Maddux et al. 1987; Baucom, Douglas, and Hendrickson, pers. comm.). Putative hybrids between flannelmouth and razorback suckers (hereinafter termed hybrids) have been recorded from the Paria River area (Suttkus et al. 1976; Suttkus and Clemmer 1979).

Moffett (1943) noted razorback suckers in Lake Mead but gave no indication of the population size. Wallis (1951) reported the species "abundant in Lake Mead" and exceptional among native fishes in "holding its own and reproducing abundantly." It had apparently remained common since closure of the dam (Jonez and Sumner 1954; Minckley and Deacon 1968). Deep-water detonations of explosives in 1962 yielded eight razorback suckers (3.1% of adult fish recorded; Jonez and Wood 1963), and numerous adults were observed but not collected in 1962–1967

(Deacon, pers. comm.). McCall (1980) recorded eight razorback suckers electrofished by AZGFD from Lake Mead in 1967, 1.0% of a total sample of 810 fish. The reported total lengths (22.4 cm average for six fish, and two others 24.3 and 26.0 cm) are almost certainly in error. Such small fish were not known from the system even at that time, and associated weights (mean 2.3 kg for the six, 3.0 and 3.3 kg, respectively) correspond to fish of greater than 50 or 60 cm TL.

By the early 1970s the Lake Mead population had become noticeably reduced (Anonymous 1973; Minckley 1973). McCall (1980) caught only three among more than ten thousand fishes recorded at fifty-seven sampling sites in 1978 and 1979; eighteen were caught between 1976 and 1980 (Minckley 1983); and none was taken by NDOW personnel between 1981 and 1988 (Withers, pers. comm.). In April 1990, however, an adult male was caught by a fisherman, and fifteen others in an apparent spawning aggregation were observed by the NDOW. Two ripe males (54.8 and 61.0 cm TL) and a gravid female (59.2 cm) were captured and others were observed in May 1990 (Sjoberg, pers. comm.).

Razorback suckers were recorded below Boulder (Hoover) Dam in the 1940s, but there were no indications that they were abundant (Moffett 1942). Wallis (1951) considered them less common in Lake Mohave, then in an early stage of impoundment, than in Lake Mead. In 1950, however, Jonez et al. (1951) reported razorback suckers "very common" in the reach soon to be flooded by Lake Mohave (Davis Dam was variously closed from 1946 through 1951, and finally completed and closed in 1954), but of variable and lesser abundance below Davis Dam to the Mexican boundary. Other workers also reported it abundant in the Lake Mohave reach in the early 1950s, and spawning and production of larvae were noted (Jonez and Sumner 1954; Winn and Miller 1954). Egg predation

by common carp was observed, and few larvae were expected to survive because they were intermingled with predatory centrarchids (Jonez and Sumner 1954). The Lake Mohave stock nonetheless remained common (LaRivers 1962; Minckley 1973, 1983; Allan and Roden 1978; Marsh and Langhorst 1988). In fact, there has been no evident change in abundance between 1974 and 1990 (Minckley and Marsh, unpub. data), and Lake Mohave fish comprised in 1990 the last major population in existence.

Kimsey (1957) noted that the razorback sucker was able to maintain "good numbers" in the lower Colorado River while other native species had "virtually disappeared." Yet actual records downstream from Davis Dam were sporadic after the 1950s. One adult was taken and others were seen in Topock Marsh upstream from Needles, California, in 1963 (Minckley, unpub. data), and one large fish was seen in Topock Gorge in 1968 (Minckley 1973). The CADFG recorded nine adults between Davis Dam and Lake Havasu in 1972–1976 (Minckley 1983; Ulmer and Anderson 1985). G. B. Edwards (pers. comm.) regularly saw adults between Davis Dam and Bullhead City while studying striped bass (*Morone saxatilis*) in 1966–1972 (Edwards 1974). J. Warnecke (pers. comm.) observed and captured a few individuals in that same reach in 1977–1979, and two large adults were taken from a slough near Laughlin, Nevada, in 1987 (Marsh and Minckley 1989).

Documentation is less complete for Lake Havasu, the second major reservoir constructed on the lower mainstream (formed by Parker Dam, closed in 1938). In 1950, Jonez et al. (1951) noted the species, and P. A. Douglas (1952) observed spawning by numerous adults. Guenther and Romero (1973) considered it rare. Grabowski et al. (1984) recorded two from a cove of the main lake in 1978 and another from the Bill Williams arm in 1979. Minckley (1979a, 1983) caught none in trammel-net surveys between 1976 and 1982. According to Ulmer and Anderson (1985), only fifteen individuals were observed or captured by CADFG personnel between 1962 and 1984. Most recently, five adults (two determined to be greater than twenty years of age by otolith examination; Marsh, unpub. data) that could only have been pumped from Lake Havasu were caught in the Central Arizona Project Granite Reef Aqueduct (USBR 1986; Mueller, pers. comm.). Larvae in samples from the aqueduct and Lake Havasu (USBR 1988; Marsh and Papoulias 1989) documented at least some reproduction in 1985 and 1986.

Below Parker Dam, Dill (1944) netted razorback suckers only once during an extensive survey in 1942 from Needles to Yuma. Of thirteen caught near the town of Parker, ten measured 34–53 cm TL (average 42 cm). Dill wrote: "Certainly it was most plentiful at one time throughout the lower river and in the agricultural ditches according to residents. It is now very scarce." He further offered the following local explanation for the decline of native fishes in the lowermost river: "They believe that during one of the droughts in the pre-Boulder [Hoover Dam] period, most of the fishes swam far up the river seeking cooler water. Here, so they say, one can still find them—trapped by the dam."

More recently, single specimens were taken near Blythe in 1969, from Imperial Reservoir in 1973, and in Palo Verde Valley in 1976 and 1977. One or two adults per year were reported by anglers from the Imperial-Coachella irrigation system between 1978 and 1982 (Minckley 1983). Ulmer and Anderson (1985) recorded at least a dozen adults, probably including some of those just noted, between Blythe and Imperial Reservoir from 1969 to 1985. Despite these occurrences, none was caught in general surveys from Davis Dam to Mexico in 1974–1976 (Minckley 1979a, 1983) nor in 1983–1987 between Laguna

Dam and the U.S.–Mexican boundary (Marsh and Minckley 1985, 1987). A specific survey for the fish from Parker Dam to Mexico in 1980 and 1981 also failed to capture any (Loudermilk and Ulmer 1985; Ulmer and Anderson 1985). No records are known from Mexico since the late 1800s (Follett 1961).

Razorback suckers were also absent from collections in areas other than the main-stem Colorado River. None was taken from 1967 to 1977 in the Salt River or its reservoirs in central Arizona (Minckley and Deacon 1968; Minckley 1969a, 1973; J. E. Johnson 1969; Bersell 1973; Rinne 1973). Surveys in 1976–1978 where the species formerly occupied the Gila River also failed to take any (Minckley and Clarkson 1979).

Upper Colorado River basin

Similar to the situation downstream, rarity of razorback suckers in the upper basin was beginning to be seriously discussed in the late 1960s. In response to accelerating development of water resources, fishery surveys were begun in areas that included the forbidding whitewater canyons. Razorback suckers, anticipated to be common, proved scarce.

As in Grand and Marble canyons, razorback suckers may never have been abundant in Glen Canyon, which was flooded by Lake Powell with the closure of Glen Canyon Dam in 1963. The only preimpoundment survey produced two young (each ca. 38 mm TL), one each from a creek mouth and a backwater (G. R. Smith 1959; G. R. Smith et al. 1959; McDonald and Dotson 1960).

To our knowledge, Lake Powell was not systematically sampled for razorback suckers in the early years after impoundment, although, based on other studies, the species never became common there. Five or six adults were taken in 1980 and eleven more in 1981 in Gypsum Canyon at the extreme upper end of the reservoir (Persons and

Bulkley 1980; Persons et al. 1982). Valdez et al. (1982a) also reported on some of the 1980 specimens (four were indicated in their fig. 30) and one taken at Hite Marina on Lake Powell. Gustaveson (pers. comm.) provided records for six ripe males (55–56 cm TL) collected in April 1982, and one each in November 1984 and 1985 (both ripe males, 55 and 54.5 cm TL), all from the San Juan arm. One male was caught in Lake Powell about 2.4 km downstream from the Dirty Devil River in 1983, and seven (including recapture of the latter fish) were netted in 1984 from the Dirty Devil arm (USBR 1984); four additional fish were caught from this area in 1990 (Williams, pers. comm.).

The Utah Division of Wildlife Resources began sampling for native fishes in 1987, when they caught, tagged, and released twelve adult razorback suckers near Piute Farms Marina on the San Juan arm (Meyer and Moretti 1988). Nine were males (eight ripe) measuring between 54.6 and 60.5 cm TL, and three were females 62.5–67.5 cm long. In 1988 ten adults (including six of the twelve tagged in 1987) were taken in the same area (Moretti, pers. comm.): five males 55–67.8 cm, four females 59.5–67.5 cm, and one fish of undetermined sex that was 64.6 cm TL.

The San Juan River basin presents an enigma. Based on local testimony, D. S. Jordan (1891) reported movements of adult razorback and flannelmouth suckers and "white salmon" (Colorado squawfish) into the Animas River, a major tributary of the San Juan in southern Colorado, yet no specimens of razorback suckers are known from the state of New Mexico (Koster 1957, 1960; Sublette 1977; Sublette et al. 1990), through which the river flows from Colorado into Utah (Fig. 17-3). No razorback suckers were taken in surveys of the San Juan River in New Mexico and Utah in 1987, and only one 57-cm, tuberculate male was captured in the river

near Bluff, Utah, in 1988 (Meyer and Moretti 1988; Brooks, unpub. data).

Nonetheless, one of the largest concentrations of razorback suckers reported in the upper basin, estimated by Behnke and Benson (1980) at 250 fish, was stranded in a drying irrigation reservoir connected to the San Juan near Bluff, Utah, in 1977. Behnke and Benson gave no authority for their report. McAda and Wydoski (1980) attributed the report to "N. Armentrout [sic], personal communication," noting "abundance was not determined." Meyer and Moretti (1988) cited the USBLM (1981) as reporting razorback suckers seined in 1976 not only from an irrigation pond near Bluff but also from backwaters of the river downstream from the pond. They cited "Neil Armantrout" as the authority for these data. All reports were almost certainly for the same incident. In 1990 Armantrout (pers. comm.) recollected the existence of two ponds. One had dried before he arrived, and only unidentified fish carcasses remained; he was told they included razorback suckers. At least two adult razorbacks were seined from the other pond. He also was told of razorbacks in river backwaters downstream from the irrigation ponds, but he did not observe them.

We discovered records for only five razorback suckers from the Cataract Canyon area, through which the Colorado River flows downstream from its confluence with the Green River and upstream from Lake Powell (Fig. 17-3). Of these, two were taken just upstream from the Green-Colorado confluence in the Colorado in 1980 (Persons et al. 1982), two were caught in Cataract Canyon in 1981 (Valdez et al. 1982a), and the other was captured there in 1987 (Valdez 1988); all were adults longer than 50 cm.

GREEN RIVER SUB-BASIN. Native fishes have been studied intensively in the Green River drainage since the early 1960s, mostly as a result of the controversial poisoning of upper parts of that system (Holden, *this volume*, chap. 3) and other conflicts between water developments and fisheries resources (Wydoski and Hamill, *this volume*, chap. 8). Broadly overlapping studies resulted in multiple reports dealing with the same or parts of the same data sets, and numbers and distributions of fish captures become confused. The pertinent literature regarding this area is voluminous and less familiar to us than that for the lower basin, and our coverage may be less than complete. Nonetheless, collections made before major alterations provide a baseline unavailable for lower-basin waters. Furthermore, most information is from studies specifically designed to assess the status of native fishes, contrasting with the anecdotal types of data available elsewhere.

Bosley (1960) recorded a razorback sucker in 1959 from Hideout Forest Camp on the upper Green River main stem, a locality now flooded by Flaming Gorge Reservoir. McDonald and Dotson (1960), G. R. Smith (1960), and Gaufin et al. (1960) each reported an adult from the same area and year; all referred to the same fish. C. L. Hubbs and Miller (1953) further reported five hybrids near this same place in 1950, and G. R. Smith (1960) noted those plus another taken in 1959 near the Utah-Colorado boundary. Azevedo (1962a) caught one razorback sucker between Jensen and Green River, Utah, in April 1962, and another was netted (Azevedo 1962b, c) from the Yampa River at Echo Park in October 1962. A 39.3-cm SL male from the Green River at Lodore Ranger Station (Branson and McCoy 1966) was taken during rotenone treatment in September 1962.

According to Hagen and Banks (1963), no razorback suckers were observed in Dinosaur National Monument during survey operations in 1961 and 1962 until rotenone passed through (7 September 1962), when "many of

the rare forms were picked up from Lodore [Ranger Station] all the way to Split Mountain campground [a total of about 70 km by river]." Banks's field notes, which we paraphrase with reference to razorback suckers, were reproduced by Hagen and Banks (1963):

> Rotenone reached the [Brown's Park] bridge at 7:00 A.M., one razorback sucker collected; between 50 and 60 razorback suckers were observed by Jack [C. J.] McCoy floating down the middle of the river [site unknown, but in the Brown's Park area]; six washed on shore near the [Lodore] ranger station; two observed in distress [site unknown]; one decomposed individual found [site unknown]; and two dead specimens were noted at Split Mountain campground.

Banks (1964) collected only ten razorback suckers, all from the rotenone operation (a total of eleven was taken if Branson and McCoy's [1966] specimen is included). Its overall rarity was further indicated by Binns's (1965, 1967a) listing of razorback sucker as the species least commonly seen during the eradication project. Five putative hybrids were secured: one in May 1962 at Echo Park and two each from the Gates of Lodore and Echo Park during the September poisoning. Neither razorback suckers nor hybrids were caught at Castle Park, Colorado, on the Yampa River (Banks 1964).

The additive impacts of rotenone and the closure of Flaming Gorge Reservoir in 1962 and Fontenelle Reservoir upstream in 1963 eliminated razorback suckers from Wyoming (G. T. Baxter and Simon 1970; Anonymous 1977), where they apparently were rare even before those alterations (Simon 1946, 1951; Bosley 1960; Binns 1965, 1967a). We found no evidence that the species ever reappeared upstream from Flaming Gorge Dam.

The species persisted downstream, since Vanicek (1967) caught sixty-five razorback suckers and fifteen putative hybrids, all of adult size, in 1964–1966 between Flaming

Gorge and the mouth of the White River, Utah. They comprised only 0.3% of almost 24,000 fishes taken from eight stations. These same data were treated by Vanicek et al. (1970), who added information from 1963 to record seventy-three razorback suckers (Echo Park, Island Park, Split Mountain, and Ouray) and sixteen hybrids (Lodore Canyon and all the above localities except Ouray). No young-of-year or juveniles were taken. The first 100 km of river below Flaming Gorge Dam was devoid of most native fishes, including razorback suckers. Species composition of native fishes below inflow of the undammed Yampa River was not appreciably changed from that prior to closure of Flaming Gorge (Stalnaker and Holden 1973), a pattern that remains today (Holden, *this volume*, chap. 3).

A comprehensive survey of upper-basin fishes by Holden (1973) between 1967 and 1972 encompassed twenty-two reaches in both the Green and Colorado sub-basins. He caught fifty-three razorback suckers and considered the species rare. Forty putative hybrids, "readily distinguished by an intermediate lateral line scale number and a much abbreviated, although clear, keel," were also taken. Again, all were adults. These same data were reported in part by Holden and Stalnaker (1975a), who listed both razorback suckers and hybrids as rare in Dinosaur National Monument, Desolation Canyon, and Canyonlands National Park. Both were absent in collections from the upper Yampa River.

Seethaler et al. (1979) summarized collections from the Green and Colorado rivers in 1974–1976, noting that razorback suckers were common (although not numerous) at the mouth of the Yampa River in early spring and late autumn; no numbers were indicated. Despite the presence of ripe fish of both sexes in spring, no young were collected. McAda (1977) and McAda and Wydoski (1980) reported one fish from Sand Wash, three from

Island Park, and twenty-nine razorback suckers and five hybrids from Echo Park.

Based on other collections, razorback suckers were common in 1978 near the mouth of Duchesne River, Utah; an unknown number was found between Horseshoe Bend (below Jensen) and the Duchesne River in 1978; and one was caught from the Yampa River in 1979, near the Little Snake River (McAda and Wydoski 1980). The last individual was also noted by Wick et al. (1981). Additional sampling between the towns of Jensen and Green River in 1975 and 1977 by Holden (1977a, 1983) caught two adults and six tentatively identified juveniles from near Ouray. He also reported another adult collected by the USFWS from that area in 1976. Another hybrid was taken 96 km up the White River in 1979 (Lanigan and Berry 1981).

Between 1978 and 1980 Holden and Crist (1981) collected sixty-seven razorback suckers and four hybrids in the Green and Yampa rivers. Collection sites extended from the Yampa River a few kilometers above the Green-Yampa confluence to a lowermost site at Jensen, where fifty-six (83.6%) of the fish were caught. In 1979–1981 Tyus et al. (1982b) recorded ninety-two specimens below Split Mountain Canyon, forty-two (45.6%) of which were taken near Jensen. All were from flat-water sections: ninety-one from between Split Mountain and Desolation canyons, and one upstream from Labyrinth Canyon.

Tyus (1987) and Tyus et al. (1987) summarized records for 323 adults, including the 92 just listed, from between the Yampa and the junction of the Green and Colorado rivers in 1981–1986. Most were taken between the lowermost Yampa and the mouth of the Duchesne; only 6 were from farther downstream. They ranged from 42.6 to 60.9 cm TL, ripe males averaging 50.3 cm ($N = 37$) and ripe females 54.4 cm ($N = 23$). Razorback suckers comprised only 1.8% of the number of federally endangered Colorado squawfish caught in a standardized sampling program, and 33% of catches of squawfish in springtime electrofishing, a time of year the sucker was most vulnerable to capture. Lanigan and Tyus (1989) then analyzed 50 recaptures of 360 razorback suckers, combining those tagged by the USFWS and the Utah Division of Wildlife Resources between 1981 and 1986, to statistically estimate 978 ± 232 (95% confidence limits) adults in the Green River above Desolation Canyon. None was taken in Desolation and Gray canyons over the same period, and too few (13) were available from the lower Green River for a population estimate.

UPPER COLORADO RIVER SUB-BASIN. Studies in this sub-basin began in the early 1960s. One of the few reports of wild juvenile razorback suckers is that of Taba et al. (1965), who caught eight specimens 90–115 mm long (presumably TL) in the period 1962–1964 from backwaters of the 40-km reach of mainstream between Moab and Dead Horse Point, Utah. Holden (1973) found adults scarce and juveniles lacking in surveys made in 1967–1972. He caught both razorback suckers and putative hybrids in the Colorado River in Canyonlands National Park and between Grand Junction and Fruita, Colorado, and hybrids only at Moab, Utah. Neither was taken in the main stem near Rifle, Colorado, nor in the Gunnison or Dolores rivers (Holden and Stalnaker 1975a, b). Intensive work in 1980 and 1981 also failed to collect the species in the latter two streams (Valdez et al. 1982a). However, a specimen from the Gunnison between Delta and the Escalante Bridge in 1976 was 49.5 cm (TL?) long and estimated (presumably through examination of scales) as nine years old (Kidd 1977; Wiltzius 1978), and a gravid female was caught by other workers in 1981 about 90 km upstream in the Gunnison River (Valdez et al. 1982b).

Kidd (1977) reported 49, 166, and 19 razorback suckers captured in 1974, 1975, and 1976, respectively, all from backwaters of the Colorado River main stem near Grand Junction, Colorado. Others were observed in 1976 near Grand Junction and upstream near the mouths of Roan and Asbury creeks. Joseph (1978) recorded another site in the Colorado mainstream about 35 km below the inflow of the Gunnison River. McAda (1977) and McAda and Wydoski (1980) recorded 74 razorback suckers and three hybrids in the Walker Wildlife Area; all but one of the former was in an artificial gravel pit adjacent to the river channel. It is likely that some interchange of data occurred among these last investigators; we cannot sort out duplications.

Sampling in 1979 and 1980 by Wick et al. (1981) yielded one fish from backwaters in the Walker Wildlife Area, seven from a flooded backwater upstream, eight from a gravel pit, and only one from the main stem. They also noted "several" hybrids between Loma and Ruby Canyon, Colorado. Valdez et al. (1982b) reported forty-seven adults from between Rifle, Colorado, and Lake Powell in 1980 and 1981. Twelve localities were represented, but only four fish were in the channel. Most (thirty-seven) were again from backwaters and gravel pits. W. H. Miller et al. (1983) reported twenty-one adults from five gravel pits in this same area in 1982. Kaeding and Osmundson (1988a) caught five adults from the river about 25 km upstream and downstream from the Gunnison in 1986 and 1987, and saw another in the lower Gunnison River in 1987.

Where are the little fish?

Few small razorback suckers have ever been taken in either the upper or lower basin (C. L. Hubbs and Miller 1953; McAda and Wydoski 1980; Valdez and Magnan 1980; Minckley 1983; Tyus 1987). In fact, Holden (1977a, 1978, 1983) reported his 1977 collection as the first ever of postlarval young-of-year from the upper basin. As we have already noted, G. R. Smith (1959) captured two immature razorback suckers from Glen Canyon, and Taba et al. (1965) recorded eight juveniles from Colorado River backwaters downstream from Moab, Utah. We know of only four collections of juveniles in the lower basin prior to the 1980s: one from below Laguna Dam, two from below Davis Dam (Jonez and Sumner 1954; Minckley 1983), and one at Cottonwood Landing, Nevada (Winn and Miller 1954; R. R. Miller 1961; W. F. Sigler and Miller 1963), all in 1950. P. A. Douglas's (1952) "four-inch humpback sucker" from below Davis Dam was a speckled dace (*Rhinichthys osculus*; Winn and Miller 1954).

The only substantial numbers of juveniles resulting from natural spawning in the 1980s were caught from irrigation canals and ponds downriver from Parker Dam (St. Amant et al. 1974; Ulmer and Anderson 1985; USFWS 1981a; Marsh and Minckley 1989; Milstead and Yess, pers. comm.). Reintroductions after 1984 (see below) confuse the issue, yet of twenty-four juveniles recorded from 1973 to 1986 (15.0–37.5 cm TL), at least sixteen were captured before stocking or from places that reintroduced individuals could not have reasonably attained, and were unquestionably wild fish. Sixty-eight other juveniles were caught from canals near the town of Parker in 1987 (thirty-eight specimens, twenty-one preserved, averaging 45.1 cm TL), 1988 (three fish averaging 28.8 cm TL), and 1990 (twenty-seven fish averaging 49.4 cm TL).

The intake for the canals, the only point of ingress for fishes, is upstream from any known reintroduction sites (Marsh and Minckley 1989); ten of the twenty-one specimens and all three from 1988 were determined by otoliths as being two to seven years old (Marsh, unpub. data) and must also represent natural recruits. The mainstream Colorado River has yielded only one juvenile

since 1950, a 31.5-cm individual caught 16 km below the town of Parker in 1987 (Langhorst 1988) that was two years old based on otoliths. Because it also occurred above a barrier isolating reintroduction sites, this fish must have been naturally spawned. The geographic origin of these wild fish is unknown, but Marsh and Minckley (1989) suggested spawning areas either within or immediately downstream from Lake Havasu.

Early records of larvae are similarly scarce. A single specimen was recorded from Lake Havasu in 1950 (P. A. Douglas 1952), and Jonez and Sumner (1954) noted larvae in Lake Mohave at about the same time. A few have been caught in plankton tows (Mueller et al. 1982), by hand in fine-meshed nets during daylight (Minckley and Marsh, unpub. data), and infrequently in midwater larval trawls from Lakes Mohave (Marsh and Langhorst 1988) and Havasu (Marsh and Papoulias 1989). Now that spawning areas have been identified and sampling techniques perfected, larval razorback suckers have not proven difficult to obtain. They are phototactic to bright light at night, and hundreds may be collected if sufficient time (usually a few hours) is allowed (Bozek et al. 1984; Langhorst and Marsh 1986; Langhorst 1987; Papoulias 1988). Mueller et al. (1982) and Mueller (1989) used SCUBA and fine-meshed nets to observe and collect larvae directly from gravels in and adjacent to spawning depressions in Lake Mohave. Tyus (1987) and Tyus et al. (1987) tentatively identified thirty-three 10.6–13.6 mm TL larvae from fine-meshed seine samples downstream from spawning areas in the Green River in 1984. Razorback suckers were in thirteen (31%) of forty-two samples, comprising 3% of 1085 larvae collected.

No recruitment to reservoir populations has been detected between 1963 and 1990 in the lower basin, despite collecting with appropriate equipment (Minckley 1983, 1985b;

Ulmer and Anderson 1985) and even when successful larval production was demonstrated in Lakes Mohave and Havasu and Senator Wash Reservoir (Bozek et al. 1984; Langhorst et al. 1985; Loudermilk 1985; Langhorst 1986, 1987; Marsh and Langhorst 1988; Marsh and Papoulias 1989; Mueller 1989). Paulson et al. (1980a) and Liles (1981) reported "25–30 cm" razorback suckers in upper Lake Mohave that may have been an exception, but none was collected and they may have resulted from the escape of some 350 hatchery fish from Willow Beach NFH in 1978 (Minckley 1983). The only actual specimens of juveniles were collected in July 1987, when four young-of-year (three preserved; 25, 32, and 42 mm SL [ASU nos. 11567-8]; Marsh and Minckley 1989) were seined from a bay in upper Lake Mohave. A possibility exists, although remote, that these fish also resulted from another experimental situation (see below).

McCarthy and Minckley (1987) analyzed polished otoliths of seventy Lake Mohave razorback suckers sacrificed between 1981 and 1983. Their sample collectively exhibited "variations in size and condition typical of the . . . population in the period 1980–84." None was estimated to be less than twenty-four years of age, and the oldest individual had forty-four apparent annuli. They concluded: "If these represent absolute ages, individuals comprising the sample of 70 fish . . . originated between 1937 and 1958, and 62 (88.6%) hatched prior to or coincident with construction and filling (1942–54) of Lake Mohave." Ages obtained from structures other than otoliths—by examination of scales, for example (McAda 1977; McAda and Wydoski 1980; Valdez et al. 1982b)—are almost certainly underestimated (Minckley 1983, 1989b; McCarthy and Minckley 1987). Annulus formation on scales may be incomplete or absent, and presumed annuli in individuals older than ten years are too near one another to be reliably separated and identified.

Development of a Recovery Program

Three basic problems plagued the planning for a recovery program for the razorback sucker: (1) information on its life history and habitat under natural conditions was fragmentary and superficial, (2) specific knowledge of why the fish was declining was inferential, and (3) there was substantial and justifiable lack of confidence that it could be reestablished in waters from which it had already disappeared for unknown reasons.

The first two problems could only be resolved through additional study, and in the case of the razorback sucker the Lake Mohave population seemed large enough to support a research effort. Biological aspects of the third problem could perhaps most readily and directly be approached by reintroducing the fish and tracing the results. The third problem was also a political issue, however, necessitating education toward recognition and acknowledgment of the plight of imperiled fishes in general, development of an ethic regarding their conservation, and lobbying for formulation of informal (public) and formal (institutional) resolve to perpetuate them. In this section we review data pertaining to the first two problem areas and describe some of the political activities that led to development of a recovery program in the lower Colorado River basin.

Biology

Reproductive period and reproduction

TIMES OF YEAR. Razorback suckers exhibit marked sexual dimorphism (Gustafson 1975a, b; McAda and Wydoski 1980; Minckley 1983). Males are smaller than females and slimmer bodied, with larger fins; strong, dense nuptial tubercles on surfaces of the anal and caudal fins, caudal peduncle, and posterolateral body; and a more exaggerated predorsal keel (Fig. 17-1). Females are larger, thicker bodied and have relatively smaller fins and an often lower and broader predorsal keel. Some females develop tubercles on the anal and caudal fins and caudal peduncles, but such tubercles are smaller and far less profuse than on males. Tuberculate males are not necessarily an indication of spawning. They are present in Lake Mohave much of the year, predominating in all but July–September, but most common in January–May. Ripe males— those from which milt flows with gentle pressure—have been caught from November through June but are most common in December through March. Ripe females, which shed mature ova with slight pressure, have been recorded in Lake Mohave from December through early June. Females develop enlarged, fleshy, edematous urogenital papilli when spawning.

Breeding adults of both sexes are dark brown to black above and creamy yellow below, with a variably developed yellowish, orange, reddish to reddish brown, or sometimes violet lateral band. As with other secondary sex characteristics, these colors may be found at almost any time of year but are most common in late winter and spring.

Male razorback suckers in Lake Mohave begin to congregate, or stage, in loosely knit aggregations as early as November or December. Females join them in January or February to begin spawning, which may extend into April, or rarely into May (P. A. Douglas 1952; Jonez and Sumner 1954; Minckley 1973, 1983; Mueller et al. 1982, 1985; Bozek et al. 1984; Langhorst and Marsh 1986; Marsh and Langhorst 1988). Spawning was observed once in the first week in June 1988 in the perennially cold water (11.0°C) im-

mediately downstream from Hoover Dam (Marsh, unpub. data). That activity occurred over a gravel outwash fan associated with local inflow of warm water from thermal springs. Spawning has been recorded in other places in Lake Mohave at water temperatures varying from 10.5° to 21°C, and the long reproductive season may relate to an availability of diverse temperatures within the reservoir.

Spawning may take place later in the upper than in the lower basin, although Gustaveson (pers. comm.) recorded ripe males in November 1984 and 1985, and Meyer and Moretti (1988) netted ripe males in late March 1987 in Lake Powell. Reproductive data may be incomplete farther north since ice cover and inclement weather preclude intensive fieldwork there in late winter and early spring. Tuberculate males have been noted in the Green River from March through June (Tyus, pers. comm.), although one record in September (Branson and McCoy 1966) indicates that secondary sex characteristics may be developed at other times of year. Ripe razorback suckers of both sexes were captured from the Green River near Ouray (including the lowermost Duchesne River), Jensen (Split Mountain to Ashley Creek), and at Island and Echo parks (including the lower kilometer of the Yampa River) from early May through mid-June in 1981 and 1983–1986, and mid-April through June in 1987 (Tyus 1987, unpub. data; Tyus et al. 1987). Aggregations thought to represent pre-reproductive staging in the Green River were seen in early May and June (Holden and Crist 1981; Tyus et al. 1982b), and ripe or recently spent fish were recorded from gravel pits and the Colorado River near Grand Junction at about that same time (McAda and Wydoski 1980; Wick et al. 1982; Kaeding and Osmundson 1988a).

Ripe fish were taken in the Green River at temperatures varying from 10.5° to 18°C, and averaging about 15°C. Ripe females in the

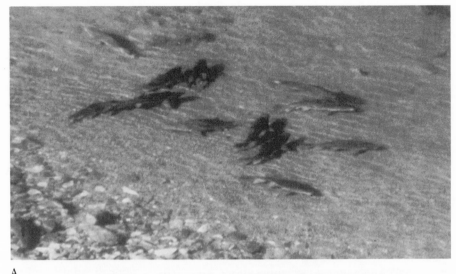

A

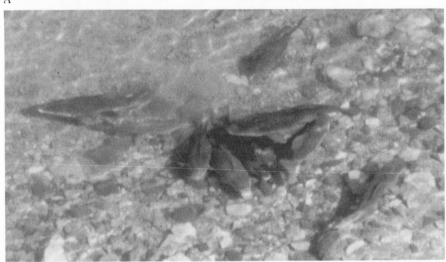

B

Fig. 17-4. Aggregation (A) and the spawning act (B) by wild razorback suckers in Arizona Bay, Lake Mohave, Arizona-Nevada. Photographs by E. S. Gustafson, March 1988.

Duchesne River in 1984 were taken at 15°C, and four were caught on a warm (17°–19°C), flooded, alluvial plain in May 1986 (Tyus 1987). McAda and Wydoski (1980) recorded water temperatures from 7° to 16°C where ripe, and presumably spawning, razorback suckers were caught in May 1975 on the Yampa River. Valdez et al. (1982b) reported ripe adults in Colorado gravel pits at 20°C in June 1980, and a pair yielded viable gametes that hatched and grew to juvenile size in the laboratory.

SPAWNING HABITATS AND BEHAVIOR. In Lake Mohave, males stage over coarse, wave-washed cobble in water 0.5–5 m deep. Groups of up to several hundred fish (Fig. 17-4) move slowly a meter or less from the bottom or lie immobile near or on the substrate for hours. Based on trammel netting, females remain in deeper water until ripe, then appear singly on the spawning grounds. Major aggregations then break apart to swim along shorelines in groups of three to thirty or more, most often representing a number of males following a female. About twice as many males as females are caught near the spawning grounds (Minckley 1983).

When she is ready to spawn, a female, flanked by two or more males, separates from a group and moves to the bottom. The males press closely against the female's posterior abdomen and caudal peduncle, and all contact and agitate the substrate for three to five seconds in apparent spawning convulsions, after which they typically return to a larger group (P. A. Douglas 1952; Jonez and Sumner 1954; Minckley 1973; Mueller et al. 1982). The entire sequence lasts from a few seconds to three minutes, usually the former. Females recognizable because of an injury or some other distinctive feature have been observed to spawn repeatedly in a given hour and day, and on successive days within a week (Minckley and Marsh, unpub. data). A female presumably releases a small fraction of her eggs with each spawning act. Fish spawn sporadically throughout the day and night, with no evident diel pattern. Severe wave action during storms results in abandonment of spawning areas for a few hours to several days, and in one instance until the following year (Bozek et al. 1984; Langhorst and Marsh 1986).

Large concentrations of spawning fish (hundreds or more) have been recorded at only a few places in Lake Mohave, although groups of ten to fifty fish may be found anywhere that suitable habitat is available. The major known spawning areas (e.g., Cottonwood Cove, Arizona Bay, Six-mile Cove, and Eldorado Canyon; Bozek et al. 1984) have been occupied

annually at least since 1974 (Minckley and Marsh, unpub. data), but it is not known if the same or different fish use an area each year. Tyus (1987) noted far fewer reproductive adults on known spawning grounds in the Green River in two of five years of study, and McCarthy and Minckley (1987) noted that some year classes seemed absent in Lake Mohave, suggesting that reproductive effort and success varied from year to year.

Mueller et al. (1982, 1985) presented data on a spawning site in Arizona Bay (Fig. 17-5) that is typical of such areas in Lake Mohave. Wave action at different lake levels in the fluctuating reservoir had formed terraces of wave-sorted substrate, each 2–4 m wide and parallel to shore. The shoreline was cobble, with offshore terraces of gravel and cobble alternating with slopes dominated by gravel, sand, and silt. Spawning razorback suckers use bottoms composed of large gravel and cobble relatively free of fine materials, and numerous fish spawned at the same place time after time.

There is no evidence that razorback suckers construct an actual nest before spawning, but spawning sites in deeper water are clearly marked by their activities, which sweep gravel and cobble substrates clear of fine materials and create depressions 20 cm or more deep. Spawning is most common near shore in water less than 0.6 m deep, where depressions are rapidly obscured by wave action or overlapping activities by spawners. Some individuals spawn with their bodies breaking the surface in water only a few centimeters deep. About 8% of the bottom between depths of 0.6 and 3.45 m was covered by spawning depressions (40 m^2 of 500 m^2 surveyed), but an estimated 30% of the bottom was used in the zone less than 0.6 m deep. Reproductive activity was observed to a depth of 2.75 m. A few depressions were evident at 3.45 m; none was seen in deeper water. Depression densities may not necessarily reflect spawning effort, since some were more heavily used than others

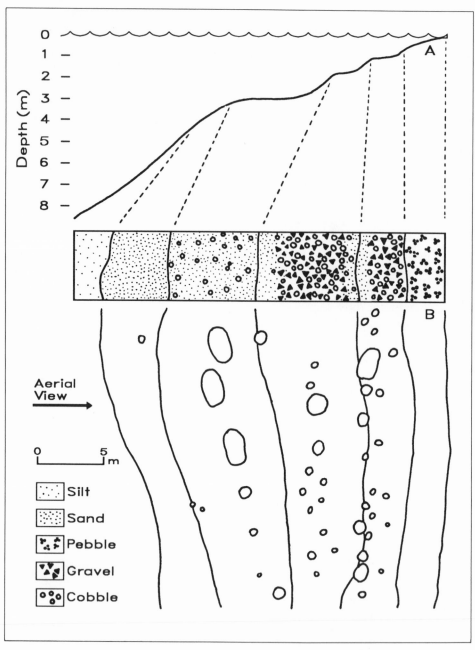

Fig. 17-5. Schematic representation of a section of razorback sucker spawning habitat in Lake Mohave, Arizona-Nevada: (A) cross-sectional profile, vertical exaggeration about 1:4; and (B) aerial view, showing distribution, shapes, and sizes (to scale) of spawning depressions.

(Mueller et al. 1982). The depressions (N = 63, including some outside the surveyed area, but excluding those in less than 0.6 m of water) were at an average depth of 2.2 ± 0.4 m (± one standard error, unless otherwise noted). They averaged 0.74 ± 0.19 m^2 in area (0.21–3.34 m^2) and were round or oval in shape (Fig. 17-5).

Concentrations of mature adults in gravel pits along the upper Colorado River have also been interpreted as staging or breeding groups. McAda and Wydoski (1980) and Wick et al. (1982) alluded to ripe fish and spawning behavior similar to that seen in Lake Mohave over cobble bottoms regularly agitated by wave action. Kidd (1977) reported spawning activities in the same gravel pits. Valdez et al. (1982b) theorized that natural backwaters were historic breeding sites, as earlier proposed by Behnke (1980) but questioned by Holden (1980b). Razorback suckers experimentally isolated in a cutoff bay of Lake Mohave successfully reproduced in two of three years (Marsh and Langhorst 1988). As we noted earlier, Indians harvested the species along lake shorelines with weirs apparently designed to capture breeding aggregations, confirming lentic spawning then as now. An abundant population of fish large enough to secure by hand ax or .22-caliber rifle in a Salton Sea canal in 1909 (Odens 1989) indicates reproduction and rapid growth in that water body after it filled in 1904–1905. In wetter times the fish may also have bred in tributaries to natural lakes in the lower basin. C. L. Hubbs (1960) reported archaeological material along now-dry channels west of the Salton Sea that may have indicated such an event.

Known and suspected spawning sites in the Green and other upper-basin rivers are all in broad, flat-water segments, as opposed to reaches in canyons and with whitewater rapids (Tyus 1987). Most ripe fish were caught over coarse sand bottom, but some were on or near gravel and cobble bars. Ripe fish also occupy creek mouths and natural and artificial backwaters. Actual and apparent riverine spawning was most commonly recorded over mixed cobble and gravel bars on or adjacent to riffles (McAda and Seethaler 1977; McAda 1977; McAda and Wydoski 1980; Tyus et al. 1982b, 1987; Tyus 1987). Twelve males and two females, all ripe and apparently spawning in May 1975, were caught from the Yampa River in water averaging 80 cm deep and flowing 92 cm s^{-1} (N = 5) over cobble 20–50 mm in diameter (McAda and Wydoski 1980). W. H. Miller et al. (1982a) caught two ripe males and one ripe female on the same bar in May 1981 (velocity 10–60 cm s^{-1}, average 40 cm s^{-1}; N = 4).

Under natural conditions, moving sand bottoms must have predominated downstream from Grand Canyon; coarse substrates may have been at a premium. Only two permanent tributaries, the Gila and Bill Williams rivers, entered that section (both are now dammed), and they were also sandy bottomed in their lower reaches. Under such conditions, Loudermilk (1985) hypothesized that razorback suckers spawned on coarse materials washed into the channel from ephemeral arroyos.

Mueller (1989) observed riverine spawning over such a 300-m^2 alluvial fan of coarse sand and gravel deposited by flash flooding into the Colorado River channel. Although within a canyon, the river reach was at the extreme upstream influence of Lake Mohave and had features of a flat-water segment. The unconsolidated material was unique to the channel's otherwise armored nature about 6.5 km downstream from Hoover Dam. The area also may have been influenced by warm water entering the river from thermal springs. Spawning depressions were 0.25–1.0 m^2 in size, oval in shape, and at depths of 1.2–1.9 m (mean 1.5 ± 0.2 m, N = 18). Water velocities in the area varied from 0 to 37 cm s^{-1} (mean 23 ± 16, N = 22), and velocities 10 cm above the substrate at five specific spawning sites averaged 15 ± 3 cm s^{-1} ($12-18$ cm s^{-1}).

Most individuals were relatively small, appeared male, and, unlike those in the body of Lake Mohave, remained stationary at the downstream end of the site when not spawning. A maximum of thirty-seven fish was present at any given time. Larger females moved occasionally into the area from the adjacent river, attracting stationary males to form groups of three to eight fish. Spawning behavior was similar to that in the lake, except that groups always oriented with heads into the current. Small groups sometimes appeared to seek shelter downstream from boulders, where they performed "rolling" movements for several seconds; their significance is unknown.

HABITS AND HABITATS IN NON-REPRODUCTIVE PERIODS. Older area residents with experience with razorback suckers in the lower basin in the early 1900s (interviewed by Minckley [1973]) indicated that "large adults [of razorback suckers] tended to remain in eddies and backwaters, lateral to the strongest currents and often concentrated behind obstructions or in deep holes near cut banks or fallen trees." Mearns's 1894 notes of adults being "snared in deep holes among the rocks" (J. O. Snyder 1915) tend to corroborate these observations.

Records of razorback suckers in backwater habitats date from Mearns's notes on "a laguna of the Salton River" (Follett 1961). The Spanish word *laguna* was applied, in part, to open waters of oxbows and other marshlands associated with streams in the Southwest (Hendrickson and Minckley 1985). Mearns reported that the "Salton River is a long slough of the Colorado River, which latter overflows its banks periodically so that the water backs up occasionally as far as the so called 'Salton Sea' north of the boundary. New

River is similar, farther west" (Follett 1961). Dill (1944) noted the species in both the river and in agricultural drains, but observed and caught it only in a "bay" below a dam in 1942. It may also be significant that the last known adults from two other streams were from a natural oxbow of the Verde River (Wagner 1954) and from an irrigation pond and backwater along the San Juan (Meyer and Moretti 1988). Relatively large numbers of razorback suckers in gravel pits and other artificial and natural backwaters associated with streams of the upper basin also speak for a proclivity for other than river channels. A large percentage of the fish taken by Tyus et al. (1982b), Valdez et al. (1982b), and Tyus (1987) in natural habitats of the upper Green and Colorado rivers were in backwaters or eddies $0.5-2.5$ m deep over sand and silt bottoms, where water velocities were less than 10 cm s^{-1}. C. L. Hubbs and Miller (1953), Marsh (1987a), and McCarthy and Minckley (1987) further commented that the razorback's closely spaced and structurally complex gill rakers, highly protrusible mouth (to a subterminal position), and the food habits of adults in reservoirs (largely zooplankton; Marsh 1987a) all may indicate that it is better adapted for backwaters than river channels.

On the other hand, a data set for six radio-tagged adults (Tyus 1987) yielded vastly different habitat relations than those just indicated. These fish occupied the channels of the Green and Duchesne rivers at depths varying from 0.6 to 3.4 m (mean 1.3 m) and velocities between 0 and 60 cm s^{-1} (mean 40 cm s^{-1}), most commonly in nearshore runs in spring and on mid-channel sandbars in summer. The summer selection of mid-channel bars of unconsolidated, coarse sand was especially unexpected, yet all six fish moved to such areas in July, where they apparently occupied the lee of current-formed dunes. Such habitats have not been investigated and were not examined by earlier workers because large, strongly flowing rivers are difficult to sample, even with modern equipment. Preliminary results from radio-tagged razorback suckers reintroduced into the Gila River, Arizona, indicate use of both sand-bottomed, flat-water, main-channel habitats and quieter pools and eddies adjacent to stronger currents (Marsh and Minckley 1991).

ADULT MOVEMENTS. Historically, adult razorback suckers apparently moved upstream in spring in main-stem rivers, and into major tributaries as well (C. L. Hubbs and Miller 1953). D. S. Jordan (1891) reported local testimony of migration into the Animas River, presumably to spawn. Ellison (1980) wrote of dense aggregations on riffles in the Salt River of Arizona in the early 1900s. C. O. Minckley and Carothers (1980) and Carothers and Minckley (1981) observed springtime movement from the Colorado into the Paria River, and Hendrickson (pers. comm.) reported the species in Bright Angel Creek, a tributary to the Colorado River in Grand Canyon, in spring 1987. Jonez and Sumner (1954) described extensive shoreward movements of razorback suckers, similar to those now occurring in Lake Mohave, during the spawning season in Lake Mead, especially in the vicinity of inflowing streams, and T. C. McCall (in Carothers and Minckley 1981) saw razorback suckers near the mouth of the Virgin River, from which, except during flood, only subsurface flow presently enters Lake Mead.

In spite of these observations, fairly extensive studies of marked razorback suckers have not demonstrated directed seasonal movements. There seems to be one group that is sedentary and another that moves extensively, a common pattern in riverine fishes (Funk 1957). As the following review will show, relationships among movements of marked fish,

reproductive condition, sex, season, river stage, time between tagging and recapture, and so on, merit further and more detailed research and analysis.

In Lake Powell, a male razorback sucker tagged near the Dirty Devil River arm in 1983 was taken in that same area in 1984, and six of twelve fish tagged at Piute Farms Marina in 1987 were recaptured along with four untagged fish at the same locality in 1988 (Meyer and Moretti 1988; Moretti, pers. comm.). In the Green River, Vanicek (1967) recaptured four of thirteen razorback suckers marked between 1964 and 1966 within 1.6 km of their points of release. McAda (1977) and McAda and Wydoski (1980) reported eleven recaptures of ninety-eight fish tagged at various locales in the upper basin. Of the eleven, eight fish tagged in 1974 in a gravel pit on the Colorado River were recaptured at the same locality in 1975; one tagged in autumn 1975 was recaptured the following spring along with forty untagged razorback suckers in a backwater 26 km upstream; a second, marked at Horseshoe Bend on the Green River in December 1974, was recaptured 130 km upstream from the point of release three and a half years later; and the last, tagged in April 1975 at Island Park on the Green River, moved 21 km upstream and was thought to have spawned in the Yampa two weeks later. In an effort that may have overlapped McAda's (1977) studies, Kidd (1977) tagged seventy fish in backwaters of the Colorado River (Grand Junction area) in 1974–1976. He recaptured nineteen fish, but only one had moved from the area of tagging to another backwater a few kilometers distant. Five fish tagged by Valdez et al. (1982b) at about the same place remained there for periods of a week to a year. Three fish successfully tracked by Kaeding and Osmundson (1988a) in the river channel near Grand Junction in 1986 and 1987 remained within 28 km of their

points of tagging over periods of four to eighteen months.

Ultrasonic transmitters proved successful in
short-term tracking of razorback suckers in
a Colorado gravel pit (McAda and Wydoski
1980). A comparable study in Senator Wash
Reservoir, California, in 1980 and 1981 produced similar but far more extensive results,
which have not yet appeared in print (Ulmer,
pers. comm.). In the Colorado study, one
female, most active in late evening and late
morning, typically remained in deep (3 m)
parts of the pool but moved to shallows in
early morning. The following spring, the same
female and two males never moved together,
but all remained in deep water; two fish
moved throughout the pond and the third remained sedentary. Five fish were also ultrason-
ically tagged by McAda and Wydoski (1980)
in the Yampa River in spring 1975, but turbulence, depth, and obstructions resulted in
poor signal detection. No fish moved more
than a kilometer in periods varying from a day
to two weeks. Data based largely on one fish,
which was intensively tracked, indicate that
they mostly remained in quiet water near
shore, infrequently moving to swifter water
near gravel bars.

Tyus (1987) and Tyus et al. (1987) reported
on movements of sixty-one marked fish recaptured seventy-four times. Of these, thirty-five
were recaptured within a year, and the remainder roamed free for one to eight years.
Excluding nineteen recaptured within two
weeks of tagging and three identified only by
threads from which tags were lost, twenty-
four individuals (32%) traveled less than 10
km from the point of capture. The remaining
twenty-eight (38%) moved a net average of
59.3 km (13–206 km). One of the latter exhibited net movement of 192 km in four years
and another moved 206 km in five years. The
longest gross movement was 266 km by a fish
moving from Island Park to the Duchesne
River between 1978 and 1982, and returning

to Island Park between 1982 and 1986.

During spawning seasons, twenty-one
tagged fish were found at two or more suspected spawning sites, some moving from one
locale to another within a season, and others
moving between sites in succeeding seasons;
for example, eleven fish moved about 21 km
downstream from a spawning site in Dinosaur
National Monument to Ashley Creek, and
three reciprocated. Five others moved from
Ashley Creek to the Duchesne River 77 km
downstream, while one made the reverse
movement. In contrast, travels of one fish
tracked intensively between April and August
1980 were between the Duchesne and Green
rivers and within the latter, all within a linear
distance of about 15 km (Tyus 1987).

HABITS AND HABITATS OF YOUNG FISH.
Under natural conditions young razorback
suckers may remain along shorelines, in embayments along sandbars, or in tributary
mouths, and then disperse into channels or
larger backwaters. However, this scenario is
based mostly on observations of hatchery-
produced fish, since the few records for wild
larvae and juveniles provide essentially no information on their historic habitat relations.
G. R. Smith (1959) caught two young—one
each in a backwater and a creek mouth in
Glen Canyon—and Taba et al. (1965) found
juveniles only in backwaters of the Colorado
River downstream from Moab, Utah. A substantial percentage of other small fish recorded were caught below dams or along
shorelines of reservoirs (Minckley 1983). In
1950 R. R. Miller (in W. F. Sigler and Miller
1963) caught 6600 young-of-year in two seine
hauls along warm (21.7°–24.4°C), shallow
margins of the Colorado River at Cottonwood
Landing, Nevada. The temperature of the adjacent river, where no young were seined, was
colder (14.4°C) due to outflow of hypolimnetic
water from Lake Mead.

Recent studies of razorback sucker larvae in

reservoirs documented exclusive occupation of littoral zones for a few weeks after hatching (Langhorst et al. 1985; Langhorst 1987), then a tendency to move offshore (Langhorst and Marsh 1986). Papoulias (1988) described the same behavior in hatchery ponds. The few wild juveniles found in canals indicate that larvae or postlarvae achieved those habitats by moving downstream into intake structures (Marsh and Minckley 1989). Most recoveries of juveniles reintroduced into backwaters or along quiet edges of streams have been downstream from their points of stocking (Hendrickson, AZGFD, letter [dated 19 October 1987] to P. C. Marsh, ASU; B. L. Jensen 1988; Marsh and Brooks 1989). Langhorst (1988, 1989) also noted alongshore orientation and movement by reintroduced juveniles (12.7–19.6 cm TL) in Colorado River backwaters, which ultimately led them to disperse into the channel.

Reasons for Decline

Although it has been variously attributed to dam building, habitat destruction, habitat alteration, impairment of water quality, and interactions with non-native fishes, the reasons for the overall decline of endemic fishes in the Colorado River basin remain speculative. With few exceptions, direct cause-and-effect relationships are yet to be established.

Physical-Chemical Factors

It is irrefutable that some physical and chemical changes in natural systems have directly affected native fishes. At the extreme, some rivers have been dewatered and cannot support fish life of any kind. Other segments are in chemical or physical states incompatible with fishes in general, or with some stage in the life cycle of one or more species. Perennially cold temperatures of hypolimnetic water released from reservoirs may preclude reproduction or completion of some other essential life-history activity or stage in tailwaters (Vanicek et al. 1970; Stalnaker and Holden 1973; Marsh 1985). Beland (1953) attributed loss of native fishes and other changes in fish populations of the lower Colorado River to channelization. Water storage and delivery structures and operations (Wydoski 1980; Hickman 1983) may also curtail reproductive or other critical activities by masking chemical or other cues, by killing eggs or larvae of some species but not others, or by adversely affecting some other life-history stage. Fluctuating water levels in reservoirs can strand adherent or interstitial razorback sucker eggs and larvae, and result in substantial mortalities (Gustafson 1975a; Bozek et al. 1984). A species that spawned deeper would not be so affected.

Declines of some native species in systems that retain natural attributes sufficient to support other indigenous species (Tyus et al. 1982a; Minckley 1983) are not so readily explained. One inferred possibility for selective extirpation is the existence of dams and diversions that physically block spawning or other movements necessary for some species and not for others. However, the limited information on movements of razorback suckers and their successful spawning in reservoirs and even hatchery ponds (Jensen, unpub. data) indicate that long-distance migrations are not mandatory for completion of that step in the life cycle. Barriers (as well as dewatered or otherwise physicochemically altered reaches) nonetheless preclude recolonization of upstream reaches depopulated by natural or human-induced catastrophes (Moyle and Nichols 1973; Hubbs and Pigg 1976). Diversions also may direct migrating larvae or other life-history stages to their deaths in agricultural canals or fields (Marsh and Minckley 1989).

Evaluations of responses of hatchery-produced razorback suckers to temperature, salinity, and different concentrations, gradients, and extremes of pesticides and other potential pollutants have also received some attention (Bulkley et al. 1982; Beleau and Bartosz 1982; Bulkley and Pimentel 1983a, b; Marsh 1985). But the effects of such factors in nature are generally limited to relatively short reaches of streams and should not account for a lack of recruitment and decline of the species throughout its extensive range.

Biological Factors

Dill (1944) was among the first to propose the idea that non-native fishes introduced into the lower Colorado River were in some way causing declines of native species. His interviews with local residents revealed "absolute agreement . . . that the indigenous fishes were quite rare now [as compared with earlier, and that] the decline became most evident during the 1930s." Among other factors forwarded as explaining this event, Dill penned the following observations:

> Before the dams were built the native fishes were at the mercy of an adverse physical environment, but the deleterious effect of predaceous exotic fishes must have been slight. That is, the population of the latter fishes was small before the creation of Boulder Dam, and floods and droughts must have worked just as severe a hardship— and probably more—on them. Because of the unfavorable water conditions around the early thirties it seems possible that the population of native fishes sank to one of its low points, and that the coincidental advent of clear water following Boulder Dam brought about a heavy production of bass and other alien fishes which preyed upon the already reduced natives. Competition as well as direct predation may have played a large part in this supposed destruction.

Only a few new ideas have been added to this thesis in almost fifty years. Native fishes are displaced by non-natives through direct or indirect competition for food, space, or other factors, or through predation on eggs, larvae, or juveniles. Hybridization between razorback suckers and other catostomids and transmission of parasites or diseases from non-native fishes are ideas that have been recently forwarded (Tyus et al. 1982a; Wick et al. 1982). Competition, although often mentioned, has not been clearly defined or carefully examined, nor has the potential genetic impact (Molles 1980; Meffe 1986, 1987) of fragmentation of species' formerly continuous ranges by dams or desiccated reaches.

The vast numbers of eggs and larvae produced by fishes like the razorback sucker would seem to preclude loss of an entire annual production to any but catastrophic circumstances. The fecundity of five Lake Mohave razorback suckers was volumetrically estimated by Minckley (1983) as averaging 1812 ± 91 ova cm^{-1} SL, and ten upper-basin fish yielded 1166 ± 491 ova cm^{-1} (calculated from McAda and Wydoski 1980). Three-year-old hatchery-reared females (average 30.5 cm SL, $N = 70$) yielded a mean of 2086 ova cm^{-1} SL by gravimetric analysis after hormone injection and manual stripping (calculated from Hamman 1985a), and twenty-five wild females (mean 56.5 cm SL) from Lake Mohave produced 2179 ova cm^{-1} SL following the same procedure (Rinne et al. 1986). In winter and spring 1983 and 1984, Marsh (unpub. data) volumetrically estimated 36,200–136,300 ova (790–2519 ova cm^{-1} SL), for an average of 1704 ± 447 ova cm^{-1} SL in ovaries dissected from twenty ripe 44.1–57.7 cm (average 49.1 ± 3.7 cm) females from Lake Mohave. Ovaries of twenty-two others of the same sizes were spent or appeared atrophied and produced at most 1900 ova per fish. As we have already noted, razorback suckers spawn each year in Lake Mohave, yet no recruitment has been detected for almost three decades.

Reproduction by razorback suckers in Lake Mohave has been studied for almost a decade without finding a larva larger than ~12 mm TL. Furthermore, the mean length of wild-caught larvae remains the same, 10.6 ± 0.3 mm TL, from January through March or April. Larvae in hatchery ponds grew from about 9 mm TL at hatching to 22 mm TL or more in about fifty days without artificial feeding (Papoulias 1988). Those captured from Lake Mohave can only represent continuing emergence of hatchlings, which at larger sizes either disappear or somehow manage to escape all our sampling efforts.

The search for escapees has been extensive, including SCUBA, larval traps, plankton nets, ichthyoplankton trawls, seines, and other means, but only a single larva was taken offshore, and it was the same size (10.5 mm TL) as those found inshore (Marsh and Langhorst 1988). If larvae exist in the body of Lake Mohave, they are remarkably rare, cryptic, evasive, or transient. We concluded that they disappear, as substantiated by the lack of recruitment to the adult population, and we offer three hypotheses to explain the phenomenon: (1) transport from the system, (2) nutritional constraints resulting in starvation, and (3) loss of early life-history stages to predation.

We examined the first possibility—that larvae moving from shorelines into the body of the reservoir were entrained by subsurface currents and carried from the system—through field studies. Thermal stratification of the reservoir begins in March or April, when larvae migrating offshore at depth would encounter hypolimnetic temperatures that average $10°–12°C$, plus density currents (Priscu et al. 1981, 1982) that ultimately lead to the turbulence, pressure changes, and other physical abuses of passage through penstocks of Davis Dam. Lake Mohave has a small volume compared with the amounts of water entering at Hoover Dam and exiting through Davis Dam, and the theoretical retention time is only seventy-nine days. Subsurface currents were $4–8$ cm s^{-1} at one transect studied by Priscu et al. (1981) uplake from Cottonwood Basin, and were estimated from discharge volumes and cross-sectional profiles to vary between 1 and 25 cm s^{-1} (Langhorst and Marsh 1986). Direct measurement of subsurface cur-

rents by drift of weighted drogues at depths of $5–15$ m in that same area varied from 5 to 25.5 cm s^{-1} (19.8 ± 8.8, $N = 7$; Marsh, unpub. data). Most larval fishes are inexorably entrained by currents exceeding 10 cm s^{-1} (Grabowski et al. 1984), and 10-mm to 12-mm razorback suckers are not a likely exception. However, we could not confirm passage of larvae through Davis Dam. No razorback sucker larvae were captured immediately below the dam at the times of year they might be expected (Marsh, unpub. data), and the rare occurrences in the river a distance below Davis Dam and in Lake Havasu (Marsh and Papoulias 1989) were of individuals too young to have reasonably originated in Lake Mohave.

The possibilities for qualitative or quantitative food constraints on larval razorback suckers' survival were studied experimentally at Dexter NFH, New Mexico. Absence of adequate or appropriate foods or key nutritional components is often implicated in mass mortalities of fish larvae (Kashuba and Matthews 1984; Theilacker 1986; Leggett 1986). Larval fishes pass through a "critical period" when they shift from endogenous to exogenous food supplies (May 1974), and catostomids experience another crisis as ontogenetic morphological changes accompany (or cause) shifts from larval surface or mid-water foraging to adult benthic habits (N. H. Stewart 1926). Razorback suckers are adaptable to variable habitat conditions, but impoundment and other alterations may have changed nutritional conditions in the Colorado River in unknown ways.

At ambient water temperatures, larval razorback suckers absorb their yolk by seven to twenty-one days after hatching, when they are between 10 and 13 mm TL, about the same size at which they disappear from samples in Lake Mohave. Larvae in hatchery ponds all had yolk to six days after hatching; 77 of 172 (45%) had absorbed their yolk in seven to thirteen days (65 of these [84%] contained food, and of the 95 fish retaining yolk, 64 [67%] also had food in the gut); and none carried yolk after fourteen days (Papoulias 1988). Initial foods were diatoms, other phytoplankton, and fine detritus, but animal materials began to increase in frequency by the fifth day after yolk absorption, and by the seventh day, animals (rotifers, cladocerans, chironomids, invertebrate eggs, and undetermined nauplii) clearly dominated gut contents.

After fourteen days, digestive tracts of larvae were consistently full. Volume of foods increased linearly with size, while the numbers of organisms each larva consumed remained about the same. Larvae ate progressively larger organisms to maintain constant fullness. Taxonomic trends in food organisms were mostly explained by the sizes of foods available, with cladocerans dominating at larger larval sizes (Papoulias 1988). Wild larvae in Lake Mohave also ate cladocerans most abundantly, along with rotifers and copepods (Marsh and Langhorst 1988), but a large percentage of wild-caught larvae had empty guts (76% in Lake Mohave and 28% in an isolated bay), and those with food in their guts had fewer organisms per larva than in hatchery ponds (8 versus 25 in larvae 10.8 mm TL, and 50 versus 100 per gut in those 13.9 mm long).

If larvae received no food at all under experimental conditions, they died twenty to thirty days posthatching at mean total lengths of 9.6 ± 0.2 mm (Papoulias and Minckley 1990). Starvation for up to nineteen days did not increase overall mortality, but mortality was greater than 85% if larvae went without food for more than twenty-seven days. Larvae fed ten brine shrimp (*Artemia salina*) per day had a significantly higher mortality rate than those fed more than fifty per day. Both a delay in feeding and restricted rations resulted in slower growth. Plankton populations in ponds were positively correlated with fertili-

zation rates, but survivorship of larvae was not significantly different in fertilized and unfertilized systems. Total biomass and individual growth were, however, greater at higher fertilization (Papoulias 1988).

Lake Mohave has high primary productivity because of constant input of nutrients from the hypolimnion of Lake Mead (Priscu et al. 1982), yet zooplankton densities are spatially and seasonally variable and relatively low (Paulson et al. 1980a). Nonetheless, based on analyses of Papoulias's (1988) and Papoulias and Minckley's (1990) data, year class failure in Lake Mohave could only be attributed to nutritional deficiencies at the lowest recorded levels of reservoir zooplankton.

Predation on eggs and larvae, the third hypothesis forwarded as limiting recruitment of razorback suckers, seems most significant, as it is proving to be in other fishes (de Lafontane and Leggett 1988). Razorback suckers are large enough after one year to be immune to most piscine predators, except historically for Colorado squawfish or today's striped bass or flathead catfish (*Pylodictis olivaris*), but larvae and especially eggs have been repeatedly observed to be eaten by non-native piscine predators (common carp, channel catfish [*Ictalurus punctatus*], green sunfish [*Lepomis cyanellus*], and other centrarchids) under both natural and experimental conditions (Jonez and Sumner 1954; Medel-Ulmer 1983; Minckley 1983; Loudermilk 1985; Pister 1985c; Langhorst and Marsh 1986; Marsh and Langhorst 1988; Langhorst 1988, 1989).

On the other hand, razorback suckers have high fecundity and spawn long before substantial numbers of predatory species move inshore in spring to occupy, feed, and breed in the littoral zone, and it is difficult to visualize predation pressures adequate to remove all their progeny over a period of years. The waters of Lake Mohave are clear year-round (Fig. 17-4), and fishes may be readily observed during wave-free periods. Only rainbow trout

(*Oncorhynchus mykiss*), stocked throughout the year for a sport fishery, and carp, along with aggregated razorback suckers, are common along shorelines between November and March. Green sunfish 30–80 mm TL are sometimes found in bays and within flooded stands of riparian plants, but their populations seem too small to eat entire cohorts of larvae. Spawning by razorback suckers usually ends before young-of-year centrarchids or populations of mosquitofish (*Gambusia affinis*) develop along shorelines and become potential threats. We have no evidence for invertebrate predators of the size, species, or density necessary to play a role in this phenomenon. Crayfish (*Orconectes* sp., *Procambarus clarki*, and likely others) have spread from introductions as bait and forage to populate much of the Colorado River basin, including Lake Mohave, but their population sizes and impacts are unknown. We nonetheless conclude that predation by non-native fishes is the single most likely factor precluding recruitment of razorback suckers in nature, based on the following.

Small populations, or body sizes of predators, may not be accurate measures of the impacts a predator has on its prey. Apparently small numbers of green sunfish have been demonstrated to suppress populations of other fishes in streams (Lemly 1985). Smaller predators may also be highly efficient. Mosquitofish that rarely exceed 30 mm SL as adults can eliminate equally small Sonoran topminnows (*Poeciliopsis o. occidentalis*) by eating their young as well as by attack, injury, and direct or indirect killing of adults (Meffe 1983a, 1985b). Mosquitofish also have been known to suppress population growth of much larger fish species by feeding on their young (G. S. Myers 1965).

If numbers of potential larvivores are any indication, predation in Lake Mohave should be far more significant when and if razorback sucker larvae move offshore. Red shiners (*Cyprinella lutrensis*), stocked in the period 1953–1956 (C. L. Hubbs 1954; USFWS 1980a, 1981a), live in the littoral zone. Threadfin shad (*Dorosoma petenense*), introduced and established in 1953 (Kimsey et al. 1957), occupy lower littoral and mid-water habitats deeper than 5 m in winter (based on echolocation surveys; Minckley, unpub. data), becoming epilimnetic and upper littoral in habit after temperatures increase in spring. They feed all year in desert reservoirs (Haskell 1959), and although feeding on larval fishes is not commonly reported (Burns 1966; Ingram and Ziebell 1983), it should not be unexpected. Threadfin shad might also compete with razorback sucker larvae if planktonic foods are in short supply. SCUBA observations and gill-net sampling indicate that young channel catfish, bluegill (*Lepomis macrochirus*), green sunfish, and largemouth bass (*Micropterus salmoides*) of sizes expected to feed on larval suckers also are near bottom at depths greater than 3 m. Beds of aquatic macrophytes, excluded from shallow waters by reservoir fluctuations, grow at 3–7 + m depths in this clear impoundment, providing cover for predators.

The strongest evidence that predation is the major factor in loss of larval razorback suckers is simply that larvae persist and grow, to maturity if given adequate time, in habitats from which predators are excluded. This was demonstrated by Langhorst (unpub. data) through exclosure experiments in Lake Mohave that resulted in 80% survival of fifty larvae that grew to an average of 11.5 mm TL in fourteen to twenty-one days, while those free in the reservoir consistently remained 10.5 mm TL. In a later attempt, seven larger larvae (ca. 18 mm TL) produced in a pond experiment (see below) survived twenty-one days in cages to attain a mean total length of 34 mm.

A more extensive study involved removal of non-native fishes from an isolated habitat adjacent to the reservoir and stocking it with adult razorback suckers. The habitat was a small (0.85–2.1 ha, depending on lake elevation), relatively deep bay (1.7–3.7 m maximum depth), isolated behind a spit formed by wave action across a tributary arroyo (Langhorst and Marsh 1986; Marsh and Langhorst 1988). The resulting pond had been separated from Lake Mohave for several years, although it is sometimes breached by storm-driven waves during high water levels.

Non-native fishes (threadfin shad, carp, channel catfish, mosquitofish, largemouth bass, green sunfish, and bluegill) were removed by ichthyocide in autumn 1984. The pond was stocked in January 1985 with 30 female and 150 male razorback suckers seined nearby. The adults spawned successfully in March, and abundant larvae survived until April, when high lake levels breached the spit, allowing reinvasion by exotic fishes. Green sunfish preyed heavily on larval razorback suckers, as confirmed by direct observation and stomach analyses, and within a month no larvae could be found. Growth to more than 16 mm TL was nonetheless greater than that in the adjacent lake, where larvae did not exceed 10.6 mm TL when they disappeared (Marsh and Langhorst 1988).

The habitat was again treated with ichthyocide in October 1985, and only non-native fishes were recovered. Predatory fishes had apparently destroyed the entire 1986 cohort, excepting those that may have escaped to the reservoir. The adults presumably evacuated when connection to the lake was achieved.

Another stocking of sixty-four male and twenty-five female razorback suckers was made in January 1986. They had reproduced by late March, when fifty-seven larvae 18.0 ± 1.4 mm TL (16–22 mm) were collected. Subsequent growth rates were comparable to those under hatchery conditions. One 10-cm TL specimen was caught in June, forty averaged 15.7 ± 1.7 cm TL (range 10.6–19.7 cm) when seined in July, and by December the fish

had attained 18.2–20.4 cm TL (average 18.9 ± 2.0 cm, N = 12). Adults in the pond were augmented in March 1987 with an additional thirty-two females and eighteen males, an effort that proved futile since high water allowed non-native fishes to invade and the suckers to escape in April. The 1987 year class, if produced, probably failed to survive.

The 1986 year class must have moved into the reservoir, just as adults had done on each opportunity. It is unlikely that they fell prey to invading piscivores because of their size. A single adult razorback sucker and abundant exotic fishes comprised later collections.

These manipulations demonstrated that razorback suckers can successfully spawn and recruit to the juvenile stage in a seminatural, predator-free habitat of the Lake Mohave system. Hatchery-produced larvae stocked in earthen hatchery ponds, livestock watering ponds, ornamental lakes in golf courses, and urban recreational lakes persist and mature (see below) so long as exotic predators are excluded. The Lake Mohave embayment was rendered unusable for study because its isolating spit was so severely eroded in 1987 that connection with the reservoir persisted for several months and invasion by predators was ensured. A cooperative program is now under way to rebuild the berm and rear razorback suckers to substantial sizes (25–35 cm) in the reclaimed backwater.

Introduced fishes, enhanced by each newly constructed lentic habitat in the lower basin, appeared too rapidly and with too great a diversity for us to pinpoint a pattern of relationship between decline of razorback suckers and any one of the many established exotic species. It is more than evident, however, that Dill's (1944) observation that closure of Boulder Dam enhanced non-native fishes was accurate. That event in 1935 was closely followed by increases in abundance and rapid expansions in ranges of exotic forms. Razorback suckers essentially disappeared from two lower-basin reservoirs, Lakes Mead (closed in 1935) and Havasu (closed in 1938), about forty years after they were filled (Minckley 1983). The fish was gone by the 1950s from the Salt River chain of lakes, again about forty years after Roosevelt Dam was closed in 1913. Four decades may therefore mark the maximum time razorback suckers can maintain a population without recruitment, and those in Lake Mohave, almost all hatched before 1954, may disappear in the 1990s (Minckley 1983; McCarthy 1986; McCarthy and Minckley 1987).

Hybridization

Another biological constraint on razorback suckers could be that small population sizes, for whatever reason, promoted hybridization with a more common species (C. L. Hubbs et al. 1942; C. L. Hubbs 1955), and genetic contamination resulted in decline of the rarer form. Hybridization between flannelmouth and razorback suckers has long been known or inferred. D. S. Jordan's (1891) *Xyrauchen uncompahgre*, collected in 1889, was based on this combination, and rarer and more local hybrids between Sonora (*Catostomus insignis*) and razorback suckers also are known (C. L. Hubbs and Miller 1953). A number of authors have suggested that hybridization is increasing (Joseph et al. 1977; Behnke 1980; Wick et al. 1982; Loudermilk 1985), although evidence for this is less than convincing for a number of reasons.

First, apparent hybrids were collected very early in the history of razorback sucker studies. C. L. Hubbs and Miller (1953) reported on a hybrid specimen collected by D. S. Jordan (1891) in 1889, and others were collected in 1927, 1947, and 1950, all before the system was highly modified. G. R. Smith (1960) reported another in the Green River in 1959, and the first major sample from the upper basin by Banks (1964) included 33% putative hybrids (five of fifteen fish) of sizes that clearly

had hatched prior to modifications associated with Flaming Gorge and Fontanelle reservoirs.

Second, if fish now occupying the upper Green River basin are of year classes dating to or before closure of Flaming Gorge, as suggested by limited aging based on otoliths (Minckley 1989b), variations in hybrid incidences may simply reflect sampling bias at various times in different habitats.

Last, of reports known to us, only C. L. Hubbs and Miller (1953) presented data to support their identifications of putative hybrids, although some others (Holden 1973; Holden and Stalnaker 1975a; Lanigan and Berry 1981) noted some diagnostic morphological features. The flannelmouth × razorback sucker hybrid combination seems distinctively intermediate between its parental species (C. L. Hubbs and Miller 1953; Minckley and Marsh, unpub. data), and one might expect most identifications to be accurate. However, of eighteen specimens selected from Lake Mohave as morphologically suspect (Buth and Murphy 1984; Buth et al. 1987), only one was electrophoretically identified as a hybrid, and it proved to be an introgressed individual. Furthermore, sexual dimorphism may not have been well understood by some workers. Some individual females have low, broad predorsal keels that cause them to appear quite different from males and other females.

Thus, we can only take on faith the substantial number of "hybrids" reported from the Green River sub-basin, viz.: 10 razorback suckers and 5 hybrids (33%) cited by Hagen and Banks (1963) and Banks (1964); 73 and 16 (18%) from Vanicek et al. (1970); 53 and 40 (43.0%) reported by Holden (1973) and Holden and Stalnaker (1975a); 57 and 8 (12.3%) noted by Seethaler et al. (1979); 33 and 5 (13.2%) from McAda and Wydoski (1980); and the 67 razorback suckers and 4 hybrids (5.6%) listed by Holden and Crist (1981). Hybrids were also recorded from the Colorado River sub-basin: 74 razorback suckers and 8 hybrids (10.8%) by McAda and Wydoski 1980; and 17 and "several" hybrids by Wick et al. (1981). Yet, Kidd (1977) and Valdez et al. (1982a) recognized none in their respective samples of 234 and 52 fish. Based on these data, and even if possible problems are ignored, a pattern of increasing incidence of flannelmouth × razorback suckers does not exist.

Kidd (1977) further noted hybridization between razorback suckers and bluehead (*Pantosteus discobolus*), white (*Catostomus commersoni*), and longnose (*C. catostomus*) suckers without supporting data. This must have been an error (see also Joseph et al. 1977), and we suspect that Kidd intended to record miscegenation between non-native (white and longnose suckers) and native catostomids (bluehead and flannelmouth suckers), which is known in the upper basin (Holden 1973; Tyus et al. 1982a). Loudermilk (1985) similarly erred in citing Holden (1973) as reporting white × razorback sucker hybrids.

In the lower basin, Gustafson (1975a, b) reported 5 (2.6%) putative hybrids in collections of 189 catostomid fishes from Lake Mohave in 1975. Only 15 more were identified in the following fifteen years of research on that population, and the cumulative overall incidence of putative hybrids among more than 6500 individuals from the reservoir was about 0.3% by May 1990 (Marsh and Minckley, unpub. data). All 20 suspected catostomid hybrids we have examined from Lake Mohave appear to involve flannelmouth and razorback suckers (Minckley 1983; Minckley and Marsh, unpub. data).

An interesting problem exists with Gustafson's records that emphasizes some drawbacks inherent in the use (or misuse) of gray literature and has resulted in an initial error being perpetuated in the literature. A single, unique specimen from Lake Mohave in 1975 was first identified (Gustafson 1975a) as a

white sucker, then later (Gustafson 1975b) as a Utah sucker (*Catostomus ardens*). R. R. Miller and G. R. Smith (in Minckley 1983) confirmed the latter identification, which, except for specimens from bait tanks (R. R. Miller 1952), is the only record of a Utah sucker from Arizona. The specimen was caught in the same net with spawning razorback suckers and hybrids. A final report (Gustafson 1975b), accepted by USFWS Region 2 (Minckley, unpub. data), recorded "hybridization between *X. texanus* and an undetermined species of *Catostomus*," which was later identified as *C. latipinnis* (see above; Minckley 1983). Unfortunately, however, two versions of Gustafson's (1975b) report exist; one is a preliminary draft that Gustafson intended to destroy, which tentatively (and erroneously) identified Lake Mohave hybrids as Utah × razorback suckers. Rather than being destroyed, however, in some manner the report beame available to be cited by Joseph et al. (1977), and the error was perpetuated at least by Seethaler et al. (1979), Wick et al. (1982), Buth and Murphy (1984), Buth et al. (1987), and perhaps others.

Parasites and Disease

Parasites and diseases studied to date have little apparent overall impact on razorback suckers. Pathogens and parasites include bacteria, protozoans, cestodes, trematodes, nematodes, and the parasitic copepod crustacean *Lernaea cyprinacea* (Flagg 1980, 1982; Wydoski et al. 1980; USFWS 1981; Mpoame 1983; Valdez et al. 1982b; Minckley 1983; Mpoame and Rinne 1983). However, none appears to effect major damage on the host, nor has their presence yet been directly related to the occurrence of introduced vectors (see, however, B. L. Wilson et al. 1966; A. E. James 1969; Deacon 1979; Brienholt and Heckman 1986). Razorback suckers in both the upper and lower basins also show a high incidence of blindness, tumors, and other maladies (Valdez et al. 1982b; Minckley 1983; Bozek et al. 1984; Minckley and Marsh, unpub. data), which may reflect either great age, high susceptibility to injury and disease, or both. Lake Mohave fish appear more afflicted than others, yet Minckley (1983; Marsh and Minckley, unpub. data) has detected no decrease in that population's size through mortality between 1974 and 1990, despite an apparent lack of recruitment. Flagg (1980) pointed out that the uniqueness of native Colorado River fishes was paralleled by their equally distinctive parasites, but later concluded (Flagg 1982) that parasites and diseases were unlikely agents in the decline of native fishes in the upper basin.

Political Bases for Recovery Efforts

Political actions dealing specifically with threatened and endangered wildlife, including fishes, were uncommon until the 1960s, when a Committee on Rare and Endangered Wildlife Species was formed by the USFWS. By 1964 a tentative list of species had been circulated among advisers. This list, along with the Endangered Species Preservation Act of 1966, resulted in compilation of *Rare and Endangered Fish and Wildlife of the United States* (USBSFW 1966), in which the razorback sucker appeared in the list of "status undetermined fishes." The Endangered Species Conservation Act of 1969 expanded formal recognition (listing) to imperiled foreign organisms but added little to protection for species in the United States (J. D. Williams 1981).

By the early 1970s biologists in Arizona, California, and elsewhere had begun a search for information on the razorback sucker, which led to a concerted effort toward its protection and recovery. It was protected by the Game and Fish Code by 1970 in California. R. R. Miller (1972a) listed it as rare in Utah

and California, and presumably extinct in Wyoming. Behnke (1973) produced an early review of its status. An ad hoc lower Colorado River basin fishes recovery team organized in 1973 (Leach et al. 1974) dealt mostly with the razorback sucker and produced and circulated a brochure advertising its plight (Fig. 17-6). Minckley and Kobetich (1974) drafted a never-completed recovery plan for the species that recommended, among other things, a program of propagation and reintroduction within its native range. The USFWS funded initial studies on basic biology of the species in Lake Mohave in 1974 (Gustafson 1975a, b; Minckley and Gustafson 1982; Minckley 1983) with the goal of assessing potentials of that stock for use in a propagation effort. The first brood fish were captured from Lake Mohave in 1974 and transported to Willow Beach NFH, Arizona, where they were successfully spawned.

The passage of the Endangered Species Act of 1973 was pivotal in focusing attention on imperiled plants and animals and providing a vehicle for protection and recovery of habitats as well as species. It produced legal provisions and funding that stimulated individual states as well as federal agencies to increase or initiate efforts to define the status of rare species. Nonetheless, USFWS attempts in the late 1970s to encourage, coordinate, and expedite reintroductions of endangered fishes into historic habitats (J. E. Johnson 1977, 1980a) initially failed due to political opposition. Apprehension on the part of states and other governmental agencies over possibilities for curtailment of sport fishing, hunting, or other water uses, as well as potentials for use restrictions through designation of critical habitat, led to marked resistance (J. E. Johnson 1979, 1980b; J. E. Johnson and Rinne 1982); essentially all reintroduction efforts stopped.

Scarcity of the razorback sucker throughout its range was nonetheless clearly recog-

WANTED

FOR FUTURE GENERATIONS

THE RAZORBACK SUCKER
(ALIAS THE HUMPBACK SUCKER)

DO NOT CATCH!
THIS SPECIES PROTECTED BY CALIFORNIA AND NEVADA STATE LAW.

DESCRIPTION

The razorback sucker *(Xyrauchen texanus)* was one of the most abundant fishes in the lower Colorado River, but today it is rarely seen and many fisheries biologists fear it will disappear. For this reason it has been listed as an endangered species by the states of California and Nevada, and a joint federal-state recovery team has been established to determine its status and take steps for its preservation.

You can recognize the adult razorback sucker by the high, sharp-edged keel-like hump behind the head. On younger fish this hump is less prominent. The head is flattened on top. The body is rather stout and the color is olive-brown above to yellowish on the belly. Head and keel are quite dark in breeding males. The razorback sucker grows to a large size, reaching three feet in length and a weight of 16 pounds.

IF YOU ACCIDENTALLY CATCH A RAZORBACK SUCKER PLEASE RETURN IT TO THE WATER ALIVE AND NOTIFY YOUR FISH AND GAME DEPARTMENT OF TIME AND PLACE TAKEN AND APPROXIMATE SIZE.

Fig. 17-6. Part of a 1974 poster intended to enhance public awareness and understanding of the plight of imperiled native fishes of the American Southwest. The photograph of an adult male razorback sucker, 46 cm SL, replaces a picture of lesser quality. The original poster was designed and produced by G. C. Kobetich, USFWS.

nized by 1977. The AZGFD considered the species uncommon and "possibly in jeopardy in the foreseeable future"; the CADFG listed it as endangered, and the Colorado Division of Wildlife as threatened (both the latter states had placed it under full legal protection); Nevada considered razorback suckers locally common but under legal protection; no records existed for New Mexico; Utah protected the species and considered it of limited distribution; and Wyoming listed it as extinct in the state (Anonymous 1977). A plan for recovery of razorback suckers by the state of Colorado in the late 1970s was not strongly supported and did not pass the draft stage (Bennett, pers. comm.). A formal proposal to officially list the razorback sucker as threatened (USFWS 1978c) was withdrawn (USFWS 1980c) because critical habitat had not been identified within two years of the original proposal date as required by 1978 amendments to the Endangered Species Act (Valdez et al. 1982a). In the interim, Deacon et al. (1979) considered the fish threatened by habitat destruction, competition, and hybridization in all six basin states in which it historically occurred.

In 1981 a decision was made to proceed toward recovery of razorback suckers in the lower basin. The AZGFD and the USFWS signed a ten-year memorandum of understanding that enabled razorback reintroduction and monitoring in historic habitats of the Gila, Salt, and Verde rivers. Under this agreement, the USFWS was to produce up to 100,000 razorback suckers a year, and the AZGFD was to assist in stocking and monitoring the progress of reintroduction efforts. The proposal succeeded where others failed because it committed the USFWS not to list the species so long as the reintroduction effort continued and appeared successful. Recovery through means such as propagation and reintroduction had ironically become more feasible for an unprotected (nonlisted) species than for those protected under the act (J. E. Johnson 1985). The

state of California initiated a similar effort in the Colorado River main stem in 1986, which, unfortunately, was implemented in earnest for only two years, then essentially abandoned by 1989. The razorback sucker had been afforded legal protection by 1987 in all five states where it persisted (Arizona, California, Colorado, Nevada, and Utah; J. E. Johnson 1987a).

Research and management efforts toward recovery of razorback suckers in the upper basin, mostly linked to funding for other, higher-profile species, proceeded independently of lower-basin programs. Studies of Colorado squawfish and humpback chub (*Gila cypha*) were emphasized. The apparent assumption was that razorback sucker requirements would be met by habitat protection and manipulation, and other protocols would be applied to other species. Current efforts rely on the commitments put forth in an interagency *Recovery Implementation Plan* (USFWS 1987a, b, c), which lists on behalf of razorback suckers: (1) research activities, for example, to determine range, distribution, and abundance (W. H. Miller et al. 1982d; Tyus et al. 1987; Tyus 1987); (2) hatchery feasibility studies (Inslee 1982a, b; Hamman 1985a, 1987; Rinne et al. 1986); (3) continuation and expansion of radio-tracking studies (Tyus 1987); and (4) development of identification techniques for larvae (D. E. Snyder 1983). All these projects have been proceeding for a number of years. In 1987 the Colorado River Fishes Recovery Team, formed by the upper-basin USFWS Region 6 office, again recommended that the razorback sucker be federally listed as threatened (St. Amant, pers. comm.). A petition for listing the species as endangered was delivered to the USFWS in March 1989 by the Sierra Club Legal Defense Fund, representing a number of conservation organizations (Potter and Cheever 1989), and the proposal to list as endangered was published (USFWS 1990a). At this writing, the federal listing process is under way.

Hatchery Production and Reintroduction as Recovery Options

Options for recovering a species that has already declined to critical population levels, as was the case for razorback suckers in the lower Colorado River basin, are limited. Habitat alteration was already extensive, an apparent lack of reproduction and recruitment were of special concern, and the fish was rare where it had once been common, all of which indicated that protection of existing habitats and remnant populations might fail to perpetuate the species. Nonetheless, direct acquisition of brood fish, development of a breeding program, and reintroduction were deemed the least hazardous and most direct means to prevent extinction.

Practical reasons also existed for this decision. Suitable hatchery and research facilities and expertise were already in place, with little need for preparation or major costs. Research personnel were available and had already begun to evaluate the status of the species, and a large, apparently viable population of adult razorback suckers persisted in Lake Mohave, immediately adjacent to research and hatchery facilities at Willow Beach NFH.

Acquisition and Evaluation of Brood Stock

Field operations

The first priority for the propagation program was the assessment of availability and quality of brood fish, since success was contingent on producing large numbers of fish suitable for reintroduction. The Lake Mohave stock, although thought (and now known; see McCarthy and Minckley 1987) to be of substantial age, was deemed abundant enough to provide brood stock.

The first acquisition of wild fish was in March 1974, when 40 adults were seined from a breeding aggregation in Lake Mohave and transported to Willow Beach NFH (Kobetich, pers. comm.). They produced eggs that hatched into 3259 fry, some of which were successfully reared to adulthood. Sixty-nine additional brood fish were transferred from Lake Mohave to Willow Beach in 1975 (Gustafson 1975b). Investigation of the fecundity and survivability of gametes (Toney 1974; Gustafson 1975b) indicated that both were adequate to justify an intensive effort to establish, hold, and maintain brood fish as a source of progeny to reestablish one or more wild populations (J. E. Johnson 1985). A total of 281 adults was transferred from Lake Mohave and Willow Beach NFH to Dexter NFH, New Mexico, in January 1981 and 1982 to commence a propagation and reintroduction effort. The ultimate fates of these individuals and their progeny are shown in Figure 17-7.

Qualitative evaluation of different capture methods indicated that seining of aggregations was least injurious to individual fish. Electrofishing resulted in relatively high mortalities, and trammel netting, which had proven highly effective for this large species (Minckley 1983), tended to abrade and cut fish entangled more than briefly (Inslee 1982a). The fish is comparatively unperturbed by human intrusion, fleeing only when closely approached by boat (to within 3–5 m) or divers (to less than 1 m), or essentially encircled by seines. Disturbed fish typically move slowly away from a site but return within a few minutes. Although capable of remarkably strong and rapid movements, razorback suckers typically remain docile, even when captured and restrained.

A few hours between capture and transport produced little obvious holding stress or physical damage, but restraint in tanks or in-lake live-cars for more than twenty-four hours prior to transport increased long-term mortality. Realized and potential problems with injuries, lesions, and external bacterial and parasitic copepod infestations on otherwise healthy fish were countered by topical applica-

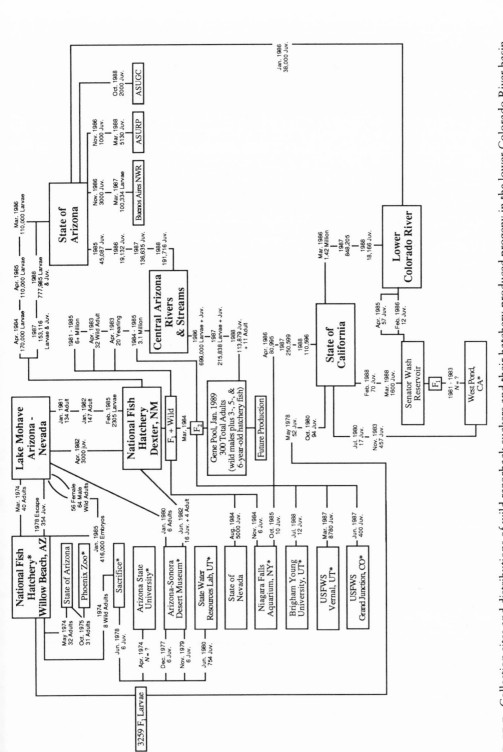

Fig. 17-7. Collection sites and distributions of wild razorback suckers and their hatchery-produced progeny in the lower Colorado River basin, 1974–1988. Asterisks indicate populations no longer in existence. Acronyms not otherwise identified in text are: ASURP, Arizona State University Research Park; and ASUGC, Arizona State University Golf Course.

tion of antiseptic solutions of Betadine or potassium permanganate, intramuscular injections of antibiotics (combiotics and tetramycin), and removal of ectoparasitic copepods by hand.

Hatchery handling and propagation

Adult razorback suckers proved readily adaptable to handling and hatchery conditions (Inslee 1982b). Work at Dexter NFH with fish collected in 1981 and 1982 enabled the development and refinement of techniques for holding adults, hormone injection to enhance gamete collection, hand stripping of eggs and milt, incubation, and treatment of fungi on developing embryos (Inslee 1982b; J. E. Johnson 1985). Techniques for rearing large numbers of young in earthen ponds—for example, pond preparation, stocking, and feeding rates—were also developed. Additional refinements to increase production and survival from spawning through transport and distribution completed the program (Hamman 1985a, 1987). Production became routine by 1983, and young were available to support reintroductions far exceeding the 100,000 fish per year committed in the AZGFD and USFWS agreement (J. E. Johnson and Rinne 1982; Minckley 1983; Rinne et al. 1986; Marsh 1987b; Johnson and Jensen, *this volume*, chap. 13).

Growth rates of razorback suckers proved remarkably high in the first few years of life, under both natural and artificial conditions. Based on otolith analysis, half or more of the maximum known adult size of about 70 cm TL in Lake Mohave was attained by wild fish in their fifth year (McCarthy and Minckley 1987). Some individuals of the 1974 year class produced and reared in relatively cold water (as low as 10°–12°C) at Willow Beach NFH matured at six years of age and 35–39 cm TL (Minckley 1983), while siblings less than 35 cm TL showed no sexual development. In seasonally warmer water (summer

temperatures exceeded 24°C) at Dexter NFH (Hamman 1985a), F_1 progeny of the 1981 and 1982 year classes attained 35 cm TL and became sexually mature in two (males) or three (females) years. They were added in 1984 to the brood stock, which numbered approximately 300 fish in 1988 (Jensen, unpub. data).

Genetic controversies and their resolution

Razorback suckers destined to become brood fish were culled in the field and at the hatchery on the basis of apparent condition and morphology. Diseased, injured, or otherwise high-risk individuals were removed, as were suspected hybrids, to minimize the possibility of genetic contamination. Assurances of genetic integrity and maintenance of heterozygosity of hatchery fish (Echelle, *this volume*, chap. 9) were major concerns. However, no screening other than visual examination by trained personnel was applied.

In part because of a lack of genetic data, Wilde (in Bozek et al. 1984), Ulmer (in Buth and Murphy 1984), and Loudermilk (1985) expressed concern that genetic contamination of razorback suckers in Lake Mohave through introgression with flannelmouth sucker might make their progeny undesirable for reintroductions.

> The issue is not one of generating a stock of razorback sucker that is any more "pure" than the native, presumably partially introgressed, population. Rather, it is introgressive hybridization, coupled with sampling error, which can cause the introgressed genes of flannelmouth sucker to be grossly overrepresented in the hatchery stock of razorback sucker. The release of massive numbers of such stock would do irreparable damage to the gene pool of wild razorback suckers, flooding it with introgressed genes far above the low level currently maintained by occasional hybridization. (Buth et al. 1987)

Wilde emphasized the probability that most reintroduced razorback suckers would die. A scenario was proposed in which a few survivors would hybridize with other native suckers because of a scarcity of conspecifics (C. L. Hubbs et al. 1942; C. L. Hubbs 1955), which might result in extensive genetic contamination of another species.

These concerns revived resistance to the reintroduction program. The state of California withdrew from a 1980 agreement to begin restocking the Colorado River mainstream until the genetic integrity of Lake Mohave brood fish was evaluated (Ulmer, pers. comm.), and a ten-year project similar to that already under way in Arizona was delayed until 1986 (Ulmer and Anderson 1985; Ulmer 1987). A 1981 memorandum of understanding between the USFWS and the New Mexico Game and Fish Department was also deferred, due not only to concern over genetic quality of brood fish but also the possibility for hybridization between reintroduced razorback and flannelmouth suckers (Hatch, pers. comm.); it remained suspended in 1991.

The risk of some hybridization of reintroduced razorback suckers with other native suckers in the lower basin had been deemed acceptable in Arizona, especially if the alternative was extinction of the monotypic genus *Xyrauchen*. Furthermore, the Sonora sucker, with which razorback suckers were known to hybridize (C. L. Hubbs and Miller 1953), remained one of the most widespread and persistent native fishes in the lower basin (Minckley 1973, 1985b), occupying many habitats where razorback suckers had never occurred and were not expected to invade. Other catostomids are either extirpated and unavailable in the lower basin (flannelmouth sucker) or are not known to hybridize with the razorback sucker (desert sucker, *Pantosteus clarki*; Minckley 1973).

These questions were addressed in an electrophoretic study that compared allozymes of Lake Mohave brood fish, their progeny from Dexter NFH, and a number of populations of flannelmouth suckers (Buth and Murphy 1984; Buth et al. 1987). Suspect fish selected from Lake Mohave proved to be razorback suckers (with one exception), with an estimated incidence of alien genes (of *C. latipinnis*) proving to be no greater than indicated by analyses for comparable genic interchange between other catostomid species. The single individual that proved different was tentatively identified as an introgressed hybrid (backcrossed to razorback sucker). Buth et al. (1987) concluded that parental stocks could be taken randomly from Lake Mohave without electrophoretic screening, since measured genetic introgression was rare. The genic composition of wild-hatched and hatchery-produced fish was identical, and the latter showed no evidence of alteration under artificial culture.

Concern that brood fish might underrepresent genetic diversity in Lake Mohave nonetheless prompted a December 1984 collection of eggs from fifty-six females fertilized in the field by sperm from sixty-four males. Loss of that stock due to a technical mistake (Fig. 17-7) stimulated capture and transport to Dexter NFH of 2400 larvae that were hand caught along 5 km of Lake Mohave shoreline in February 1985. When they were mature, survivors of that collection (about a hundred individuals in 1988; Jensen, unpub. data) were incorporated into the brood stock.

Reintroductions

The lower parts of many major streams and their backwaters and oxbows in central and western Arizona were beheaded by impoundments (Corle 1951; Rea 1983; Fradkin 1984; Graf 1985). Loss of headwaters was exacerbated by groundwater pumping, and lowland habitats maintained by return flows of irrigation and domestic wastewaters were physically or chemically suspect or inhabited by remarkable numbers of non-native fishes (Minckley 1973; Marsh and Minckley 1982).

Thus, practicalities of water availability were major considerations in deciding where to stock hatchery-produced fish. Potential habitat remained extensive only in and associated with channels upstream from impoundments in the Verde, Salt, and Gila rivers and their numerous tributaries.

A second problem involved the politics of water, which included a marked and continuing aversion on the part of water users and water-development agencies to the presence of threatened or endangered fishes, and led to strong resistance to reintroduction (J. E. Johnson 1979, 1980b; J. E. Johnson and Rinne 1982). Agencies and organizations involved in development were suspicious of stocking due to legal protection of species and habitats listed under the Endangered Species Act and the potential for ultimately including razorback suckers under that legislation.

The 1981 memorandum of understanding between the AZGFD and the USFWS reduced opposition to reintroduction of nonlisted species, and a similar agreement between those agencies and the U.S. Forest Service to reintroduce endangered Sonoran topminnows (Brooks 1985) provided similar assurances for a listed fish. In 1982 the act was amended (USFWS 1984f) to include reduced protection for experimentally reintroduced populations. By 1985, federally endangered Colorado squawfish were being reintroduced into historic habitats in Arizona under an "experimental, nonessential" category (J. E. Johnson 1987b). Little effort occurred outside Arizona until 1986, when California became involved (Langhorst 1988, 1989; Marsh and Minckley 1989). Proposed introduction of razorback suckers into the San Juan River, New Mexico, started and stopped no fewer than three times in 1986 alone; the USFWS and the New Mexico Game and Fish Department aligned to favor the program, and the USBR and the Utah Division of Wildlife Resources remained opposed.

Site selection

Original decisions on where, when, and how to stock hatchery-produced razorback suckers were based largely on speculation. Historic records suggested that mainstreams and backwaters of major rivers offered the most favorable sites (J. E. Johnson 1985), and fish were stocked in 1981–1984 (Figs. 17-7, 17-8) in channels of the Salt, Verde, and Gila rivers, and in or near mouths of their major tributaries (Brooks et al. 1985). These large habitats were abandoned in favor of smaller tributaries in 1986 and 1987 because predation by non-native fishes was heavy in larger streams, reintroduced razorback suckers moved rapidly downflow and thus into reservoirs where they were presumably subject to even heavier predation by alien species, and access to large streams for stocking and monitoring was limited and difficult.

Non-native piscivorous fishes known and expected to prey on young suckers (Marsh and Brooks 1989) were less abundant in the upper reaches of tributaries (Minckley 1973; Minckley and Clarkson 1979; Propst et al. 1985a, b). Stocking in tributaries, particularly above temporary barriers such as dry reaches, was also anticipated to offset downstream movement of hatchery-reared fish, a behavior predicted from experience with other species (see E. H. Brown 1961; Moring and Buchanan 1978) and demonstrated for razorback suckers (B. L. Jensen 1988; Marsh and Brooks 1989). Spreading stocked fish into tributary networks might also decrease losses due to other than regional droughts or floods. Stocking sites in these areas were more numerous, and although many reaches remained impossible to reach by motor vehicle, access was less difficult on tributaries than along large streams. This practice improved poststocking monitoring and thus the probability for recaptures. Smaller habitats have fewer barriers such as dams, irrigation diversions, and in-

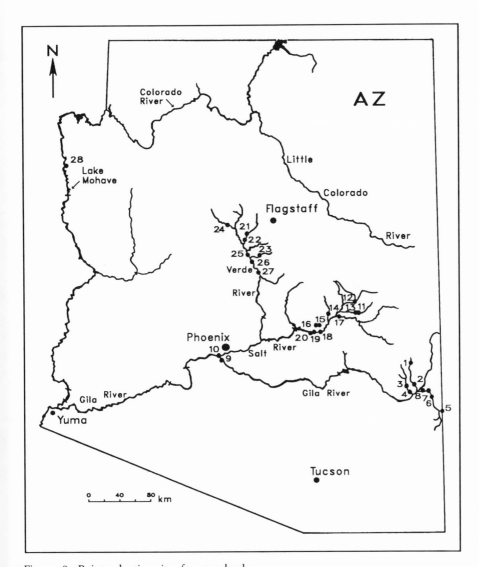

Fig. 17-8. Reintroduction sites for razorback
suckers in the Colorado River basin, Arizona,
1981–1988. Numbers: 1–2, Eagle Creek;
3–4, Bonita Creek; 5–8, mainstream Gila River;
9–10, gravel pits along Gila and Salt rivers; 11
and 13, Cedar Creek; 12, Carrizo Creek;
14, Canyon Creek; 15, Cherry Creek; 16, Coon
Creek; 17–20, mainstream Salt River and
backwaters; 21–22, Fossil Creek; 23, West Clear
Creek; 24–27, mainstream Verde River and
backwaters; 28, Lake Mohave. Recapture sites are
listed in Table 17-1.

takes (Minckley 1985b), and water supplies tend to be more reliable (that is, less variable; Minckley and Brown 1982) and more diversified (greater pool and riffle development). Most permanent tributaries of the Gila, Salt, and Verde rivers drain from relatively undeveloped federal, state, or Indian lands and are sometimes less influenced by environmental pollutants and other alterations. Disease problems, although scarcely studied and perhaps insignificant, may also be less pronounced in tributaries than in mainstreams. Last, primary and secondary production in low desert streams is remarkably high (Minckley 1979b, 1981; Gray 1981; S. G. Fisher 1986), and food supplies are probably superior in tributaries (as they are in harder-bottomed reaches of rivers). Shifting, sandy bottoms of large streams tend to support far smaller standing crops of algae and invertebrates (Minckley 1979a, b; Marsh and Minckley 1987).

Transport and stocking

Transport of hatchery-reared razorback suckers by truck, following USFWS (1978a) recommendations, was successful. Larvae were transported and stocked at or near the time of transition from yolk dependence to active feeding (Minckley and Gustafson 1982; Papoulias 1988), and before receiving food. Fish more than 30 mm TL were denied food for four to five days before transport to allow evacuation of guts to reduce potential ammonia stress during hauling. Larvae and juveniles were transported in 17°–18°C well water at densities of 0.1 and 0.2 kg l⁻¹, respectively; no bactericide or anaesthetic was used. Transport was usually at night and did not exceed twelve hours from loading to release, and mortality was consistently less than 0.5% (Jensen, unpub. data).

Razorback suckers were stocked as larvae in March, small juveniles in June and July, and larger juveniles in September and October, a sequence that followed maturation, spawning, and hatching of an annual cohort and the concomitant need to progressively reduce numbers in limited rearing space at Dexter NFH.

Fish stocked in the Verde and Gila rivers in 1981–1984 were marked by coded wire tags embedded in the snout (Jefferts et al. 1963; Bergman et al. 1968; Wydoski and Emery 1983), a practice that was originally deemed advisable to differentiate hatchery fish from wild individuals that might persist despite the more than twenty-five years since the species was recorded in central Arizona. Tagging was discontinued in 1985 for a number of reasons. Stocking of unmarked larvae was begun in the Salt River in 1982 for a projected evaluation of the relative success of stocking fish of different sizes, and no practicable method is yet known for long-term marking of larval fishes (A. C. Jensen 1962; D. E. Arnold 1966; Wydoski and Emery 1983). Also, larvae hatched at Dexter NFH were reared to juvenile size at the AZGFD Page Springs SFH and stocked unmarked in the Verde River in 1984 and 1985 because nose-tagging equipment was prohibitively expensive. The presence of both marked and unmarked fish in two of the three rivers by 1984 precluded possibilities of differentiating wild from hatchery fish. Furthermore, tag loss was determined to be 4% and 18% for 397 and 49 fish (Jensen, unpub. data), respectively, over a period of six months. In the interim, otolith aging was found to be reliable for identifying year classes of fish recovered.

Fish were stocked into the slowest currents available to facilitate their acclimation to local conditions. Larvae were transferred from the truck to buckets and placed in quiet backwaters, along stream margins, near algal mats and macrophyte beds, and in braids of small channels. Juveniles were planted by buckets or dip nets, or directly from hatchery trucks through an exit tube into backwaters, low-velocity runs, and pools. When necessary, transport water was brought to within 3°–5°C of stream temperature. During floods, stocking was often exclusively in backwaters. No allowances were made for turbidity. All fish scheduled for a given locality were released within 50 m of an access point. Undesirable habitat conditions during low flow were countered by not stocking a site, distributing fish at several adjacent places within a site, or spreading fish scheduled for a single locality to a number of localities.

Different-sized fish were stocked in different streams in an attempt to evaluate size-related survivorship: larvae in the Salt River, intermediate sizes in the Verde, and the largest fish in the Gila River. Monitoring soon indicated that stocking of larval razorback suckers should be abandoned in favor of larger fish. The few razorback suckers recaptured more than a few weeks after reintroduction all were from fish stocked as juveniles; none came from stocks of larvae (Brooks et al. 1985; Brooks and Marsh 1986; B. L. Jensen 1988). Ample data are available to indicate the same pattern for other fishes. Poststocking survival of greater numbers of "catchable" salmonids (longer than 25 cm TL) than of juveniles (Schuck 1948; Nielson et al. 1957; Sholes 1979) may be attributable to reduced predation pressure on larger fish. Marsh and Brooks (1989) provided comparative data suggesting that predation was less severe on razorback suckers stocked at 11.3 cm TL than on smaller fish (mean length 68 mm). The 15–25 cm attained by razorback suckers in thirteen or fewer months (McCarthy and Minckley 1987; Marsh 1990; Marsh and Minckley, unpub. data) exceed the predation size range of most resident piscivores in the Gila River basin.

Fish well past the larval stage were stocked exclusively after 1985, but demands for larger fish, along with needs for greater production to serve the CADFG program on the Colorado River mainstream, exceeded available rearing space at Dexter NFH. This problem was alleviated in part by using ponds at Page Springs SFH (AZGFD) and Niland SFH (CADFG). The adaptability of razorback suckers to a diversity of holding and rearing facilities (Inslee 1982b; Minckley 1985b; Hamman 1987) greatly increased the available options, and arrangements were also made for use of livestock watering ponds and water bodies on municipal golf courses. Livestock ponds are in use at Buenos Aires National Wildlife Refuge, Arizona (Marsh 1990), and golf course ponds were successfully utilized in 1984–1987 for razorback suckers and bonytail in California (in part, Ulmer 1987; Langhorst 1988) and razorback suckers in 1986 and 1987 in Nevada (Withers, pers. comm.). Urban lakes in Arizona have also been used for rearing razorback suckers, bonytail, and Colorado squawfish. Growth rates and survival are similar to those under hatchery conditions (Marsh 1987b, 1990).

Stocking razorback suckers in winter—a time of reduced activity and lesser food demands of many warm-water piscivores—should reduce predation on newly reintroduced fish. Survival has been demonstrably increased for salmonids stocked in spring rather than autumn (Millis and Kanaly 1958; Cuplin 1967).

Producing larger fish also physically limits the numbers that can be transported and stocked at a given site and time to groups of hundreds or at most a few thousand. We expect this to enhance survival, especially if fish are released over a period of time rather than abruptly. Data of Marsh and Brooks (1989) suggest a surge in feeding by non-native catfishes, seemingly elicited by the sudden abundance of large numbers of newly stocked and vulnerable razorback suckers (see also Keith

Table 17-1. Recaptures of reintroduced razorback suckers in Arizona, 1981–1987, excluding data for fish recaptured at their point of release less than two weeks after stocking.

Number recaptured	Recapture site, date, and SL (mean, range)	Reintroduction site, date, and SL (mean, range)
Gila River Drainage		
5	Highline Canal near San Jose Crossing, Graham County, 9/81 (not measured)	Bonita Creek mouth, Graham County, 9/9/81 (~98 mm)
1	Salt or Gila River, downstream from Phoenix, Maricopa County, 8/7/85 (~29.2 cm)	Same locality, 7/13/83 (~59 mm)
60	Upper Bonita Creek, Graham County, 10/15/85 (75 ± 19 mm, 49 mm– 14.1 cm)	Same locality, 7/8/85 (~59 mm)
14	as above, 8/14/86 (17.3 ± 2.7 cm, 14–23.3 cm)	as above
1	as above, 5/15/88 (18.5 cm)	as above
2	Eagle Creek, Greenlee County, near Morenci, 8/13/86 (21.1 and 22.3 cm)	Eagle or Willow creeks, unknown dates 1981–1985, various sizes[1]
10	Eagle Creek, Greenlee County, 7/14– 15/87 (58 ± 8 mm [8], 47–72 mm)	Upper Eagle Creek, Greenlee County, 7/22/87 (~60 mm)
Salt River Drainage		
12	Cedar Creek, Gila County, 7/23/85 (65 ± 15 mm, 42–95 mm)	Cedar Creek, Gila County, 5/22/85 (~30 mm)
2	Carrizo Creek, Gila County, 7/23–24/85 (64 and 74 mm)	Same locality, 5/22/85 (~30 mm)
6	Cherry Creek above Ellison Ranch, Gila County, 6/1/87 (13.7 ± 1.7 cm, 12–16 cm)	Same locality, 5/19/86 or 6/9/86 (~47 mm)
1	Canyon Creek below OW Ranch, Gila County, 5/31/87 (83 mm)	1 km upstream from OW Ranch, Gila County, 5/19/86 or 6/9/86 (~47 mm)
1	Salt River at Roosevelt Diversion, Gila County, 3/16/88 (20.4 cm)	Cherry Creek or Salt River at Cherry Creek 10/5/86 or 10/8/86 (10.9–19.9 cm)[2]
Verde River Drainage		
1	Verde River at Beasley Flat, Yavapai County, 1985 (~34 cm)	Unknown[3]

[1]Original stocking of these individuals cannot be ascertained because Eagle Creek and a tributary, Willow Creek, received fish ranging in mean SL from 5.9 to 23.9 cm on eleven occasions from 1981 to 1985, any of which could have contributed to the downstream recapture.

[2]Although date and location of stocking of this specimen cannot be ascertained with certainty, it likely was a survivor of larger (10.9–19.9 cm SL) fish stocked upstream into Cherry Creek or the Salt River on 5 and 8 October 1986, respectively.

[3]Numerous stockings upstream from 1981 through 1985, or immediately downstream in 1981 through 1983, could have given rise to this individual.

Recapture habitat	Distance and direction moved	Source
agricultural canal	~20 km, downstream	Campbell, pers. comm.
gravel pit (connected to river)	near same locality	Kepner, pers. comm.
oxbow backwater	same locality	original data
oxbow backwater	same locality	original data
oxbow backwater	same locality	original data
on-stream impoundment	~43–49 km, downstream	original data
canyon-bound reach	~2.3–20 km, downstream	original data
stream margin	same locality	Winham, pers. comm.
stream margin, pool	same locality	Winham, pers. comm.
isolated backwater	same locality	original data
stream margin, pool	3 km downstream	original data
run, above diversion dam	20–28 km, downstream	Leibfreid, pers. comm.
presumably pool	unknown	Burton, pers. comm.

1969; Berry and Kaeding 1986a). Lower winter temperatures might also reduce undetected stress and further increase survival.

Status of Reintroduced Populations: An Interim Report

Catostomid fishes produce remarkable numbers of eggs and larvae, but only a few survive to adulthood. Information on survival of sucker larvae in nature is difficult to find, but fishes such as these typically suffer extremely high larval mortality (Dahlberg 1979). Geen et al. (1966) estimated survival rates of 0.3% to the yolk sac stage for white and longnose suckers in Canadian streams. Estimated survival from egg to yolk sac larvae for cui-ui (*Chasmistes cujus*) in the Truckee River, Nevada, was much higher (1.4% and 16.0% in two different years) and attributed to local absence of predators (Buchanan and Strekal 1988). After one year, survival increases dramatically; for example, Verdon and Magnin (1977) estimated an annual survival rate of 79% for two-year-old white suckers.

Both egg and larval mortality are partially circumvented by hatchery propagation. Hatch for razorback sucker eggs at Dexter NFH commonly exceeded 90% (Jensen, unpub. data). Survival of unfed larvae in variously fertilized hatchery ponds was from 20.5% to 100% for 50 days (mean 78.1 ± 8.4%; Papoulias 1988), while fed larvae survived at rates of 92.7%–96.3% for about 120 days (Hamman 1987). It is not known, however, if hatchery-reared young survive as well in the wild as those that hatch under natural conditions. Mortality after stocking may be as great or greater than that suffered by naturally spawned fish.

Even in light of low expectations, efforts to determine survival during the first seven years of reintroductions were disappointing. Through autumn 1988, 118 fish were recovered from nearly 12 million larvae and juveniles stocked into central Arizona (Table 17-1). The Gila River and its tributaries yielded 93 recaptures on seven separate occasions, the Salt River and its tributaries produced 24 fish among five samples, and the Verde River yielded only one (Fig. 17-8). None was recovered from the Colorado main stem, which was stocked with 2.9 million larvae and juveniles in 1987 and 1988. Although large numbers were caught while fish remained concentrated just after release, none was taken after being at large more than two weeks (Langhorst 1988, 1989). Earlier smaller, less intensive plants into the Colorado mainstream and associated waters (Marsh and Minckley 1989) similarly produced no known recaptures.

In order to keep these numbers in perspective, one should consider that 120 wild, 55-cm females could spawn at least 12 million eggs. Further, an estimated 275 linear km of stream channels (550 surface ha if our estimated width of 20 m was near average) and nearly 17,000 surface hectares of reservoir are available in the Gila River basin. Recapture of 118 individuals from the 175.5 km² of water available to reintroduced fish (stocking averaged only about 100 individuals per ha per year, including larvae, in central Arizona) may, in fact, be a reasonably high return. An additional 274 river km (perhaps 3000 ha at an estimated average width of 100 m) comprises the lower Colorado River reintroduction area downstream from Lake Havasu (USBR 1980).

Monitoring began immediately after the first reintroductions in central Arizona in 1981 and is to continue through 1990, and beyond if an assessment of recent stockings of larger fish is to be accomplished. Sampling has been by seines, hoop, gill, and trammel nets, and electrofishing equipment mounted on backpacks, boats, rubber rafts, all-terrain cycles, and banks, as appropriate to the complexity, size, and accessibility of habitat. Collecting was specific to the monitoring effort or included as part of other studies. In addition to streams, reservoirs on larger rivers (San Carlos Reservoir, Roosevelt Lake, Bartlett Reservoir, and an impoundment on Eagle Creek) were also sampled in 1985 and later (in part Minckley 1985a). Fish stocked in 1981 or 1982 could have achieved mature sizes (greater than 35–40 cm) by 1985 and begun the nearshore aggregation that makes them highly susceptible to capture in winter and spring. Only one recapture has as yet been recorded from an impoundment.

Future Needs in the Recovery Program

Changes in emphasis during razorback sucker recovery efforts have been based on practical considerations, inference, and new data derived from research and experience, in order of greater to lesser importance, which reflects the perceived urgency of the program. In an ideal sequence, the order would have been reversed. Hypotheses would have been generated and systematically tested to support a progression of positive results. Inferences would have been applied on the basis of data derived from such tests rather than from scanty information in a historic record or anecdote. And each problem would have been solved by application of funds, expertise, and an innovative management scheme.

At first, time and resources did not allow such a scenario; existing facilities, resources, and personnel were respectively modified, reallocated, and retrained. With no apparent recruitment into wild populations and the apprehension that adults were approaching death from old age, along with continued reduction and deterioration of habitats and lack of information on why the species suffered such problems in the first place, immediate action was required to prevent extinction. The program of hatchery propagation allowed research—both descriptive and systematic—on life-history stages not otherwise available for

study, and the data base expanded rapidly.

Additional changes in the razorback sucker recovery effort will undoubtedly be proposed as needs arise. We perceive three major areas where additional emphasis should be applied: (1) means to improve survival of reintroduced fish, including insurance against extinction should the present program fail; (2) information exchange; and (3) critical research needs. All three pertain to efforts to conserve long-lived western fishes in general, as well as razorback suckers.

Ways to Increase Survival

Habitat alteration and enhancement

Reduction of predation and competition pressures may be effected by removal and transfer of non-native fishes, or through their eradication or reduction in discrete reaches coupled with prevention of immigration by installation of barriers. The first alternative is rarely economically sound, especially with warm-water species. The second option, however, is particularly attractive. One approach might involve relaxation of regulations for harvest of non-native predators by anglers or others. Relaxed creel limits, diversification of legal methods of capture, or commercial harvest of species such as channel and flathead catfishes could reduce local populations temporarily or indefinitely and enhance the probability of survival for reintroduced razorbacks.

Another, more direct means is the use of ichthyocides to remove unwanted populations. Reduction or removal of resident fishes before sport fishes are introduced has been practiced for years in the American Southwest (Hemphill 1954; AZGFD 1960–1973; in this volume, Holden, chap. 3; Rinne and Turner, chap. 14) and elsewhere. Such action to favor reintroduced native species has been attempted only recently, on a relatively small scale (Meffe 1983b), and most frequently to benefit threatened or endangered salmonids (see Rinne et al. 1982). If ichthyocides are used, reaches on the order of a few hundred meters would suffice to allow newly reintroduced fish to remain unmolested while acclimating to their new environment; for example, adjusting to novel flow conditions after stresses of handling and transport, and shifting from artificial foods and pond foraging activity to seeking natural foods in a stream. Reaches should include both pools and riffles to allow fish to select among various kinds of habitats, and an isolated segment should be large enough to preclude crowding. Of practical importance is a need to block the reach from egress by reintroduced fishes, both upstream and downstream.

Reclamation of longer reaches or portions of whole watersheds would be more rewarding, especially if ichthyocides were applied at points upstream from the farthest penetration by target species. Such treatment would ideally end at a natural or artificial barrier to undesired fishes (e.g., Apache Falls on the Salt River, or any of several dams or diversions). Complete removal of non-native fishes could not be ensured, but reduction in numbers could certainly be accomplished. Stocks of nontarget species should be collected and held until detoxification is complete, then reintroduced to reestablish them along with hatchery-produced razorback suckers. Rapid recovery of native western fishes after natural and artificial catastrophes has been demonstrated (Rinne 1975; Minckley and Meffe 1987), and resident native species may provide a buffer between remaining predators and reintroduced fish.

Environmental preparation before stocking has not been tried for razorback suckers in streams, but, as we described before, it has been successfully performed with ichthyocide in an isolated embayment adjacent to Lake Mohave. In another instance, the state of California began stocking juvenile razorback suckers in 1988 into backwaters separated from the Colorado River by nets and depleted of non-native fishes through netting and electrofishing. The intent, which was realized in part (Langhorst 1988, 1989), was to rear young suckers to sizes relatively immune to predation prior to releasing them into the channel. Stocking of young Colorado squawfish in backwaters also appears to hold promise, especially when predatory fishes are removed first (Berry and Kaeding 1986a, b; Osmundson 1987; Kaeding and Osmundson 1988a). Juvenile razorback suckers stocked with young Colorado squawfish and fathead minnows (*Pimephales promelas*) grew almost as rapidly as in warmer lower-basin ponds, from 58 ± 6 mm TL at stocking in June to 30.7 ± 18.8 cm in November (Kaeding and Osmundson 1988a).

Even when predators and competitors cannot be eliminated, razorback suckers grow far more rapidly than most other species, and a few months may be adequate to establish a population that persists even in the face of substantial invasion by non-native fishes. Two specific examples are available. First, an 8.9-ha urban lake complex in Tempe, Arizona, was stocked with 1000 seven-month-old razorback suckers (average 12.7 cm TL; range 8–20 cm; $N = 50$) in November 1986, just after it was filled and before a large non-native fauna established. They grew to 25.5 ± 1.9 cm ($N = 77$) in April 1987, 36.3 ± 2.7 cm ($N = 60$) in September, and 43.9 ± 2.2 cm ($N = 36$) in March 1988. In the same period, the non-native fish fauna increased from two to sixteen species through invasion from the local canal system. In the second case, 100 young-of-year razorback suckers stocked in 1984 in a pond otherwise devoted to the production of goldfish (*Carassius auratus*) as food for Colorado squawfish at Dexter NFH grew to 43.7 ± 3.9 cm TL ($N = 77$, range 34.5–50.0 cm) in November 1987. They matured to reproduce in spring 1987, and their young survived to achieve more than 10 cm

TL in midsummer despite the presence of the thriving goldfish population. We pointed out earlier that survivorship and growth rates of razorback suckers in diverse habitats from livestock tanks to ornamental ponds are comparable to those under hatchery conditions.

We strongly support these efforts since, despite regional aridity and the highly modified nature of the basin, natural, seminatural, and artificial habitats amenable to native fishes are common on most lower-basin floodplains (Ohmart et al. 1975; Ohmart and Anderson 1982; L. B. Brown 1983). Natural oxbows and swales, seeps behind levees and spoil banks, meanders isolated by channelization, and dredged backwaters originally created for sport fisheries or other purposes may all be used for native fish management. More than 5100 surface ha of water is already set aside within existing national wildlife refuges along the lower Colorado River (Fowler-Propst, pers. comm.), and similar holdings exist in the upper basin. Management of refuges for native fishes along with waterfowl as the principal target group is a compatible multiple use that should be mutually beneficial; we recommend it highly.

Use of naturally isolated habitats, or isolating them with heavy equipment, followed by eradication of undesired species and stocking of native fishes, are inexpensive management techniques. Such a project is currently under way to enhance the razorback sucker population in Lake Mohave. Valdez and Wick (1981) earlier reviewed natural versus man-made backwaters as native fish habitat in the upper basin and concluded that enhancement of natural features could be a valuable management technique. Minckley (1987) and Jackson et al. (1988) discussed rationales for construction and maintenance of seminatural habitats for razorback suckers or other native species on the San Pedro River floodplain, Arizona. In the lower basin, aquatic habitats are also fed by treated municipal wastewater that flows through natural and artificial channels into abandoned gravel pits and marshy depressions of the Gila, Santa Cruz, and Hassayampa rivers, and elsewhere. These waters are among the few that may be expected to increase in numbers and volume as human populations increase in the region. They are now mostly inhabited by introduced fishes but could as well be managed for native species with minimal outlay of money and effort.

Insurance against the disappearance of razorback suckers may further hinge upon their marked longevity. Modifications and repairs of a number of major dams in the lower basin are in progress or anticipated. Lake Carl Pleasant on the Agua Fria River is being enlarged as part of the Central Arizona Project, Roosevelt Dam is to be heightened and strengthened in the near future, Coolidge Dam on the Gila River is in need of repair, and other such projects are anticipated in the next decade (Rinne, pers. comm.). In each instance, nursery areas manageable for native fishes should be built while reservoirs are drawn down, to be used for ongoing production when the system is again in operation. If a reservoir is dried, it should be heavily stocked with razorback suckers (and perhaps other appropriate species) as it is refilled, anticipating that survival and rapid growth rates in the initial absence of predators will allow development of large populations of individuals of body sizes that preclude predation and circumvent other detrimental interactions with non-native fishes. Each population will thus provide a thirty- to forty-year buffer, "hedging the bet" against the species's extinction.

Manipulations of restocked fish

As we noted above, for largely unexplained reasons, razorback suckers (B. L. Jensen 1988) and many other warm-water fishes (Wickliff 1938; Funk 1957) move downstream after transplantation or stocking. Restraining fish until they are acclimated to a new habitat might alter this behavior. We held juvenile razorback suckers for twenty-four to thirty-six hours in live-cars before stocking them in the Gila River, however, and saw no differences in their behavior compared with that of fish placed directly from a transport truck into the stream. Nonetheless, stocking in isolated backwaters or in stream reaches cleared of other fishes and isolated by netting might help hatchery-reared fish to establish residency, assuming they tend to remain once they are familiar with an area. Groups of fish could be released after days to weeks of restraint. Inasmuch as survival of salmonids acclimated to raceways or streams has been shown to exceed that of pond-reared stocks (Schuck 1948; R. B. Miller 1953), conditioning of razorback suckers in raceways might be a useful substitute for in-stream holding. Again, a preliminary assessment (B. L. Jensen 1988) indicated no discernible differences in behavior of raceway-acclimated versus pond-acclimated individuals.

In-stream orientation and homing is known in numerous fishes (Gunning and Shoop 1962; Harden-Jones 1968; Hasler 1971; Leggett 1977) and is often demonstrably performed through the gusto-olfactory senses (Gunning 1959; Kleerkoper 1969). Although we know of no specific accounts in catostomids (see, however, Bangham and Bennington 1939; Gerking 1950, 1953, 1959; Siebert 1980), razorback suckers may have highly developed chemosensory capabilities (R. B. Miller and Evans 1965; Minckley 1983) that might be exploited through conditioning. Some fishes respond to chemicals secreted by conspecifics (Dizon et al. 1973; Asbury et al. 1981), and Tyus (1985) suggested that Colorado squawfish use cues such as the chemistry of spring inflows to "home" for reproduction. Chemical "leading" of salmonids to natal habitats by artificial substances has been successful (Hasler and Scholz 1978). A short period of

exposure to a unique chemical, or to foods to which they were conditioned, could be an inexpensive way to stimulate hatchery-reared fish to establish residency.

Monitoring and documentation

MONITORING EFFORTS. Projected sampling of central Arizona streams into which razorback suckers had been stocked (the Salt, Verde, and Gila rivers), and downstream reservoirs where fish might eventually become established, was far too extensive to accomplish with the manpower, equipment, and funding that were available. Access to most of the mainstream Salt and Verde rivers is only by whitewater rafting in high-water periods or canoe at lower flows. Either of these involves a large commitment of resources, especially if monitoring is to be done in different seasons, since different equipment and techniques are required not only for access but also for collecting. Parts of the Gila River are similarly difficult. On the other hand, concentrating efforts on a single river, with emphasis on one or a few tributaries, may not be satisfactory since each system is unique and information obtained from one may not be applicable to others.

We believe that concentrating efforts in selected, accessible reaches in each system is the most reasonable approach to monitoring. Intensive effort with a diversity of gear, year after year at the same localities, also provides long-term data bases on other fishes useful in the overall management of such systems. We predict that a major proportion of surviving razorback suckers have moved into downstream reservoirs (Fig. 17-3), where, if they follow the pattern elsewhere, adults should aggregate and attempt to spawn. These fish should be mature when they are between three and six years old, and fish stocked in 1981 through perhaps 1986 should thus be susceptible to capture no later than 1989.

Public participation is another way to monitor, and, with some effort, it could be promoted throughout the entire lower basin. The posting of appropriate advertisements for assistance (e.g., Fig. 17-6) stimulates public response and assistance. The razorback sucker is large, seasonally colorful, and spectacular in appearance, and public sentiment now favors the plight of disappearing native animals. Further, channeling funds, personnel, and interest into development of an informed, educated, and more sympathetic public may do more in the long term to save endangered species than more localized efforts.

To date, all places where razorback suckers have been reintroduced (Fig. 17-8) have been monitored at least once after stocking, and more than half have been sampled three or more times. Only a few fish have been recovered (Table 17-1), however, and there is no evidence that populations have reestablished. Only after resident adults persist and stocking is curtailed will the occurrence of their young-of-year provide direct evidence of recruitment.

RESPONSIBILITIES FOR DOCUMENTATION. Documentation of the results of a recovery program is a responsibility rarely realized. Yet documentation has ramifications far exceeding the simple recording of information. Dozens of persons representing at least six agencies or institutions have been involved in recovery and management programs for razorback suckers, and many others have responsibilities for the habitats into which the species has been stocked or may eventually occur. There are substantial gaps in information, mostly fostered by lack of communication or failure to realize the importance and value of individual activities. Moreover, as active management toward recovery of razorback sucker spreads to adjacent states such as California (Ulmer 1987), New Mexico (Propst, pers. comm.), and elsewhere, as we anticipate it

will, and stocking and monitoring of this and other species expands, the need for accurate, up-to-date, and readily available information will increase.

It is generally recognized that centralized repositories for data, memoranda, and reports are invaluable for planning, executing, and evaluating recovery programs. Agencies and individuals circulate memoranda, reports, and reprints, and each maintains an internal file, but general mailings are exceptional. Redundancy is relatively high, and funding is a problem, but greater circulation of information and stable storage facilities are sorely needed.

Minckley (1985b) recommended centralizing documentation for USFWS Region 2, for example, at Dexter NFH; other federal alternatives exist. State offices dealing with endangered species (e.g., nongame fisheries and wildlife branches of game and fish departments) also might be used but are limited by their political boundaries as well as by facilities, personnel, funds, or stability of purpose. Private organizations such as The Nature Conservancy have programs of data storage and retrieval that suffer far less from political constraints or instability, but they are rarely free from a general lack of funds. Academic institutions enjoy the distinct advantage of an established commitment to library facilities, relatively high stability, and at least modest funding. They also enjoy intercommunications, such as interlibrary loan services, that are well developed, reliable, nationwide, and often computerized. The availability of storage space and information-retrieval systems could be ensured through paid subscription by concerned agencies and organizations, perhaps with a part of the "fee" taking the form of a commitment from each to store information. We do not strongly recommend one of these options over another, but some program of this sort must be established if data bases are to be stabilized and information flow maintained.

We do not propose to further a program that preserves gray literature. Persons holding research and management positions must be urged to publish their results and recommendations. Agencies should make publication in peer-reviewed, public outlets a part of individual performance evaluations, ensuring that agency personnel and contractors alike perform such tasks by allocating time, money, and appropriate incentives and acknowledgments to the preparation and publication of manuscripts.

Subjects for Additional Research

Research needs for endangered species can be broadly separated into two major categories: those necessary if recovery is to be effected, and those that would be desirable but are not directly applicable to perpetuation of the species. For organisms nearing extinction, studies of the second category are often ill advised. Most of these species are too rare to allow sampling, because each individual is more valuable for its potential genetic contribution than for the biological information it might provide. Furthermore, such research tends to siphon funds away from more urgently needed studies. We emphasize needs in the first category.

Fortunately, a large population of wild razorback sucker adults persisted in Lake Mohave, so basic information was at least obtained on the adult stage of its life cycle. The propagation, culture, and reintroduction provided data on other life-history stages and made feasible the necessary studies of basic biology of the species, at least under seminatural or artificial conditions.

Future research programs for recovery of the razorback sucker fall into three major areas: (1) resolution of the recruitment problem, (2) individual and populational movements and ecology, and (3) genetics. Until the problem of recruitment failure is resolved and circumvented, recovery through natural means seems unlikely. Habitat alteration, competition, parasites and disease, nutrition, and predation all have been proposed as factors in this failure, but how one or a combination of factors influenced populations over the entire geographic range of the species remains unexplained. How does local or even regional habitat alteration, other than absolute drying or contamination beyond tolerance levels, cause the disappearance of a whole species?

Other than the reduced temperatures below impoundments that may result in local avoidance of cold water by adults or preclude hatching of eggs or survival of larvae, the physical reasons for the general disappearance of razorback sucker remain speculative. Cold-water reaches resulting from hypolimnetic releases from dams make up only a small percentage of the streams available. Food conditions may be a factor, but nutritional states in Lake Mohave (based on zooplankton standing crops), where razorback sucker larvae disappear before reaching the juvenile stage, are quantitatively and qualitatively similar to those in isolated backwaters in nature and in ponds at Dexter NFH where they successfully survive and grow to adulthood. A predation hypothesis has yet to be ruled out, although it remains problematic for a species like the razorback sucker, which has high fecundity and apparent survival potential. Razorback sucker larvae nonetheless survive, grow to adulthood, and even mature in the absence of predators, while they almost invariably disappear when predatory fishes are present, in both seminatural and experimental situations.

Perhaps this last simple fact—that predator exclusion results in successful recruitment—constitutes an adequate level of information for management toward recovery of the razorback sucker. One can argue that no more re-search is needed at this level, since the technol-ogy already exists to produce vast numbers of fertile and fecund, and thus viable, razorback suckers under hatchery or seminatural condi-tions. One might also ask if the political and management problems that arise in selection, preparation, and maintenance of habitat suit-able for recovery and perpetuation of this species are the biologist's responsibility. We believe they may not be, and we urge adminis-trators to examine and define their roles in this phase of expediting and complying with the goals and requirements of the Endangered Species Act.

The second major area of needed research hinges on a general lack of quantitative data on daily, seasonal, and annual movements and habitat preferences of the various life stages. A beginning has been made through use of conventional tagging and telemetry in streams of the upper basin. As yearlings are produced for stocking, larger ones should be radio-tagged to facilitate monitoring. Subadults and adults should also be equipped with transmit-ters to determine their activities, dispersal, and overall behaviors.

Tracking individuals of known origins should expedite monitoring efforts by helping to define areas of concentration of reintro-duced fish, thus allowing reduction of effort in the long term. Survivorship data can also be obtained, as can precise information on diel, seasonal, and annual habitat selection, movements for reproduction, and so on, all of which are clearly pertinent to management in perpetuity. Nowhere in the Colorado River basin can "natural" conditions be found or, if present, maintained; human water demands in this water-poor region are simply too great. Only through definition of habitats suitable for life for razorback suckers and other large-river species can they be perpetuated, and how better to determine and define habitat

than through direct telemetry of the animals themselves?

The third major area of research needs—genetics—exists for essentially all endangered fishes. Razorback suckers from Lake Mohave are variable (heterozygous) and relatively free of introgressed genes derived through hybridi-zation with other species. However, no com-parisons have been made among remnant populations in the upper and lower Colorado River basins. Such will be necessary to assess historic isolation (or lack of same) and the fragmentation effects of water diversions and dams, and to evaluate the suitability of vari-ous stocks for recovery attempts.

We reemphasize that the highest priority in all research on razorback suckers must lie in a program that ensures establishment of resident populations of reproductive age to compen-sate for the loss of the Lake Mohave popula-tion predicted to occur within the next decade. The ecology of remnant riverine populations of the upper basin must be thoroughly studied as well, both toward the goal of reestablishing healthier populations there and to provide in-ferences for management of reintroduced stocks in lower-basin streams. Research com-paring natural and reintroduced riverine stocks will be especially informative.

Criteria and Potentials for Recovery

Reestablishing self-sustaining populations of a long-lived fish presents the unique problem that most workers on the project in 1989 will probably pass on (figuratively or actually) be-fore the species is secure. Criteria for deciding when the razorback sucker is successfully reestablished have yet to be defined. More questions than answers remain. How much natural reproduction and how frequently, and what levels of recruitment are necessary to maintain a population?

As we have already noted, technology is

available to establish substantial populations of hatchery-produced razorback suckers, either during the ten-year AZGFD/CADFG and USFWS stocking programs or through some later permutation of the present efforts. If the program is successful, and a life span of forty to fifty years is assumed, fish established by 1990 could survive until 2030 or 2040. But there are no data on recruitment success of razorback suckers under natural conditions. Perhaps recruitment occurred annually, but it also might have succeeded only once every decade, or longer, by simple variations in survivorship. Data provided by McCarthy and Minckley (1987), although not extensive, imply sporadic production of year classes by razorback suckers in Lake Mohave. Such a pattern has been convincingly demonstrated for cui-ui in Pyramid Lake, Nevada (Scoppettone and Vinyard, *this volume*, chap. 18).

Hatchery-produced razorback suckers must be marked in some manner—for example, with oxytetracycline (Wydoski and Emery 1983)—to distinguish them from those produced naturally. Growth of razorback suckers is remarkably variable. A single cohort may include fish from 3 to 20 cm long at the end of their first summer (Minckley 1983, unpub. data; McCarthy and Minckley 1987), and such disparities appear to persist at least through the first few years of life (Minckley unpub. data). Size-frequency distributions are thus unreliable for year class estimation. Otolith examination seems to be a valid aging technique and may be used to identify year classes. However, unless all recaptured fish are aged, it will be impossible to evaluate natural reproduction by unmarked, stocked fish, even for a number of years after reintroductions are terminated.

Sexual maturity is attained in the second (males) or third year (females) in razorback suckers under conditions at Dexter NFH, or in the fifth or sixth year under other captive regimes. Thus, reproduction by fish stocked in 1981, which might have occurred by 1983 or 1984, could have been delayed until 1985 or 1986, or may require even longer under conditions in the field. Production of a viable year class by hatchery-reared fish thus might not even occur in the ten-year period provided by the 1981 AZGFD/USFWS memorandum of understanding, or reproduction might only occur after two or even three or more decades. Monitoring obviously must be continued for a long time. The present program clearly involves a far longer commitment than the projected decade, no matter what the ultimate result.

Appendix: Names and Affiliations of Individuals Providing Personal Communications Cited in Chapter 17.

N. Armantrout, USBLM, Portland, Ore.

F. Baucom, USFWS, Phoenix, Ariz.

J. Bennett, Colorado Division of Wildlife, Denver, Colo.

J. Burton, AZGFD, Phoenix, Ariz.

E. Campbell, USBLM, Huachuca City, Ariz.

J. Deacon, University of Nevada, Las Vegas, Nev.

B. DeMarais, ASU, Tempe, Ariz.

M. Douglas, ASU, Tempe, Ariz.

G. Edwards, USFWS, Washington, D.C.

J. Fowler-Propst, USFWS, Albuquerque, N.M.

A. Gustaveson, Utah Department of Wildlife Resources, Salt Lake City, Utah

M. Hatch, New Mexico Department of Game and Fish, Santa Fe, N.M.

D. Hendrickson, University of Texas, Austin, Tex.

W. Kepner, U.S. Environmental Protection Agency, Las Vegas, Nev.

G. Kobetich, USFWS, Sacramento, Calif.

W. Leibfried, Flagstaff, Ariz.

E. Milstead, Calipatria, Calif.

M. Moretti, Utah Department of Wildlife Resources, Price, Utah

G. Mueller, USBR, Boulder City, Nev.

D. Propst, New Mexico Department of Game and
   Fish, Santa Fe, N.M.

W. Rinne, USBR, Boulder City, Nev.

J. St. Amant, CADFG, Long Beach, Calif.

J. Sjoberg, NDOW, Las Vegas, Nev.

H. Tyus, USFWS, Vernal, Utah

L. Ulmer, CADFG, Bonnie Doon, Calif.

J. Warnecke, AZGFD, Mesa, Ariz.

R. Williams, USBR, Salt Lake City, Utah

T. Winham, USFWS, Dexter, N.M.

D. Withers, USFWS, Reno, Nev.

S. Yess, USFWS, Winona, Minn.

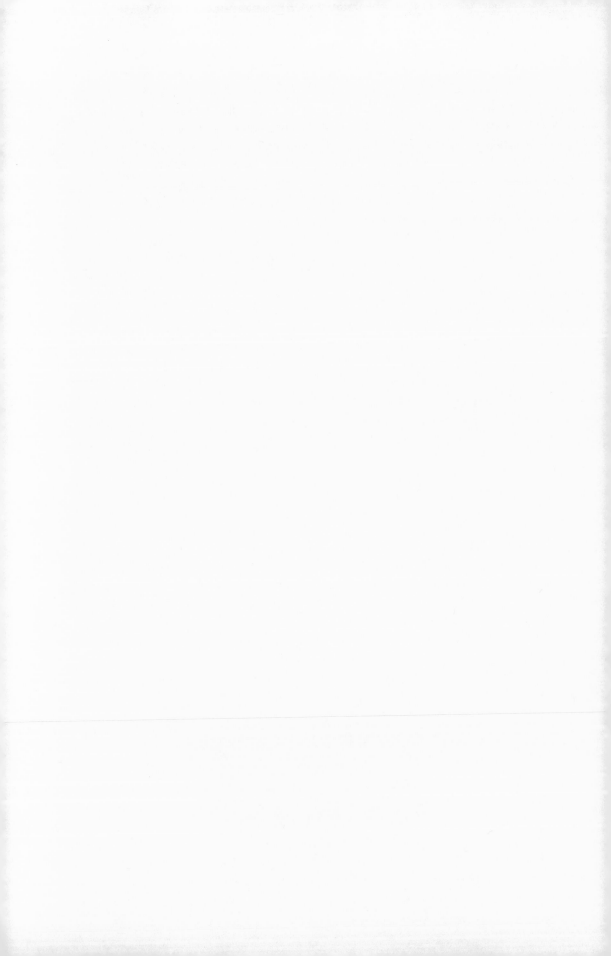

Chapter 18

Life History and Management of Four Endangered Lacustrine Suckers

G. Gary Scoppettone and Gary Vinyard

Introduction

Among the catostomids of western North America, all three living species of the genus *Chasmistes* and the single member of the genus *Deltistes* are obligatory lake dwellers. Here we refer to them collectively as lakesuckers (Fig. 18-1). These fishes reached their greatest abundance and widest geographic distributions in Pliocene through Pleistocene times, when *Chasmistes* inhabited lakes throughout the Snake River system, much of the Great Basin, and extended south into what are now the southeastern California deserts (R. R. Miller and Smith 1981). The range of *Deltistes* included the Klamath River basin east to the present Snake River plain of southwestern Idaho (G. R. Smith 1975), and to southeastern Oregon (R. R. Miller and Smith 1981) and the Lahontan Basin (Honey Lake, California; D. W. Taylor and Smith 1981).

With desiccation of the American West since the last pluvial period, the range and abundance of lakesuckers declined to isolated remnants. The June sucker (*Chasmistes liorus* Jordan) retreated to Utah Lake, Utah; the cui-ui (*C. cujus* Cope) was confined to Pyramid and Winnemucca lakes, Nevada; and shortnose (*C. brevirostris* Cope) and Lost River suckers (*Deltistes luxatus* Cope) occupied lakes of the upper Klamath basin, Oregon, and the Lost River system, Oregon and Cali-

fornia. A fifth species, the Snake River sucker (*C. muriei* Miller and Smith), described after its recent extinction, presumably occupied lakes of the upper Snake River basin near Jackson Hole, Wyoming (R. R. Miller and Smith 1981).

Chasmistes and *Deltistes* are closely related to the older, more diverse, widespread genus *Catostomus* (R. R. Miller and Smith 1981). *Chasmistes* species are distinctive in having branched gill rakers and a terminal mouth, while *Deltistes* has unique triangular gill rakers and a more ventral mouth with papillose lips (Seale 1896). The exceptional terminal mouth in *Chasmistes* is a presumed adaptation for feeding on zooplankton. Life-history patterns are similar among lakesuckers; they are long-lived lake dwellers and obligatory stream spawners, and typically begin to reproduce at five to ten years of age.

All four lakesuckers are federally listed as endangered (U.S. Department of the Interior 1973; U.S. Fish and Wildlife Service [USFWS] 1986e, 1988a). In each case their declines seem to be from similar causes. Most populations were subject to intensive fishing in the late nineteenth and early twentieth centuries (D. S. Jordan and Evermann 1903; Carter 1969; Howe 1981). The habitats of each have been greatly altered by water-development projects, which have resulted in the loss of Winnemucca Lake and reductions in the sur-

359

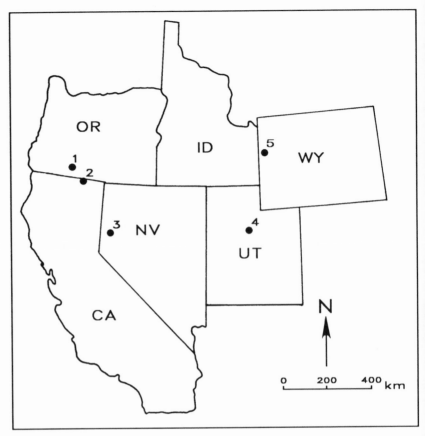

Fig. 18-1. Map of the western United States, showing distributions of lakesuckers (genera *Chasmistes* and *Deltistes*) as follows: 1, Upper Klamath basin; 2, Lost River system; 3, Lower Truckee basin; 4, Utah Lake; and 5, Jackson Hole.

face areas in others (Pyramid, Lower Klamath, and Tule lakes), as well as in fluctuating water levels in those now operated as storage reservoirs for irrigation (Upper Klamath, Clear, and Utah lakes). In all basins, diversion dams and reduced downstream flows resulting from water use have blocked spawning migrations (LaRivers 1962; Golden 1969; Radant et al. 1987). Degraded water quality from increased agricultural and municipal use has adversely affected their habitats. The introduction of non-native species is also thought to have affected lakesuckers, and all except the cui-ui

have experienced significant hybridization and introgression with other catostomid species (Moyle 1976; R. R. Miller and Smith 1981).

Life spans exceed forty years in the cui-ui, June sucker, and Lost River sucker, and thirty years in the shortnose sucker (Scoppettone 1988), and analyses of age structure suggest exceptional longevity as the key to their persistence. Lakesuckers, along with the razorback suckers (*Xyrauchen texanus*) of the Colorado River basin (McCarthy and Minckley 1987), are among the longest lived of all western

North American catostomids. We here assess age structure and review the life history, current status, and management options for this unique group.

Methods

Opercular bones have been verified to provide an accurate and reliable means of aging lakesuckers (Scoppettone 1988). Opercles preserve a record of the individual's age; also, by correlating opercle and fish size, one can reconstruct a growth curve based on the spacing of opercular annuli. Because growth rates decline at the onset of reproduction (Royce 1972), it is further possible to estimate age at first reproduction from an abrupt decrease in annulus width. Annulus formation results from the merging of two distinct regions. A zone of dense, relatively translucent, high-calcium-content bone is deposited during periods of slow growth (winter), while bone formed during periods of rapid growth (summer) contains higher concentrations of proteins and is more opaque. These two zones together compose an annulus (Casselman 1974).

We present new age-frequency data for *Chasmistes brevirostris*, *C. liorus*, and *Deltistes luxatus*; data for *C. cujus* are derived from Scoppettone et al. (1986). Opercles were removed and cleaned by boiling, and the fenestrated support of the hyomandibular socket was ground smooth to expose hidden annuli. Annuli counts were made with a binocular dissecting microscope.

More than 1000 cui-ui were examined from spawning runs of 1978 through 1987. Mortalities resulted from handling at the Marble Bluff Fish Handling Facility, Nevada, and from white pelican (*Pelicanus erythrorhynchos*) attacks on fish during spawning aggregations (Scoppettone et al. 1986).

During an unexplained fish kill in Upper Klamath Lake, Oregon, in summer 1986, 191 Lost River and 7 shortnose suckers were collected and measured, and operculae were removed for aging (Scoppettone 1988). Length-frequency comparisons of these with a sample of adult spawners indicated that the die-off provided a representative sample of the adult population of Lost River suckers. An additional sample of 19 shortnose suckers was obtained from Copco Reservoir, California, for the purpose of determining taxonomic status; lengths and opercular bones were later provided to us by Beak Consultants (1987).

Eighteen June suckers collected in 1984 and twenty-eight in 1988, netted from the Provo River, Utah, and held for artificial culture, were maliciously killed by vandals (R. Radant, Utah Department of Natural Resources, pers. comm.). These fish, preserved in formalin, were measured, and right opercles were removed and cleaned by scraping rather than boiling. Length-frequency measurements of other river fish indicated that the sample was representative of the adult population.

Results

Cui-ui

Historically, cui-ui inhabited two sister water bodies, Pyramid and Winnemucca lakes (J. O. Snyder 1917; LaRivers 1962), both of which received their primary water supply from the Truckee River (Fig. 18-2). As terminal sumps of a single stream, both lakes were sensitive to water manipulation and management, and by 1938 the diversion of the Truckee River for irrigation resulted in drying of the shallower (24 m) Winnemucca Lake (F. H. Sumner 1940). Similarly, the level of Pyramid Lake has declined by as much as 25 m (to a depth of approximately 116 m) between 1906 and 1989. The primary cause was diversion at Derby Dam, about 64 km upstream from Pyramid Lake. The first U.S. Bureau of Reclamation (USBR) irrigation project, the Newlands Project (completed in 1906), involved construction of a transbasin canal, which has diverted an annual average of 3.1×10^8 m^3 of water from the Truckee River (50% of the average flow). This water flows into Lahontan Reservoir, where it joins inflow from the Carson River to be used for irrigation (Scoppettone et al. 1986).

Life history

The cui-ui is the best known and most abundant lakesucker, and it is typical of the group in most life-history characteristics (Table 18-1;

Scoppettone et al. 1986). In spring, cui-ui congregate at the south end of Pyramid Lake near the mouth of the Truckee River. Initiation of spawning migration seems to be correlated with increasing water temperatures, and begins in early April through mid-May and ends in early to mid-June. Peak migrations have occurred at river temperatures varying between 9° and 17°C and mean daily temperatures from 12° to 15°.

Historically, cui-ui spawned in the lower 70 km of the Truckee River, the only perennial

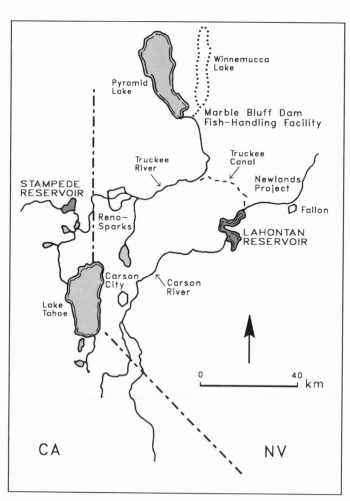

Fig. 18-2. The Truckee River–Pyramid Lake system, California-Nevada, and its artificial connection to the Carson River system.

Table 18-1. Comparisons of lakesucker life-history parameters.

Parameters	Cui-ui[1,2]	Lost River[3,4,5,6]	Shortnose Upper Klamath Lake[4,6]	Shortnose Copco Reservoir[5,7,8]	Shortnose June[9,10]
No. spawners ($\times 10^3$)	36	11.6	2	Unknown	< 0.5
Year estimated	1986	1985	1985		1984
Spawning habitat					
Depth (cm)	21–110	21–80	15–61	—	30–76
Velocity (cm s^{-1})	27–140	18–82	43–131	—	6–137
Substrate	gravel	cobble	cobble	cobble	cobble
Temperature (°C)	9–17	11–15	10–15	13–17	11–15
Fecundity ($\times 10^3$)	24–196	102–236	34–72	37–56	52–89
Age					
Maximum	45	43	25	33	42
At first reproduction	6–12	5–9	—	5–7	5–7
Maximum sizes					
Total length (cm)	70	100	—	52	55
Total weight (kg)	2.4	4.5	—	1.5	2.0

[1]Scoppettone et al. 1986
[2]Coleman et al. 1987
[3]Coleman et al. 1988
[4]Andreasen 1975
[5]Moyle 1976
[6]Bienz and Ziller 1987
[7]Beak Consultants 1987
[8]Coots 1965
[9]Radant et al. 1987
[10]John 1984

tributary to the lake (V. K. Johnson 1958). Recent data from radio-tagged fish indicate that river migration encompassed less than 10 of the 19 km now available, and adults remain in the river for 4 to 16.5 days (Scoppettone et al. 1986). Spawning occurs in relatively fast water (average 50 cm s^{-1}), at depths of 21–140 cm on predominantly gravel substrates, and individual fish may spawn several hundred times in 3 to 5 days (Scoppettone et al. 1983). Hatching success is highest at mean daily temperatures lower than 17°C (M. E.

Coleman et al. 1987), and embryonic development is temperature dependent (Koch 1973). At diel water temperature fluctuations from 14° to 17°C, eggs hatch in 10 days and larvae swim up after an additional 4 days (Coleman et al. 1987).

Most larvae move downstream at night. In 1982 and 1984, maximum larval densities were observed in the river twenty-six and twenty-nine days, respectively, after the peak of adult upstream migration (Scoppettone et al. 1986). Their mouths do not open until about sixteen days after hatching (Bres 1978), and emigrating larvae usually retain their yolk sacs. The timing of mouth opening corresponds with entry into the lake. Cui-ui have been observed spawning in Pyramid Lake itself (V. K. Johnson 1958; Koch 1973), but extreme alkalinity and elevated salinity preclude successful reproduction there (Chatto 1979).

Cui-ui have specialized food habits and consume primarily bottom-oriented zooplankton and macroinvertebrates such as ostracods, *Cyclops*, and chironomid dipteran larvae and pupae (Scoppettone et al. 1986; N. Vucinich, Pyramid Lake Fisheries, unpub. data).

Adult cui-ui are rarely encountered except during spawning migrations, but they have been gill-netted in Pyramid Lake in water less than 23 m deep (V. K. Johnson 1958; Vigg 1980). They may first spawn at six years of age (36–38 cm fork length [FL]), although most do not reproduce until eight to twelve years old (46–52 cm). Cui-ui have been aged to forty-one years (Scoppettone 1988). Mortality rates for males appear higher than for females, since females dominate older age classes. The sex ratio (females to males) was 1.3:1 in thirteen-year-old fish, 2:1 at eighteen years, and 5:1 at ages greater than thirty years.

Current status

Historically, cui-ui were an important food resource and an integral part of the Pyramid Lake Paiute Indian culture (Wheeler 1974). They were caught during spawning migrations, and many were dried for later consumption. In more recent times the population was exploited by both westerners and Indians, largely through snagging of fish from prespawning aggregations near the mouth of the Truckee River. Creel census data collected from 1954 to 1968 by the Nevada Department of Wildlife showed a precipitous decline in catch, from a high of 790 fish in 1955 to only 10 in 1968 (Koch 1976). Sport fishing

was eliminated in 1969 and the Indians voluntarily halted subsistence fishing in 1978. The primary predators on cui-ui then became white pelicans, which in 1982 killed or consumed several hundred fish (Scoppettone et al. 1986).

Clearly, the decline of the cui-ui resulted from a lack of successful reproduction and recruitment. During the drought of the 1930s the level of Pyramid Lake dropped rapidly, and a large, frequently impassable, delta formed at the mouth of the Truckee River (LaRivers 1962). By the early 1940s the famous Pyramid Lake strain of Lahontan cutthroat trout (*Oncorhynchus clarki henshawi*) had disappeared, probably due to reproductive failure (LaRivers 1962). In most years after the late 1930s, neither cui-ui nor cutthroat trout had been able to gain access to the river for spawning. Additionally, in many years virtually the entire spring runoff was diverted for agricultural use.

Cui-ui populations have greatly declined in the last fifty years (V. K. Johnson 1958; LaRivers 1962; Scoppettone et al. 1986). In 1983, 187,000 cui-ui were estimated in the prespawning aggregation. Examination of population age structure in that group, however, suggested the species was in jeopardy (Scoppettone et al. 1986) because about 92% were derived from the 1969 year class, and most of the remainder were hatched in 1950. There was apparently no significant recruitment between 1950 and 1969 (Fig. 18-3).

Before this alarming situation was recognized, cui-ui were thought to be reproducing in springs and other freshwater interfaces of Pyramid Lake (Koch 1973), and the absence of young from lake samples was attributed to their secretive behavior (V. K. Johnson 1958; Wheeler 1974). Before 1984, only one fish smaller than 35 cm FL had been collected, a 32-mm standard length specimen obtained in 1938 (R. R. Miller, University of Michigan, pers. comm.).

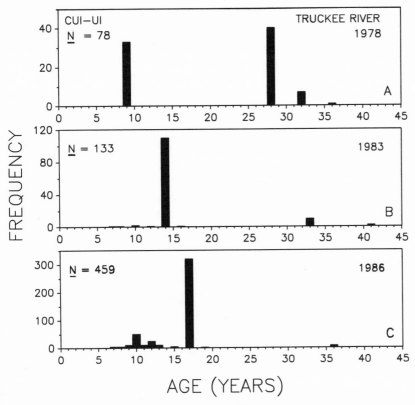

Fig. 18-3. Cui-ui age structure, derived from fish collected during spawning periods: (A) Fish collected in 1978 from the prespawning aggregation in Pyramid Lake; and (B and C) fish collected from spawning migrations.

Management

In response to the lack of cui-ui spawning success, Marble Bluff Dam on the Truckee River and a 5.1-km-long channel were completed in 1976 to allow fish to bypass the delta. The dam maintains flow in the bypass channel and serves as a hydraulic control to reduce erosion caused by declining lake level (Scoppettone et al. 1986). The bypass channel contains four modified, Ice Harbor–type fish ladders, through which a flow of 1.1 m³ s⁻¹ is maintained during the spawning season. A fish-handling building is located at the upstream end of the channel, adjacent to Marble Bluff Dam. Fish gain access to the river above the dam through this channel, or they may be collected in a trap at the base of the dam and transferred over the structure. The 4-km reach of river downstream of the dam is extremely erosive. Its degradation coupled with elevated lake level essentially obliterated the infamous delta from 1984 through 1987, and cui-ui entered the river trap during this time period.

Most fish are allowed to continue their upstream migration; 100–300 females are spawned artificially, with appropriate numbers of males, and fertilized eggs are reared in a hatchery operated by the Pyramid Lake tribe. Most larval fish from the hatchery are released shortly after swim-up. Between 1980 and 1987 the hatchery annually released approximately eight million larval cui-ui into the lower Truckee River and Pyramid Lake (S. Cerocke, Pyramid Lake Tribal Fisheries, pers. comm.). In 1986 facilities for extended rearing were constructed at Sutcliffe, Nevada, with the intent of releasing larger fish in the future.

In 1984 Pyramid Lake Fisheries tribal biologists began collecting young cui-ui (10–35 cm FL) in gill-net samples in Pyramid Lake (L. Carlson, Pyramid Lake Tribal Fisheries, pers. comm.). The appearance of juveniles was an encouraging sign, and several new age classes have appeared in the spawning run since 1983. In 1978, 59% of the fish in the spawning run were from the 1950 year class, 41% were from the 1969 year class, and a small remainder were born before 1950 (Fig. 18-3A). In 1983 the 1969 year class comprised 85%, the 1950 year class or older made up 10% of the run, but approximately 4% were fish hatched during the 1970s (Fig. 18-3B). In 1986 the 1969 year class comprised 70% of the run, while 28% consisted of younger fish (Fig. 18-3C).

This change in relative proportions reflects both the results of successful recruitment in the 1970s and accelerated mortality within the 1969 year class. Although the additional year classes do not appear particularly strong, they are the only recorded additions to a population that had virtually no recruitment for eighteen years. It is clearly exceptional for any species to tolerate such an extended period without successful reproduction, and without its longevity the cui-ui would have been driven to extinction. The combination of hatchery propagation and the wet years of 1983–1986 reduced the precariousness of their situation.

Water demands in the Truckee River basin are inexorably increasing, and there is continuing concern about maintaining the limited supply necessary to preserve the cui-ui. Extended drought would almost certainly affect it adversely (Buchanan and Strekal 1988). Water quality in the Truckee River may continue to be a problem because of an expanding urban population. Only one non-native fish species, Sacramento perch (*Archoplites interruptus*), has successfully invaded the highly alkaline environment of Pyramid Lake (Galat et al. 1981), and its population has recently declined.

The threat of increased salinity is another concern arising from mass water diversion from the Truckee River. Galat and Robinson (1983) predicted that an increase in salinity could decrease crustacean zooplankton diversity and change community structure, as occurred in Walker Lake, Nevada (Koch et al. 1979), perhaps negatively affecting the forage base for cui-ui. The ownership of water rights in the Truckee basin has been in dispute and subject to litigation for many years (Knack and Steward 1984). Negotiations continue between agricultural and municipal water users, the Pyramid Lake tribe, and the U.S. government. Until water issues are settled, ambiguity clouds the future of the cui-ui.

Lost River Sucker

Lost River suckers are endemic to the Klamath and Lost River drainages of the Klamath Basin, Oregon and California (Fig. 18-4). Their historical distribution is uncertain, but Gilbert (1898) stated that they primarily occupied deep waters of Tule and Upper Klamath lakes. The most substantial tributary feeding Upper Klamath Lake is the Williamson River, which receives most of its flow from a series of cold-water springs 25 km north of

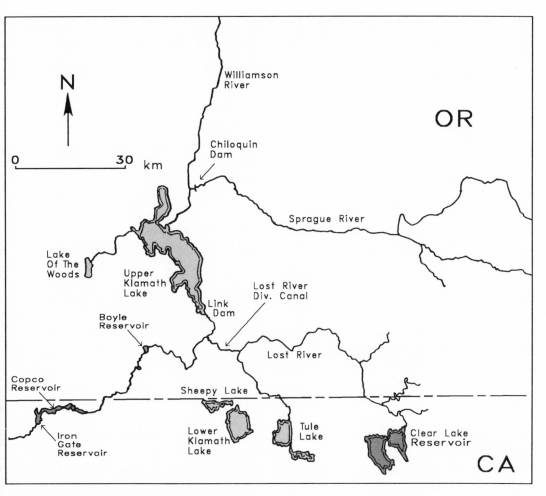

Fig. 18-4. Map of upper Klamath River basin and Lost River system, California-Oregon, and the transbasin canal connecting the two.

the lake. Its major tributary, Sprague River, extends 80 km eastward. These streams join approximately 17 km from their entrance into the lake.

There has long been concern regarding the decline of Lost River suckers in Upper Klamath Lake (Vincent 1968; Andreasen 1975). Angler catch fell from more than 10,000 fish in the Williamson and Sprague rivers in 1968 (Golden 1969) to only 630 in 1985 (Bienz and Ziller 1987). An estimated

11,900 entered the Williamson and Sprague rivers to spawn in 1985, a number much reduced from those reported historically (Golden 1969). Lost River suckers were once sufficiently abundant in Tule Lake to support a commercial fishery and serve as an important food resource for the Modoc and Klamath Indians (Howe 1981).

A 1975 survey of the system indicated the only nonhybridizing Lost River suckers remaining were in a small population in Clear

Lake Reservoir (Contreras 1973; Koch et al. 1975). In other sections of the river they hybridized with shortnose and Klamath large-scale suckers (*Catostomus snyderi*; Moyle 1976). The present status of the fish in this system is unknown, but it is unlikely to have improved since 1975. Lost River suckers have colonized J. C. Boyle Reservoir, Oregon, and Copco Reservoir, California, in the Klamath River system. Both populations are thought to be small, and their reproductive status is unknown (Moyle 1976).

Life history

Little information exists regarding the life history of Lost River suckers; that available is from Upper Klamath Lake (Table 18-1). The fish spawn in Sprague River and at several spring inflows in Upper Klamath Lake (Fig. 18-4). Upstream migration appears to be triggered by warming water temperatures (Golden 1969); spawning migrations begin in the Sprague River in March and April. In 1987 Lost River suckers spawned over compacted cobble in velocities between 18 and 82 cm s^{-1} at depths varying from 21 to 80 cm. They preferentially spawned on loose gravel when it was available. Fecundity estimates range from 101,000 to 235,000 eggs per female (Golden 1969; Andreasen 1975). Larvae moved downstream at night immediately after swim-up, with maximum numbers recorded about twenty-one days after peak spawning (M. E. Coleman et al. 1988).

Limited data suggest that adults consume primarily bottom-oriented macroinvertebrates and zooplankton (Coleman et al. 1988). The maximum life span is at least forty-three years (Scoppettone 1988). Examination of the age structure of fish from Upper Klamath Lake in 1986 indicated that 95% were nineteen to thirty years old (Fig. 8-5), and there had been little successful recruitment over the past fifteen years. Spacing of opercular annuli suggest that fish first become reproductive at six

to seven years of age, with most individuals doing so at eight or nine years.

Current status

Upper Klamath Lake and its watershed have been greatly affected by agriculture, and water use has reduced volume and changed flow patterns into the lake. Chiloquin Dam, constructed in 1928, blocks upstream migration and precludes accumulation of suitable spawning gravels below the dam. Open diversions and pump stations may entrain larvae in their downstream migration, and the lake itself has been markedly altered.

In 1912 Link Dam (2.5 m high) was constructed across the outlet of Upper Klamath Lake, and the system has since been operated as an irrigation reservoir. Since construction, sedimentation has greatly increased, nutrient loading has risen through return of agricultural water (U.S. Army Corps of Engineers 1979), and available spawning habitat has been markedly reduced. Increasingly eutrophic conditions are indicated by summer blooms of the blue-green alga *Aphanizomenon*, and at peak densities pH may reach 10.5 (Vincent 1968; Coleman et al. 1988). When algal populations crash, local depletion of dissolved oxygen occurs. *Aphanizomenon* densities were high in summer 1986 and may have contributed to fish mortality then, as has been suspected in the past (Vincent 1968). Measurements after the bloom collapsed in 1986, and shortly after a major fish mortality, indicated that dissolved oxygen may have been locally absent from water deeper than 2 m.

Introductions of non-native fishes may have affected the Upper Klamath Lake population. Yellow perch (*Perca flavescens*) were stocked in the 1930s (J. Ziller, Oregon Department of Fish and Wildlife [ORDFW], pers. comm.) and are now abundant, while fathead minnows (*Pimephales promelas*) have become the dominant fish since their introduction in about 1980 (Bienz and Ziller 1987).

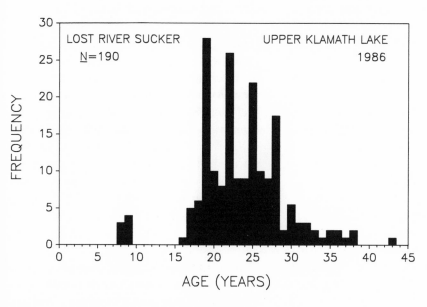

Fig. 18-5. Lost River sucker age distribution in 1986, derived from fish collected from Upper Klamath Lake, Oregon.

Spawning habitat has been reduced to less than 4% of that historically available. Inflow from the spring-fed Williamson River is probably too cold (5°C) for sucker reproduction (Golden 1969). Temperatures in Sprague River are far less extreme, but only 1.6 km of channel now exists below Chiloquin Dam. Spawning substrate below the dam is of poor quality, consisting mostly of armored, cobble bottom. Although the dam has a fish ladder, it is virtually unused by suckers.

General alteration and destruction of aquatic habitats in upper parts of the system also must have had an impact. One of the earliest USBR projects (the Klamath Project, authorized in 1902) involved draining Tule and Sheepy lakes for agriculture and constructing a transbasin canal linking the Lost and Klamath rivers (Fig. 18-4). During high water, flow from the Klamath system entered Lost River through Lost River Slough and eventually discharged into Tule Lake (formerly called Rhett Lake; J. O. Snyder 1908). Lost River Slough was dredged to form a canal uniting the two systems in 1912. Both Lost River and shortnose suckers may have entered the Lost River system via this route, either through the natural connection or following construction. Also as part of that project, the outflow of Clear Lake (headwaters of the Lost River system) was dammed to create an irrigation reservoir. Both Tule and Sheepy lakes were deemed unsuitable for agriculture after draining and were allowed to partially refill with irrigation runoff.

Management

In 1983 the Klamath Indian Tribe, the ORDFW, and the USFWS began studies of Lost River, shortnose, and Klamath largescale suckers. In 1986 the Klamath Basin Interagency Working Group (KBIWG) was established for

preservation of Lost River and shortnose suckers. This group includes representatives of the
Klamath tribe, ORDFW, USFWS, California Department of Fish and Game, U.S. Forest Service, and USBR. Concern about the decline of
Klamath Basin suckers prompted Klamath Indian tribal members to curtail their catch in
1985, and in 1987 the state of Oregon closed
the sport fishery.

Klamath suckers are an integral part of the
culture of the Klamath Indians (C. Kimball,
Klamath tribal chairman, pers. comm.), and
the tribe has developed a hatchery and rearing
facilities as a short-term measure to preserve
them. The hatchery program is intended to
maintain the genetic integrity of both species
until the specific causes of their decline are
determined and the problems are remedied.

Shortnose Sucker

Historically, shortnose suckers were known
only from Upper Klamath Lake and Lake of
the Woods, Oregon. They also may have occurred in the Lost River system, but this is
undocumented by specimens (Moyle 1976).
Putative hybrids of unknown origin in the
Lost River system (Koch et al. 1975; Moyle
1976; R. R. Miller and Smith 1981) may have
originated from either natural or postmodifi-
cation dispersal (see above).

The taxonomic status of shortnose suckers
is unclear throughout the Klamath Basin. Andreasen (1975) stated that some "pure" (unhybridized) shortnose suckers remain in Upper
Klamath Lake; however, R. R. Miller and
Smith (1981) suggested that no pure individuals exist. Hybridization has clearly taken place
in Upper Klamath Lake, where approximately
a third of the shortnose suckers appear to be
hybrids with either Lost River or largescale
suckers. Fish resembling shortnose suckers
nonetheless remain (J. Ziller, ORDFW, pers.
comm.), and it was estimated that in 1985,
about two thousand apparently pure short-

nose suckers remained in the spawning run
(Bienz and Ziller 1987).

Putative hybrids involving shortnose suckers also occur in artificial impoundments (J. C.
Boyle and Copco reservoirs) downstream
from Upper Klamath Lake. The largest of
these populations, consisting of putative hybrids with Klamath smallscale suckers (*Catostomus rimiculus*; R. R. Miller and Smith
1981) is in Copco Reservoir. The KBIWG has
placed highest priority on determining the
taxonomic status of the shortnose sucker
(J. E. Williams, U.S. Bureau of Land Management, pers. comm.).

Life history

The shortnose sucker is the smallest in body
size and appears to have the shortest life span
among the lakesuckers (Table 18-1). They seldom exceed 52 cm FL and 1.5 kg (Moyle
1976). The oldest specimen known was a
thirty-three-year-old hybrid collected from
Copco Reservoir in 1987 (Scoppettone 1988).
A small amount of data on fish in Upper
Klamath Lake suggest that they feed almost
exclusively on cladocerans (Coleman et al.
1988). Two 49-cm FL shortnose suckers produced 36,000 and 57,000 eggs (Coots 1965).

In Upper Klamath Lake, shortnose suckers
begin migrating to spawn as water temperatures warm in April and May—generally later
than Lost River and largescale suckers (C.
Bienz, Klamath tribe biologist, pers. comm.).
In 1988 the peak spawning migration of
largescale suckers was in March, Lost River
suckers in April, and shortnose suckers in
May (Coleman et al. 1988). In 1987 shortnose suckers migrated from Copco Reservoir
up the Klamath River in late April (Beak Consultants 1987); spawning was recorded in the
lower 4.8 km. Cope (1884) was informed by
an Indian chief that *xooptu* (the Indian name
for shortnose sucker) did not ascend the Williamson River with Lost River and largescale
suckers.

Current status

We examined nineteen shortnose suckers from Copco Reservoir in 1987 that were sixteen to thirty-three years old, with a mean age of twenty-three years (Fig. 18-6). There was no evidence of recent recruitment. Extensive gill-netting and electrofishing in 1987 also suggested that juveniles were rare or absent (Beak Consultants 1987). During the 1986 fish kill in Upper Klamath Lake, only seven shortnose suckers (ages four to twenty years) were found.

Management

Because shortnose suckers are generally sympatric with Lost River suckers, they are subject to similar habitat destruction and influences from introduced fishes. Management practices for shortnose suckers have paralleled those for Lost River suckers. In the upper Klamath Lake basin, harvest has been curtailed and the Klamath Indians have developed a hatchery. Research is being conducted on Upper Klamath Lake in an attempt to determine factors that limit populations.

June Sucker

Life history

June suckers spawn in the Provo River, the primary source of water for Utah Lake (Fig. 18-7). Flows are controlled by Deer Creek Reservoir and by agricultural diversions, which restrict spawning to 6.1 km of the lowermost river. Water manipulation virtually dewatered the stream during past spawning seasons (Radant et al. 1987).

The lacustrine habitat has also been greatly altered. Utah Lake is now operated as a reservoir, and water levels fluctuate extensively. It is increasingly eutrophic and saline (Fuhriman

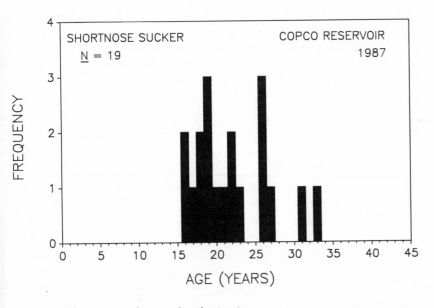

Fig. 18-6. Shortnose sucker age distribution in 1986, derived from fish collected from Copco Reservoir, California.

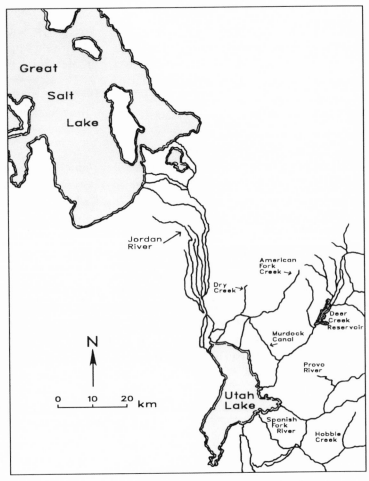

Fig. 18-7. The Provo River–Utah Lake–Great Salt Lake system, Utah.

et al. 1981), and introduced fishes have virtually replaced the native fauna. Catostomids were once the dominant fishes, forming the base of a major fishery (Carter 1969); today they comprise only a small fraction of total biomass (Radant and Sakaguchi 1981; Radant and Hickman 1985). Common carp (*Cyprinus carpio*) constituted almost 93% of the total biomass in 1981, while more than 5% was white bass (*Morone chrysops*), walleye (*Stizostedion vitreum*), and black bullhead (*Ameiurus melas*). Native fishes comprised less than 1% of the biomass.

The spawning migration of June suckers up the Provo River typically occurs in the month for which the species is named. Among the several hundred entering the stream from 1979 to 1987, spawning was observed annually over a five- to eight-day period, on substrates varying from coarse gravel to small cobble, and at water temperatures ranging from 11° to 15°C (Table 18-1). Depths at spawning sites were 30–76 cm, and water velocities were 6–137 cm s^{-1} (Shirley 1963; Radant and Hickman 1985; Radant et al. 1987). Larvae moved downstream immedi-

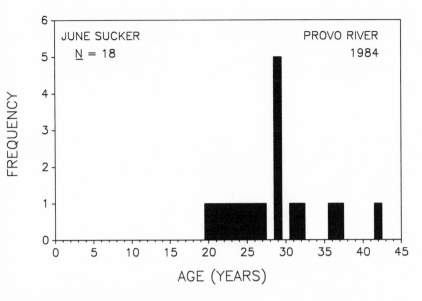

Fig. 18-8. June sucker age distribution in 1984, derived from fish removed from the Provo River spawning migration.

ately after swim-up (N. Muirhead, Brigham Young University, pers. comm.). The branched, filamentous gill rakers of June suckers suggest that they are plankton feeders (R. R. Miller and Smith 1981), although little is known of their food habits.

During a drought in the 1930s, Utah Lake was nearly drained to accommodate irrigation needs. Fishes were crowded, and a freeze caused mass mortality (V. M. Tanner 1936). The June sucker subsequently hybridized with more abundant Utah suckers (*Catostomus ardens*). Since the 1950s, closely following introductions of white bass and walleye, the numbers of all forms of pure and hybrid June suckers have declined. Predation by the introduced species may have contributed to the decline (Radant et al. 1987).

Current status

The total population of June suckers may be fewer than a thousand (Radant et al. 1987).

Eighteen fish collected in 1984 were between twenty and forty-two years of age (Fig. 18-8). Thus, June suckers are probably near extinction, and no genetically pure individuals are believed to exist. The introgressed form was designated a new subspecies, *Chasmistes liorus mictus*, by R. R. Miller and Smith (1981).

Management

A management plan completed in 1984 suggested that the Utah Lake ecosystem has been so highly altered that restoration of a self-sustaining population of June suckers may not be feasible (Radant 1984). Artificial culturing is planned to prevent extinction. Stocking hatchery-reared fish into Utah Lake at a large size (>15 cm) would enhance their probability of survival. An alternate site containing suitable habitat and located within the Bonneville Basin is being sought as a refuge. In 1987, 208 juveniles were stocked into Camp Creek Reservoir, Utah, and several survivors were netted a year later. Only 2 ha in surface area, with an inlet stream that may be prohibitively small for reproduction, this reservoir's only utility may be in maintaining a brood stock (D. Knight, Utah Division of Wildlife Resources, pers. comm.) or serving as a rearing area for juveniles to be stocked elsewhere. It may nonetheless be an important, immediate link in the future for survival of the species. Meanwhile, research has begun into the life history of the June sucker to determine the causes of its decline.

Summary and Discussion

Comparison of life history patterns in lake-suckers disclosed striking similarities among species (Table 18-1). This specialized group of North American catostomids consists of fecund, late-maturing, long-lived fishes that are all potadromous spring spawners in fast, cool water, and over similar substrates. Larvae emigrate into lakes immediately after swim-up. Each species has suffered a significant amount of habitat degradation, and all populations have been drastically reduced.

Within this group, all species have survived nearly complete interruptions of recruitment for periods of ten to twenty years. Consequently, all species have declined, most to precariously low levels; the remaining fish are quite old. Lost River suckers in Upper Klamath Lake and shortnose suckers in Copco Reservoir have undergone ten-year periods without significant recruitment. Utah suckers have not successfully reproduced since the mid-1960s, and the cui-ui had no known recruitment from 1950 to 1969. This same phenomenon has been observed in the razorback sucker in Lake Mohave, Arizona-Nevada (Minckley 1983; McCarthy and Minckley 1987; Minckley et al., *this volume*, chap. 17), and in Green River, Utah-Colorado (Lanigan and Tyus 1989), for even longer periods of time (more than twenty to thirty years).

Environmental and ecological changes potentially responsible for recruitment failure include the following: (1) reduced habitat quantity and quality in spawning areas (Table 18-2), (2) changed discharge patterns in spawning habitats, (3) decreased water quality, (4) hybridization, (5) competition and predation from non-native species (Table 18-3), and (6) overharvest. Many of these problems can be linked directly to water-management practices associated with agricultural and urban development.

Most adverse conditions developed over relatively long periods, and their impacts were unappreciated. All the lakesucker populations were historically large, and their exceptional longevity, plus slow rates of growth after maturation, masked changes in population structure. Amelioration of these conditions poses

Table 18-2. Historic and present spawning habitat relations for western lakesuckers.

Species	Spawning site(s)	Available stream km		
		Historic	Today	% remaining
Cui-ui	lower Truckee River, NV	80	19.0	24.0
Shortnose and Lost River suckers	Sprague/Williamson rivers, OR	48	1.6	3.3
June suckers	Provo River, UT	32	6.1	20.0

Table 18-3. Relative abundance of native and non-native fishes in three lakesucker habitats.

Habitat	Native species			Non-native species	
	Historic no.	Present no.	% relative abundance	No.	% relative abundance
Pyramid Lake	6	6	99.7	4	0.3
Upper Klamath Lake	10	10	15.5	5	84.5
Utah Lake	5	4	0.4	14	99.6

difficult problems because of the decades-long development of each situation. Declines of all the lakesuckers were well along before they were identified, and it has taken more years to assess their magnitude. In each instance there was a similar suite of environmental perturbations, and similar responses of the lakesucker populations—a failure to recruit.

Populations of June suckers, shortnose suckers in Copco Reservoir, and Lost River suckers in Upper Klamath Lake consist entirely of ever-diminishing numbers of very old adults. These populations are in clear and imminent danger of extinction. The cui-ui provides an exception to this pattern, we hope, because of extensive management efforts. It too experienced an extended period without recruitment, but there was modest recruitment from fish hatched in the 1970s, and numerous juveniles were captured in the 1980s.

The most important long-term measures necessary to perpetuate the cui-ui include providing sufficient water of adequate quality. Enough water must be maintained in the Truckee River to allow cui-ui access to spawning grounds during spring migrations, and it must remain at the appropriate temperature and chemistry for successful hatch and survival of young. Water quality in Pyramid Lake must remain adequate to support a suitable food base for growth of juveniles and maintenance of adults.

Buchanan and Strekal (1988) developed a population model for cui-ui that related spawning success to Truckee River flow characteristics. This model has recently been used by the USBR to develop new operational criteria and operating procedures (OCAP) for implementation by water users. A primary consideration in this program's development was survival of cui-ui, and the proposal would increase flows into Pyramid Lake to benefit the species. However, the Truckee-Carson Irrigation District is contesting OCAP implementation on the grounds that it will increase the frequency of water shortages.

There is also concern about the quality and quantity of water to be supplied to Stillwater National Wildlife Refuge at the terminus of the irrigation system. This concern is based on unacceptable amounts of materials such as mercury, selenium, and boron leached from irrigated soils. As it now stands, irrigation water is removed from the Truckee River to the detriment of the native fishes. This water accumulates pollutants, which are discharged into the refuge, where waterfowl and other wildlife are adversely affected. An obvious solution would be purchase of water rights in the Newlands Project to ensure an adequate volume to maintain Pyramid Lake and also to provide sufficient high-quality water for the refuge.

As an advocate of the new OCAP, the

Pyramid Lake Paiute tribe contends that implementation is necessary for survival of the cui-ui. Interested parties in the dispute are involved in negotiations attempting to resolve the issue. However, even with resolution of the OCAP, threats to cui-ui will continue. Expanding urban water demands in Reno, Nevada, will continue to threaten the allocation of upstream reservoir waters.

In 1980 a U.S. federal court ruled that water from Stampede Reservoir in the upper Truckee Basin would be used to promote recovery of the endangered cui-ui and threatened Lahontan cutthroat trout. But what if recovery is judged successful and both fish are delisted? Reallocation of water from Stampede Reservoir to other purposes would be a devastating blow. Such a change in status must not be allowed to jeopardize continued water availability. The resolution of this problem requires that water from Stampede Reservoir remain committed to fisheries, regardless of cui-ui population status. The presence of Lahontan cutthroat trout constitutes another variable in discussions of water allocation. A desirable game fish, this species is also important in the ecology and management of the system, and the trout has more advocates than do cui-ui. Fortunately, water-management practices that enhance either of these fishes should benefit the other.

The other major threat to cui-ui is deterioration of Pyramid Lake itself. This problem is again clearly related to water supply. The factors most detrimental to reproduction and survival are increased levels of dissolved solids, especially various inorganic nitrogen compounds.

Human population growth in the Reno area has significantly altered water quality in the Truckee River. The Reno-Sparks Municipal Wastewater Treatment Facility discharges about 15,000 m³ of treated effluent into the stream each day. Although the plant is equipped with tertiary treatment capacity that removes much of the nutrient content, the total volume still represents considerable loading. Some have proposed that effluents be used in land application to counteract impacts of nutrient loading, with little or none directly entering the river. Although the effluent quality might limit its potential uses, such application could substitute for agricultural withdrawal and improve water quality in the lower river. Another factor of potentially great but unknown importance is the possible importation of water by Sierra Pacific Power Company. Such efforts are likely to increase in future years, and it is impossible to predict the outcome. Although cui-ui have the most viable population of any lakesucker, their survival is thus contingent upon the favorable resolution of pending water-supply issues. They cannot be considered secure.

Unlike the situation with cui-ui, the problems facing Klamath River basin suckers are not at all clear-cut. The specific causes of decline and necessary remedies for recovery have not been identified. Clearly there have been extensive disruptions of both lake and river habitats, and such changes are immediately suspect. The current hatchery program (M. Coleman, Seattle National Fishery Center, pers. comm.) has only recently begun, is limited in scope, and must be regarded as a short-term measure against extinction.

The most evident action to promote reproduction is modification or elimination of Chiloquin Dam. That structure reduces access to spawning habitat and is an impediment to survival of Lost River and shortnose suckers. Removal or alteration of the dam to allow fish passage would also enlarge the potential spawning habitat by at least a factor of fifty, which may be more efficient and cost-effective than any hatchery program. Water quality in Upper Klamath Lake is poor; it is suspected that sucker survival is directly affected by low dissolved oxygen and high pH, at least, and there is considerable fear that the lake will continue to deteriorate. The cause of hyper-eutrophication needs investigation. Sources and impacts of nutrients must be identified, along with impacts of the extensive destruction of marsh habitat and damming of its outlet.

Reservoir populations of native suckers are similarly declining throughout the Klamath River basin. Less information is available about the causes of these declines than for those in Upper Klamath Lake, and many more data must be accumulated before management strategies are formulated.

June suckers are precariously near to extinction. They remain only as a rapidly shrinking and aging remnant population, without recent successful reproduction. This demographic observation, combined with the overwhelming dominance of non-native fishes in Utah Lake and current water-management practices, may preclude their survival in nature. The situation in Utah Lake is similar to that of the Klamath Basin—there is virtually no information on factors responsible for the decline of the fish. Again, there have been substantial reductions in spawning habitat and major alterations in limnological characteristics of the lake. It is imperative that hatchery propagation be vigorously pursued to avert extinction. Preservation of the species requires intensive short-term efforts to save a viable population, and long-term efforts to effect its recovery (Radant and Hickman 1985).

Conclusion

The outlook for lakesuckers is bleak; saving them will be challenging. Recent attempts to artificially spawn Lost River and shortnose suckers, and increasing research being directed at these species, are positive steps toward ensuring their survival. Clearly, an intensive and sustained effort will be required. The June sucker population may have declined so low that extraordinary efforts will be required to avert extinction. Sustained research efforts and hatchery rearing programs must be put into place.

The growing human populations of the western United States, and their apparently ever-increasing abilities (and proclivities) to disrupt existing ecosystems, have produced a conflict between fishes and humans that will not be resolved in favor of the fishes unless we choose to make it so. It will be a sad commentary on management of fisheries resources when lakesuckers are eliminated and their survival is ensured only through hatchery propagation, and even sadder if we permit them to spiral to extinction.

Of the five species of lakesuckers present when Europeans entered the scene, one is extinct, one no longer exists in its original form due to hybridization, two are clearly at imminent risk of extinction, and only one has had significant, successful reproduction in the last two decades. Without our commitment to their preservation, this unique group of western fishes is truly in peril of vanishing as we watch.

Chapter 19

Ecology and Management of Colorado Squawfish

Harold M. Tyus

Introduction

The Colorado squawfish (*Ptychocheilus lucius*) is the largest minnow (family Cyprinidae) native to North America, reaching total lengths of 1.8 m and weights of 36 kg (R. R. Miller 1961). Popular names of "salmon," "white salmon," "whitefish" (Evermann and Rutter 1895), and "Colorado River salmon" (Measeles 1981) were applied to this attractive species (Fig. 19-1), reflecting its large size, body shape, coloration in springtime, and migratory nature. American Indians and early settlers knew about squawfish and used it for food (Minckley 1965, 1973; Seethaler 1978). Ichthyologists of the later nineteenth century documented its widespread distribution, and it was reported common in most larger rivers of the Colorado River system from Wyoming to Mexico (Girard 1856; Evermann and Rutter 1895; D. S. Jordan and Evermann 1896b, 1923; R. R. Miller 1961).

Although squawfish were common in the lower Colorado River basin in Arizona and California until the 1930s (R. R. Miller 1961), records were few after then, and efforts to collect them there in the mid-1960s met with failure (Minckley 1973). By the early 1970s the species was essentially gone from the lower basin (Moyle 1976; Minckley 1985b). It fared better in the upper basin in Colorado and Utah (Fig. 19-2) and now exists in largest numbers in the Green River sub-basin (Holden and Wick 1982; Tyus et al. 1982a, 1987; Archer et al. 1986). It was listed as an endangered species by the U.S. Bureau of Sport Fisheries and Wildlife (now U.S. Fish and Wildlife Service [USFWS]) in 1967 (*Federal Register* 32[43]:4001) and received protection under the Endangered Species Act (ESA) of 1973 (U.S.C. 1531, et seq.) in early 1974 (*Federal Register* 39[3]:1175).

Historical information on the Colorado squawfish is largely taxonomic and distributional, and little was known of its life history until the 1960s (R. R. Miller 1964a). Interest in extant populations increased at that time, and pre- and postimpoundment studies of Flaming Gorge Reservoir began under section 8 of the Colorado River Storage Project Act (Binns et al. 1963). These studies, conducted by various governmental agencies (Bosley 1960; McDonald and Dotson 1960; Azevedo 1962a, b, c; Binns 1967a, b) and graduate students at Colorado State and Utah State universities (Banks 1964; Vanicek 1967; Holden 1968, 1973; Seethaler 1978), contributed much toward early understanding of the biology of squawfish and other endemic Colorado River fishes.

The environmental movement, energy crisis, and water-development issues of the 1970s stimulated intensified research on rare and imperiled fishes. This was reflected by a sudden

Fig. 19-1. Juvenile Colorado River squawfish, about 20 cm TL, from the Green River, Utah. Photographs by H. M. Tyus, summer 1986.

proliferation of literature (reviewed by Joseph 1977; Joseph et al. 1977; Wydoski et al. 1980; Ferriole 1987; USFWS 1989a) and special symposia (e.g., W. H. Miller et al. 1982c; Spofford et al. 1980). By the late 1970s an increasing need for information to evaluate water-resource development under the ESA resulted in formation of the Colorado River Fish Project (CRFP), an interagency cooperative effort involving three field research stations (W. H. Miller et al. 1982d; Shields 1982; Wydoski and Hamill, *this volume*, chap. 8). The CRFP served as a focus for research activities in the upper Colorado River basin for the next decade.

Attempts at management of Colorado squawfish and other endemic fishes by implementation of flow-related and non-flow-related measures are recent developments and

center largely on authorities pursuant to section 7 of the ESA (Lambertson 1982). Squawfish management in the upper Colorado basin is now centered on a multiagency consortium coordinated by the USFWS (1987a), and there is optimism for its recovery by the end of this century (USFWS 1987a; Rose and Hamill 1988; Wydoski and Hamill, *this volume*, chap. 8). However, little information on the planned activities, or their ecological bases, has been published.

This chapter summarizes current knowledge of the life cycle, status, and management options for Colorado squawfish, using published and previously unpublished data. Much of the background information was developed from unpublished agency reports that contain important information not otherwise available. Emphasis is placed on management to-

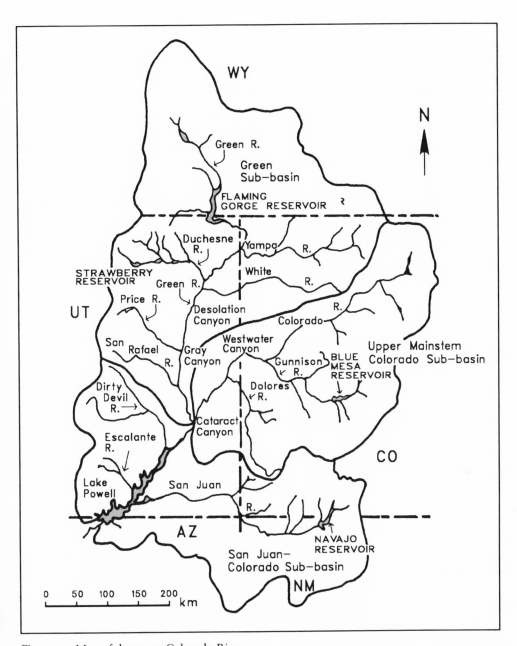

Fig. 19-2. Map of the upper Colorado River
basin, showing sub-basins and river systems
mentioned in text.

ward recovery of the species in the Green River sub-basin, where it is most widely distributed and abundant, and its ecology is best known.

Background Information

Management options for Colorado squawfish were difficult to develop because its life cycle was so poorly understood. Remote habitats and difficulty in collecting data in the swift, canyon-bound, hostile Colorado River basin also were factors. As late as 1978, the *Colorado Squawfish Recovery Plan* (USFWS 1978b) stated that spawning migrations had not been reported in ten years.

Knowledge of spawning areas, nursery habitats, and concentration areas was almost nonexistent, and they remained poorly documented until the 1980s (Tyus et al. 1981, 1982b, 1987; Wick et al. 1981, 1983, 1985; Haynes et al. 1984; Tyus and McAda 1984; Archer et al. 1985, 1986; Tyus 1985, 1986; Nesler et al. 1988), when new and better technology and sufficient funds enabled accelerated research. Relative densities for adults, juveniles, and young have now been mapped, and management strategies have been drafted (Archer et al. 1986; USFWS 1987a). Behavior and responses of the fish to environmental factors have been investigated in hatcheries (Toney 1974; Hamman 1981, 1986, 1989) and laboratories (Berry and Pimentel 1985; Black and Bulkley 1985a, b; Marsh 1985; Pimentel et al. 1985; Karp and Tyus 1990). In addition, efforts were made to accumulate habitat information on endangered Colorado River fishes, including squawfish, for use in stream-flow recommendations (W. H. Miller et al. 1982b; Valdez et al. 1987; Tyus and Karp 1989).

The Colorado squawfish has a complex life cycle. Coevolving as it did with the depauperate ichthyofauna of the Colorado River, this large, predaceous fish is a generalist adapted to large seasonal water fluctuations, low food

bases, and changing riverine subsystems (M. L. Smith 1981; G. R. Smith 1981b; Tyus 1986). Cyprinid fishes invaded the New World from Asia in the Oligocene or early Miocene (Cavender 1986), and fossil *Ptychocheilus* are reported from as early as Miocene times (Uyeno and Miller 1965; G. R. Smith 1981a; Minckley et al. 1986). We must assume that the fish survived by incorporating life strategies to deal with changing climates varying from pluvial to arid, using migration and long-distance movement for exploiting changing habitats and environmental conditions of the late Cenozoic, and developing adaptations that enabled it to compete and survive until recent times (G. R. Smith 1981a, b; Tyus 1986). As a result, adult, juvenile, and young Colorado squawfish acquired different life-history attributes for survival, and they adapted to utilize virtually every habitat available.

The adaptability of Colorado squawfish must form the basis for their management toward recovery. However, habitat preferences are difficult to ascertain in nature, and fish distributions and abundances are typically used to infer them. As an example, the USFWS has used densities of different life-history stages in designating "sensitive areas" to guide protective and management measures (Archer et al. 1986; USFWS 1987a). In the following sections particular emphasis is placed on statistics for abundance, since the development of appropriate management objectives and the success of any management tool can only be judged on the basis of population response. Detailed descriptions of specific habitat parameters are outside the scope of this presentation, but references are supplied where needed.

Life Cycle

Longevity, age, and sexual maturity

Colorado squawfish are capable of living long lives and attaining large sizes. Adult squawfish were the largest predaceous fish in the Colorado River, and large size presumably was an important attribute with apparent selective advantages. Thus, a long growth period accompanied by delayed reproduction would favor large adults. Rinne et al. (1986) cited slow growth (to 50.8 cm total length [TL] in nine years) and long life in hatchery-reared Colorado squawfish, and speculated that a 1.8-m specimen might be fifty or more years of age. In the Green River, Seethaler (1978) indicated that Colorado squawfish less than 42.8 cm TL were immature, and those more than 50.3 cm were mature.

Work on spawning grounds from 1981 to 1988 found 14 ripe females (average 65.4 cm TL) and 194 ripe males (55.5 cm TL; Tyus and Karp 1989). These observations supported the earlier work of Vanicek (1967), who speculated that females grew larger and perhaps older than males. However, it is probable that females do not remain ripe (with expressible sex products) as long as males do, and unripe fish captured during the spawning season may also be females. If all fish captured are divided on the basis of ripeness, and secondary sex characteristics such as presence of dense nuptial tubercles in males (or their absence in

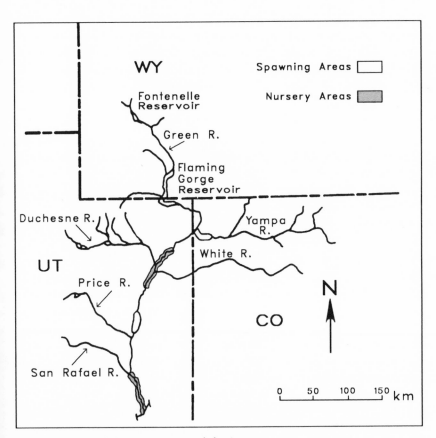

Fig. 19-3. Map of the Green River sub-basin, showing spawning areas to which annual migrations of adult Colorado squawfish have been confirmed and nursery areas in which squawfish larvae are common.

females) are also used to determine sex, the percentage of females increases from 7% (ripe fish only) to 20% (ripe, plus fish identified by secondary sex characters as females; Tyus and Karp 1989).

Growth rates for fifty-nine tagged adults subsequently recaptured in the Green River (range 48.2–77.0 cm TL) averaged 11.2 mm per year (Tyus 1988). Colorado squawfish have been aged to eleven years by examination of scales (Seethaler 1978), but more recent results using otoliths and vertebrae (conducted by D. Schultz, W. L. Minckley, and myself) suggested ages for wild Green River adults from seven to thirty years.

Foods and feeding

Adult Colorado squawfish are piscivores (Vanicek and Kramer 1969; Minckley 1973; Holden and Wick 1982), although some other foods are also consumed. Large fish have been caught with various baits, including Mormon crickets (*Anabrus migratorius*; Tyus and Minckley 1988) and carcasses of mice, birds, and rabbits (Beckman 1953). I have also interviewed fishermen who caught adults on hooks baited with earthworms, night crawlers, cut bait of sucker flesh; chicken, moose, and sage grouse livers; and various artificial lures, including spinners, plugs, and spoons (E. Wick, M. Hughes, J. Johnson, and L. Masslich, pers. comm.). W. L. Minckley (pers. comm.) saw a large brood fish at Dexter National Fish Hatchery, New Mexico, consume young American coots (*Fulica americana*). This voracious appetite was further illustrated by a large individual that surfaced and proceeded to strike my plastic float and take it underwater.

Reproduction

Colorado squawfish make extensive migrations, and homing of adult fish of 100 km and more (Fig. 19-3) to spawning sites was documented through radio-tracking and repeated recaptures of the same fish (Wick et al. 1981,

1983, 1985; Tyus 1985, 1990; Tyus et al. 1987; Tyus and Karp 1989). Large size presumably aids in undertaking long movements, which must require significant energy expenditures. The strategies involved are not fully understood, but habitat selection and recruitment success (fitness) are no doubt important. Migration is a logical adaptation of intermontane desert fishes to seasonally low discharges and is related to selection of optimal spawning sites as well (G. R. Smith 1981b).

Only isolated captures of ripe squawfish were reported from the Green River in the late 1960s (Vanicek and Kramer 1969) and early 1970s (Holden and Stalnaker 1975b; Seethaler 1978). More detailed studies in the 1980s, however, indicated that reproductively active adults seek faunally depauperate whitewater canyons for deposition of gametes in summer. Two major migrations and spawning areas have been identified by tracking radiotagged fish (Fig. 19-3). One migration to Yampa Canyon (Yampa River km 0–32) was identified in 1981 and studied annually through 1988 (Tyus et al. 1982b; Wick et al. 1983; Tyus and McAda 1984; Tyus et al. 1987; Tyus and Karp 1989). The other migration, to Gray Canyon of the Green River (km 232–256), was suspected in 1982, but insufficient fish were radio-tracked to confirm it until 1983 (Tyus 1985). The area was monitored continuously from 1983 through 1988 (Tyus 1990), and ripe adults have been caught at both sites each year. Successful reproduction was confirmed by collection of larvae immediately downstream, but not upstream, in both areas, from 1980 to 1987 (Haynes et al. 1984; Tyus et al. 1987; Nesler et al. 1988). It is possible that other spawning sites exist, including one in Labyrinth and Stillwater canyons of the lower Green River (Tyus et al. 1987). If so, few radio-tagged fish used them and the populations are presumably small.

Homing behavior was inferred from long-distance movement patterns and recaptures of the same Colorado squawfish on the same spawning grounds over a number of different years; fidelity to these specific areas has been demonstrated (Wick et al. 1983; Tyus 1985, 1990; Tyus and Karp 1989). There is no record of a fish moving from one spawning reach to the other (e.g., interchanges of Yampa and Gray Canyon fish). Lack of suitable spawning substrate or other factors is not judged to be limiting. Migrating squawfish pass through many kilometers of apparently suitable spawning habitat to reach other sites, and an olfactory homing mechanism was proposed to account for this behavior (Tyus 1985, 1990).

Turbidity in the Green and Yampa rivers has precluded direct observation of Colorado squawfish spawning behavior. However, radiotelemetry observations of wild spawning fish were provided by Archer and Tyus (1984), and Hamman (1981) noted behavior of spawning fish in hatchery conditions. Fish behavior was similar to that described for a congener, the northern squawfish (*Ptychocheilus oregonensis*), by Patten and Rodman (1969) and Beamsderfer and Congleton (1982). Northern squawfish remained in deep pools or eddies, moved abruptly to cobble bars to spawn, then returned to the pools or eddies. This same behavior in Colorado squawfish prompted division of habitats into two types: a resting-staging area in pools or large, shoreline mixing (eddying) currents, where fish find suitable resting and feeding conditions between spawning bouts or where males gather around females until they are ready to deposit eggs; and a deposition-fertilization habitat in riffles and shallow runs where they congregate for actual reproduction (Archer and Tyus 1984; Tyus and Karp 1989; Tyus 1990).

It has been difficult to identify spawning grounds for Colorado squawfish in other locations, in part because the fish is so rare. The presence of small larvae in the upper mainstem Colorado and San Juan rivers indicates successful reproduction, but their low numbers suggest limited reproduction or recruitment. No conclusive migratory patterns have been detected in the Colorado main stem despite years of study, although some migration-like movements have been noted (Archer et al. 1985, 1986). Perhaps spawning behavior documented for the Green and Yampa rivers was disrupted in the Colorado by years of flow and habitat alterations, or perhaps that stream was never optimal for the fish (Kaeding and Osmundson 1988b). The San Juan River has been studied far less, and squawfish ecology is less known in that system.

Habitat Relations

Larval ecology

Colorado squawfish larvae hatch in 3.5–6.0 days at 20°–22°C (Hamman 1981). Drift netting and seining indicate that larvae emerge from the cobble soon after hatching and move downstream out of the Yampa River in three to fifteen days (Haynes et al. 1984; Nesler 1986; Nesler et al. 1988). Drifting young are predominantly protolarvae, and those from shoreline backwaters are metalarvae (Haynes and Muth 1982, 1984), suggesting that the fish seek warmer and more productive habitats as they grow (Tyus et al. 1987). Young-of-year (postlarval) squawfish are rare in Yampa and Gray canyons but are distributed 100–250 km downstream a few weeks after the spawn (Fig. 19-3; Tyus et al. 1982b, 1987). They occupy shallow, alongshore, ephemeral embayments (backwaters) formed in late summer by receding water levels (Holden 1977b; Tyus et al. 1982b, 1987; Haynes et al. 1984; Tyus and Haines 1991). This observation led to development of a downstream transport hypothesis (W. H. Miller et al. 1982b; Tyus et al. 1982b). Although the downstream movement of young away from spawning sites was once debated (Holden 1977b; Holden and Wick 1982), the hypothesis is now generally accepted (USFWS 1978b,

1989a; Wick et al. 1983; Haynes et al. 1984; Nesler et al. 1988), and field (Hendrickson and Brooks 1987; Tyus and Haines 1991) and laboratory experiments (unpub. data) suggest active downstream movements for larvae six weeks old and older.

Juveniles

Juvenile Colorado squawfish from about 60 to 400 mm TL (Tyus et al. 1982b) include age-1 to about age-5 fish. Little is known of smaller juveniles (size range 100–300 mm) because they are apparently difficult to capture and too small to radio-tag and track. Downstream concentrations of postlarvae indicate that upstream movement of juveniles must occur to repopulate upstream reaches. Such probably occurs in the late juvenile or early adult stage (i.e., subadult, 30–50 cm TL), as suggested by highest concentrations of these larger juveniles in lowermost sections of the Green River and highest numbers of adults in upstream sections (Tyus et al. 1987). Collections from the White and Yampa rivers (W. H. Miller et al. 1982a, b) tend to support this hypothesis (Tyus 1986), and such a partial separation of life stages may reduce cannibalism. Small juveniles are nonetheless sometimes captured in the same backwater habitat used by younger life stages (Holden 1977b; Tyus et al. 1982b), whereas larger juveniles are not uncommon in shoreline habitats similar to those occupied by adults (Tyus et al. 1987).

Adults

After spawning, adults in the Green River return to the areas they previously occupied in spring (Wick et al. 1983; Tyus and Karp 1989; Tyus 1990), where they also reside in winter (Valdez and Masslich 1989; Wick and Hawkins 1989). There are no indications that adults stray any significant distances from the specific reaches they occupy during the non-migratory period. Large adults used compara-

bly large areas of habitat, often moving about within a 5-km or longer reach of stream in their day-to-day activities. Radiotelemetry of twenty-two fish (38.5–70.7 cm TL, $N = 2329$ observations) indicated that they were most often associated with shorelines (Tyus et al. 1984). Comparable data on habitat use was presented by Holden and Wick (1982), Tyus et al. (1982b, 1984), Wick et al. (1983, 1985), and Valdez et al. (1987).

Distribution and Abundance

General Observations

Colorado squawfish were recorded from the Green River as early as 1825, when Colonel W. H. Ashley's party subsisted on fish caught by angling (Morgan 1964). Dellenbaugh (1908) reported their capture during the 1871 Powell Expedition, and D. S. Jordan (1891) caught them near the town of Green River, Utah, in 1889. Residents of Vernal, Utah, remarked that squawfish were abundant in the Green River; individuals taken there reached a documented 52 pounds (23.6 kg), and a photograph of a large specimen was provided by Vanicek (1967). Seethaler (1978) summarized interviews with local people in Green River, Wyoming; Brown's Park, Colorado; and other places, who reported fish of thirty pounds (13.6 kg). He also reported an interview with Mr. Rial Chew, who observed a squawfish 5.5–6 ft (1.7–1.8 m) long, caught just below the confluence of the Green and Yampa rivers in 1911 in what is now Dinosaur National Monument, and presented a photograph of a 25-pound (11.4-kg) adult taken in the same area in 1928. Mr. Chew also said that he caught several of the fish by hand during a flood. Figure 19-4 is a recently discovered photograph of a large specimen captured in the lower Yampa River in the late 1930s (Burton 1987), where the species was, and is, regularly taken. Older residents of Vernal and surrounding communities reported

Fig. 19-4. Adult Colorado squawfish from the
lower Yampa River, Colorado, with Charles, Jr.,
and Pat Mantle, about 1935. Photograph from
Burton (1987), with permission.

squawfish caught in substantial numbers during periods of low flow, when semi-isolated pools in the channel were haul seined. I also interviewed Mr. Sylvan N. Arrowsmith, who said that he had caught many "white salmon" (75–100 large fish) prior to their legal protection and ate them in preference to any other species. Mr. Arrowsmith also reported seeing several (12 or more) in the 13.6–15.9 kg (30–35 pounds) size class that were captured by seines.

Abundance in the Last Decade

Distribution and abundance of Colorado squawfish were poorly understood until systematic surveys of the late 1960s and early 1970s (Holden and Stalnaker 1975a, b). Squawfish are found in small numbers in the main stems of larger rivers of the upper basin, but only the Green and Yampa rivers population and perhaps that in the Colorado main stem are thought large enough to be of viable, self-sustaining sizes.

Although the exact size of the Green River population is unknown, data based on catch rates and the size of occupied range indicate that it is larger than others. The CRFP provided substantial data for the years 1979–1981; these are summarized in Table 19-1. The same sampling methods were used in various locations, including seines, trammel nets, and electrofishing. Some methods were locally more efficient, and electrofishing was considered more effective in the Colorado and Yampa rivers than in the Green River, where turbidity was twice as great or more (U.S. Geological Survey records). However, these data are the most representative available for comparisons. Summing the four nonoverlapping studies in Table 19-1 for each sub-basin, catch per year was 2218 young, 221 juveniles, and 221 adults for the Green River (2.32, 0.23, and 0.23 fish km^{-1}, respectively, for the reaches studied) and 169 young, 36 juveniles, and 59 adults for the Colorado main stem (0.26, 0.06, and 0.09 fish km^{-1}, respectively). These values suggest that the Green River sub-basin has about an order of magnitude more squawfish of each life-history stage than the upper Colorado River sub-basin.

Table 19-1. Comparisons of Colorado squawfish catch data in the Green and upper main-stem Colorado rivers sub-basins, 1979–1981.

| | | River | | Annual catches | | |
Author[1]	River	miles	Years	Young	Juveniles	Adults
		Green River				
A	Green	22–319	2	2211	202	93
B	Green	319–345	1	7	0	16
C	White	0–150	1	0	19	39
B	Yampa	0–124	1	0	0	73
		Colorado River				
D	Colorado	−48–241	2	169	36	53
E	Gunnison	0–42	2	0	0	6
E	Dolores	0–68	1	0	0	0

[1]A, Tyus et al. 1982b; B, W. H. Miller et al. 1982b; C, W. H. Miller et al. 1982c; D, Valdez et al. 1982b; E, Valdez et al. 1982a.

Table 19-2. Six approximations and an average for population size for adult Colorado squawfish in the Green River, Utah.

Method (see text)	Estimated population sizes		
	Range	Average or mean	Fish km^{-1}
1	—	7728	14.0
2	775–3875	2325	4.2
3	—	8611	15.6
4	7369–8840	8105	14.7
5	4226–12,134	9764	17.7
6	3091–44,160	11,150	20.2
Averages	3865–17,252	7947	14.4

Relative proportions of these values are supported by interagency collections of adults and young, respectively, by spring electrofishing and autumn seining in 1986 and 1987 (USFWS 1987e, 1988c). Collections in reaches of densest populations of these two life stages (Archer et al. 1986) averaged 1.01 fish per hour electrofishing time for adults in the Green, and 0.31 fish per hour in the Colorado rivers. Seining caught 26.5 and 7.0 young per 100 m² in the Green and Colorado rivers, respectively (USFWS 1987e, 1988c). Although perhaps not directly comparable because of different habitat size and turbidities, the Yampa and White rivers of the Green sub-basin yielded electrofishing catches of 0.7 and 0.2 adults per hour.

Absolute numbers of Colorado squawfish are unknown, since rarity and lack of information on movements, recruitment, and mortality have precluded use of standard population estimation methods. However, many years of study allow some meaningful interpretations regarding standing stock, and crude approximations of adult population size (Table 19-2) were therefore developed from (1) comparisons with catch rates of razorback sucker (*Xyrauchen texanus*), for which a population estimate is available (Lanigan and Tyus 1989);

(2) use of electrofishing catch rates, converted to number of fish per kilometer and corrected for an assumed catch-rate efficiency; (3) use of electrofishing catch rates corrected by an estimated efficiency for razorback suckers; (4) a crude capture-recapture estimate, generally disregarding usual population estimation constraints; (5) an opinion survey of experienced fishery workers to determine relative numbers of razorbacks and squawfish observed per unit distance electrofished; and (6) a similar survey using the number of squawfish estimated per unit distance based on electrofishing catches. A computed overall average summed the outputs of all the above methods. The rationale and computations for these approximations are given in the appendix to this chapter.

Despite strong reservations, I believe the resulting range of averages—3875–17,252 adult fish—is realistic for the Colorado squawfish population of the main-stem Green River, and the grand average of all approximations—about 8000 adults—is also reasonable. Accuracy remains unknown; however, the similarity among results of sampling techniques implies either an inherent bias common in all methods attempted or a reasonably accurate grand average. The abundance of other life stages could also be estimated, but it would be far more difficult and the results less meaningful. Smaller squawfish are not vulnerable to sampling methods usually and practically employed (Holden and Wick 1982), and the relative abundance of fish less than 60 mm TL fluctuates greatly with environmental conditions.

No population estimate has been published for Colorado squawfish in the upper main-stem Colorado River. However, catch statistics in Table 19-1 indicate that absolute numbers are lower than in the Green River and about an order of magnitude less abundant; perhaps less than 1000 adult fish remain in the upper Colorado mainstream. The impacts of recent stockings of hatchery fish are unknown but apparently have barely affected this approximation. Recent work in Cataract Canyon and upper Lake Powell (Valdez 1988, 1990), and in the San Juan River (Meyer and Moretti 1988; Platania and Bestgen 1988; Platania et al. 1990), indicated that adult Colorado squawfish persist in even smaller numbers in those areas.

Management toward Recovery

General Review

Environmental changes in the Colorado River basin have been dramatic since the turn of the century, and far too rapid for genetic adaptation by its native fishes. As shown in Figure 19-5, much of the system was significantly altered by main-stem impoundments, diversions, and other water-resource development. The lower Colorado River was largely converted from a natural, fluctuating, turbid system to a modified (dammed), channelized, water-delivery system. Native fishes in the lower basin have been mostly extirpated from the mainstream and replaced by a new fauna sorted out from forty-four or more introduced forms, twenty of which have become locally or regionally abundant (Minckley 1973,

1982). Fish habitat in the upper basin has also been altered (Joseph et al. 1977), but more than 2000 km of riverine habitat is estimated to remain. Although about the same number of non-native fishes have been introduced (forty-two species; Tyus et al. 1982b), replacement of natives has not been complete, and populations persist in the reaches least affected by humans.

Losses of reproducing populations of Colorado squawfish and other endemic fishes have been mostly attributed to conversion of riverine habitat to artificial impoundments, replacement of warm-water habitat by cold tailwaters of dams, and erection of migration barriers (Vanicek 1967; Joseph et al. 1977; Seethaler 1978; Holden and Wick 1982; Tyus 1984). More insidious impacts, including introductions of non-native fishes, small but cumulative water depletions, and downstream effects of water manipulations, have also been suggested (reviewed by USFWS 1978b, 1989a), but these are more difficult to document or prove. Preimpoundment poisoning of the Green River is also sometimes implicated in this decline (in this volume, see Holden, chap. 3; Rinne and Turner, chap. 14). There is no doubt that squawfish were eradicated in the reservoir area, and above, when it was treated with rotenone in 1962, and fish were killed as far downstream as the mouth of the Yampa River. However, conversion of the treated area to a reservoir with its cold tailwaters, and blockage of potential migratory routes for adults and young by the dam, make the issue academic.

With passage of the ESA, federal and state agencies were provided with policy direction and monies to protect and recover remaining stocks of endangered fishes. Federal agencies, who were required to consult with the USFWS regarding effects of water-development projects, began funding research to learn more about squawfish and other species. Provisions of the ESA focused national attention on man-

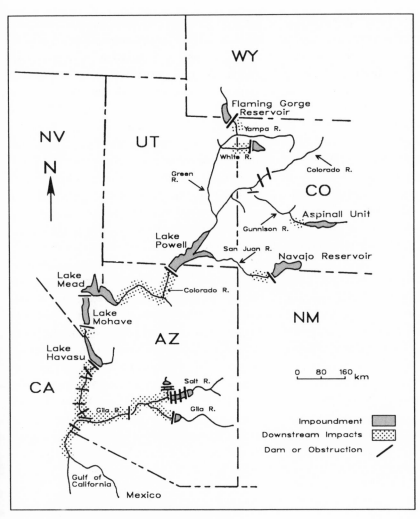

Fig. 19-5. Map of the Colorado River basin, western United States, showing relative concentrations of modifications and their impacts on the system.

agement options for these species for the first time. As a result, species-specific recovery plans (USFWS 1978b, 1989a) and regional recovery implementation programs (USFWS 1987a) were drafted by cooperating federal, state, and other agencies and individuals. Under the ESA, management options must, by law, place the highest priority on removing threats to extinction. Hence, downlisting species from endangered to threatened (par-

tial recovery) or delisting (full recovery) are primary goals. Only after these goals are met should other options be considered.

Frankel (1983) noted three elements to conservation: (1) the target (in this case the Colorado squawfish), (2) "time scale of concern" (Frankel 1984), and (3) management. Management affects the distribution, abundance, and numerical relations of and between species. Whereas "preservation" is often touted by "environmentalists," preservation of a species in the strictest sense can be accomplished by a zoo, aquarium, or other such facility. In contrast, conservation may only be accomplished within an ecosystem context. It is critical that ecological insights prevail in endangered species management, and an understanding of biotic and abiotic factors limiting the target organisms must be incorporated into any management scheme.

The preceding discussion has dealt with the life cycle and ecology of Colorado squawfish as a background for management. What follows summarizes my interpretations of important management activities to date and provides recommendations for future initiatives. I incorporated the terminology and order of the five recovery elements presented by the upper Colorado River Recovery Implementation Program (RIP) for rare and endangered fishes (USFWS 1987a) for consistency, and to aid the reader in making comparisons between this chapter and the RIP. The five elements are: (1) habitat management, (2) habitat development and maintenance, (3) stocking of native species, (4) non-native species and sport fishing management, and (5) research, monitoring, and data management. Further information on these elements, including development and assessments of the RIP, may be found in USFWS (1987a), Rose and Hamill (1988), and Wydoski and Hamill (*this volume*, chap. 8).

Habitat Management

Effective management implies protection of the native riverine habitat necessary to maintain viable populations of Colorado squawfish and prevent further jeopardy to its continued existence. Recovery can proceed only when existing populations are secure. Of first consideration is provision and maintenance of in-stream flows of proper quality, timing, duration, and magnitude at the proper location

for each life stage. The attainment of sufficient quantities of water requires determination of in-stream flow needs based on habitat requirements. Because water resources in the arid West are hotly contested, identified requirements for water must be biologically defensible. Flow issues are national and international in scope, potentially affecting allocations among and between all seven Colorado River basin states and the United States and Mexico.

Determination of in-stream flow needs of nonendangered fishes need not be as exacting as for endangered fishes because mistakes can be corrected in time through reestablishment of populations. If the stream in question is, for example, a put-and-take recreational trout fishery, more hatchery-reared fish can be stocked. In the case of the endemic Colorado River fauna, in-stream flow determination is deadly serious because limited gene pools cannot be sacrificed, and no successful self-sustaining population of Colorado squawfish, once lost, has yet been reestablished. No ecologist or fish manager wishes to participate in the eradication of a species (or its associated fauna) through an error in professional judgment. And no water-resource planner wishes to give up water rights if not absolutely required to do so. Therefore, determination and implementation of in-stream flows for habitat management are nearly impossible. Biologists are unwilling to make recommendations without some safety factor, and development interests are loathe to accept recommendations unless provision of each volume allocated is clearly supported by proof of need.

Although large reservoirs upstream of endangered fish habitat can aid in providing flow releases, the appropriation or acquisition of water rights is needed to provide in-stream flow for recovery and to offset impacts associated with continued development. Section 7 of the ESA provides for consultation between the USFWS and other federal agencies, and has resulted in flow-related conservation measures necessary to prevent loss of endangered species but not for recovery per se. This is an important difference, because it is conceivable that protection under section 7 may ensure against further losses, but in doing so may limit options for recovery.

The ecology and life cycles of fish are best studied where they remain abundant, in the least altered and most historic location, with the premise that conditions to which fish are best adapted are those in which they are most likely to maintain an advantage. Recovery needs may thus be formulated by comparing this "optimal" condition with habitats in areas where the species is in decline. In the following sections, protocols for habitat management in the Green River, where the first condition obtains, will be compared where possible with those in the upper Colorado and San Juan rivers, where the species is rarer.

Green River

Many interagency studies since 1979 (reviewed by Ferriole 1987; USFWS 1989a) have been concerned with determination of instream flow needs for endangered Colorado River fishes. Based on their results, the USBR and USFWS are cooperating to provide and evaluate the discharges from Flaming Gorge Dam necessary to form and maintain nursery habitats for young squawfish, an activity based on empirical verification of recommended flow scenarios. Associated with these efforts, the USBR and USFWS are completing studies of the requirements of rare and endangered fishes with respect to operations of Flaming Gorge, results of which will be provided as part of a biological opinion and supporting documents under section 7 of the ESA.

In 1984 the USFWS recommended interim flows to maintain nursery habitat for young (age-0) Colorado squawfish based on catch records (Jones and Tyus 1985). Survival of young was evaluated by analyzing catch data collected as part of an autumn sampling program. Recruitment was high in 1979 and 1980 when the August–September discharge varied from 44.8 to 53.2 m^3 s^{-1} (1600–1900 ft^3 s^{-1}), and low in 1983 and 1984 when discharges were 84.9–118.9 m^3 s^{-1} (3000–4200 ft^3 s^{-1}. A regression analysis of catches on means for August–September discharges at Jensen, Utah, demonstrated a high negative correlation ($r = -0.89$, $p < 0.05$). In addition, a similar relationship appeared between larval growth and mean monthly discharge ($r = -0.87$, $p < 0.05$). The Green River is historically and predictably low in discharge in late summer, and field observations between 1979 and 1988 suggested that low abundance and slow growth of larvae were due to an abnormally high summer discharge (releases of water stored during flood years of 1983 and 1984), which inundated prime nursery habitat (backwaters) for the species (Jones and Tyus 1985; Tyus and Haines 1991). This hypothesis was supported by research conducted by the USBR in 1987 and 1988, in which aerial photography at different, controlled discharge levels indicated that the availability of backwater habitat was negatively correlated ($r = -0.91$, $p < 0.05$) with late summer flows. Specifically, the surface area of backwater habitat was maximized at 47.4 m^3 s^{-1} (1687 ft^3 s^{-1}; average of Jensen, Ouray, and Sand Wash study areas), and numbers of backwaters were maximized at 39 m^3 s^{-1} (1381 ft^3 s^{-1}; Pucherelli et al. 1988).

Instream flow incremental methodology (IFIM), in association with physical habitat simulation modeling (PHABSIM; Bovee 1982), was also evaluated for use in determining habitat availability for young fish. Peaks and dips in model outputs made acceptable flows difficult to determine, and discharge levels at which recruitment, growth, and habitat were collectively maximized in field studies conflicted with the model's prediction. Subsequent evaluations revealed that aerial photographic mapping was a more appropriate means of evaluating these habitats.

Flow recommendations must also consider temperature effects on larvae, because increased flows of colder water also lowered average temperatures in the Green River upstream from Desolation Canyon. We noted a temperature differential of 10°C at the junction of the Green and Yampa rivers in 1983, whereas in better recruitment years the differential was usually less than 2°C (1979 = 0°C, 1980 = 1.5 °C, 1981 = 1.5°C; Tyus et al. 1987). Berry (1988) demonstrated that 10°–15°C cold shock adversely affected larval Colorado squawfish (fourteen days old, 9.0 ± 0.3 mm TL). No such effects were noted for larger, forty-day-old fish (24.4 ± 0.4 mm TL). Cold temperatures during larval drift could be implicated in the partial loss of the 1983 year class. However, temperatures in 1984 (3.0°–6°C) did not approach this differential and thus do not explain loss of that year class. Also, similar losses occurred in the Green River downstream of the Gray Canyon spawning area, a reach that was not affected by low temperatures from the dam, but was affected by high discharges as in the upper river. Temperatures in the few backwaters observed during USFWS autumn sampling in 1984 averaged 22.8°C; main-channel temperatures were 19.5°C (Tyus et al. 1987). Thus, field data supported the concept of a loss of recruitment due to high discharges that flooded ephemeral backwater habitats. Low temperature can have an influence, but it was not the primary factor limiting larval production in 1983 and 1984.

Yampa River

Habitat use and stream-flow needs of rare and endangered fishes in the Yampa River were presented by Tyus and Karp (1989), who evaluated requirements of fishes by interpreting empirical biological and discharge data, and then recommended flow regimens. Protection of flow and temperature regimens for initiation of spring spawning migrations and flows, temperatures, and sediment transport conditions during spawning were emphasized. Based on habitat use and behavior, they demonstrated that high spring discharges in concert with increasing water temperatures were predictably associated with initiation of spawning migration. Decreasing discharges in early summer to midsummer were associated with successful spawning and downstream drift of larvae to nursery habitats. Existing winter base flows appeared adequate for maintenance of winter habitat conditions (Wick and Hawkins 1989).

Tyus and Karp (1989) also reviewed outputs of IFIM/PHABSIM modeling of Colorado squawfish requirements developed in 1984 for staging and spawning (Archer and Tyus 1984; Rose 1984), and in 1987 for spawning, for possible use in developing discharge recommendations. Recommended discharges at which 90% simulated spawning habitat was obtained varied between 9.9 and 31.1 $m^3 s^{-1}$ (350–1100 $ft^3 s^{-1}$) for staging and 8.5–42.5 $m^3 s^{-1}$ (300–1500 $ft^3 s^{-1}$) for egg deposition, using habitat utilization curves developed in 1984. However, curves produced in 1987 produced different results. Adult habitat was optimized as flows approached zero, and spawning habitat was optimized at 42.5 $m^3 s^{-1}$ (1500 $ft^3 s^{-1}$). Furthermore, some discharges presumably optimizing habitat for other endangered species conflicted with those preferred by squawfish. The model outputs were rejected in deference to interpretations of empirical results. Many constraints have been placed on the use of IFIM habitat-suitability curves (Valdez et al. 1987), and our findings indicated that a simple three-variable model (depth, velocity, and substrate) did not accurately or adequately describe relations between fish habitat and river flows. An apparent incompatibility between heterogeneous habitats used by squawfish and hydrologic data collected as input to the modeling effort was presented by Tyus (1988). IFIM/PHABSIM modeling must be considered experimental until input variables reflect factors limiting populations of this or other warm-water species.

Upper Main-stem Colorado River

Much research has been conducted on habitat needs of Colorado squawfish in the Upper Colorado main-stem from Lake Powell to Palisades, Colorado (Valdez et al. 1982b; Archer et al. 1985, 1986; Wick et al. 1985; Kaeding and Osmundson 1989). Low numbers of fish and apparently marginal habitats have hampered development of in-stream management options, as indicated by Valdez et al. (1982b) and Kaeding and Osmundson (1988a, b, 1989). Habitat alteration by construction of water-development projects has been implicated as a major factor in low numbers of squawfish in the stream (Joseph et al. 1977; Holden 1979); average water temperatures there were judged only 71% as suitable for the species as those in the Green River (Kaeding and Osmundson 1988b). Blockage and other subsequent alterations of upstream habitats in the Gunnison and Colorado rivers above the presently occupied reach may have already removed habitats requisite to support a large adult population.

At present, the USFWS is evaluating the need for habitat management in 24 km of the upper Colorado above Grand Junction, Colorado, and has provided preliminary flow recommendations for that reach for enhancement of summer discharges for adults. No other specific habitat management plans have been developed. More optimal conditions for spawning and rearing are desirable because abundance and growth of age-0 fish are low (Kaeding and Osmundson 1989).

San Juan River

Colorado squawfish in the San Juan River have recently been evaluated for their significance to the overall recovery effort (Meyer and Moretti 1988; Platania and Bestgen 1988; Platania et al. 1990). However, the factors limiting those stocks are unknown, and development of habitat management options awaits future studies.

Habitat Development and Maintenance

Development and maintenance of new habitat and provision of access to new areas are considered high-priority recovery goals (USFWS 1987a). The first category includes application of experimental techniques to existing areas to determine if artificially created habitat will be acceptable for use. This includes construction of artificial spawning channels, or conversion of existing (but not currently suitable) stream reaches to spawning habitat, construction of nursery areas, and other efforts. The second category (e.g., construction of fish passageways) includes opening up new areas or restoring access to reaches now blocked by dams, diversions, and so on.

Colorado squawfish migrations have been observed for more than a hundred years, and we have documented travel of hundreds of kilometers in which adults occupy flat-water sections of rivers in autumn, winter, and spring but migrate to whitewater canyons to spawn in summer. Construction of dams, including Flaming Gorge Dam on the upper Green River and Taylor Draw Dam on the White River, blocked the passage of migrating fish, as evidenced by their congregation below the obstructions prior to the spawning season (McDonald and Dotson 1960; Seethaler 1978; Martinez 1986). These dams were not fitted with passageways for fishes. The only feasibility study was conducted in 1986 by the U.S. Army Corps of Engineers for the Redlands Diversion Dam on the Gunnison River. The lower Gunnison was historical habitat for Colorado squawfish, and the USFWS wished to evaluate the feasibility of retrofitting the dam with a passageway as a means of reestablishing connection with parts of its range. The work was placed on hold because of the projected high cost of the structure and a lack of convincing arguments that the fish would use it.

Extirpation of squawfish from the lower Colorado River basin was almost certainly caused by habitat alterations. It is therefore somewhat alarming to encounter assumptions that further alterations would be helpful to viable populations. Attempts to provide new, better, or more productive aquatic habitats must be approached with caution because they may result in the opposite effect. The "Frankenstein effect" (Moyle et al. 1986)—immediate and local solutions to a problem that cause adverse effects over a long-term or widespread scale—is a possibility that must be avoided. In the case of Colorado River fishes, a narrow species (or life stage) approach may well have negative effects on other species or life stages of the same species. Far too often habitat improvement projects have been ill planned, politically expedient approaches to management, with little or no follow-up or project maintenance (Rinne and Turner, *this volume*, chap. 14).

Provision of new or improvement of existing habitats can only be successful if strict ecological principles are applied. This challenge is recognized in the RIP, and the following stipulations are to be met (USFWS 1987a) before habitat development and management studies occur: (1) testing and implementation of management techniques will not be conducted in confirmed spawning and nursery areas, or in river reaches in which modifications might adversely affect use of confirmed spawning or nursery areas; and (2) genetic integrity of wild populations must be protected

when using hatchery-reared animals. There are obviously other stipulations that could be added, including a need to develop adequate follow-up and provide for maintenance of habitats that are developed. Also, delineation and protection of sensitive areas (USFWS 1987a) demand further attention relative to criteria for their formation and the types of protection afforded.

Stocking Rare Fishes

Stocking for recovery of squawfish has not occurred in the main-stem Green River because there is a naturally reproducing population there. Some stocking has been performed in Kenny Reservoir on the White River, Colorado, in an experimental attempt to establish a sport fishery. Colorado squawfish also have been stocked in other locations, including the upper main-stem Colorado and in the lower basin.

Squawfish stocked in the upper main-stem Colorado River included 1474 six-year-old fish, planted in 1980 (Valdez et al. 1982b) and tagged with Carlin dangler tags. Seventeen were recaptured in the last eight years. About 76,000 age-0 fish (50–125 mm TL) were also stocked between 1982 and 1984 (Kaeding, USFWS, pers. comm.). An additional 71,576 were stocked in streamside ponds in 1983 and 1984, and an undetermined number escaped into the Colorado River as a result of high water (D. Osmundson, USFWS, pers. comm.). Reintroduction of squawfish into parts of the lower basin from which they had been extirpated was initiated in 1985 by the Arizona Game and Fish Department and the USFWS (1989a). More than 100,000 fish, fingerlings through adults, have been stocked into the Salt and Verde rivers, and some have been recaptured (Brooks 1986b; Hendrickson and Brooks 1987). The relative success of all these introductions needs more evaluation.

Stocking can benefit management for recovery, but only in a comprehensive program in which life requirements of the target species are well understood and are provided at the sites proposed for stocking. Sites must be monitored to determine this. Habitat loss was presumably the prime factor associated with extirpation of squawfish in the lower basin, and if habitats have not improved, reintroductions are at best a stopgap measure (Rinne et al. 1986).

Fish culture has dominated fishery science in the United States for at least fifty years and has been regarded (by some) as a panacea to stop the loss of fisheries. It was not until the decade 1930–1940 that research highlighted the need for habitat protection and management (Radonski and Martin 1986). In the past, fish culture was not directed at restoring endangered fishes. Rather, it was directed at maintaining fisheries for sport or commercial harvest. There are many questions to be answered regarding use of hatcheries and stocking in management for recovery of wild populations before the methodology is used in accelerated or broadly applied efforts. Without habitat maintenance or improvement, hatcheries should be considered temporary refuges for genetic material and research facilities for laboratory studies.

The Colorado squawfish is a case in point. It has a complex life cycle in which there are spatial separations of life-history stages, incorporation of energetically costly migrations, and no doubt other, as yet unknown, components. If management for recovery is to succeed, these and other life-history attributes must be considered in the hatchery/stocking program. Some individuals migrate *downstream* to reach spawning grounds, and others migrate *upstream* to reach the same place (Tyus 1985, 1990). This phenomenon could be based on environmental cues (e.g., through olfaction or other means), or it might be under genetic control (Bowler 1975). Imprinting or other forms of fixation of cues may be disrupted under hatchery conditions, and a

genetic basis could preclude successful establishment of a population.

These migration behaviors suggest mechanisms for which we only have hypotheses. Ideally, areas proposed for reintroductions should be evaluated and ranked for availability of suitable habitat for all life stages. For example, shallow nursery habitat should be present below areas suitable for spawning, and availability of both spawning and nursery areas should coincide with proper water temperatures and discharges.

Non-native Species and Sport Fishing

Impacts of introduced fishes on native faunas (reviewed by J. N. Taylor et al. 1984) are seldom understood and have been largely inferred from major alterations of species composition. Competition (Connell 1983; Ross 1986) between native and introduced forms is difficult to document experimentally, and perhaps impossible to prove in nature. Direct effects of fish introductions on native fishes were categorized and reviewed by Moyle et al. (1986) as follows: elimination, reduced growth and survival, changes in community structure, and no effect. They stressed that many fishes are introduced into habitats that are already changing due to the impacts of humans, making it difficult to determine if habitat change, competition, both, or other factors are responsible for decline of the native fauna.

This already complex issue is further confounded by introductions into waters in which species interactions were not well understood anyway. Stocking of new species into an area to promote sport fishing has often created more problems than it solved, mostly because methods for predicting outcomes are few (Li and Moyle 1981). Stocking of non-native warm-water species has been reduced in recent years, and under the RIP will not occur in watersheds occupied by Colorado squawfish unless it would not interfere with recovery initiatives. The reverse of stocking—elimination or removal of non-natives—is worth considering but has not yet been meaningfully addressed. Introductions of non-native fishes in the Colorado basin have been documented by many, and data on their present status was provided by Minckley (1973, 1982) and Tyus et al. (1982a). Competition of non-native fishes with Colorado squawfish has been proposed by various workers (Holden 1977a, 1979; Joseph et al. 1977; USFWS 1978b, 1989a; Holden and Wick 1982; Behnke and Benson 1980; Osmundson 1987).

The impacts of such introductions on native species remain problematic because they differ in their effects on various life stages, both temporally and spatially. For example, larval squawfish are eaten by other fishes, and probably suffer competition from the young of other species as well. Studies of predation are difficult because the fragile larvae are rapidly digested by predators (Marsh and Langhorst 1988).

Postlarval Colorado squawfish attain 40–50 mm TL in their first year of life, and they may begin to eat other fishes at that time. It is possible that larger or more aggressive individuals experience better growth in backwaters where small non-native fishes are used as food. Some studies indicate that squawfish postlarvae consume appreciable volumes of unidentified fishes (Jacobi and Jacobi 1982); however, our stomach analysis of hundreds of small squawfish collected in the Green River in 1986–1988 (USFWS unpub. data) indicated that age-0 fish subsisted mainly on invertebrates, and seldom other fishes, in agreement with findings of Vanicek and Kramer (1969).

On the other hand, aggressive behavior toward squawfish larvae by some introduced species has been observed in aquaria (Karp and Tyus 1990) and ostensibly occurs in nature as well. Postlarvae may also serve as prey for some larger fishes, particularly centrarchids, which commonly enter backwaters. Osmundson (1987) recorded predation on Colorado squawfish by largemouth bass (*Micropterus salmoides*), green sunfish (*Lepomis cyanellus*), black crappie (*Pomoxis nigromaculatus*) and black bullhead (*Ameiurus melas*).

Fish in the juvenile life stage are large enough to navigate freely along most river channels, and presumably do so. There are possible advantages and disadvantages to them from introduced species, the squawfish eating smaller species and individuals (Vanicek and Kramer 1969) and being eaten by larger ones until adulthood is attained. However, stomachs of 123 northern pike (*Esox lucius*), 61 walleye (*Stizostedion vitreum*), and 755 channel catfish (*Ictalurus punctatus*) taken from the Green and Yampa rivers from 1984 to 1988 (Tyus and Beard 1990; Tyus and Nikirk 1990) provided no evidence that these predators ate squawfish (some fish remains were not identified). The fish remains most commonly identified were other introduced forms.

Growth of adult Colorado squawfish may be more rapid than formerly if their prey base has increased due to introduction of non-native fishes, but lack of precise data on prey abundance makes such comparisons difficult. Some introduced prey may be lethal; large squawfish have been known to choke on channel catfish (Vanicek 1967; McAda 1983; Pimentel et al. 1985). Larger introduced fishes such as northern pike and walleye are potential competitors for food and space.

Large squawfish readily take fishing baits or artificial lures and can frequently be observed and fished at the water surface. Therefore sport fishing must be considered a factor limiting future recovery. Sport fishing mortality of adults (intentional or not) is a serious problem in areas of concentration, particularly in the Yampa River, a popular fishing spot (Saile 1986; Tyus and Karp 1989). In the

Green River a 10% angler take has been documented for some years (USFWS, unpub. data). This situation needs more evaluation; however, loss of large, reproductive fish, particularly females, may be significant. More and better angler contact and education and a more rigorous creel program could be beneficial.

Research, Monitoring, and Data Management

The CRFP (W. H. Miller et al. 1982c; Shields 1982; Wydoski and Hamill, *this volume*, chap. 8) is a cooperative USFWS project involving the USBR, the USNPS, the USBLM, the states of Colorado and Utah, and others. The project was designed for rapid incorporation of information into ESA activities and recovery programs. These goals were further supported by development of the RIP in 1987, and annual reviews are carried out in which needed studies and information are given priorities and funds are allocated. The USFWS, under the RIP, is computerizing an extensive data base. In addition, joint annual research meetings have been conducted by agencies and cooperators each year since the early 1980s to aid in reducing isolationism (Wydoski and Hamill, *this volume*, chap. 8).

Monitoring is required as a follow-up to research to maintain surveillance on the system and to further refine management protocols. Long-term monitoring may also be geared to testing hypotheses, thus providing a dynamic input and fine-tuning management. Monitoring of Colorado squawfish involves, at a minimum, periodic electrofishing surveys for adults as well as annual sampling of larval production related to habitat conditions.

Research, monitoring, and concomitant development and refinement of management options are essential if recovery is to succeed. However, these activities are just as valuable for other freshwater species and other rivers.

The systematic damming of rivers of the world has reduced other fauna, particularly large endemic forms with complex life cycles like the Colorado squawfish. Included are other North American species like paddlefish (*Polyodon spathula*) and sturgeons (Acipenseridae), and Eurasian fishes like sturgeons and the cyprinids zheltosek (*Elopichthys bambusa*), pikeasp (*Aspiolucius esocinus*), and others (Pavlov et al. 1985). These large riverine fishes are rapidly disappearing, and the predicted extinction of many of the world's large migratory fishes (G. Howes, British Museum [Natural History], pers. comm.) may soon become a reality. Knowledge gained for Colorado squawfish will be useful in understanding life cycles and habitat requirements of these animals as well. The knowledge must be shared with others, however, which requires a commitment from agencies and individuals to publish their data and studies in widely circulated periodicals.

Conclusions

Loss of the Colorado squawfish, top carnivore of the pristine Colorado River ichthyofauna, would signal a final collapse of the most endemic riverine fish community in North America (R. R. Miller 1959), and perhaps foretell the doom of many of the large migratory fishes of the world as well. If the lower Colorado River ecosystem is indeed a model that predicts the outcome of water-resource development and the introduction of non-native competitors and predators in the upper basin (Molles 1980), the future is grim. The loss of species was not a national concern until recently, as indicated by passage of the ESA fewer than twenty years ago and the formation of active, concerned groups like the Desert Fishes Council. Recovery of the big-river fishes of the Colorado system is now a major issue in the battle against extinction. Although recovery will be costly and will involve difficult research and management activities, the Colorado squawfish and its environment can be saved.

Most of the habitat alterations that decimated these fishes were caused, directly or indirectly, by construction of federal reservoirs, which produce large revenues. It is only proper that some of these funds aid in the recovery of the fauna that is left, and, indeed, most of the funding for upper-basin recovery efforts has been contributed by the USBR (about $6.5 million to date; R. Williams, USBR, pers. comm.); this is only a small percentage of the money realized (about 0.05%) from hydroelectric operations of mainstream dams in the same period. Biologists have been retained whose functions are to conduct and manage research, cooperate with and foster interest by others in the academic and private sectors, and interpret and apply results to biological conservation. Without continuing cooperation, management toward recovery of Colorado river fishes is impossible.

Will the battle be worth it? Ehrlich and Ehrlich (1981) presented convincing arguments for saving endangered species. Indirect benefits to humanity were cited as most important but least understood. Rare species must be viewed as part of the fabric upon which humans depend for maintenance of the biosphere. The Colorado squawfish is just too valuable to lose. Historically, the fish furnished food and livelihood for Native and non-native Americans, and was important in settlement of the West. Its table qualities are evident, since "Colorado River Salmon" was an entrée for apparently lavish July Fourth and Christmas dinners served at Lee's Ferry, Arizona, in 1899 (Measeles 1981). The spectacular potadromous migrations and unique life cycle of the squawfish are certainly important from the scientific point of view. Philosophically, do humans wish to lose this large predator along with a reduction of perhaps 20% of the world's biota by the year

2000? Economically, in view of the increasing expense of maintaining cold-water fisheries, is not conservation of one of the largest warm-water species on the continent prudent? What of the distant future? Will humans again depend upon subsistence fishing for their livelihood?

The final curtain has yet to come down on the Colorado squawfish. Although reduced to perhaps 25% of its former range, the fish survives. A sizable population exists in the Green River, demonstrating its fitness by coping with a plethora of environmental insults. We have shown that research and management can achieve some recovery goals in the Green River, and chances for developing such options exist elsewhere.

As we learn more about this species and work to provide for its needs, the concept of "preservation" must be replaced with a long-term conservation ethic (Soulé and Wilcox 1980), a process of continuing evolution (Frankel 1983) that includes management on an ecosystem scale. This ethic is the only practical and possible solution for this unique life form. Conservation of the Colorado squawfish will require a new and more ecologically appropriate approach to fishery management— management for recovery—which obviously includes recovery not only of the fish but of its environment as well. Success will require sincere efforts on the part of scientists, managers, environmentalists, water-resources developers, governments, and private citizens, all with sometimes conflicting interests. If we are successful for this fish, and for its ecosystem, success will follow more quickly for other endangered fishes.

Financial support for the CRFP was provided by the U.S. Bureau of Reclamation (USBR), USFWS, U.S. National Park Service (USNPS), U.S. Bureau of Land Management (USBLM), and others. The states of Colorado and Utah cooperated and provided logistical support. I thank L. Mills for many years of support and encouragement. R. Williams, C. Karp, J. Hamill, and H. Williamson improved an early version of the manuscript. C. Karp, K. Paulin, L. Kaeding, M. Moretti, S. Cranny, and S. Lanigan participated in abundance approximations. I especially thank the more than fifty permanent and seasonal employees who assisted with data collection between 1979 and 1989.

Appendix

Approximations of abundance for adult squawfish in the mainstream Green River are summarized in the text and presented in Table 19-2. I intentionally excluded tributaries, the Yampa, White, and Duchesne rivers, and the Green River upstream from Echo Park. Similar exercises could be conducted for those streams, but they have not been sampled thoroughly or with the same methods for longer than a year. Most tributary studies lasted a year or less (e.g., W. H. Miller et al. 1982b, c), extended only to state boundaries (e.g., Radant 1982, 1986), or were conducted in limited geographic areas (various workers). The following exercises used five to seven years of catch/effort data from between river kilometers 43 and 552 on the Green River, all during the same time of year and with the same equipment.

Method 1

Electrofishing is a common tool used by fishery workers to catch large fishes (J. B. Reynolds 1983). Lanigan and Tyus (1989) used fish collected using this technique to estimate population size of razorback suckers in the upper Green River with mark-recapture methodology. A population estimate for that species was considered valid because assumptions of geographic and demographic closure appeared satisfied, whereas data for Colorado squawfish do not meet those necessary assumptions. However, the two species were commonly collected together; shorelines were occupied by squawfish most of the year (Tyus et al. 1984), and by razorback suckers in spring (Tyus 1987). Although electrofishing efficiency was probably different for the two species, I used the estimated size of the razorback population to approximate standing stock of squawfish.

As indicated by Tyus (1987), the catch rate for razorback suckers comprised 20.6% of that for squawfish (0.20 suckers and 0.97 squawfish hr^{-1}) in the 171 km of Green River downstream of the Yampa River. Assuming the population estimate for razorbacks (948 individuals) by Lanigan and Tyus (1989) is correct, there would be about $984/0.21 = 4515$ squawfish in the reach. This figure includes both larger juveniles (30–40 cm TL) and adults (> 40 cm TL). Based on an adult/juvenile ratio for Green River provided by Tyus et al. (1987), the adult component amounts to $(0.79)(4515) = 3567$ fish, or about 21 adults km^{-1}. The estimate for the upper Green River was extended to the entire main stem (552 km) by substituting an average catch rate from the entire main stem of 0.68 squawfish hr^{-1} in place of the average rate of 0.97 hr^{-1} in the uppermost part (Tyus 1987). This provides an estimate of about 14 fish km^{-1}, and $(14 \text{ fish } km^{-1})(552 \text{ km}) = 7728$ adult squawfish in the main stem.

Method 2

If the average electrofishing catch rate for the Green River is 0.56 adult Colorado squawfish hr^{-1} (Tyus et al. 1987), and if a boat electrofishes one shoreline at a rate of 4.0 km hr^{-1}, the average catch would be 0.14 fish km^{-1} for one shoreline. The number for both shorelines would be double that, or 0.28 fish km^{-1}. The total number of adult fish would then be approximated by $(0.28)(552) = 155$ fish, if electrofishing efficiency was 100%. However, efficiency is affected by many factors and varies from 4% to 54% (Ruhr 1957;

Jacobs and Swink 1982). High conductivity and turbidity of the Green River reduce electrofishing efficiency. I therefore used an arbitrary range of 4%–20% to obtain maximum and minimum estimates of $155/0.04 = 3875$, and $155/0.2 = 775$ adult squawfish; average $= 2325$ fish.

Method 3

Despite the above estimations, electrofishing efficiency for capture of squawfish in the Green River is unknown. However, an efficiency can be calculated from the population estimate for razorback suckers of 948 fish from 171 km. I used a catch of 0.2 fish hr^{-1} from Tyus (1987) rather than the 0.49 given by Lanigan and Tyus (1989), because the latter was estimated for effort concentrated in a reach of greatest razorback sucker abundance. A capture rate of 0.2 razorbacks hr^{-1}, assuming travel time of 4 km hr^{-1} and two shorelines, gives a catch per kilometer of 0.1 fish. Calculation of catch for the upper Green River would thus be $(171) (0.1) = 17$ fish. Electrofishing efficiency would then be calculated by $17/948$ fish in the population, or 0.018. If one assumes that Colorado squawfish and razorback suckers are equally vulnerable in spring, the population size of squawfish would be approximated by: 155 fish (method 2 calculation)/0.018 = 8611 adult individuals.

Method 4

The total number of Colorado squawfish tagged in the Green River by the USFWS from 1979 through 1987 was 625; 7 of 77 fish captured in 1988 were recaptures. A crude approximation of number of fish at large, disregarding constraints of geographic and demographic closure, would be $7/66 = 625/x$, or 5893 fish. In addition to other problems, this approximation does not consider tag loss (Wydoski and Emery 1983), an important factor in active, migrating fishes. Tag loss may be as high as 42% (Dunning et al. 1987) for some external tags. If limits of 25% and 50% tag loss are assumed, the value of 5893 would be converted to 7369 and 8840 adults, average = 8105.

Methods 5 and 6

I employed a modified Delphi technique (Crance 1988), using six biologists experienced in electrofishing for squawfish in the Green River. Their experience averaged about four and a half years. Each was asked two questions: (1) what was the ratio of squawfish to razorback suckers; and (2) what was the maximum number of squawfish per mile of river, based on their best estimate of number of adult fish electrofished and relative electrofishing efficiency. The responses consisted of a range of numbers per individual. Group averages were then provided to all participants, who were asked to supply best average numbers. Their estimates varied from 2 to 50 squawfish per razorback; group averages varied from 4.5:1 to 12.8:1, with a grand mean of 10.3:1. Estimated Colorado squawfish population sizes were $(4.5) (948) = 4266$, $(12.8) (948) = 12,134$, and $(10.3) (948) = 9764$, respectively (948 is the razorback population size estimated by Lanigan and Tyus [1989]).

The number of squawfish estimated per mile varied from 2 to 500, the range of averages was 5.6–80 fish km^{-1}, and the grand average was 20.2 fish km^{-1}. An estimate of squawfish abundance using the above range of averages is $(5.6) (552$ km$) = 3091$, and $(80) (552) = 44,160$ fish, respectively. The grand average was calculated as $(20.2) (552) = 11,150$ adult squawfish.

SECTION VII

Epilogue: Swords of the Fathers, Paying the Piper, and Other Clichés

This last contribution attempts to bring the remarkable diversity of skirmishes in the "battle against extinction" of western American freshwater fishes into "real world" perspective. The authors are convinced that the knowledge already exists with which to save most endangered species, if the will can be mustered and concentrated into an effective effort. Habitat preservation and management are feasible, and with the exception of biological pollutants like introduced fishes, technology clearly exists to identify, set aside, and renovate (or re-create) habitats to conserve much of the natural aquatic diversity. In addition, legislation already in place usually is adequate to do the conservation job, so long as its interpretation and application is commensurate with its original goals. Problems of definition and compliance often hinge on the discretionary powers of local or regional administrators.

The establishment of national priorities by a federal administration profoundly affects the enthusiasm with which conservation goals are pursued. Under U.S. law, as we have seen, especially during the past decade, it is possible to promote practices that cause or accelerate the major ecosystem disruptions that are increasingly evident in western deserts and throughout the world. Let us not forget that it is equally possible under the law to promote policies leading toward sustainable coexistence of human and nonhuman communities.

Efforts in the current battle against extinction have been successful in tangible ways. Knowledge, technology, and legislation for conservation are in place. We must now make these tools work for the common good. Public education is the key, and until a worldview of limits and a viable land ethic, both of which serve to perpetuate the biosphere, become ingrained in the public, we must continue an uphill fight against losses in biodiversity.

These cui-ui had started to run upstream from Pyramid Lake into the Truckee River, Nevada, but were stranded by a drop in river level when upstream water was diverted for irrigation, May 1940. Photograph taken and provided by T. Trelease.

Chapter 20

Western Fishes and the Real World: The Enigma of "Endangered Species" Revisited

James E. Deacon and W. L. Minckley

Introduction

More than twenty years ago we framed, polished, and finally published a paper in *Science* titled "Southwestern Fishes and the Enigma of 'Endangered Species'" (Minckley and Deacon 1968). Then, as today, we were alarmed that the actions of humans were directly and indirectly exterminating native freshwater fishes in western North America. R. R. Miller (1946a) was among the first to predict, then document (R. R. Miller 1961), the extent to which fishes and fish habitats were disappearing in the American West. In our 1968 paper we reviewed some specific problems, using examples to illustrate generalizations about the disappearance of various components of the fauna, then tried to place the problem in its broader, worldwide context, the history of which had already been reviewed by White (1967). We alluded to most of the same fish species, aquatic habitats, and ecologic, geographic, and sociologic factors discussed in this volume. As time went by, new problems appeared; a few were resolved, many remain, and native fishes continue to disappear (R. R. Miller et al. 1989). The question of how to preserve these aquatic animals and other biotic elements of the West in the face of continued human population increases and development is more pressing than ever before.

This is not to say that progress has not been made. A review of the information presented in this book demonstrates a significant and positive transition from attempts to promote awareness of the plight of endangered species to development of comprehensive programs aimed at preventing their further decline.

Beginnings, and the Path of Conservation

Many state and federal managers were made aware of their role in speeding the decline of native western fishes as a result of the Green River poisoning in Colorado, Wyoming, and Utah in 1962 (Holden, *this volume*, chap. 3). That event, considered a debacle by some, but necessary and justified by others, spawned a debate that deeply divided fishery managers from more conservation-minded individuals, a schism that remains. As academicians, we view this as one consequence of specialized training for managers in degree programs isolated from basic science (Soulé 1986), and also as an excellent illustration of the "A-B cleavage" described by Leopold (1949; Pister, *this volume*, chap. 4). Others, we are confident, view it differently. Nonetheless, the considerable damage done to native nongame fishes by management activities, while evident, may never be known in full (Rinne and Turner, *this volume*, chap. 14).

Toward the end of the 1960s, isolated efforts were under way by state game and fish departments, in cooperation with universities and the U.S. Fish and Wildlife Service (USFWS), to preserve some nongame fishes in Arizona, California, and Nevada (see, in this volume, Pister, chap. 4; Williams, chap. 11). These efforts marked beginnings of a shift from the utilitarian policies of earlier times. Recognition of the nonutilitarian value of native plant and animal resources and ecosystems is working its way into state and federal management agencies, university curricula, and the public mind. Intrinsic value is becoming recognized even more slowly.

To our knowledge, twenty years ago no western state employed a biologist with primary responsibilities for native nongame species of fishes. Colorado was the first to do so in 1976. Today in the western United States, only Idaho, Oregon, and Wyoming are without at least one biologist responsible for a program for native nongame fishes (Pister, *this volume*, chap. 4, table 4-1). Federal agencies have also expanded in this area, despite budgetary constraints. The development of state and federal programs for endangered species was encouraged by public awareness of environmental problems and supported by the resulting legislation of the 1960s and 1970s. The Endangered Species Act (ESA) and Clean Water Act presently may be most important for western fishes, although other legislation is also potentially useful (Williams and Deacon, *this volume*, chap. 7).

During the past two or three decades, the philosophical basis for environmental policy in the United States has undergone scrutiny, resulting in an expanding foundation for environmental attitudes, legislation, and litigation (White 1967; Callicott 1987a; Hargrove 1989; Rolston 1989, *this volume*, chap. 6). Society and social institutions are more willing to recognize their duties to endangered species, and in some measure recognize the species' rights (Varner 1987; Rolston 1988; Williams and Deacon, *this volume*, chap. 7).

In spite of the development of a substantial legal commitment and expanded agency programs, the status of fishes continues to deteriorate (R. R. Miller et al. 1989; J. E. Williams et al. 1989; Minckley and Douglas, *this volume*, chap. 1; Contreras Balderas, chap. 12), reflecting the same alarming trend in the earth's biota (Wilson 1988). J. E. Williams et al. (1989) could not justify removal of a single taxon of North American fish from a list of rare species on the basis of recovery efforts from 1979 to 1988. The status of 7 taxa had improved, while 22 declined and 136 were added to the 251 kinds already listed. Twenty-five taxa were removed from the list, 10 due to extinction and 15 because of new information on their taxonomy or status. Regionally, the largest number of listed taxa was from the southwestern United States, even though that ichthyofauna is far less speciose than in any other part of the continent.

Why is this so? Land in the western United States is mostly under federal or state ownership, and thus is public domain, upon which the national resource of endangered species should be managed, by law. Williams and Deacon (*this volume*, chap. 7) suggest that most agencies have sufficient authority, if they choose to apply it, to maintain species diversity on lands they administer. The deteriorating status of fishes on federal lands in the West therefore is usually a consequence of failure to exercise authority vested in the responsible agency. Similar conditions apply to state lands. Private owners may, of course, do largely as they wish, including using their land in ways that promote stability of habitat and its contained biota. Unfortunately, like federal agencies, state and private owners often make decisions that do not uphold the integrity, stability, and beauty of the land.

The inescapable fact is that institutional decisions regarding land use in general and endangered fishes in particular reflect the individual worldview of the person(s) responsible. There is considerable diversity of opinion regarding the value of maintaining biotic diversity, as reflected in actions taken by public servants and private landowners alike. Much of the agency authority to protect biotic integrity is discretionary, leaving ample opportunity for managers to express their individual views. Economic, political, social, and biotic factors are all involved in such decisions, even when the ESA directs that they be based only on biological considerations.

Strategies of Protection

Isolated Endemics

Strategies for protecting isolated endemic fishes are relatively simple and straightforward. Since habitats are usually distinct and well defined, the preferred strategy is to save the habitat, which usually consists of a single spring or spring system. Other choices are to transplant the animal or to place it under intensive culture.

Moyle and Sato (*this volume*, chap. 10) emphasize protection of entire, naturally functioning communities as the only way to ensure survival of a species in its evolutionary context. This is easier to do for isolated endemics and the communities upon which they depend than for wide-ranging taxa. Probably because of this fact, refuges for western fishes most commonly are established for isolated endemics; twenty-two of the thirty-three discussed by Williams (*this volume*, chap. 11) provide protection for such special species.

Echelle (*this volume*, chap. 9) points out that small, isolated populations of western fishes frequently have a lower genetic diversity than large, widespread populations. The cause appears to be frequent bottlenecking— the passage of a genetic lineage through a small number of individuals. Management strategies thus should avoid creating or allowing conditions that would occasionally require severe reductions in either habitat or population. This is another reason why natural habitats are more desirable than artificial ones as refugia. Artificial habitats, because they depend on human maintenance, are less reliable.

Protecting natural habitat is the most satisfactory solution for protection of isolated endemic fishes. It has been successful for the Death Valley pupfishes (R. R. Miller 1967; Deacon and Williams, *this volume*, chap. 5), the Borax Lake chub (*Gila boraxobius*), the Moapa dace (*Moapa coriacea*), and others (Williams, *this volume*, chap. 11). In spite of the fact that restricted distributions and small population sizes of these fishes dictate their virtually permanent status as endangered or threatened, there can be no doubt that their prospects for survival and continued evolution are better now than in the recent past.

Opportunities to use this strategy are becoming rarer as natural habitats continue to be modified or destroyed. Furthermore, even when complete protection is accomplished, there frequently will be continued damaging perturbations from external causes, the "eternal external threat" (Janzen 1986). Introduction of exotic species, groundwater withdrawals or contamination of water supplies (often distant from a site), floods, and innumerable other possibilities require constant vigilance even for the most "secure" habitat. Damage requires habitat reconstruction, and in those cases it is necessary to understand both the natural habitat conditions and the ecology of the native species protected. For example, it is essential to recognize that a spring system in the desert most often consists of multiple outflows draining through frequently changing channels into marshes that fluctuate seasonally (Miller et al., *this volume*, chap. 2). Such systems are dynamic and should be encouraged to remain so. Reconstructive management should not create a stabilized springhead draining through a fixed channel into a regulated reservoir, albeit designed to look like a natural marsh.

Efforts have been less successful when only a part of the natural habitat is protected. The first two attempted—Hot Creek, Nevada, and Owens Valley, California—demonstrated that continual vigilance coupled with frequent management intervention is essential (Hardy 1980; C. Allan 1983; Courtenay et al. 1985; in this volume, Pister, chap. 4; Minckley et al., chap. 15). Episodic population bottlenecking has been the rule. Both White River springfish (*Crenichthys baileyi thermophilus*) at Hot Creek and Owens pupfish (*Cyprinodon radiosus*) survive in good numbers, but their long-term prospects depend precariously on the reliability of humans rather than that of the habitats supporting them. Similar conditions exist at the San Bernardino National Wildlife Refuge (Rinne and Turner, *this volume*, chap. 14) and elsewhere.

Transplanting endemic fishes into natural habitats outside their native ranges raises questions that must be addressed before the transplant is attempted (J. E. Williams et al. 1988). The strategy can perpetuate taxa that would otherwise disappear along with doomed habitats, but adverse impacts on the host habitat and its indigenous biota may be serious.

Nonetheless, the genus *Empetrichthys* continues to exist only through such action. *Empetrichthys l. latos* was transplanted into Corn Creek Spring, a human-modified natural habitat on the Desert National Wildlife Refuge, Nevada, in 1971, just before its native spring system at Manse Ranch was destroyed. The founding population of twenty-nine individuals may have represented a severe bottleneck. Second and third bottlenecks occurred when mosquitofish (*Gambusia affinis*) appeared, resulting in dramatic declines in poolfish populations (Minckley et al., *this volume*, chap. 15). Since then, two additional stocks have been established (Baugh et al. 1987; in

this volume, Minckley et al., chap. 15; Hendrickson and Brooks, chap. 16).

Although all three seem secure at present, long-term survival depends on habitat maintenance by humans. In addition, there are substantial differences in physical and other characteristics of each refuge that could lead to shifts in adaptive peaks. Corn Creek Spring forms three ponds, each with wider annual thermal variation as a function of distance from a constant-temperature source. Flow from the lowermost pond spreads over a seasonally variable, wetted marshland. Spring Mountain Ranch State Park is a small water-supply reservoir that exhibits wide seasonal temperature and other fluctuations. Shoshone Ponds has an artesian source that maintains a relatively constant temperature in a small, isolated pond throughout the year. If the species persists long enough in all three places, one would predict evolution to eventually produce three distinct products, each different from the one that would have been at Manse Ranch Spring.

Maintenance of endemic fishes in artificial habitats is the least satisfactory of the options available, but it is not to be discounted. Difficulties with artificial habitats have been discussed by Baugh and Deacon (1988), J. E. Johnson and Hubbs (1989), Moyle and Sato (*this volume*, chap. 10), Johnson and Jensen (*this volume*, chap. 13), and others. This strategy may be successful in saving fishes for a short time while more secure natural habitats are located or prepared, but it should not be accepted as a long-term solution under any but the most extreme circumstances.

Wide-ranging Species

Complexities of conservation strategies increase as the geographic range of target species becomes larger. Some species, like the Colorado squawfish (*Ptychocheilus lucius*), demonstrate distinctive ecologies for each of a number of life-history stages and collectively occupy (and presumably require) hundreds of kilometers of river (Tyus, *this volume*, chap. 19). Others appear to have naturally fragmented ranges, distributed in tributaries of a large river system, with infrequent genetic contact among subpopulations.

The earliest efforts to protect native fishes of the West focused on trouts (see, in this volume, Williams, chap. 11; Rinne and Turner, chap. 14), a group long popular with humans because of its sporting, culinary, and aesthetic values. Many of the now-isolated western trouts were widespread at suitable elevations or latitudes. Habitat loss and disruption, especially of streams needed for reproduction by lake-dwelling populations, took major tolls. Non-native trouts were introduced and proved in many cases to outcompete, prey upon, or hybridize with native stocks, resulting in their demise. Losses also resulted from failure until too late to recognize distinct taxa or stocks.

In general, protection of trouts has been effective. The strategies involved propagation, protection of portions of natural ranges by artificial barriers or use of natural ones, and eradication of introduced trout (and, coincidentally, other native taxa) followed by reintroduction of natives. However, non-native trouts are repeatedly reintroduced by unknowing persons or vandals, barriers are breached by floods, and so on. Time and time again, managers have learned the necessity of constant vigilance and management efforts. Interestingly, the natural fragmentation that encouraged the development of the remarkable diversity of native trouts in the arid West is exaggerated by these programs. Native stocks are far more isolated in even smaller populations than ever before, creating a situation similar to that for pupfishes and other isolated species. These trout will differentiate in time to produce populations different from the original.

The same situation prevails in some small nongame species from lowland habitats, such as the Gila topminnow (*Poeciliopsis o. occidentalis*). This live-bearer lived in a variety of habitats, including the margins of large rivers. It must have formed an almost continuous population at low elevations throughout the Gila River system, at least in wetter periods. Propagation and widespread reintroduction have reestablished it as disjunct populations in small habitats (Simons et al. 1989; Minckley et al., *this volume*, chap. 15), an extension of the strategy used to preserve restricted endemics and trouts. Recovery, if it is to resemble the natural state, will require provision for gene flow among populations. The long-term genetic implications of isolating populations of formerly widespread taxa are beginning to be addressed (Meffe and Vrijenhoek 1988; in this volume, Echelle, chap. 9; Minckley et al., chap. 15).

Large species, such as lakesuckers (species of *Chasmistes* and *Deltistes*), formerly restricted to single, large (but rare) habitats of natural western lakes, have presented other problems, as well as diverse opportunities for recovery. Scoppettone and Vinyard (*this volume*, chap. 18) demonstrate that a strategy of intervention designed to circumvent problems caused by interruption of critical life-history events for cui-ui (*Chasmistes cujus*) can be successful.

The other lakesuckers have similar life-history attributes and problems of survival, and successful recovery could similarly be accomplished for all. The fact that an apparently successful recovery program has only been developed for cui-ui, and not for other lakesuckers, is a consequence of political and social conditions rather than a lack of knowledge. Cui-ui are fortunate to live in a lake and spawn in a river also used by the desirable Lahontan cutthroat trout (*Oncorhynchus clarki henshawi*), the presence of which adds significant local political impetus and justification to development of facilities and technology. Cui-ui are also important to the Pyramid Lake Paiute Indian tribe, which increases their political and social value.

The widely variable climatic and hydrologic conditions of the Great Basin certainly give creative managers the opportunity to encourage natural reproduction of lakesuckers in wetter years, provided the technology and planning are in place. For species that live thirty to fifty years (Scoppettone 1988), society could surely accommodate two or three opportunities to reproduce! Some supplemental artificial propagation might also be necessary, as well as attempts to find opportunities to minimize predation, competition, and declining water quality and quantity. These are not inconsequential problems, but they are solvable ones, especially as we come to grips with the reality of living in a world of limits.

Perhaps the most complex conservation effort during the past two decades has been directed toward recovery of main-stem Colorado River fishes (Wydoski and Hamill, *this volume*, chap. 8). All are large, long-lived species that once roamed undammed rivers. In the lower basin, from which two of the three species disappeared early, hatchery propagation and reintroduction have been the primary management strategies (Johnson and Jensen, *this volume*, chap. 13). Research has been extensive on remnant populations living under artificial and seminatural conditions. As a result, as with lakesuckers, present technology and knowledge can ensure perpetuation of these fishes if social and political hurdles can be overcome (Minckley et al., *this volume*, chap. 17).

These endangered big-river fishes persist in wild populations in the upper Colorado River basin, where emphasis has been on perpetuation of existing stocks. A complex interagency program was developed, involving purchase of water rights to preserve in-stream flows, habitat modifications aimed at enhancing native species, control of competitive and predatory non-native fishes, and hatchery propagation and stocking, along with the negotiated accommodation of diverse water-development interests (USFWS 1987a). In our view, the last purpose of this Endangered Fishes Recovery Implementation Program (Wydoski and Hamill, *this volume*, chap. 8)—allowing states, agencies, and private organizations to proceed with water development while attempts are made to recover the endangered fishes—fails to recognize a world of limits. It does, however, recognize the existence of the "real world," where economic development, political expediencies, and biological imperative interact, and sometimes clash, to forge a definition of social values.

Perhaps this program will succeed; we certainly hope so. However, we would be remiss if we did not express our fear that it will fail, or succeed only partially. The water developments will be in place, and the fishes will have suffered the consequences. At a minimum, it is clear that the projected fifteen-year life of the Recovery Implementation Program must be extended in perpetuity if it is to succeed, and we urge its proponents and administrators to plan for that contingency.

Prospects

During the past two decades it has become evident that knowledge no longer limits our ability to protect native fishes. Most endangered species can be recovered, if we choose. When we lack knowledge, it can be developed quickly enough to permit an adequate management response, provided the will exists to pursue the answers. In the past, fishes were lost as inevitable consequences of development, with no thought that it could be otherwise, and frequently without awareness of the loss, even by the people directly responsible for it. Failure of springs caused by pumping of groundwater resulted in the unheralded extinction of Las Vegas dace (*Rhinichthys deaconi*; R. R. Miller 1984) and Raycraft Ranch and Pahrump Ranch poolfishes (*Empetrichthys latos concavus, E. l. pahrump*) prior to 1958 (Minckley and Deacon 1968). Construction and operation of irrigation works contributed to the demise of Pahranagat spinedace (*Lepidomeda altivelis*; R. R. Miller and Hubbs 1960) and the Pyramid Lake strain of Lahontan cutthroat trout (LaRivers 1962). In all these cases, even with the trout, there is little evidence to suggest that the fishes were given more than a passing thought.

Specific efforts by state and federal management agencies to prevent extinction of native nongame fishes began in the West in 1967 (Williams, *this volume*, chap. 11). R. R. Miller et al. (1989) reported that at least eight western taxa (whether or not formally described) became extinct between 1967 and 1989. J. E. Williams et al. (1989) noted improvement in the status of only six others between 1979 and 1989, while that of fifteen more declined. Efforts were made to prevent extinction of only three of the eight that disappeared (in part, Minckley et al. *this volume*, chap. 15). Four of the six taxa that showed improvement in status did so as a result of deliberate conservation efforts. The other two improved due to circumstances unrelated to conservation. The Fish Creek Springs tui chub (*Gila bicolor euchila*) apparently improved because of increased habitat resulting from either a greater spring flow or changes in water use (Baugh et al. 1986). The Zuñi bluehead sucker (*Pantosteus discobolus yarrowi*) probably improved because of a wet climatic cycle. Some conservation activities were also directed toward improvement of conditions for four of the fifteen taxa whose status declined. To our knowledge, the remaining eleven that declined were not subjected to any concerted conservation efforts, although some may have experienced status

surveys or other activity directed toward gathering information.

It is apparent that most western fishes have declined or become extinct because institutional or social forces to prevent such events failed to materialize. The necessary response should originate within federal or state agencies charged with responsibilities for resource management. These responsibilities have, however, been embraced by western states with varying degrees of enthusiasm. Many federal agencies (or subunits within them) have responded with about the same degree of enthusiasm, sometimes apparently viewing administration of or even compliance with the ESA as an added chore rather than a legal and moral responsibility.

Individual citizens or conservation organizations have been forced to attempt to compel agencies to discharge their obligations toward endangered species and habitats. In this volume, Pister (chap. 4) and Deacon and Williams (chap. 5) provide details of such struggles concerning fishes in the Death Valley region, and Williams and Deacon (chap. 7) broaden the picture to other areas. Private organizations such as The Nature Conservancy (TNC) have responded to the need to set aside habitat for endangered biota. In some instances TNC purchased lands that were passed on to the USFWS or some other governmental agency for management, but some reserves continue to be operated privately (Williams, *this volume*, chap. 11).

As is pointed out in the chapters just cited, and also by Hendrickson and Brooks (chap. 16), Minckley et al. (chap. 17), and Tyus (chap. 19), some recent amendments to the ESA have complicated conservation efforts. We question the advisability, for example, of designating *any* population of an endangered species as "experimental, *nonessential*" (emphasis ours), and thus essentially removing them from protection under the ESA. We originally supported establishing the "experimen-tal" designation for its potential to further research, a critical part of recovery, and we continue our support in that context. How-ever, in practice there are some hard questions to be answered: What is the relationship of "nonessential" populations to recovery? Is an established, viable, "nonessential" stock of an endangered species acceptable as evidence that a species is recovered and thus eligible for downlisting or delisting?

In practice, some agencies appear to inten-sify their intransigence in dealing with endan-gered species *unless* they are classed as experi-mental, nonessential. After the classification is achieved, they participate, for example, by aiding in surveys for suitable sites and allow-ing reintroductions to occur after habitat is identified. In most instances, however, no funding is provided for studies or even moni-toring of nonessential populations; while in others, nonessential reintroduced populations seem to absorb a substantial proportion of funds available. We suppose (or perhaps hope), in the final analysis, that if other efforts at recovery should fail, success with such a stock would contribute to the salvation of a species. Thus, benefits would accrue without expenditure of funds or assumption of respon-sibility, "something for nothing," which, in anyone's view, is the best of all available op-tions. We question the ethics, legality, and most of all the advisability of such courses of action when dealing with the possibility for extinction.

Conclusions

Local and regional battles against extinction have become major factors in the larger war guiding society toward recognition of the need to change the premises upon which we base our present social, economic, and politi-cal systems. These systems produced the trends leading toward destabilization of the

Earth's ecosystem. Disappearing species are harbingers of this problem. People are coming to realize the enormous consequences of global warming, acid rain, depletion of the ozone layer, proliferation of toxic wastes, and loss of biodiversity. More slowly, the populace is becoming aware that these unacceptable trends are caused by the twin demons of popu-lation growth and resource use, and that the only solution lies in reducing both. The ines-capable conclusions are: (1) we must convert our growth economy to a steady state in which conservation is more important than material consumption; (2) population reduc-tion to long-term, worldwide carrying capac-ity is a desirable goal; and (3) international cooperation must be recognized as essential to achieving ecosystem stability.

Such sweeping changes require understand-ing of the human role as a plain member and citizen of the biotic community. Just as we have learned that it is right to have ethical standards guiding our relationships with other people, we must now develop ethical stan-dards to guide our relationships with the natu-ral world, in which *Homo sapiens* is only one of several million species having rights. To in-sist on exclusive rights for a single species, which causes perturbations in life-support sys-tems of all species, is a bankrupt ethic.

A new environmental ethic must be forged and assimilated into societies' world view. Leopold's (1949) observation that "the land-relation is still strictly economic, entailing privileges but not obligations" remains gener-ally true. Most people in the Western world will affirm some degree of stewardship re-sponsibility toward land, plants, and animals. The somewhat stronger idea of duties to the natural world is less easily acknowledged. Those who have acquired some understand-ing of the world's environmental problems are able to develop an environmental ethic ac-knowledging rights for natural objects. While the literature in this area is becoming extensive and more available, a consensus has by no means been achieved. Nevertheless, worldwide demands for environmentally sensitive, sustainable development have a solid philosophical foundation.

The battle against extinction for western fishes is playing a role in the evolution of the formal (Rolston, *this volume*, chap. 6) and the popular (Pister, *this volume*, chap. 4; Deacon and Williams, chap. 5) conceptions of stewardship, duties, and rights regarding natural objects. Legislation and results of litigation support and define these duties, and move toward according rights (Williams and Deacon, *this volume*, chap. 7). While progress is being made, both in acceptance of environmental values and in programs to reduce loss of biodiversity, we still lose more battles than we win. The proximate cause for failure is more often attributable to bureaucratic intransigence than to inadequate legislation or knowledge. The ultimate cause is growth of resource use by a growing population that fails to recognize or acknowledge the rights of and its duties toward natural objects. It is obvious that this situation cannot continue indefinitely. How long it does go on will depend on how vigorously we participate. The alternative is a system irreversibly changed to a new, unknown state, a legacy to be endured by our descendants through no fault of their own.

It is already recognized that biotic diversity cannot be maintained without constant attention in a world so completely dominated by humans. But what else should humans do but take care of their environment? They must either rationally nurture and conserve resources or be subject to stochastic events resulting from natural and self-induced factors that control human populations. Constant attention to the details of maintaining biotic diversity—the earth's life-support system—is not an option; it is mandatory.

The enigma of endangered species remains with us. Rare and disappearing animals and plants are deterrents to development in some people's eyes, but there are now laws that must be dealt with. Other people consider these species important and essential to preserve. In fact, rare and endangered species have become the standard-bearers of the conservation movement, signals of our seriously deteriorated ecosystem. Loss of species, or, better stated perhaps, the continuing environmental degradation their extinctions document, are a constant and vivid reminder of where and when self-imposed limits must be applied so that biological diversity can be maintained. We are convinced that some form of conservation biology will become an essential element in human social institutions. Otherwise, humans will proceed with the unacceptable degradation of earth's ecosystem, with continued losses and catastrophic results for the biosphere upon which we depend. If so, our immediate and long-term future as well will be far different and darker than the one we visualize today.

Literature Cited

Abbott, C. C. 1861. Descriptions of four new species of North American Cyprinidae. *Proceedings of the Philadelphia Academy of Natural Sciences* 12 (1860), 473–474.

Abbott, I. 1983. The meaning of *z* in species/area regressions and the study of species turnover in island biogeography. *Oikos* 41, 385–390.

Abele, L. G., and E. F. Connor. 1979. Application of island biogeography theory to refuge design: Making the right decision for the wrong reasons. In *Proceedings of the First Conference on Scientific Research in the National Parks*, ed. R. M. Linn, pp. 89–94. Washington, D.C.: U.S. National Park Service Transactions and Proceedings Series 5.

Adler, J. 1984. Edge of extinction. *Newsweek*, 30 January 1984, 7.

Allan, C. 1983. Interbasin report. *Proceedings of the Desert Fishes Council* 6 (1974), 151.

Allan, R. C., and D. L. Roden. 1978. Fish of Lake Mead and Lake Mohave. *Nevada Department of Wildlife Biological Bulletin* 7, 1–105.

Allendorf, F. W. 1986. Genetic drift and the loss of alleles versus heterozygosity. *Zoo Biology* 5, 181–190.

Allendorf, F. W., and R. F. Leary. 1988. Conservation and distribution of genetic variation in a polytypic species, the cutthroat trout. *Conservation Biology* 2, 170–184.

Allendorf, F. W., and S. R. Phelps. 1981. Isozymes and the preservation of genetic variation in salmonid fishes. In *Fish gene pools*, ed. N. Ryman, pp. 37–52. Ecological Bulletins 34. Stockholm, Sweden: The Editorial Service.

Allendorf, F. W., and N. Ryman. 1987. Genetic management of hatchery stocks. In *Population genetics and fishery management*, ed. N. Ryman and F. Utter, pp. 141–160. Seattle: University of Washington Press.

Allendorf, F. W., N. Ryman, and F. M. Utter. 1987. Genetics and fishery management. In *Population genetics and fishery management*, ed. N. Ryman and F. Utter, pp. 1–19. Seattle: University of Washington Press.

Almada, P., and S. Contreras Balderas. 1984. El Bolsón de Cuatro Ciénegas, Coah., Méx. *Reunion Registrata Ecologia, Norte* 1, 125–129. Mexico, D. F., Mexico: Secretaria de Desarrollo Urbano y Ecologia.

Altukhov, Y. P., and E. A. Salmenkova. 1987. Stock transfer relative to natural organization, management, and conservation of fish populations. In *Population genetics and fishery management*, ed. N. Ryman and F. Utter, pp. 322–354. Seattle: University of Washington Press.

Ammerman, L. K. 1988. Biochemical genetics of endangered Colorado squawfish (*Ptychocheilus lucius*) populations. U.S. Fish and Wildlife Service, San Marcos National Fish Hatchery and Research Center, San Marcos, Tex.

Ammerman, L. K., and D. C. Morizot. 1989. Biochemical genetics of endangered Colorado squawfish populations. *Transactions of the American Fisheries Society* 118, 435–440.

Andreasen, J. K. 1975. Systematics and status of the family Catostomidae in southern Oregon. Ph.D. diss., Oregon State University, Corvallis.

Anonymous. 1973. Lake Mead bass study. U.S. Bureau of Reclamation, Lower Colorado River Region, Boulder City, Nev.

Anonymous. 1977. Endemic fishes of the Colorado River system, a status report. Colorado River Wildlife Council, Las Vegas, Nev.

Arai, N. P., and D. R. Mudry. 1983. Protozoan and metazoan parasites of fishes from the headwaters of the Parsnip and McGregor rivers, British Columbia: A study of possible parasite transfaunations. *Canadian Journal of Fisheries and Aquatic Science* 40, 1676–1684.

Archer, D. L., L. R. Kaeding, B. D. Burdick, and C. W. McAda. 1985. A study of the endangered fishes of the upper Colorado River. Final Report for Northern Colorado Water Conservation District and U.S. Fish and Wildlife Service Cooperative Agreement 14-16-0006-82-959R. U.S. Fish and Wildlife Service, Grand Junction, Colo.

Archer, D. L., L. R. Kaeding, and H. M. Tyus. 1986. Colorado River fishes monitoring project. Final Report for U.S. Bureau of Reclamation Contract 2-07-40-L3083. U.S. Fish and Wildlife Service, Lakewood, Colo.

Archer, D. L., and H. M. Tyus. 1984. Colorado squawfish spawning study, Yampa River. U.S. Fish and Wildlife Service, Salt Lake City, Utah.

Archer, D. L., H. M. Tyus, L. R. Kaeding, C. W. McAda, and B. D. Burdick. 1984. Colorado River fishes monitoring project. U.S. Fish and Wildlife Service, Salt Lake City, Utah.

Arizona Game and Fish Department. 1960–1973. Statewide fishery investigations. Federal Aid to Fisheries Restoration Project Completion Report, F-7-R, variable pagination. Arizona Game and Fish Department, Phoenix.

Arnold, D. E. 1966. Marking fish with dyes and other chemicals. *U.S. Fish and Wildlife Service Technical Paper* 10, 1–44.

Arnold, E. T. 1972. Behavioral ecology of pupfishes (genus *Cyprinodon*) from the Cuatro Ciénegas Basin, Coahuila, México. Ph.D. diss., Arizona State University, Tempe.

Asbury, K., W. J. Matthews, and L. G. Hill. 1981. Attraction of *Notropis lutrensis* (Cyprinidae) to water conditioned by the presence of conspecifics. *The Southwestern Naturalist* 25, 525–528.

Atapattu, S., and C. S. Wickremasinghe. 1974. Sri Lanka's Gal Oya National Park: Aspects and prospects. *Biological Conservation* 6, 219–222.

Avery, E. L. 1978. The influence of chemical reclamation on small brown streams in southwestern Wisconsin. *Wisconsin Department of Natural Resources Technical Bulletin* 110, 1–35.

Axelrod, D. I. 1979. Age and origin of Sonoran Desert vegetation. *Occasional Papers of the California Academy of Science* 132, 1–74.

———. 1983. Paleobotanical history of the western deserts. In *Origins and evolution of deserts*, ed. S. G. Wells and D. R. Haragan, pp. 113–129. Albuquerque: University of New Mexico Press.

Azevedo, R. 1962a. Fishery Management Report, Green River Survey. U.S. Bureau of Sport Fisheries and Wildlife, Albuquerque, N.M.

———. 1962b. Fishery Management Report, Dinosaur National Monument—Green River fish collections. U.S. Bureau of Sport Fisheries and Wildlife, Albuquerque, N.M.

———. 1962c. Log: Numbers of fish collected before and after eradication, Green River. U.S. Bureau of Sport Fisheries and Wildlife, Springerville, Ariz. Interoffice memorandum to regional director, U.S. Bureau of Sport Fisheries and Wildlife, Albuquerque, N.M.

Bailey, R. M. 1951. Division of fishes. In *Annual report of the director, Museum of Zoology, University of Michigan*, pp. 20–23, 52 (72). Ann Arbor: University of Michigan Press.

Bailey, R. M., and C. E. Bond. 1963. Four new species of freshwater sculpins, genus *Cottus*, from western North America. *Occasional Papers of the Museum of Zoology, University of Michigan* 634, 1–27.

Baird, S. F., and C. Girard. 1853. Fishes. In *Report of an expedition down the Zuñi and Colorado rivers*, by Capt. L. Sitgreaves, pp. 148–152. Washington, D.C.: U.S. Senate Executive Document 59, Thirty-third Cong., second sess.

Bangham, R. V., and N. L. Bennington. 1939. Movement of fish in streams. *Transactions of the American Fisheries Society* 68, 256–262.

Banks, J. L. 1964. Fish species distribution in Dinosaur National Monument during 1961–1962. Master's thesis, Colorado State University, Fort Collins.

Barbour, C. D., and J. H. Brown. 1974. Fish species diversity in lakes. *American Naturalist* 108, 473–489.

Barlow, G. W. 1958a. Daily movements of desert pupfish, *Cyprinodon macularius*, in shore pools of the Salton Sea, California. *Ecology* 39, 580–587.

———. 1958b. High salinity mortality of desert pupfish, *Cyprinodon macularius*. *Copeia* 1958, 231–232.

———. 1961. Social behavior of the desert pupfish, *Cyprinodon macularius*, in the field and in the aquarium. *American Midland Naturalist* 65, 339–358.

Bartlett, J. P. 1854. *Personal narrative of exploration and incidents in Texas, New Mexico, California, Sonora, and Chihuahua, connected with the U.S. and Mexican Boundary Commission during the years of 1850, '51, '52, and '53*, vols. 1 and 2. New York: D. Appleton.

Bassett, H. M. 1962. Statewide fishery investigations: A manipulation of environmental conditions pertaining to minor jobs of a developmental nature in District II. Federal Aid to Fisheries Restoration Project Completion Report, F-7-R-5, 1–5. Arizona Game and Fish Department, Phoenix.

Bateman, R. L., A. L. Mindling, and R. L. Naff. 1974. Development and management of ground water in relation to preservation of desert pupfish in Ash Meadows, southern Nevada. *University of Nevada, Desert Research Institute, Technical Report Series H-W, Hydrology Water Resources Publication* 17, 1–39.

Baugh, T. M., and J. E. Deacon. 1983. Daily and yearly movement of the Devils Hole pupfish *Cyprinodon diabolis* Wales in Devil's Hole, Nevada. *Great Basin Naturalist* 43, 592–596.

———. 1988. Evaluation of the role of refugia in conservation efforts for the Devils Hole pupfish, *Cyprinodon diabolis* Wales. *Zoo Biology* 7, 351–358.

Baugh, T. M., J. E. Deacon, and P. Fitzpatrick. 1988. Reproduction and growth of the Pahrump poolfish (*Empetrichthys latos latos*) in the laboratory and in nature. *Journal of Aquariculture and Aquatic Science* 5, 1–5.

Baugh, T. M., J. E. Deacon, and D. Withers. 1985. Conservation efforts with the Hiko White River springfish, *Crenichthys baileyi grandis* Williams and Wilde. *Journal of Aquariculture and Aquatic Science* 4, 49–53.

Baugh, T. M., J. W. Pedretti, and J. E. Deacon. 1986. Status and distribution of the Fish Creek Springs tui chub, *Gila bicolor euchila*. *Great Basin Naturalist* 46, 441–444.

Baxter, G. T., and J. R. Simon. 1970. Wyoming fishes. *Wyoming Game and Fish Department Bulletin* 4, 1–168.

Baxter, R. M. 1977. Environmental effects of dams and impoundments. *Annual Review of Ecology and Systematics* 8, 255–283.

Beak Consultants. 1987. Shortnose and Lost River sucker studies: Copco Reservoir and the Klamath River. Final Report to the City of Klamath Falls, Oregon. Beak Consultants, Portland, Ore.

Beamsderfer, R. C., and J. L. Congleton. 1982. Spawning behavior, habitat selection and early life history of northern squawfish, with inferences to Colorado squawfish. In Part 3, *Colorado River Fisheries Project, final report contracted studies*, ed. W. H. Miller, J. J. Valentine, D. L. Archer, H. M. Tyus, R. A. Valdez, and L. R. Kaeding, pp. 27–110. Final Report for U.S. Bureau of Reclamation Contract 9-07-40-L-1016, Memorandum of Understanding CO-910-MU9-933, U.S. Bureau of Land Management. U.S. Fish and Wildlife Service, Salt Lake City, Utah.

Bean, M. J. 1978. Federal wildlife law. In *Wildlife and America*, ed. H. B. Brokaw, pp. 279–289. Washington, D.C.: Council for Environmental Quality.

———. 1983. *The evolution of national wildlife law.* New York: Praeger Scientific.

Beatley, J. C. 1971. Vascular plants of Ash Meadows, Nevada. University of California Laboratory of Nuclear Medicine and Radiation Biology, Report 12–845.

———. 1977. Ash Meadows. *Mentzelia* 3, 20–35.

Becker, G. 1975. Fish toxification: Biological sanity or insanity? In *Rehabilitation of fish populations with toxicants: A symposium*, ed. P. H. Eschmeyer, pp. 41–53. Bethesda, Md.: North Central Division, American Fisheries Society, Special Publication 4.

Beckman, W. C. 1953. Guide to the fishes of Colorado. *Colorado State Museum Leaflet* 11, 1–110.

———. 1963. *Guide to the fishes of Colorado.* Boulder: Colorado State Museum.

Bednarz, J. C. 1979. Ecology and status of the Pecos gambusia, *Gambusia nobilis* (Poeciliidae) in New Mexico. *The Southwestern Naturalist* 24, 311–322.

Behnke, R. J. 1968. Rare and endangered species: The native trouts of North America. *Proceedings of the Annual Meeting of the Western Association of State Game and Fish Commissioners* 48, 530–533.

———. 1972. The systematics of salmonid fishes of recently glaciated lakes. *Journal of the Fisheries Research Board of Canada* 29, 639–671.

———. 1973. Humpback sucker, *Xyrauchen texanus*. Threatened and Endangered Fish Report. Colorado Cooperative Fisheries Unit, Colorado State University, Fort Collins.

———. 1979. *Monograph of the native trouts of the genus* Salmo *of western North America.* Lakewood, Colo.: U.S. Department of Agriculture, Forest Service.

———. 1980. The impacts of habitat alterations on the endangered and threatened fishes of the upper Colorado River basin: A discussion. In *Energy development in the Southwest: Problems of water, fish and wildlife in the upper Colorado River basin*, vol. 2, ed. W. O. Spofford, Jr., A. L. Parker, and A. V. Kneese, pp. 182–192. Research Paper R-18. Washington, D.C.: Resources for the Future.

Behnke, R. J., and D. E. Benson. 1980. Endangered and threatened fishes of the upper Colorado River basin. *Colorado State University Cooperative Extension Service Bulletin* 503A, 1–34.

Behnke, R. J., and M. Zarn. 1976. *Biology and management of threatened and endangered trouts.* Washington, D.C.: U.S. Department of Agriculture, Forest Service General Technical Report RM-28, 1–45.

Beland, R. D. 1953. The effect of channelization on the fishery of the lower Colorado River. *California Fish and Game* 39, 137–139.

Beleau, M. H., and J. A. Bartosz. 1982. Acute toxicity of selected chemicals: Data base. In Part 3, *Colorado River Fisheries Project, final report contracted studies*, ed. W. H. Miller, J. J. Valentine, D. L. Archer, H. M. Tyus, R. A. Valdez, and L. R. Kaeding, pp. 242–254. Final Report for U.S. Bureau of Reclamation Contract 9-07-40-L-1016, Memorandum of Understanding CO-910-MU9-933, U.S. Bureau of Land Management. U.S. Fish and Wildlife Service, Salt Lake City, Utah.

Berg, W. J. 1987. Evolutionary genetics of rainbow trout, *Parasalmo gairdneri* Richardson. Ph.D. diss., University of California, Davis.

Berg, W. J., and G. A. E. Gall. 1988. Gene flow and genetic differentiation among California coastal rainbow trout populations. *Canadian Journal of Fisheries and Aquatic Science* 45, 122–131.

Berger, B. L., R. E. Lennon, and J. W. Hogan. 1969. Laboratory studies on antimycin-A as a fish toxicant. *U.S. Bureau of Sport Fisheries and Wildlife, Investigations in Fish Control* 26, 1–19.

Bergman, P. K., K. B. Jefferts, H. F. Fiscus, and R. C. Hager. 1968. A preliminary evaluation of an implanted coded wire fish tag. *Washington Department of Fisheries, Fisheries Research Paper* 3, 63–84.

Berry, C. R., Jr. 1988. Effects of cold shock on Colorado squawfish larvae. *The Southwestern Naturalist* 33, 193–197.

Berry, C. R., Jr., and L. R. Kaeding. 1986a. Growout ponds useful for Colorado squawfish culture. *U.S. Fish and Wildlife Service Research Information Bulletin* 86-42, 1.

———. 1986b. Largemouth bass predation on stocked Colorado squawfish. *U.S. Fish and Wildlife Service Research Information Bulletin* 86-45, 1.

Berry, C. R., Jr., and R. Pimentel. 1985. Swimming performance of three rare Colorado River fishes. *Transactions of the American Fisheries Society* 114, 397–402.

Bersell, P. O. 1973. Vertical distribution of fishes relative to physical, chemical and biological features in two central Arizona reservoirs. Master's thesis, Arizona State University, Tempe.

Bestgen, K. R., S. P. Platania, J. E. Brooks, and D. L. Propst. 1989. Dispersal and life history traits of *Notropis girardi* (Cypriniformes: Cyprinidae), introduced into the Pecos River, New Mexico. *American Midland Naturalist* 122, 228–235.

Bienz, C. S., and J. S. Ziller. 1987. Status of three lacustrine sucker species (Catostomidae). Klamath Indian Tribe and Oregon Department Fish and Wildlife, Chiloquin.

Binns, N. A. 1965. Effects of rotenone treatment on the fauna of the Green River, Wyoming. Master's thesis, Oregon State University, Corvallis.

———. 1967a. Effects of rotenone treatment on the fauna of the Green River, Wyoming. *Wyoming Game and Fish Commission, Fisheries Technical Bulletin* 1, 1–114.

———. 1967b. Fishery survey of Fontenelle Reservoir and the Green River downstream to Flaming Gorge Reservoir; Addendum, Green River and Flaming Gorge Reservoir post-impoundment investigations. Wyoming Game and Fish Department, Cheyenne.

———. 1978. Habitat structure of Kendall Warm Spring, with reference to the endangered Kendall Warm Springs dace, *R. o. thermalis*. *Wyoming Game and Fish Commission, Fisheries Technical Bulletin* 4, 1–145.

Binns, N. A., F. Eiserman, F. W. Jackson, A. F. Regenthal, and R. Stone. 1963. The planning, operation, and analysis of the Green River fish control project. Wyoming Game and Fish Department, Laramie, and Utah Department of Fish and Game, Salt Lake City.

Bishop, A. B., M. D. Chambers, W. O. Mace, and D. W. Mills. 1975. Water as a factor in energy resource development. *Prjero* (Water Research Laboratory, Utah State University, Logan) 28-1, 1–102.

Bisson, P. A., and R. E. Bailey. 1982. Avoidance of suspended sediment by juvenile coho salmon. *North American Journal of Fisheries Management* 4, 371–374.

Black, G. F. 1980. Status of the desert pupfish, *Cyprinodon macularius* (Baird and Girard), in California. *California Department of Fish and Game Special Publication* 80-1, 1–42.

Black, T., and R. V. Bulkley. 1985a. Growth rate of yearling Colorado squawfish at different water temperatures. *The Southwestern Naturalist* 30, 253–257.

———. 1985b. Preferred temperature of yearling Colorado squawfish. *The Southwestern Naturalist* 30, 95–100.

Boecklen, W. J. 1986. Optimal design of nature reserves: Consequences of genetic drift. *Biological Conservation* 38, 323–338.

Boecklen, W. J., and N. J. Botelli. 1984. Island biogeographic theory and conservation practice: Species-area or specious-area relationships? *Biological Conservation* 29, 63–80.

Booth, W. 1988. Reintroducing a political animal. *Science* 241, 154–158.

Bosley, C. E. 1960. Pre-impoundment study of the Flaming Gorge Reservoir. *Wyoming Game and Fish Commission Technical Report* 9, 1–81.

Bouma, R. C. 1984. Contribution to the management of *Gambusia nobilis* at Bitter Lake National Wildlife Refuge. Final Report for U.S. Fish and Wildlife Service Contract 14-16-002-81-214. Cornell University, Ithaca, N.Y.

Bovee, K. D. 1982. A guide to stream habitat analysis using the instream flow incremental methodology. Instream Flow Information Paper 12. *U.S. Fish and Wildlife Service Report FWS/OBS-82/26*, 1–248.

Bowler, B. 1975. Factors influencing genetic control in lakeward migrations of cutthroat trout fry. *Transactions of the American Fisheries Society* 104, 474–482.

Bowman, T. H. 1981. *Thermosphaeroma milleri* and *T. smithi*, new sphaeromatid isopod crustaceans from hot springs in Chihuahua, Mexico, with a review of the genus. *Journal of Crustacean Biology* 1, 105–122.

Bozek, M. A., L. J. Paulson, and J. E. Deacon. 1984. Factors affecting reproductive success of bonytail chubs and razorback suckers in Lake Mohave. Final Report for U.S. Fish and Wildlife Service Contract 14-16-0002-81-251. University of Nevada, Las Vegas.

Bradley, W. G., and E. L. Cockrum. 1968. A new subspecies of meadow vole (*Microtus pennsylvanicus*) from northwestern Chihuahua, Mexico. *American Museum Novitiates* 2325, 1–7.

Branson, B. A., and C. J. McCoy, Jr. 1966. Observations on breeding tubercles in *Xyrauchen texanus* (Abbott). *The Southwestern Naturalist* 11, 301.

Branson, B. A., M. E. Sisk, and C. J. McCoy, Jr. 1966. *Ptychocheilus lucius* from Salt River, Arizona. *The Southwestern Naturalist* 11, 300.

Bratton, S. P. 1984. Christian ecotheology and the Old Testament. *Environmental Ethics* 7, 117–133.

Braun, R. H. 1986. Emerging limits in federal land management discretion, livestock, riparian ecosystems, and clean water law. *Environmental Law* 17, 43–79.

Bres, M. 1978. The embryonic development of the cui-ui, *Chasmistes cujus* (Teleostei, Catostomidae). Master's thesis, University of Nevada, Reno.

Brewer, W. H. 1904. William H. Brewer Papers, Yale Historical Manuscript Collection: "Forest Physiography," Lecture 24, 24 March, Box 40/Folder 203 [not seen].

Brienholt, J. C., and R. A. Heckman. 1980. Parasites from two species of suckers (Catostomidae) from southern Utah. *Great Basin Naturalist* 40, 149–156.

Briggs, J. C. 1986. Introduction to the zoogeography of North American fishes. In *The zoogeography of North American freshwater fishes*, ed. C. H. Hocutt and E. O. Wiley, pp. 1–16. New York: John Wiley and Sons.

Brittan, M. R. 1967. The Death Valley fishes—an endangered fauna. *Ichthyologica, Aquarists' Journal* 39, 81–92.

Brönmark, C., J. Herrman, B. Malmqvist, C. Otto, and P. Sjöström. 1984. Animal community structure as a function of stream size. *Hydrobiologia* 112, 73–79.

Brooks, J. E. 1985. Factors affecting the success of Gila topminnow (*Poeciliopsis o. occidentalis*) introductions on four Arizona National Forests. Annual Report to the Office of Endangered Species, U.S. Fish and Wildlife Service. Arizona Game and Fish Department, Phoenix.

———. 1986a. Status of natural and introduced Sonoran topminnow (*Poeciliopsis o. occidentalis*) populations in Arizona through 1985. Annual Report to the Office of Endangered Species, U.S. Fish and Wildlife Service. Arizona Game and Fish Department, Phoenix.

———. 1986b. Reintroduction and monitoring of Colorado squawfish (*Ptychocheilus lucius*) in Arizona, 1985. Annual Report to the U.S. Fish and Wildlife Service. Arizona Game and Fish Department, Phoenix.

————. No date. Reservoir sampling for razorback sucker and bonytail chub. Memorandum, U.S. Fish and Wildlife Service, Dexter National Fish Hatchery, New Mexico (1986).

Brooks, J. E., and P. C. Marsh. 1986. Poststocking dispersal and predation on fingerling razorback suckers in the Gila River, Arizona. Paper presented to the Annual Meeting of the Arizona–New Mexico Chapter, American Fisheries Society (abstract).

Brooks, J. E., P. C. Marsh, and W. L. Minckley. 1985. Reintroduction and monitoring of razorback sucker in the lower Colorado River basin. *Proceedings of the Western Division, American Fisheries Society* 65, 203 (abstract).

Brooks, J. E., and M. K. Wood. 1988. A survey of the fishes of Bitter Lake National Wildlife Refuge, with an historic overview of the game and nongame fisheries. Final Report to the Office of Refuges and Fisheries Resources, U. S. Fish and Wildlife Service. Dexter National Fish Hatchery, New Mexico.

Brown, D. A. 1987. Ethics, science, and environmental regulation. *Environmental Ethics* 9, 331–349.

Brown, D. E. 1988. The challenge and the hope of species reintroduction: Return of the natives. *Wilderness* 52 (183), 40–52.

Brown, E. H., Jr. 1961. Movement of native and hatchery-reared game fish in a warm-water stream. *Transactions of the American Fisheries Society* 90, 449–456.

Brown, J. H. 1971. The desert pupfish. *Scientific American* 225 (5), 104–110.

————. 1978. The theory of insular biogeography and the distribution of boreal birds and mammals. *Great Basin Naturalist Memoirs* 2, 209–227.

Brown, J. H., and C. R. Feldmeth. 1971. Evolution in constant and fluctuating environments: Thermal tolerance of desert pupfish (*Cyprinodon*). *Evolution* 25, 390–398.

Brown, L. B. 1983. Colorado River changes, 1903–1982. Final Report for U.S. Bureau of Reclamation Purchase Order 3-PGH-30-09150. Southwest Natural History Association, Phoenix, Ariz.

Browne, R. A. 1981. Lakes as islands: Biogeographic distribution turnover rates and species composition in the lakes of central New York. *Journal of Biogeography* 8, 75–85.

Bruce, J. 1961. Statewide fishery investigations: Postdevelopment evaluation (investigations). Federal Aid to Fisheries Restoration Project Completion Report, F-7-R-7, 1–2. Arizona Game and Fish Department, Phoenix.

Bruce, R. V. 1987. *The launching of modern American science*. Ithaca, N.Y.: Cornell University Press.

Brune, G. 1975. Major and historical springs of Texas. *Texas Water Development Board Report* 189, 1–94.

Buchanan, C. C., and T. T. Strekal. 1988. Simulated water management and evaluation procedure for cui-ui (*Chasmistes cujus*). U.S. Fish and Wildlife Service, Reno, Nev.

Bulkley, R. V., C. R. Berry, R. Pimentel, and T. Black. 1982. Tolerance and preferences of Colorado River endangered fishes to selected habitat parameters. In Part 3, *Colorado River Fisheries Project, final report contracted studies*, ed. W. H. Miller, J. J. Valentine, D. L. Archer, H. M. Tyus, R. A. Valdez, and L. R. Kaeding, pp. 185–242. Final Report for U.S. Bureau of Reclamation Contract 9-07-40-L-1016, Memorandum of Understanding CO-910-MU9-933, U.S. Bureau of Land Management. U.S. Fish and Wildlife Service, Salt Lake City, Utah.

Bulkley, R. V., and R. Pimentel. 1983a. Concentrations of total dissolved solids preferred or avoided by endangered Colorado River fishes. *Transactions of the American Fisheries Society* 112, 595–600.

————. 1983b. Temperature preference and avoidance by adult razorback suckers. *Transactions of the American Fisheries Society* 112, 601–607.

Bunnell, S. 1970. The desert pupfish. *Cry California* 5, 2–13.

Burns, J. W. 1966. Threadfin shad. In *Inland fisheries management*, ed. A. Calhoun, pp. 481–488. Sacramento: California Department of Fish and Game.

Burr, B. M. 1976. A review of the Mexican stoneroller, *Campostoma ornatum* Girard (Pisces: Cyprinidae). *Transactions of the San Diego Society of Natural History* 18, 127–144.

Burton, D. K. 1987. *Blue Mountain folks, their lives and legends*. Salt Lake City: K/P Graphics.

Busack, C. A., and G. A. E. Gall. 1981. Introgressive hybridization in populations of Paiute cutthroat trout (*Salmo clarki seleniris*). *Canadian Journal of Fisheries and Aquatic Sciences* 38, 939–951.

Buth, D. G., C. B. Crabtree, R. D. Orton, and W. J. Rainboth. 1984. Genetic differentiation between the freshwater subspecies of *Gasterosteus aculeatus* in southern California. *Biochemical Systematics and Ecology* 12, 423–432.

Buth, D. G., and R. W. Murphy. 1984. Genetic characteristics of the flannelmouth sucker (*Catostomus latipinnis*) and of hatchery and native stocks of the razorback sucker (*Xyrauchen texanus*): A data base for studies of introgressive hybridization and population differentiation. Final Report for U.S. Fish and Wildlife Service Job Order 20181-0643. University of California, Los Angeles.

Buth, D. G., R. W. Murphy, and L. Ulmer. 1987. Population differentiation and introgressive hybridization of the flannelmouth sucker and of hatchery and native stocks of the razorback sucker. *Transactions of the American Fisheries Society* 116, 103–110.

Cairns, J., Jr. 1986. Restoration, reclamation, and regeneration of degraded or destroyed habitats. In *Conservation biology: The science of scarcity and diversity*, ed. M. Soulé, pp. 465–484. Sutherland, Mass.: Sinauer Associates.

———. 1988. Increasing diversity by restoring damaged ecosystems. In *Biodiversity*, ed. E. O. Wilson, pp. 333–343. Washington, D.C.: National Academy of Sciences Press.

Callicott, J. B., ed. 1987a. *Companion to a Sand County Almanac*. Madison: University of Wisconsin Press.

———. 1987b. Introductions to *Companion to a Sand County Almanac*, ed. J. B. Callicott, pp. 3–13. Madison: University of Wisconsin Press.

———. 1991. Conservation ethics and fishery management. *Fisheries* 16, 22–28.

Campbell, R. R. 1984. Rare and endangered fishes of Canada: The Committee on the Status of Endangered Wildlife in Canada (COSEWIC) Fish and Marine Mammals Subcommittee. *Canadian Field-Naturalist* 98, 71–74.

———. 1985. Rare and endangered fishes and marine mammals of Canada: COSEWIC Fish and Marine Mammals Subcommittee status report: II. *Canadian Field-Naturalist* 99, 404–408.

———. 1987. Rare and endangered fishes and marine mammals of Canada: COSEWIC Fish and Marine Mammals Subcommittee status report: III. *Canadian Field-Naturalist* 101, 165–170.

———. 1988. Rare and endangered fishes and marine mammals of Canada: COSEWIC Fish and Marine Mammals Subcommittee status report: IV. *Canadian Field-Naturalist* 102, 81–86.

Campton, D. E. 1987. Natural hybridization and introgression in fishes: Methods of detection and genetic interpretations. In *Population genetics and fishery management*, ed. N. Ryman and F. Utter, pp. 161–192. Seattle: University of Washington Press.

Carley, C. J. No date. Activities and findings of the red wolf recovery program from late 1973 to 1 July 1975. U.S. Fish and Wildlife Service, Office of Endangered Species, Albuquerque, N.M.

Carlson, C. A., and R. T. Muth. 1989. The Colorado River: Lifeline of the American Southwest. In *Proceedings of the International Large River Symposium*, ed. D. P. Dodge, pp. 220–239. *Canadian Journal of Fisheries and Aquatic Sciences*, Special publication 106.

———. In press. Endangered species management. In *Management of North American fisheries*, ed. C. C. Kohler and W. A. Hubert. Bethesda, Md.: American Fisheries Society.

Carlson, C. A., C. G. Prewitt, D. E. Snyder, E. J. Wick, E. L. Ames, and W. D. Fronk. 1979. Fishes and macroinvertebrates of the White and Yampa rivers, Colorado. *U.S. Bureau of Land Management Biological Series* 1, 1–276.

Carothers, S. W., and C. O. Minckley. 1981. A survey of the fishes, aquatic invertebrates and aquatic plants of the Colorado River and selected tributaries from Lees Ferry to Separation Rapids. Final Report for U.S. Bureau of Reclamation Contract 7-07030-C0026. Museum of Northern Arizona, Flagstaff.

Carpelan, L. H. 1961. History of the Salton Sea. In *The ecology of the Salton Sea, California, in relation to the sportfishery*, ed. B. W. Walker, pp. 9–15. California Department of Fish and Game Fish Bulletin 113.

Carter, D. 1969. A history of commercial fishing on Utah Lake. Master's thesis, Brigham Young University, Provo, Utah.

Carufel, L. H. 1964. Statewide fisheries investigations: Postdevelopment evaluation (investigations). Federal Aid to Fisheries Restoration Project Completion Report, F-7-R-7, 163–203. Arizona Game and Fish Department, Phoenix.

Casselman, J. M. 1974. Analysis of hard tissue of pike (*Esox lucius* L.) with special reference to age and growth. In *Proceedings of an International Symposium on the Ageing of Fish*, ed. T. B. Bagenal, pp. 13–27. London: Unwin Brothers.

Castro, A. 1971. Steinhart Aquarium log for Devil's Hole pupfish colony, *Cyprinodon diabolis*. *Proceedings of the Desert Fishes Council* 3 (1970), 30–31.

Cavender, T. M. 1978. Taxonomy and distribution of the bull trout, *Salvelinus confluentus* (Suckley), from the American Northwest. *California Fish and Game* 64, 139–174.

———. 1986. Review of the fossil history of North American freshwater fishes. In *The zoogeography of North American freshwater fishes*, ed. C. H. Hocutt and E. O. Wiley, pp. 699–724. New York: John Wiley and Sons.

Chakraborty, R. 1980. Gene-diversity analysis in nested subdivided populations. *Genetics* 96, 721–723.

Chakraborty, R., and D. Leimar. 1987. Genetic variation within a subdivided population. In *Population genetics and fishery management*, ed. N. Ryman and F. Utter, pp. 89–120. Seattle: University of Washington Press.

Chamberlain, F. M. 1904. Notes on fishes collected in Arizona, 1904. Unpublished manuscript, U.S. National Museum, Washington, D.C.

Chamberlain, T. K. 1946. Fishes, particularly the suckers, Catostomidae, of the Colorado River drainage and of the Arkansas River drainage, in relation to the Gunnison-Arkansas transmountain diversion. U.S. Fish and Wildlife Service, College Station, Texas [not seen].

Chambers, S. M., and J. W. Bayless. 1983. Systematics, conservation and the measurement of genetic diversity. In *Genetics and conservation: A manual for managing wild animal and plant populations*, ed. C. M. Schonewald-Cox, S. M. Chambers, B. MacBryde, and L. Thomas, pp. 349–363. Menlo Park, Calif.: Benjamin/Cummings.

Chatto, D. A. 1979. Effects of salinity on hatching success of the cui-ui. *Progressive Fish-Culturist* 41, 82–85.

Chernoff, B. 1985. Population dynamics of the Devils Hole pupfish. *Environmental Biology of Fishes* 13, 139–147.

Chernoff, B., and R. R. Miller. 1982. *Notropis bocagrande*, a new cyprinid fish from Chihuahua, México, with comments on *Notropis formosus*. *Copeia* 1982, 514–522.

Cobb, J. B. 1979. Christian existence in a world of limits. *Environmental Ethics* 1, 149–158.

———. 1988. A Christian view of biodiversity. In *Biodiversity*, ed. E. O. Wilson, pp. 481–485. Washington, D.C.: National Academy of Sciences Press.

Cokendolpher, J. C. 1980. Hybridization experiments with the genus *Cyprinodon* (Teleostei: Cyprinodontidae). *Copeia* 1980, 173–176.

Colbert, E. H. 1984. *The great dinosaur hunters and their discoveries*. Mineola, N.Y.: Dover Publications.

Cole, B. J. 1982. Colonizing abilities, island size, and the number of species on archipelagoes. *American Naturalist* 117, 629–638.

Cole, G. A., and C. A. Bane. 1978. *Thermosphaeroma subequalum*, n. gen., n. sp. (Crustacea: Isopoda) from Big Bend, Texas. *Hydrobiologia* 59, 225–228.

Coleman, G. A. 1929. A biological survey of the Salton Sea. *California Fish and Game* 15, 218–227.

Coleman, M. E., J. Kann, and G. G. Scoppettone. 1988. Life history and ecological investigations of catostomids from the upper Klamath Basin, Oregon. Klamath Indian Tribe, Chiloquin, Ore., U.S. Fish and Wildlife Service, Reno, Nev.

Coleman, M. E., D. Winkelman, and H. Burge. 1987. Cui-ui egg and larvae temperature tolerance evaluations under controlled fluctuating temperature regimes. National Fisheries Research Center, U.S. Fish and Wildlife Service, Seattle, Wash.

Collins, J. P., C. Young, J. Howell, and W. L. Minckley. 1981. Impact of flooding in a Sonoran Desert stream, including elimination of an endangered fish population (*Poeciliopsis o. occidentalis*, Poeciliidae). *The Southwestern Naturalist* 26, 415–423.

Connell, J. H. 1978. Diversity in tropical rain forests and coral reefs. *Science* 199, 1302–1310.

———. 1983. On the prevalence and relative importance of interspecific competition: Evidence from field experiments. *American Naturalist* 122, 661–691.

Connor, E. F., and E. D. McCoy. 1979. The statistics and biology of the species-area relationship. *American Naturalist* 113, 791–833.

Constantz, G. D. 1974. Reproductive effort in *Poeciliopsis occidentalis* (Poeciliidae). *The Southwestern Naturalist* 19, 47–52.

———. 1975. Behavioral ecology of mating in the male Gila topminnow, *Poeciliopsis occidentalis* (Cyprinodontiformes: Poeciliidae). *Ecology* 56, 966–973.

———. 1976. Life history strategy of the Gila topminnow, *Poeciliopsis occidentalis*: A field evaluation of theory on the evolution of life histories. Ph.D. diss., Arizona State University, Tempe.

———. 1979. Life history patterns of a livebearing fish in contrasting environments. *Oecologia* 40, 189–201.

———. 1980. Energetics of viviparity in the Gila topminnow (Pisces: Poeciliidae). *Copeia* 1980, 876–878.

———. 1989. Reproductive biology of poeciliid fishes. In *Ecology and evolution of livebearing fishes (Poeciliidae)*, ed. G. K. Meffe and F. F. Snelson, Jr., pp. 33–50. Englewood Cliffs, N.J.: Prentice-Hall.

Contreras, G. P. 1973. Distribution of the fishes of the Lost River system, California-Oregon, with a key to species present. Master's thesis, University of Nevada, Reno.

Contreras Balderas, S. 1969. Perspectivas de la ictiofauna en las zonas aridas del Norte de Mexico. Memoires Primer Simposio Internacional American Producción Alimentos, A.Z. *ICASALS* (Texas Technological University Publication) 3, 293–304.

———. 1975. Impacto ambiental de obras hidraulicas. *Informes* 1, 1–129.

———. 1977. Biota endémica de Cuatro Ciénegas, Coahuila, México. *Memoires Primer Congreso Nacional Zoologica (Chapingo, México)* 1, 106–113.

———. 1978a. Lista de peces Mexicanos en peligro y amenazados de extinción en las provincias Chihuahuense, Sonorense y Tamaulipense. *Segundo Congreso Nacional Zoologica (Monterrey, Mexico)* (resumenes).

———. 1978b. Speciation aspects and man-made community composition changes in Chihuahuan Desert fishes. In *Transactions of the Symposium on the Biological Resources of the Chihuahuan Desert Region, United States and Mexico*, ed. R. H. Wauer and D. H. Riskind, pp. 405–431. Washington, D.C.: U.S. National Park Service Transactions and Proceedings Series 3.

———. 1984. Environmental impacts in Cuatro Ciénegas, Coahuila, México. *Journal of the Arizona-Nevada Academy of Sciences* 19, 85–87.

———. 1987. Lista anotada de especies de peces Mexicanos en peligro o amenazados de extinción. *Proceedings of the Desert Fishes Council* 16 (1984), 58–65.

———. In press. Importancia, biota endemica y perspectivas actuales de Cuatro Ciénegas, Coahuila, México. *Memoires Primer Simposio Nacional Areas Protegidas 1986 (México, D.F., México)*.

Contreras Balderas, S., and A. Maeda. 1985. Estado actual de la ictiofauna nativa de la Cuenca de Parras, Coah., México, con notas sobre algunos invertebrados. *Octavo Congreso Nacional Zoologia (México)* 1, 59–67.

Conway, W. 1988. Editorial. *Conservation Biology* 2, 132–134.

Cook, L. M. 1961. The edge effect in population genetics. *American Naturalist* 95, 295–307.

Cook, S. F., and C. D. Williams. 1982. The status and future of Ash Meadows, Nye County, Nevada. Office of the Nevada Attorney General, Carson City.

Cooke, R. U., and R. W. Reeves. 1976. *Arroyos and environmental change in the American Southwest.* London: Oxford University Press.

Coon, K. L., Sr. 1965. Some biological observations on the channel catfish, *Ictalurus punctatus* (Rafinesque), in a polluted western river. Master's thesis, Utah State University, Logan.

Coots, M. 1965. Occurrences of the Lost River sucker, *Deltistes luxatus* (Cope), and shortnose sucker, *Chasmistes brevirostris* (Cope), in northern California. *California Fish and Game* 51, 68–73.

Cope, E. D. 1884. On the fishes of the Recent and Pliocene lake of the western part of the Great Basin and of the Idaho Pliocene lake. *Proceedings of the Philadelphia Academy of Natural Sciences* 35 (1883), 134–167.

Cope, E. D., and H. C. Yarrow. 1875. Reports upon the collections of fishes made in portions of Nevada, Utah, California, Colorado, New Mexico, and Arizona during the years 1871, 1872, 1873, and 1874. *Report of Geographical and Geological Explorations West of the 100th Meridian (Wheeler Survey),* vol. 5, 635–703.

Corle, E. 1951. *The Gila: River of the Southwest.* Lincoln: University of Nebraska Press.

Courtenay, W. R., Jr., J. E. Deacon, D. W. Sada, R. C. Allan, and G. L. Vinyard. 1985. Comparative studies of fishes along the course of the pluvial White River, Nevada. *The Southwestern Naturalist* 30, 503–524.

Courtenay, W. R., Jr., and G. K. Meffe. 1989. Small fishes in strange places: A review of introduced poeciliids. In *Ecology and evolution of livebearing fishes (Poeciliidae),* ed. G. K. Meffe and F. F. Snelson, Jr., pp. 319–331. Englewood Cliffs, N.J.: Prentice-Hall.

Courtenay, W. R., Jr., and J. R. Stauffer, Jr., eds. 1984. *Distribution, biology and management of exotic fishes.* Baltimore, Md.: Johns Hopkins University Press.

Cowles, R. B. 1934. Notes on the ecology and breeding habits of the desert minnow, *Cyprinodon macularius* Baird and Girard. *Copeia* 1934, 40–42.

Cox, T. J., 1972a. The food habits of the desert pupfish (*Cyprinodon macularius*) in Quitobaquito Springs, Organ Pipe National Monument, Arizona. *Journal of the Arizona Academy of Sciences* 7, 25–27.

———. 1972b. Territorial behavior of the desert pupfish, *Cyprinodon macularius*, in Quitobaquito Springs, Organ Pipe National Monument, Arizona. *Journal of the Colorado-Wyoming Academy of Sciences* 7, 73.

Crabtree, C. B., and D. G. Buth. 1987. Biochemical systematics of the catostomid genus *Catostomus*: Assessment of *C. clarki, C. plebeius* and *C. discobolus* including the Zuñi sucker, *C. d. yarrowi. Copeia* 1987, 845–854.

Crance, J. H. 1988. Results on the use of the Delphi technique for developing category I habitat suitability criteria for redbreast sunfish. *U.S. Fish and Wildlife Service Biological Report* 88, 102–129.

Crawford, A. B., and D. F. Petersen, eds. 1974. *Environmental management of the Colorado River basin.* Logan: Utah State University Press.

Crear, D., and I. Haydock. 1970. Laboratory rearing of the desert pupfish, *Cyprinodon macularius. Fisheries Bulletin* 69, 151–156.

Cross, T. F., and J. King. 1983. Genetic effects of hatchery rearing of Atlantic salmon. *Aquaculture* 33, 33–40.

Cumming, K. B. 1975. History of fish toxicants in the United States. In *Rehabilitation of fish populations with toxicants: A symposium*, ed. P. H. Eschmeyer, pp. 5–21. Bethesda, Md.: North Central Division, American Fisheries Society, Special Publication 4.

Cuplin, P. 1967. Methods of increasing returns of hatchery fish. *Proceedings of the Annual Conference of the Western Association of State Game and Fish Commissioners* 47, 296–305.

Dahlberg, M. D. 1979. A review of survival rates of fish eggs and larvae in relation to impact assessments. *Marine Fisheries Review* 41, 1–9.

Darwin, C. 1968 [1859]. *The origin of species*. Harmondsworth, U.K.: Penguin Books.

Davis, H. S. 1936. Stream improvement in national forests. *Transactions of the North American Wildlife Conference* 1, 447–453.

Davis, J. R. 1979. Die-offs of an endangered pupfish, *Cyprinodon elegans* (Cyprinodontidae). *The Southwestern Naturalist* 24, 534–536.

———. 1980. Rediscovery, distribution, and populational status of *Cyprinodon eximius* (Cyprinodontidae) in Devil's River, Texas. *The Southwestern Naturalist* 25, 81–88.

Dawson, V. K. 1975. Counteracting chemicals used in fishery operations: Current technology and research. In *Rehabilitation of fish populations with toxicants: A symposium*, ed. P. H. Eschmeyer, pp. 32–49. Bethesda, Md.: North Central Division, American Fisheries Society, Special Publication 4.

Deacon, J. E. 1967. The ecology of Saratoga Springs, Death Valley National Monument. In Studies on the ecology of Saratoga Springs, Death Valley National Monument. Annual Report for U.S. National Park Service Contract 14-10-0434-0989. Nevada Southern University, Las Vegas.

———, ed. 1968a. Ecological studies of aquatic habitats in Death Valley National Monument, with special reference to Saratoga Springs. Final Report for U.S. National Park Service Contract 14-10-0434-0989. Nevada Southern University, Las Vegas.

———. 1968b. Endangered non-game fishes of the West: Causes, prospects, and importance. *Proceedings of the Annual Conference of the Western Association of Fish and Game Commissioners* 48, 534–549.

———. 1969. Nevada's endangered species. *Nevada Outdoors* 3, 18–19.

———. 1979. Endangered and threatened fishes of the West. *Great Basin Naturalist Memoirs* 3, 41–64.

———. 1988. The endangered woundfin and water management in the Virgin River, Utah, Arizona, Nevada. *Fisheries* 13, 18–24.

Deacon, J. E., and W. G. Bradley. 1972. Ecological distribution of fishes of the Moapa River in Clark County, Nevada. *Transactions of the American Fisheries Society* 101, 408–419.

Deacon, J. E., and S. Bunnell. 1970. Man and pupfish, a process of destruction. *Cry California* 5, 14–21.

Deacon, J. E., and M. S. Deacon. 1979. Research on endangered fishes in National Parks with special emphasis on the Devils Hole pupfish. In *Proceedings of the First Conference on Scientific Research in the National Parks*, ed. R. M. Linn, pp. 9–19. Washington, D.C.: U.S. National Park Service Transactions and Proceedings Series 5.

Deacon, J. E., C. Hubbs, and B. J. Zahuranec. 1964. Some effects of introduced fishes on the native fish fauna of southern Nevada. *Copeia* 1964, 384–388.

Deacon, J. E., G. Kobetich, J. D. Williams, S. Contreras, et al. 1979. Fishes of North America: Endangered, threatened, or of special concern: 1979. *Fisheries* 4, 29–44.

Deacon, J. E., and W. L. Minckley. 1974. Desert fishes. In *Desert biology*, vol. 2, ed. G. W. Brown, Jr., pp. 385–488. New York: Academic Press.

Deacon, J. E., and J. E. Williams. 1984. Annotated list of the fishes of Nevada. *Proceedings of the Biological Society of Washington* 97, 103–118.

de Lafontaine, Y., and W. C. Leggett. 1988. Predation by jellyfish on larval fish: An experimental evaluation employing in situ enclosures. *Canadian Journal of Fisheries and Aquatic Sciences* 45, 1173–1190.

Dellenbaugh, F. S. 1908. *A canyon voyage: The narrative of the second Powell Expedition down the Green and Colorado rivers from Wyoming and the explorations on land, in the years 1871 and 1872*. New York: Knickerbocker Press.

DeMarais, B. D. 1986. Morphological variation in *Gila* (Pisces: Cyprinidae) and geologic history: Lower Colorado River basin. Master's thesis, Arizona State University, Tempe.

———. 1991. *Gila eremica*, a new species of cyprinid fish from northern Sonora, Mexico. *Copeia* 1991, 178–189.

Denniston, C. 1978. Small population size and genetic diversity: Implications for endangered species. In *Endangered birds: Management techniques for preserving endangered species*, ed. S. A. Temple, pp. 281–289. Madison: University of Wisconsin Press.

Dermott, R. M., and H. J. Spence. 1984. Changes in populations and drift of stream invertebrates following lampricide treatment. *Canadian Journal of Fisheries and Aquatic Sciences* 41, 1695–1701.

Diamond, J. M. 1976. Island biogeography and conservation: Strategy and limitations. *Science* 193, 1027–1029.

Diamond, J. M., and R. M. May. 1976. Island biogeography and the design of natural preserves. In *Theoretical ecology: Principles and applications*, 2d ed., ed. R. M. May, pp. 228–252. London: Blackwell Scientific, Oxford University Press.

Diamond, J. M., and E. Mayr. 1976. Species-area relation for birds of the Solomon Archipelago. *Proceedings of the National Academy of Sciences* 73, 262–266.

Dill, W. A. 1944. The fishery of the lower Colorado River. *California Fish and Game* 30, 109–211.

Dizon, A. E., R. M. Horrall, and A. D. Hasler. 1973. Olfactory electroencephalographic responses of homing coho salmon, *Oncorhynchus kisutch*, to water conditioned by conspecifics. *U.S. Fish and Wildlife Service Fisheries Bulletin* 71, 893–896.

Dobson, A. P., and R. M. May. 1986. Disease and conservation. In *Conservation biology: The biology of scarcity and diversity*, ed. M. E. Soulé, pp. 345–365. Sunderland, Mass.: Sinauer Associates.

Dodge, S. E., ed. 1984. *The Nature Conservancy news* 34 (5), 1–30.

Douglas, M. E., W. L. Minckley, and H. M. Tyus. 1989. Qualitative characters, identification of Colorado River chubs (Cyprinidae: genus *Gila*), and the "art of seeing well." *Copeia* 1989, 653–662.

Douglas, P. A. 1952. Notes on the spawning of the humpback sucker, *Xyrauchen texanus* (Abbott). *California Fish and Game* 38, 149–155.

Dudley, W. W., Jr., and J. D. Larson. 1976. Effect of irrigation pumping on desert pupfish habitats in Ash Meadows, Nye County, Nevada. *U.S. Geological Survey Professional Paper* 927, 1–142.

Duff, D. A. 1980. Construction and operating efficiency of intermountain area stream habitat improvement structures. In *Proceedings of the Trout Stream Habitat Improvement Workshop*, M. E. Seehorn, coordinator, pp. 153–158, Atlanta, Ga.: U.S. Forest Service.

Duff, D. A., N. Banks, E. Spartis, W. E. Stone, and R. J. Poehlmann. 1988. *Indexed bibliography on stream habitat improvement*. Ogden, Utah: U.S. Department of Agriculture, Forest Service.

Dunning, D. J., Q. E. Ross, J. R. Waldman, and M. T. Mattson. 1987. Tag retention by, and tagging mortality of, Hudson River striped bass. *North American Journal of Fish Management* 7, 535–538.

Dunshee, B. R., C. Leben, G. W. Keitt, and F. M. Strong. 1969. The isolation and properties of antimycin-A. *American Chemical Society* 71, 2436–2437.

Dymond, J. R. 1964. A history of ichthyology in Canada. *Copeia* 1964, 2–33.

Eadie, J. M., T. A. Hurly, R. D. Montgomerie, and K. L. Teather. 1986. Lakes and rivers as islands: Species-area relationships in fish faunas of Ontario. *Environmental Biology of Fishes* 15, 81–89.

Eadie, J. M., and A. Keast. 1984. Resource heterogeneity and fish species diversity in lakes. *Canadian Journal of Zoology* 62, 1689–1695.

Echelle, A. A., and P. J. Conner. 1989. Rapid geographically extensive genetic introgression after secondary contact between two pupfish species (*Cyprinodon*, Cyprinodontidae). *Evolution* 43, 717–727.

Echelle, A. A., and A. F. Echelle. 1978. The Pecos pupfish, *Cyprinodon pecosensis* n. sp. (Cyprinodontidae), with comments on its evolutionary origin. *Copeia* 1978, 569–582.

———. 1980. Status of the Pecos gambusia. Final Report for U.S. Fish and Wildlife Service Contract 14-160002-79-133. Oklahoma State University, Stillwater (also appeared as *Endangered Species Report* 10, 1–73, U.S. Fish and Wildlife Service, Albuquerque, New Mexico).

Echelle, A. A., A. F. Echelle, and D. R. Edds. 1983. Genetic structure of three species of endangered desert fishes. Final Report for U.S. Fish and Wildlife Service Contract 14-160009-1554. Oklahoma State University, Stillwater.

———. 1987. Population structure of four pupfish species (Cyprinodontidae: *Cyprinodon*) from the Chihuahuan Desert region of New Mexico and Texas: Allozymic variation. *Copeia* 1987, 668–681.

Echelle, A. A., A. F. Echelle, and L. G. Hill. 1972. Interspecific interactions and limiting factors of abundance and distribution in the Red River pupfish, *Cyprinodon rubrofluviatilis*. *American Midland Naturalist* 88, 109–130.

Echelle, A. A., and C. Hubbs. 1978. Haven for endangered pupfish. *Texas Parks and Wildlife* 36, 8–11.

Echelle, A. A., and I. Kornfield, eds. 1984. *Evolution of fish species flocks*. Orono: University of Maine Press.

Echelle, A. A., and R. R. Miller. 1974. Rediscovery and redescription of the Leon Springs pupfish, *Cyprinodon bovinus*, from Pecos County, Texas. *The Southwestern Naturalist* 19, 179–190.

Echelle, A. A., D. M. Wildrick, and A. F. Echelle. 1989. Allozyme studies of genetic variation in poeciliid fishes. In *Ecology and evolution of livebearing fishes (Poeciliidae)*, ed. G. K. Meffe and F. F. Snelson, Jr., pp. 217–234. Englewood Cliffs, N.J.: Prentice-Hall.

Echelle, A. F., and A. A. Echelle. 1986. Geographic variation in morphology of a spring-dwelling desert fish, *Gambusia nobilis* (Poeciliidae). *The Southwestern Naturalist* 31, 459–468.

Echelle, A. F., A. A. Echelle, and D. R. Edds. 1989. Conservation genetics of a spring-dwelling desert fish, the Pecos gambusia, *Gambusia nobilis* (Poeciliidae). *Conservation Biology* 3, 159–169.

Edds, D. R., and A. A. Echelle. 1989. Genetic comparisons of hatchery and natural stocks of small endangered fishes: Leon Springs pupfish, Comanche Springs pupfish, and Pecos gambusia. *Transactions of the American Fisheries Society* 118, 441–446.

Edwards, G. B. 1974. Biology of the striped bass, *Morone saxatilis* (Walbaum), in the lower Colorado River (Arizona-California-Nevada). Master's thesis, Arizona State University, Tempe.

Edwards, R. J. 1979. A report of Guadalupe bass (*Micropterus treculi*) × smallmouth bass (*M. dolomieui*) hybrids from two localities in the Guadalupe River, Texas. *Texas Journal of Science* 31, 231–238.

Ehrenfeld, D. W. 1976. The conservation of non-resources. *American Scientist* 64, 648–656.

———. 1978. *The arrogance of humanism*. London: Oxford University Press.

Ehrenfeld, D. W., and J. G. Ehrenfeld. 1985. Some thoughts on nature and Judaism. *Environmental Ethics* 7, 93–95.

Ehrlich, P. R. 1988. The loss of diversity: Causes and consequences. In *Biodiversity*, ed. E. O. Wilson, pp. 21–27. Washington, D.C.: National Academy of Sciences Press.

Ehrlich, P. R., and A. H. Ehrlich. 1981. *Extinction: The causes and consequences of the disappearance of species*. New York: Random House.

Ehrlich, P. R., and D. D. Murphy. 1987. Conservation lesson from long-term studies of checkerspot butterflies. *Conservation Biology* 1, 122–131.

Eiserman, F., F. W. Jackson, R. Kent, A. Regenthal, and R. Stone. 1964. Flaming Gorge Reservoir post-impoundment investigations. Progress Report no. 1, Utah Department Fish and Game, Salt Lake City, and Wyoming Game and Fish Commission, Cheyenne.

Eldredge, N., and J. Cracraft. 1980. *Phylogenetic patterns and the evolutionary process*. New York: Columbia University Press.

Ellis, M. M. 1914. Fishes of Colorado. *University of Colorado Studies* 11, 1–136.

Ellison, D. N. 1980. Letter to the editor. *Wildlife views* (Arizona Game and Fish Department, Phoenix) 24, 2.

Engstrom-Heg, R. 1971a. A lightweight Mariotte bottle for field, laboratory, and hatchery use. *Progressive Fish-Culturist* 33, 227–231.

———. 1971b. Comparison of field methods for measuring stream discharge. *New York Fish and Game Journal* 18, 77–96.

Environmental Defense Fund, Sierra Club, Trout Unlimited, Friends of the Earth, and Colorado Audubon Council. 1985. Joint statement before the Subcommittee on Fisheries, Wildlife Conservation, and Environment, Committee on Merchant Marine and Fisheries, U.S. House of Representatives, Hearing on Reauthorization of the Endangered Species Act, 14 March, Washington, D.C.

Eschmeyer, P. H., ed. 1975. *Rehabilitation of fish populations with toxicants: A symposium*. Bethesda, Md.: North Central Division, American Fisheries Society, Special Publication 4.

Everhart, W. R., and W. R. Seaman. 1971. *Fishes of Colorado*. Denver: Colorado Game, Fish, and Parks Division.

Evermann, B. W. 1916. Fishes of the Salton Sea. *Copeia* 1916, 61–63.

Evermann, B. W., and C. Rutter. 1895. Fishes of the Colorado basin. *Bulletin of the U.S. Fish Commissioner* 14 (1894), 473–486.

Fausch, K. D., J. R. Karr, and P. R. Yant. 1984. Regional application of an index of biotic integrity based on stream-fish communities. *Transactions of the American Fisheries Society* 113, 81–89.

Feinberg, J. 1974. The rights of animals and unborn generations. In *Philosophy and environmental crisis*, ed. W. T. Blackstone, pp. 43–68. Athens: University of Georgia Press.

Feldmeth, R., D. Soltz, L. McClanahan, J. Jones, and J. Irwin. 1985. Natural resources of the Lark Seep system (China Lake, CA) with special emphasis on the Mohave chub (*Gila bicolor mohavensis*). *Proceedings of the Desert Fishes Council* 15 (1983), 356–358.

Fenneman, N. M. 1931. *Physiography of western United States*. New York: McGraw-Hill.

Ferriole, S. 1987. Wildlife resource information system, Colorado squawfish mapping criteria for the Colorado, Dolores, Gunnison, Green, White, and Yampa rivers. Colorado Division of Wildlife, Grand Junction.

Ferris, S. D., D. G. Buth, and G. S. Whitt. 1982. Substantial genetic differentiation among populations of *Catostomus plebeius*. *Copeia* 1982, 444–449.

Feth, J. H. 1961. A new map of western coterminous United States showing the maximum known or inferred extent of Pleistocene lakes. *U.S. Geological Survey Professional Paper* 424B, 110–112.

Fiero, G. W., and G. B. Maxey. 1970. Hydrogeology of the Devil's Hole area, Ash Meadows, Nevada. Center for Water Resources Research, Desert Research Institute, University of Nevada System, Reno.

Findley, R. 1970. Death Valley, the land and the legend. *National Geographic* 137 (1), 69–103.

Fisher, J., N. Simon, J. Vincent, and IUCN Staff. 1969. *Wildlife in danger*. New York: Viking Press.

Fisher, R. F. 1971. Environmental law. *Sierra Club Bulletin* 56 (1), 24–25.

Fisher, S. G. 1986. Structure and dynamics of desert streams. In *Pattern and process in desert ecosystems*, ed. W. G. Whitford, pp. 119–139. Albuquerque: University of New Mexico Press.

Flagg, R. 1980. Disease survey of the Colorado River fishes. U.S. Fish and Wildlife Service, Fish Disease Control Center, Fort Morgan, Colo.

———. 1982. Disease survey of the Colorado River fishes. In Part 3, *Colorado River Fisheries Project, final report contracted studies*, ed. W. H. Miller, J. J. Valentine, D. L. Archer, H. M. Tyus, R. A. Valdez, and L. R. Kaeding, pp. 177–184. Final Report for U.S. Bureau of Reclamation Contract 9-07-40-L-1016, Memorandum of Understanding CO-910-MU9-933, U.S. Bureau of Land Management. U.S. Fish and Wildlife Service, Salt Lake City, Utah.

Follett, W. I. 1961. The fresh-water fishes—their origins and affinities. *Systematic Zoology* 9 (1960), 212–232.

Forde, C. D. 1931. Ethnography of the Yuman Indians. *University of California Publications in American Archaeology and Ethnology* 28, 83–278.

Forman, R.T.T., and M. Godron. 1981. Patches and structural components from a landscape ecology. *BioScience* 31, 733–740.

Forman, R.T.T., and E. W. B. Russell. 1983. Commentary: Evaluation of historical data in ecology. *Bulletin of the Ecological Society of America* 64, 5–7.

Foster, D. 1958. Statewide fishery investigations: A manipulation of environmental conditions pertaining to minor jobs of a development nature. Federal Aid to Fisheries Restoration Project Completion Report, F-7-R-1, 3–4. Arizona Game and Fish Department, Phoenix.

Fradkin, P. L. 1984. *A river no more—the Colorado River and the West.* Tucson: University of Arizona Press.

Frankel, O. H. 1974. Genetic conservation: Our evolutionary responsibility. *Genetics* 78, 53–65.

———. 1983. The place of management in conservation. In *Genetics and conservation: A reference for managing wild animal and plant populations*, ed. C. Schonewald-Cox, S. Chambers, B. MacBryde, and W. Thomas, pp. 1–14. Menlo Park, Calif.: Benjamin/Cummings.

Frankel, O. H., and M. E. Soulé. 1981. *Conservation and evolution.* Cambridge: Cambridge University Press.

Fuhriman, D. R., L. B. Merritt, A. W. Miller, and H. S. Stock. 1981. Hydrology and water quality of Utah Lake. *Great Basin Naturalist Memoirs* 5, 43–67.

Funk, J. L. 1957. Movement of stream fishes in Missouri. *Transactions of the American Fisheries Society* 85, 39–57.

Galat, D. L., E. L. Lider, S. Vigg, and S. R. Roberston. 1981. Limnology of a large, deep, North American terminal lake, Pyramid Lake, Nevada, USA. *Hydrobiologia* 82, 281–317.

Galat, D. L., and B. Robertson. In press. Interactions between *Poeciliopsis occidentalis sonoriensis* and *Gambusia affinis* (Atheriniformes: Poeciliidae) in the Rio Yaqui drainage, Arizona, USA. *Environmental Biology of Fishes.*

Galat, D. L., and R. Robinson. 1983. Predicted effects of increasing salinity on the crustacean zooplankton community of Pyramid Lake, Nevada. *Hydrobiologia* 105, 115–131.

Garrett, G. P. 1980a. Update on some of the protected and endangered fishes of Texas. *Proceedings of the Desert Fishes Council* 11 (1979), 34–36.

———. 1980b. Species specificity in the mating systems of *Cyprinodon variegatus* and *Cyprinodon bovinus. Proceedings of the Desert Fishes Council* 11 (1979), 54–59.

Gaufin, A. R., G. R. Smith, and P. Dotson. 1960. Aquatic survey of the Green River and tributaries within the Flaming Gorge Reservoir basin, Appendix A. In *Ecological studies of the flora and fauna of Flaming Gorge Reservoir basin, Utah and Wyoming*, ed. A. M. Woodbury, pp. 139–162. University of Utah Anthropological Papers 48.

Geen, G. H., T. G. Northcote, G. F. Hartman, and C. C. Lindsey. 1966. Life history of two species of catostomid fishes in Sixteenmile Lake, British Columbia, with special reference to inlet stream spawning. *Journal of the Fisheries Research Board of Canada* 23, 1761–1787.

Geike, A. 1962. *Founders of geology,* 2d ed. New York: Dover.

Gerking, S. D. 1950. Stability of a stream fish population. *Journal of Wildlife Management* 14, 193–202.

———. 1953. Evidence for the concepts of home range and territory in stream fishes. *Ecology* 34, 347–365.

———. 1959. The restricted movement of fish populations. *Biological Review* 34, 221–242.

Gerking, S. D., and D. V. Plantz. 1980. Size-biased predation by the Gila topminnow, *Poeciliopsis occidentalis* (Baird and Girard). *Hydrobiologia* 72, 179–191.

Gifford, E. W. 1933. The Cocopa. *University of California Publications in American Archaeology and Ethnology* 31, 257–334.

Gilbert, C. H. 1893. Report on the fishes of the Death Valley expedition collected in southern California and Nevada in 1891, with descriptions of new species. *North American Fauna* 7, 229–234.

———. 1898. The fishes of the Klamath Basin. *Bulletin of the U.S. Fish Commissioner* 17, 1–13.

Gilbert, C. H., and N. B. Scofield. 1898. Notes on a collection of fishes from the Colorado Basin in Arizona. *Proceedings of the U.S. National Museum* 20, 487–499.

Gilderhaus, P. A., B. L. Berger, and R. E. Lennon. 1969. Field trials of antimycin-A as a fish toxicant. *U.S. Bureau of Sport Fisheries and Wildlife, Investigations in Fish Control* 27, 1–27.

Gilpin, M. E. 1988. A comment on Quinn and Hastings: Extinction in subdivided habitats. *Conservation Biology* 2, 290–292.

Gilpin, M. E., and M. E. Soulé. 1986. Minimum viable populations: Processes of species extinction. In *Conservation biology: The science of scarcity and diversity*, ed. M. E. Soulé, pp. 19–34. Sunderland, Mass.: Sinauer Associates.

Girard, C. 1856. Researches upon the cyprinoid fishes inhabiting the freshwaters of the United States of America, west of the Mississippi Valley, from specimens in the museum of the Smithsonian Institution. *Proceedings of the Philadelphia Academy of Natural Sciences* 8, 165–213.

Golden, M. P. 1969. The Lost River sucker, *Catostomus luxatus* (Cope). *Oregon State Game Commission Report* I-69, 1–10.

Goodman, R. 1980. Taoism and ecology. *Environmental Ethics* 2, 73–80.

Gorman, G. C., and J. R. Karr. 1978. Habitat structure and stream fish communities. *Ecology* 59, 507–515.

Gorman, G. C., and L. A. Nielsen. 1982. Piscivory by stocked brown trout (*Salmo trutta*) and its impact on the nongame fish community of Bottom Creek, Virginia. *Canadian Journal of Fisheries and Aquatic Sciences* 39, 862–869.

Grabowski, S. J., S. D. Hiebert, and D. M. Lieberman. 1984. Potential for introduction of three species of nonnative fishes into central Arizona via the Central Arizona Project: A literature review and analysis. *U.S. Bureau of Reclamation, Research Engineering Center Report REC-ERC-84-7*, 1–124.

Graf, W. L. 1985. The Colorado River, instability and basin management. *Association of American Geographers, Research Publications in Geography* 1985, 1–86.

Granger, B. H. 1960. *Arizona place names*. Tucson: University of Arizona Press.

Gray, L. J. 1981. Species composition and life histories of aquatic insects in a lowland Sonoran Desert stream. *American Midland Naturalist* 106, 229–242.

Green, D. M. 1985. Biochemical identification of red-legged frogs (*Rana aurora draytoni*) at Duckwater, Nevada. *The Southwestern Naturalist* 30, 614–616.

Greenwood, P. H. 1981. Species-flocks and explosive evolution. In *Chance, change and challenge—the evolving biosphere*, ed. P. H. Greenwood and P. L. Forey, pp. 61–74. Cambridge: Cambridge University Press and British Museum (Natural History).

Greger, P. D., and J. E. Deacon. 1982. Observations on woundfin spawning and growth in an outdoor experimental stream. *Great Basin Naturalist* 42, 549–552.

Grene, M. 1987. Hierarchies in biology. *American Scientist* 75, 504–510.

Gresswell, R. E. 1991. Use of antimycin for removal of brook trout from a tributary of Yellowstone Lake. *North American Journal of Fisheries Management* 11, 83–90.

Griffith, B., J. M. Scott, J. W. Carpenter, and C. Reed. 1989. Translocation as a species conservation tool: Status and strategy. *Science* 245, 477–480.

Grinnell, J. 1914. An account of the mammals and birds of the lower Colorado valley. *University of California Publications in Zoology* 12, 51–294.

Gruenewald, R. J. 1960. Statewide fishery investigations: Minor jobs of a developmental nature in District IV. Federal Aid to Fisheries Restoration Project Completion Report, F-7-R-3, 1–2. Arizona Game and Fish Department, Phoenix.

Guenther, H. R., and J. Romero. 1973. Lake Havasu aquatic impact study, progress report, 1972. U.S. Bureau of Reclamation, Lower Colorado River Region, Boulder City, Nev.

Guilbert, T. 1974. Wildlife preservation under federal law. In *Federal environmental law,* ed. E. Dolgin and T. Guilbert, pp. 550–594. St. Paul, Minn.: West Publishing.

Guillory, V. 1980. *Gambusia georgei* Hubbs and Peden, San Marcos gambusia. In *Atlas of North American freshwater fishes,* ed. D. S. Lee, C. R. Gilbert, C. H. Hocutt, R. E. Jenkins, D. E. McAllister, and J. R. Stauffer, Jr., p. 542. Raleigh: North Carolina State Museum of Natural History.

Gunning, G. E. 1959. The sensory basis for homing in the longear sunfish, *Lepomis megalotis megalotis* (Rafinesque). *Investigations of Indiana Lakes and Streams* 5, 103–130.

Gunning, G. E., and C. R. Shoop. 1962. Restricted movement of the American eel, *Anguilla rostrata* (LeSueur), in freshwater streams, with comments on growth rate. *Tulane Studies in Zoology* 9, 255–272.

Gustafson, E. S. 1975a. Capture, disposition, and status of adult razorback suckers from Lake Mohave, Arizona. U.S. Fish and Wildlife Service Contract. Arizona State University, Tempe.

———. 1975b. Early development, adult sexual dimorphism, and fecundity of the razorback sucker, *Xyrauchen texanus* (Abbott). Final Report for U.S. Fish and Wildlife Service Contract. Arizona State University, Tempe.

Gyllensten, U. 1985. The genetic structure of fish: Differences in the intraspecific distribution of biochemical genetic variation between marine, anadromous, and freshwater species. *Journal of Fisheries Biology* 26, 691–699.

Gyllensten, U., R. F. Leary, F. W. Allendorf, and A. C. Wilson. 1985. Introgression between two cutthroat trout subspecies with substantial karyotypic, nuclear, and mitochondrial genomic divergence. *Genetics* 111, 905–915.

Haas, P. H. 1975. Some comments on use of the species-area curve. *American Naturalist* 109, 371–373.

Hagen, H. K., and J. L. Banks. 1963. Ecological and limnological studies of the Green River in Dinosaur National Monument. Final Report for U.S. National Park Service Contract 14-10-0232-686. Colorado State University, Fort Collins.

Halliday, W. R. 1955. The miner's bathtub. In *Celebrated American caves,* ed. C. E. Mohr and H. N. Sloane, pp. 90–104. New Brunswick, N.J.: Rutgers University Press.

———. 1966. *Depths of the Earth.* New York: Harper and Row.

Hamman, R. L. 1981. Spawning and culture of Colorado squawfish in raceways. *Progressive Fish-Culturist* 43, 173–177.

———. 1982a. Culture of endangered Colorado River fishes. Section I: Induced spawning and culture of the humpback chub and bonytail chub hybrids. In Part 3, *Colorado River Fisheries Project, final report contracted studies,* ed. W. H. Miller, J. J. Valentine, D. L. Archer, H. M. Tyus, R. A. Valdez, and L. R. Kaeding, pp. 128–136. Final Report for U.S. Bureau of Reclamation Contract 9-07-40-L-1016, Memorandum of Understanding CO-910-MU9-933, U.S. Bureau of Land Management. U.S. Fish and Wildlife Service, Salt Lake City, Utah.

———. 1982b. Culture of endangered Colorado River fishes. Section II: Induced spawning and culture of the humpback chub. In Part 3, *Colorado River Fisheries Project, final report contracted studies,* ed. W. H. Miller, J. J. Valentine, D. L. Archer, H. M. Tyus, R. A. Valdez, and L. R. Kaeding, pp. 158–167. Final Report for U.S. Bureau of Reclamation Contract 9-07-40-L-1016, Memorandum of Understanding CO-910-MU9-933, U.S. Bureau of Land Management. U.S. Fish and Wildlife Service, Salt Lake City, Utah.

———. 1982c. Induced spawning and culture of bonytail chub. *Progressive Fish-Culturist* 44, 201–203.

———. 1985a. Induced spawning of hatchery-reared razorback sucker. *Progressive Fish-Culturist* 47, 187–189.

———. 1985b. Induced spawning of hatchery-reared bonytail. *Progressive Fish-Culturist* 47, 35–37.

———. 1986. Induced spawning of hatchery-reared Colorado squawfish. *Progressive Fish-Culturist* 48, 72–74.

———. 1987. Survival of razorback sucker cultured in earthen ponds. *Progressive Fish-Culturist* 49, 135–140.

———. 1989. Survival of Colorado squawfish cultured in earthen ponds. *Progressive Fish-Culturist* 51, 27–29.

Hampshire, S. 1972. *Morality and pessimism.* Cambridge: Cambridge University Press.

Hanson, J. N. 1971. Investigations on Gila trout, *Salmo gilae* Miller, in Southwestern New Mexico. Master's thesis, New Mexico State University, Las Cruces.

Harden-Jones, F. R. 1980. *Fish migration.* London: Edward Arnold.

Hardy, T. 1980. The inter-basin area report. *Proceedings of the Desert Fishes Council* 11 (1979), 5–21.

Hargrove, E. 1979. The historical foundations of American environmental attitudes. *Environmental Ethics* 1, 209–240.

———. 1989. *Foundations of environmental ethics.* Englewood Cliffs, N.J.: Prentice-Hall.

Harrell, H. L. 1980. *Gambusia geiseri* Hubbs and Hubbs, largespring gambusia. In *Atlas of North American freshwater fishes*, ed. D. S. Lee, C. R. Gilbert, C. H. Hocutt, R. E. Jenkins, D. E. McAllister, and J. R. Stauffer, Jr., p. 541. Raleigh: North Carolina State Museum of Natural History.

Harris, R. E., H. N. Sersland, and F. P. Sharpe. 1982. Providing water for endangered fishes in the upper Colorado River system. In *Fishes of the upper Colorado River system: Present and future*, ed. W. H. Miller, H. M. Tyus, and C. A. Carlson, pp. 90–92. Bethesda, Md.: Western Division, American Fisheries Society.

Haskell, W. L. 1959. Diet of the Mississippi threadfin shad, *Dorosoma petenense atchafalayae*, in Arizona. *Copeia* 1959, 298–302.

Hasler, A. D. 1971. Orientation and fish migration. In *Fish physiology*, vol. 4, ed. W. S. Hoar and D. J. Randall, pp. 429–510. New York: Academic Press.

Hasler, A. D., and A. T. Scholz. 1978. Olfactory imprinting and homing in salmon. In *Zoophysiology*, vol. 14, pp. 3–38. New York: Springer-Verlag.

Hastings, J. R. 1959. Vegetation change and arroyo cutting in southeastern Arizona. *Journal of the Arizona Academy of Sciences* 1, 60–67.

Hastings, J. R., and R. M. Turner. 1965. *The changing mile.* Tucson: University of Arizona Press.

Hatch, M. D., W. H. Baltosser, and C. G. Schmidt. 1985. Life history and ecology of the bluntnose shiner (*Notropis simus pecosensis*) in the Pecos River of New Mexico. *The Southwestern Naturalist* 30, 555–562.

Haynes, C. M., T. A. Lytle, E. J. Wick, and R. T. Muth. 1984. Larval Colorado squawfish (*Ptychocheilus lucius* Girard) in the upper Colorado River basin, Colorado, 1979–1981. *The Southwestern Naturalist* 29, 21–33.

Haynes, C. M., and R. T. Muth. 1982. Identification of habitat requirements and limiting factors for Colorado squawfish and humpback chubs. Colorado Division of Wildlife Federal Aid Project Progress Report, SE-4, 1–43.

———. 1984. Identification of habitat requirements and limiting factors for Colorado squawfish and humpback chubs. Colorado Division of Wildlife Federal Aid Project Progress Report, SE-3, 1–21.

Heckmann, R., J. E. Deacon, and P. D. Greger. 1987. Parasites of the woundfin minnow, *Plagopterus argentissimus*, and other endemic fishes from the Virgin River, Utah. *Great Basin Naturalist* 46, 662–676.

Heede, B., and J. N. Rinne. 1989. Hydrodynamic and fluvial morphologic processes: Implications for fisheries management and research. *North American Journal of Fisheries Management* 10, 75–92.

Hemphill, J. E. 1953. Untitled report [toxaphene as a fish poison]. Arizona Game and Fish Department, Phoenix.

———. 1954. Toxaphene as a fish toxin. *Progressive Fish-Culturist* 16, 41–42.

Hendrickson, D. A., and J. E. Brooks. 1987. Colorado River squawfish reintroduction studies. *Proceedings of the Desert Fishes Council* 18 (1986), 207–208.

Hendrickson, D. A., and W. L. Minckley. 1985. Ciénegas: Vanishing aquatic climax communities of the American Southwest. *Desert Plants* 6, 131–175.

Hendrickson, D. A., W. L. Minckley, R. R. Miller, D. J. Siebert, and P. H. Minckley. 1981. Fishes of the Río Yaqui basin, México and United States. *Journal of the Arizona-Nevada Academy of Sciences* 15 (1980), 65–106.

Hendrickson, D. A., and A. Varela. 1989. Conservation status of the endangered desert pupfish, *Cyprinodon macularius*, in México and Arizona. *Copeia* 1989, 478–483.

Herbold, B. 1987. Patterns of co-occurrence and resource use in a non-coevolved assemblage of fishes. Ph.D. diss., University of California, Davis.

Hershler, R., and D. W. Sada. 1987. Springsnails (Gastropoda: Hydrobiidae) of Ash Meadows, Amargosa basin, California-Nevada. *Proceedings of the Biological Society of Washington* 100, 776–843.

Hickman, T. J. 1983. Effects of habitat alteration by energy resource developments in the upper Colorado basin on endangered fishes. In *Aquatic resources management of the Colorado River ecosystem*, ed. V. D. Adams and V. A. Lamarra, pp. 537–550. Ann Arbor, Mich.: Ann Arbor Science Publications.

Hillis, D. M., and C. Moritz. 1990. An overview of applications of molecular systematics. In *Molecular systematics*, ed. D. M. Hillis and C. Moritz, pp. 502–515. Sunderland, Mass.: Sinauer Associates.

Hillyard, S. D. 1981. Energy metabolism and osmoregulation in desert fishes. In *Fishes in North American deserts*, ed. R. J. Naiman and D. L. Soltz, pp. 385–410. New York: John Wiley and Sons.

Hocutt, C. H., and E. O. Wiley, eds. 1986. *The zoogeography of North American freshwater fishes*. New York: John Wiley and Sons.

Hoffman, R. J. 1988. Chronology of diving activities and underground surveys in Devil's Hole and Devil's Hole Cave, Nye County, Nevada, 1950–1986. U.S. Geological Survey Open-file Report 88-93, 1–12. Carson City, Nev.

Holden, P. B. 1968. Systematic studies of the genus *Gila* (Cyprinidae) of the Colorado River basin. Master's thesis, Utah State University, Logan.

———. 1973. Distribution, abundance and life history of the fishes of the upper Colorado River Basin. Ph.D. diss., Utah State University, Logan.

———. 1977a. A study of the habitat use and movement of the rare fishes in the Green River from Jensen to Green River, Utah, August and September, 1977. Final Report for Western Energy and Land-Use Team Contract 14-16-0009-77-050. BIO/WEST, Logan, Utah.

———. 1977b. Habitat requirements of juvenile Colorado River squawfish. U.S. Fish and Wildlife Service Report FWS/OBS/77/65, i–vi, 1–71. Fort Collins, Colo.

———. 1978. A study of the habitat and movement of the rare fishes in the Green River, Utah. *Transactions of the Bonneville Chapter, American Fisheries Society* 1978, 64–90.

———. 1979. Ecology of riverine fishes in regulated stream systems with emphasis on the Colorado River. In *The ecology of regulated streams*, ed. J. V. Ward and J. A. Stanford, pp. 57–74. New York: Plenum Press.

———. 1980a. *Xyrauchen texanus* (Abbott), humpback sucker. In *Atlas of North American freshwater fishes*, ed. D. S. Lee, C. R. Gilbert, C. H. Hocutt, R. E. Jenkins, D. E. McAllister, and J. R. Stauffer, Jr., p. 435. Raleigh: North Carolina State Museum of Natural History.

———. 1980b. The impacts of habitat alteration on the endangered and threatened fishes of the upper Colorado River basin: A discussion. In *Energy development in the Southwest: Problems of water, fish and wildlife in the upper Colorado River basin*, vol. 2, ed. W. O. Spofford, Jr., A. L. Parker, and A. V. Kneese, pp. 217–223. Research Paper R-18. Washington, D.C.: Resources for the Future.

———. 1983. Status and preferred habitat of the rare fishes of the Green River, Utah. *Proceedings of the Desert Fishes Council* 9 (1977), 314 (abstract).

Holden, P. B., and L. W. Crist. 1981. Documentation of changes in the macroinvertebrate and fish populations in the Green River due to inlet modification of Flaming Gorge Dam. Final Report for U.S. Water and Power Resources Service (U.S. Bureau of Reclamation) Contract 0-07-40-S1357. BIO/WEST, Logan, Utah.

Holden, P. B., and C. B. Stalnaker. 1970. Systematic studies of the cyprinid genus *Gila* in the upper Colorado River basin. *Copeia* 1970, 409–420.

———. 1975a. Distribution and abundance of mainstream fishes of the middle and upper Colorado River basins. *Transactions of the American Fisheries Society* 104, 217–231.

———. 1975b. Distribution of fishes in the Dolores and Yampa River systems of the upper Colorado River basin. *The Southwestern Naturalist* 19, 403–412.

Holden, P. B., W. White, G. Sommerville, D. Duff, R. Gervais, and S. Gloss. 1974. Threatened fishes of Utah. *Proceedings of the Utah Academy of Science, Arts, and Letters* 51, 46–65.

Holden, P. B., and E. J. Wick. 1982. Life history and prospects for recovery of Colorado squawfish. In *Fishes of the upper Colorado River system: Present and future*, ed. W. H. Miller, H. M. Tyus, and C. A. Carlson, pp. 98–108. Bethesda, Md.: Western Division, American Fisheries Society.

Hooper, F. F. 1955. Eradication of fish by chemical treatment. *Michigan Department of Conservation, Fisheries Division Pamphlet* 19, 1–6.

Hoover, F., and J. A. St. Amant. 1983. Results of Mohave chub, *Gila bicolor mohavensis*, relocations in California and Nevada. *California Fish and Game* 69, 54–56.

Hopkirk, J. D., and R. J. Behnke. 1966. Additions to the known native fish fauna of Nevada. *Copeia* 1966, 134–136.

Houf, L. J., and R. S. Campbell. 1977. Effects of antimycin-A and rotenone on macrobenthos in ponds. *U.S. Fish and Wildlife Service, Investigations in Fish Control* 80, 1–29.

Howe, C. B. 1981. *Ancient tribes of the Klamath country*. Portland, Ore.: Binford and Mort.

Hubbs, C. 1957. *Gambusia heterochir*, a new poeciliid fish from Texas, with an account of its hybridization with *G. affinis*. *Tulane Studies in Zoology* 5, 1–16.

———. 1959. Population analysis of a hybrid swarm between *Gambusia affinis* and *G. heterochir*. *Evolution* 13, 236–246.

———. 1963. An evaluation of the use of rotenone as a means of "improving" sports fishing in the Concho River, Texas. *Copeia* 1963, 199–203.

———. 1971. Competition and isolation mechanisms in the *Gambusia affinis* × *G. heterochir* hybrid swarm. *Bulletin of the Texas Memorial Museum* 19, 1–47.

———. 1980. The solution to the *Cyprinodon bovinus* problem: Eradication of a pupfish genome. *Proceedings of the Desert Fishes Council* 10 (1978), 9–14.

———. 1982. Occurrence of exotic fishes in Texas waters. *Pearce-Sellard Series* (University of Texas, Austin), 36, 1–19.

Hubbs, C., and H. J. Broderick. 1963. Current abundance of *Gambusia gaigei*, an endangered fish. *The Southwestern Naturalist* 8, 46–48.

Hubbs, C., and W. F. Hettler. 1964. Observations on the toleration of high temperature and low dissolved oxygen in natural waters by *Crenichthys baileyi*. *The Southwestern Naturalist* 9, 245–248.

Hubbs, C., G. Hoddenbach, and C. M. Fleming. 1986. An enigmatic population of *Gambusia gaigei*, an endangered fish species. *The Southwestern Naturalist* 32, 121–123.

Hubbs, C., and B. L. Jensen. 1984. Extinction of *Gambusia amistadensis*, an endangered fish. *Copeia* 1984, 529–530.

Hubbs, C., J. E. Johnson, and R. H. Wauer. 1977. Habitat management plan for Big Bend gambusia, Big Bend National Park, Texas. U.S. National Park Service, Santa Fe, N.M.

Hubbs, C., T. Lucier, E. Marsh, G. P. Garrett, R. J. Edwards, and E. Milstead. 1978. Results of an eradication program on the ecological relationships of fishes in Leon Creek, Texas. *The Southwestern Naturalist* 23, 487–496.

Hubbs, C., and A. E. Peden. 1969. *Gambusia georgei* sp. nov. from San Marcos, Texas. *Copeia* 1969, 357–364.

Hubbs, C., and J. Pigg. 1976. The effects of impoundments on endangered fishes of Oklahoma. *Transactions of the Oklahoma Academy of Sciences* 5, 113–117.

Hubbs, C., and V. G. Springer. 1957. A revision of the *Gambusia nobilis* species group, with descriptions of three new species, and notes on their variation, ecology, and evolution. *Texas Journal of Science* 9, 279–327.

Hubbs, C., and J. G. Williams. 1979. A review of circumstances affecting the abundance of *Gambusia gaigei*, an endangered fish endemic to Big Bend National Park. In *Proceedings of the First Conference on Scientific Research in the National Parks*, ed. R. M. Linn, pp. 631–634. Washington, D.C.: U.S. National Park Service Transactions and Proceedings Series 5.

Hubbs, C. L. 1929. Studies of the fishes of the order Cyprinodontes. VIII: *Gambusia gaigei*, a new species from the Rio Grande. *Occasional Papers of the Museum of Zoology, University of Michigan* 198, 1–11.

———. 1932. Studies of the fishes of the order Cyprinodontes. XII: A new genus related to *Empetrichthys*. *Occasional Papers of the Museum of Zoology, University of Michigan* 252, 1–5.

———. 1954. Establishment of a forage fish, the red shiner (*Notropis lutrensis*), in the lower Colorado River system. *California Fish and Game* 40, 287–294.

———. 1955. Hybridization between fish species in nature. *Systematic Zoology* 4, 1–20.

———. 1960. Quaternary paleoclimatology of the Pacific Coast of North America. *California Cooperative Fisheries Investigations Report* 7, 105–112.

———. 1963. Secretary Udall: Reviews the Green River fish eradication program. *Copeia* 1963, 465–466.

———. 1964. History of ichthyology in the United States after 1850. *Copeia* 1964, 42–60.

Hubbs, C. L., W. I. Follett, and L. J. Dempster. 1979. List of the fishes of California. *Occasional Papers of the California Academy of Sciences* 133, 1–51.

Hubbs, C. L., L. C. Hubbs, and R. C. Johnson. 1942. Hybridization in nature between species of catostomid fishes. *Contributions of the University of Michigan Laboratory of Vertebrate Biology* 22, 1–76.

Hubbs, C. L., and R. R. Miller. 1941. Studies of the fishes of the order Cyprinodontes. XVII: Genera and species of the Colorado River system. *Occasional Papers of the Museum of Zoology, University of Michigan* 443, 1–9.

———. 1943. Mass hybridization between two genera of cyprinid fishes in the Mohave Desert, California. *Papers of the Michigan Academy of Science, Arts, and Letters* 28 (1942), 343–378.

———. 1948a. Two new relict genera of cyprinid fishes from Nevada. *Occasional Papers of the Museum of Zoology, University of Michigan* 507, 1–30.

———. 1948b. The zoological evidence: Correlation between fish distribution and hydrographic history in the desert basins of western United States. In *The Great Basin with emphasis on glacial and postglacial times*, pp. 17–166. Bulletin of the University of Utah 38, Biological Series 10.

———. 1953. Hybridization in nature between the fish genera *Catostomus* and *Xyrauchen*. *Papers of the Michigan Academy of Science, Arts, and Letters* 38, 207–233.

———. 1965. Studies of cyprinodont fishes. XXII: Variation in *Lucania parva*, its establishment in western United States, and description of a new species from an interior basin in Coahuila, Mexico. *Miscellaneous Publications of the Museum of Zoology, University of Michigan* 127, 1–104.

Hubbs, C. L., R. R. Miller, and L. C. Hubbs. 1974. Hydrographic history and relict fishes of the north-central Great Basin. *Memoirs of the California Academy of Sciences* 7, 1–259.

Huenneke, L. 1988. SCOPE program on biological invasions: A status report. *Conservation Biology* 2, 8–10.

Humphries, J. M. 1984. Genetics of speciation in pupfishes from Laguna Chichancanab, México. In *Evolution of fish species flocks*, ed. A. A. Echelle and I. Kornfield, pp. 129–140. Orono: University of Maine Press.

Hynes, H. B. N. 1970. *The ecology of running waters*. Toronto: University of Toronto Press.

Ingram, W., and C. D. Ziebell. 1983. Diet shifts to benthic feeding by threadfin shad. *Transactions of the American Fisheries Society* 112, 554–556.

Inslee, T. D. 1982a. Culture of endangered Colorado River fishes. Section III: Spawning of razorback suckers. In Part 3, *Colorado River Fisheries Project, final report contracted studies*, ed. W. H. Miller, J. J. Valentine, D. L. Archer, H. M. Tyus, R. A. Valdez, and L. R. Kaeding, pp. 145–157. Final Report for U.S. Bureau of Reclamation Contract 9-07-40-L-1016, Memorandum of Understanding CO-910-MU9-933, U.S. Bureau of Land Management. U.S. Fish and Wildlife Service, Salt Lake City, Utah.

———. 1982b. Spawning and hatching of the razorback sucker (*Xyrauchen texanus*). *Proceedings of the Annual Conference of the Western Association of Fish and Wildlife Commissioners* 62, 431–432 (abstract).

International Union for the Conservation of Nature and Natural Resources. 1988. *1988 IUCN red list of threatened animals*. Morges, Switzerland: IUCN.

Jackson, W., T. Martinez, P. Cuplin, W. L. Minckley, B. Shelby, P. Summers, D. McGlothlin, and B. van Haveren. 1988. *Assessment of water conditions and management opportunities in support of riparian values: BLM San Pedro River Properties*. U.S. Bureau of Land Management, Arizona Project Completion Report. Washington, D.C.: U.S. Government Printing Office.

Jacobi, G. Z., and D. J. Degan. 1977. Aquatic macroinvertebrates in a small Wisconsin trout stream before, during, and two years after treatment with the fish toxicant antimycin. *U.S. Fish and Wildlife Service, Investigations in Fish Control* 81, 1–23.

Jacobi, G. Z., and M. Jacobi. 1982. Fish stomach content analysis. In Part 3, *Colorado River Fisheries Project, final report contracted studies*, ed. W. H. Miller, J. J. Valentine, D. L. Archer, H. M. Tyus, R. A. Valdez, and L. R. Kaeding, pp. 285–324. Final Report for U.S. Bureau of Reclamation Contract 9-07-40-L-1016, Memorandum of Understanding CO-910-MU9-933, U.S. Bureau of Land Management. U.S. Fish and Wildlife Service, Salt Lake City, Utah.

Jacobs, K. E., and W. D. Swink. 1982. Estimations of fish population size and sampling efficiency of electrofishing and rotenone in two Kentucky tailwaters. *North American Journal of Fisheries Management* 2, 239–248.

James, A. E. 1969. *Lernaea* (Copepoda) infection of three native fishes from the Salt River basin, Arizona. Master's thesis, Arizona State University, Tempe.

James, C. J. 1969. Aspects of the ecology of the Devil's Hole pupfish, *Cyprinodon diabolis* Wales. Master's thesis, Nevada Southern University, Las Vegas.

Janzen, D. H. 1983. No park is an island: Increase in interference from outside as park size decreases. *Oikos* 41, 402–410.

———. 1986. The eternal external threat. In *Conservation biology: The science of scarcity and diversity*, ed. M. E. Soulé, pp. 286–303. Sunderland, Mass.: Sinauer Associates.

Jefferts, K. B., P. K. Bergman, and H. F. Fiscus. 1963. A coded wire identification system for macroorganisms. *Nature* 198, 460–462.

Jennings, M. R. 1987. Fredrick Morton Chamberlain (1867–1921), pioneer fishery biologist of the American West. *Fisheries* 12, 22–29.

Jensen, A. C. 1962. Marking and tagging fishes. *U.S. Fish and Wildlife Service Leaflet* 534, 1–8.

Jensen, B. L. 1988. Annual report for fiscal year 1988. U.S. Fish and Wildlife Service, Dexter National Fish Hatchery, Dexter, N.M.

Jester, D. B., and H. J. McKirdy. 1966. Evaluation of trout stream improvement in New Mexico. *Proceedings of the Annual Meeting of the Association of State Game and Fish Commissioners* 46, 316–333.

Jewish Publication Society. *Tanakh: A new translation of the Holy Scriptures, according to the traditional Hebrew text.* Philadelphia: The Jewish Publication Society.

John, R. T. 1984. June sucker fecundity. Utah Division of Wildlife Resources Memorandum, 24 October 1984, Salt Lake City.

Johns, K. R. 1963. The effect of torrential rains on the reproductive cycle of *Rhinichthys osculus* in the Chiricahua Mountains, Arizona. *Copeia* 1963, 286–291.

———. 1964. Survival of fish in intermittent streams of the Chiricahua Mountains, Arizona. *Ecology* 41, 112–119.

Johnson, J. E. 1969. Reproduction, growth, and population dynamics of the threadfin shad, *Dorosoma petenense* (Guenther), in central Arizona reservoirs. Ph.D. diss., Arizona State University, Tempe.

———. 1976. Status of endangered and threatened fish species in Colorado. *U.S. Bureau of Land Management, Technical Note* 280, 1–29.

———. 1977. Realistic management of endangered species: Progress to date. *Proceedings of the Annual Conference of the Western Association of Game and Fish Commissioners* 57, 298–301.

———. 1979. The endangered species act—cooperative management? *Proceedings of the Annual Conference of the Western Association of Game and Fish Commissioners* 59, 85–88.

———. 1980a. Reintroduction of endangered fish species into historic habitats. *Proceedings of the Desert Fishes Council* 11(1979), 92 (abstract).

———. 1980b. The impacts of habitat alterations on the endangered and threatened fishes of the upper Colorado River basin: A discussion. In *Energy development in the Southwest: Problems of water, fish and wildlife in the upper Colorado River basin*, vol. 2, ed. W. O. Spofford, Jr., A. L. Parker, and A. V. Kneese, pp. 224–235. Research Paper R-18. Washington, D.C.: Resources for the Future.

———. 1980c. The resurrection of San Bernardino. *Proceedings of the Desert Fishes Council* 12 (1980), 92 (abstract).

———. 1985. Reintroducing the natives: Razorback sucker. *Proceedings of the Desert Fishes Council* 13 (1981), 73–79.

———. 1987a. *Protected fishes of the U.S. and Canada.* Bethesda, Md.: American Fisheries Society.

———. 1987b. Reintroducing the natives: Colorado squawfish and woundfin. *Proceedings of the Desert Fishes Council* 17 (1985), 118–124.

Johnson, J. E., and C. Hubbs. 1989. Status and conservation of southwestern poeciliid fishes. In *Evolution and ecology of livebearing fishes (Poeciliidae)*, ed. G. K. Meffe and F. F. Snelson, Jr., pp. 301–317. Englewood Cliffs, N.J.: Prentice-Hall.

Johnson, J. E., and G. Kobetich. 1969. A new locality for the Gila topminnow, *Poeciliopsis occidentalis*, in Arizona. *The Southwestern Naturalist* 14, 368.

Johnson, J. E., and J. N. Rinne. 1982. The endangered species act and southwestern fishes. *Fisheries* 7, 1–10.

Johnson, L., and J. Johnson, eds. 1987. *Escape from Death Valley: As told by William Lewis Manley and other '49ers.* Reno: University of Nevada Press.

Johnson, M. P., and P. H. Raven. 1973. Species number and endemism: The Galapagos Archipelago revisited. *Science* 179, 893–895.

Johnson, V. K. 1958. Fisheries management report—Pyramid Lake. Federal Aid to Fisheries Restoration Project Completion Report, FAF-4-R, 1–47. Nevada Department of Wildlife, Reno.

Jones, R. L., and H. M. Tyus. 1985. Recruitment of Colorado squawfish in the Green River basin, Colorado and Utah: 1979–1984. U.S. Fish and Wildlife Service, Division of Endangered Species, Denver, Colo.

Jonez, A., J. Hemphill, R. D. Beland, G. Duncan, and R. A. Wagner. 1951. Fisheries report on the lower Colorado River. Arizona Game and Fish Commission, Phoenix.

Jonez, A., and R. C. Sumner. 1954. Lakes Mead and Mohave investigations: A comparative study of an established reservoir as related to a newly created impoundment. Federal Aid to Fisheries Restoration Project Completion Report, F-1-R, 1–186. Nevada Fish and Game Commission, Reno.

Jonez, A., and N. Wood. 1963. Special fisheries report on Project Vela-Uniform, Lake Mead test explosions. In Minutes of the Technical Committee, Colorado River Wildlife Management Commission, Davis Camp, Arizona. Arizona Game and Fish Department, Phoenix.

Jordan, D. S. 1891. Report of explorations in Utah and Colorado during the summer of 1889, with an account of the fishes found in each of the river basins examined. *Bulletin of the U.S. Fish Commissioner* 9, 1–40.

———. 1905. *A guide to the study of fishes.* 2 vols. New York: Henry Holt and Company.

Jordan, D. S., and B. W. Evermann. 1896a. A checklist of the fishes and fish-like vertebrates of North and Middle America. *Report of the U.S. Commissioner of Fish and Fisheries, 1894–1895* 21, 278–584.

———. 1896b–1900. The fishes of North and Middle America. *Bulletin of the U.S. National Museum* 47 (4 parts), i–lx, 1–3313.

———. 1903. *American food and game fishes.* New York: Doubleday Page.

———. 1923. *American food and game fishes.* 2d printing. New York: Doubleday Doran.

Jordan, D. S., B. W. Evermann, and H. W. Clark. 1930. Checklist of the fishes and fishlike vertebrates of North and Middle America north of the northern boundary of Venezuela and Colombia. *Report of the U.S. Fish Commissioner* 1928, 1–670.

Jordan, D. S., and C. H. Gilbert. 1883. Synopsis of the fishes of North America. *Bulletin of the U.S. National Museum* 16, i–lvi, 1–1018.

Jordan, D. S., and R. E. Richardson. 1907. Description of a new species of killifish, *Lucania browni,* from a hot spring in lower California. *Proceedings of the U.S. National Museum* 33, 319–321.

Jordan, W. R. 1988. Ecological restoration: Reflections on a half-century of experience at the University of Wisconsin-Madison Arboretum. In *Biodiversity,* ed. E. O. Wilson, pp. 311–316. Washington, D.C.: National Academy of Sciences Press.

Joseph, J. W. 1977. An indexed, annotated bibliography of the endangered and threatened fishes of the upper Colorado River system. *U.S. Fish and Wildlife Service Report* FWS/OBS/77/61, 1–169.

———. 1978. Capture locations of rare fish in the upper Colorado River system. Final Report for U.S. Fish and Wildlife Service Contract 14-16-0009-77-049, Western Energy Land Use Team. Ecological Consultants, Fort Collins, Colo.

Joseph, J. W., J. A. Sinning, R. J. Behnke, and P. B. Holden. 1977. An evaluation of the status, life history, and habitat requirements of endangered and threatened fishes of the upper Colorado River system. *U.S. Fish and Wildlife Service Report* FWS/OBS/77/62, 1–168.

Kaeding, L. R., and D. B. Osmundson. 1988a. A report on the studies of the endangered fishes of the upper Colorado River as part of conservation measures for the Green Mountain and Ruedi Reservoir water sales. Colorado River Fisheries Project, U.S. Fish and Wildlife Service, Grand Junction, Colo.

———. 1988b. Interaction of slow growth and increased early-life mortality: An hypothesis on the decline of Colorado squawfish in the upstream regions of its historic range. *Environmental Biology of Fishes* 22, 287–298.

———. 1989. Biologically defensible flow recommendations for the maintenance and enhancement of Colorado squawfish habitat in the "15-mile" reach of the upper Colorado River during July, August and September. Colorado River Fisheries Project, U.S. Fish and Wildlife Service, Grand Junction, Colo.

Kaeding, L. R., H. M. Tyus, B. D. Burdick, R. L. Jones, and C. W. McAda. 1986. Sensitive areas analysis/monitoring: Preference curve development/refinement. In Annual Report for U.S. Bureau of Reclamation, Colorado River Endangered Fishes Investigations, section II, pp. 1–55. Denver: U.S. Fish and Wildlife Service.

Kaeding, L. R., and M. A. Zimmerman. 1983. Life history and ecology of the humpback chub in the Little Colorado and Colorado rivers of the Grand Canyon. *Transactions of the American Fisheries Society* 112, 577–594.

Kapuscinski, A. R., and L. G. Jacobson. 1987. *Genetic guidelines for fisheries management.* Minnesota Sea Grant. Duluth: University of Minnesota.

Karp, C. A., and H. M. Tyus. 1990. Behavioral interactions among young Colorado squawfish and six fish species. *Copeia* 1990, 25–34.

Kashuba, S. A., and W. J. Matthews. 1984. Physical condition of larval shad during spring-summer in a southwestern reservoir. *Transactions of the American Fisheries Society* 113, 199–204.

Keast, A. 1978. Trophic and spatial interrelationships in the fish species of an Ontario temperate lake. *Environmental Biology of Fishes* 3, 7–31.

Keith, W. E. 1969. Preliminary results in the use of a nursery pond as a tool in fishery management. *Proceedings of the Meeting of the Southeastern Association of Game and Fish Commissioners* 1969, 501–511.

Kennedy, D. H. 1916. A possible enemy of the mosquito. *California Fish and Game* 2, 179–182.

Kennedy, S. E. 1977. Life history of the Leon Springs pupfish, *Cyprinodon bovinus. Copeia* 1977, 93–103.

Kidd, G. 1977. An investigation of endangered and threatened fish species in the upper Colorado River as related to Bureau of Reclamation projects. Final Report for U.S. Bureau of Reclamation Contract. Northwest Fisheries Research, Clifton, Colo.

Kimsey, J. B. 1957. Fisheries problems in impounded waters of California and the lower Colorado River. *Transactions of the American Fisheries Society* 87, 319–332.

Kimsey, J. B., R. H. Hagy, and G. W. McCammon. 1957. Progress report on the Mississippi threadfin shad, *Dorosoma petenense atchafaylae* [sic], in the Colorado River for 1956. *California Department of Fish and Game, Inland Fisheries Administrative Report* 57-23, 1–48.

Kincaid, H. L., and C. R. Berry, Jr. 1986. Trout broodstocks used in management of national fisheries. In *Fish culture in fisheries management,* ed. R. H. Stroud, pp. 122–222. Bethesda, Md.: American Fisheries Society.

Kinne, O. 1960. Growth, food intake and food conversion in a euryplastic fish exposed to different temperatures and salinities. *Physiological Zoology* 33, 288–317.

———. 1965. Salinity requirements of the fish, *Cyprinodon macularius. U.S. Public Health Service Publication* 999-WP-25, 187–192.

Kinne, O., and E. M. Kinne. 1962a. Rates of development in embryos of a cyprinodont fish exposed to different temperature-salinity-oxygen combinations. *Canadian Journal of Zoology* 40, 231–253.

———. 1962b. Effects of salinity and oxygen on developmental rates in a cyprinodont fish. *Nature* (London) 193, 1097–1098.

Kirsch, P. H. 1889. Notes on a collection of fishes obtained in the Gila River at Fort Thomas, Arizona. *Proceedings of the U.S. National Museum* 11, 555–558.

Kleerkoper, H. 1969. *Olfaction in fishes.* Bloomington: Indiana University Press.

Knack, M. C., and O. C. Steward. 1984. *As long as the river shall run.* Berkeley and Los Angeles: University of California Press.

Knapp, W. T. 1953. *Fishes found in the freshwaters of Texas*. Brunswick, Ga.: Ragland Studio Litho Printing.

Kniffen, F. B. 1932. Lower California studies. IV: The natural landscape of the Colorado Delta. *University of California Publications in Geography* 5, 149–244.

Koch, D. L. 1973. Reproductive characteristics of the cui-ui lakesucker (*Chasmistes cujus* Cope) and its spawning behavior in Pyramid Lake, Nevada. *Transactions of the American Fisheries Society* 102, 145–149.

———. 1976. Life history information on the cui-ui lakesucker (*Chasmistes cujus* Cope, 1883) in Pyramid Lake, Nevada. *Occasional Papers of the Biological Society of Nevada* 40, 1–12.

Koch, D. L., J. J. Cooper, G. P. Contreras, and V. King. 1975. Survey of the fishes of the Clear Lake Reservoir drainage. *Center for Water Resources Research, University of Nevada, Reno, Project Report* 37, 1–38.

Koch, D. L., J. J. Cooper, E. G. Jacobson, and R. A. Spencer. 1979. Investigation of Walker Lake, Nevada: Dynamic ecological relationships. *University of Nevada, Desert Research Institute Publication* 50010, 1–191.

Kodric-Brown, A. 1977. Reproductive success and the evolution of breeding territories in pupfish (*Cyprinodon*). *Evolution* 31, 750–766.

———. 1978. Establishment and defense of territories in a pupfish (Cyprinodontidae: *Cyprinodon*). *Animal Behavior* 26, 818–834.

———. 1981. Variable breeding systems in pupfishes (genus *Cyprinodon*): Adaptations to changing environments. In *Fishes in North American deserts*, ed. R. J. Naiman and D. L. Soltz, pp. 205–235. New York: John Wiley and Sons.

Koehn, R. K. 1965. Development and ecological significance of nuptial tubercles of the red shiner, *Notropis lutrensis. Copeia* 1965, 462–467.

Kohn, A. J. 1967. Environmental complexity and species diversity in the gastropod genus *Conus* on Indo-West Pacific reef platforms. *American Naturalist* 101, 251–259.

Kornfield, I., and R. K. Koehn. 1975. Genetic variation and evolution in New World cichlids. *Evolution* 29, 427–437.

Kornfield, I., D. C. Smith, and P. S. Gagnon. 1982. The cichlid fish of Cuatro Ciénegas, México: Direct evidence of conspecificity among distinct trophic morphs. *Evolution* 36, 658–664.

Koster, W. J. 1957. *Fishes of New Mexico*. Albuquerque: University of New Mexico Press.

———. 1960. *Ptychocheilus lucius* (Cyprinidae) in the San Juan River, New Mexico. *The Southwestern Naturalist* 5, 174–175.

Krumholz, L. A. 1948. The use of rotenone in fisheries research. *Journal of Wildlife Management* 12, 305–317.

Kushlan, J. A. 1979. Design and management of continental wildlife reserves: Lessons from the Everglades. *Biological Conservation* 15, 281–290.

Kynard, B. E., and R. Garrett. 1979. Reproductive ecology of the Quitobaquito pupfish from Organ Pipe Cactus National Monument, Arizona. In *Proceedings of the First Conference on Scientific Research in the National Parks*, ed. R. M. Linn, pp. 625–629. Washington, D.C.: U.S. National Park Service Transactions and Proceedings Series 5.

LaBounty, J. F., and J. E. Deacon. 1972. *Cyprinodon milleri*, a new species of pupfish (family Cyprinodontidae) from Death Valley, California. *Copeia* 1972, 769–780.

Lacy, R. C. 1987. Loss of genetic diversity from managed populations: Interacting effects of drift, mutation, immigration, selection, and population subdivision. *Conservation Biology* 1, 143–158.

———. 1988. A report on population genetics in conservation. *Conservation Biology* 2, 245–247.

Lake, P. S. 1982. The relationships between freshwater fish distribution, stream drainage area and stream length in some streams of south-east Australia. *Bulletin of the Australian Society of Limnology* 8, 31–37.

Lambertson, R. E. 1982. Mitigation and section 7 on the upper Colorado River. In *Fishes of the upper Colorado River system: Present and future*, ed. W. H. Miller, H. M. Tyus, and C. A. Carlson, pp. 86–90. Bethesda, Md.: Western Division, American Fisheries Society.

Langhorst, D. R. 1986. Potential fate of larval razorback sucker in Lake Mohave, Arizona-Nevada. *Abstracts of the Annual Meeting of the Arizona–New Mexico Chapter, American Fisheries Society* 1986, unpaginated (abstract).

———. 1987. Larval razorback sucker, *Xyrauchen texanus*, in Lake Mohave, Arizona-Nevada. *Proceedings of the Desert Fishes Council* 17 (1985), 164–165 (abstract).

———. 1988. A monitoring study of razorback sucker (*Xyrauchen texanus*) reintroduced into the lower Colorado River in 1987. Final Report for California Department of Fish and Game Contract C-1888. California Department of Fish and Game, Blythe.

———. 1989. A monitoring study of razorback sucker (*Xyrauchen texanus*) reintroduced into the lower Colorado River in 1988. Final Report for California Department of Fish and Game Contract FG-7494. California Department of Fish and Game, Blythe.

Langhorst, D. R., and P. C. Marsh. 1986. Early life history of razorback sucker in Lake Mohave. Final Report for U.S. Bureau of Reclamation Contract 5-PG-30-06440. Arizona State University, Tempe.

Langhorst, D. R., P. C. Marsh, and W. L. Minckley. 1985. Reproductive biology and larval ecology of razorback sucker, *Xyrauchen texanus*, in Lake Mohave, Arizona-Nevada. *Abstracts of the Annual Meeting of the Arizona–New Mexico Chapter, American Fisheries Society* 1985, unpaginated (abstract).

Langlois, D. 1977. Colorado's endangered fish. *Colorado Outdoors* 26, 18–21.

Lanigan, S. H., and C. R. Berry, Jr. 1981. Distribution of fishes in the White River, Utah. *The Southwestern Naturalist* 26, 389–393.

Lanigan, S. H., and H. M. Tyus. 1989. Population size and status of the razorback sucker in the Green River basin, Utah and Colorado. *North American Journal of Fisheries Management* 9, 68–73.

LaRivers, I. 1962. *Fish and fisheries of Nevada.* Carson City: Nevada Game and Fish Commission.

LaRivers, I., and T. J. Trelease. 1952. An annotated check list of the fishes of Nevada. *California Fish and Game* 38, 113–123.

Leach, H. R., J. M. Brode, and S. J. Nicola. 1974. *At the crossroads: A report on California's endangered and rare fish and wildlife for 1973–1974.* Sacramento: California Department of Fish and Game.

———. 1976. *At the crossroads: A report on California's endangered and rare fish and wildlife for 1975–1976.* Sacramento: California Department of Fish and Game.

———. 1978. *At the crossroads: A report on California's endangered and rare fish and wildlife for 1977–1978.* Sacramento: California Department of Fish and Game.

Leary, R. F., F. W. Allendorf, S. R. Phelps, and K. L. Knudsen. 1987. Genetic divergence and identification of seven cutthroat trout subspecies and rainbow trout. *Transactions of the American Fisheries Society* 116, 580–587.

Lee, D. C. 1980. On the Marxian view of the relationship between man and nature. *Environmental Ethics* 2, 3–16.

Lee, D. S., and G. H. Burgess. 1980. *Gambusia affinis* (Baird and Girard), mosquitofish. In *Atlas of North American freshwater fishes*, ed. D. S. Lee, C. R. Gilbert, C. H. Hocutt, R. E. Jenkins, D. E. McAllister, and J. R. Stauffer, Jr., p. 538. Raleigh: North Carolina State Museum of Natural History.

Lee, D. S., C. R. Gilbert, C. H. Hocutt, R. E. Jenkins, D. E. McAllister, and J. R. Stauffer, Jr., eds. 1980. *Atlas of North American freshwater fishes.* Raleigh: North Carolina State Museum of Natural History.

Lee, D. S., S. P. Platania, and G. H. Burgess. 1983. Atlas of North American Freshwater Fishes: 1983 Supplement. *Occasional Papers of the North Carolina Biological Survey* 1983-3, 1–67.

Leggett, W. C. 1977. The ecology of fish migrations. *Annual Review of Ecology and Systematics* 8, 285–308.

———. 1986. The dependence of fish larval survival on food and predator densities. In *The role of freshwater outflow in coastal marine ecosystems*, ed. S. Skreslet, pp. 117–138. New York: Springer-Verlag.

Lemly, A. D. 1985. Suppression of native fish populations by green sunfish in first-order streams of piedmont North Carolina. *Transactions of the American Fisheries Society* 114, 705–712.

Lennon, R. E. 1966. Antimycin—a new fishery tool. *Wisconsin Conservation Bulletin* 31, 4–5.

Lennon, R. E., and B. L. Berger. 1970. A resume on field applications of antimycin-A to control fish. *U.S. Bureau of Sport Fisheries and Wildlife, Investigations in Fish Control* 40, 1–19.

Lennon, R. E., J. B. Hunn, R. A. Schnick, and R. M. Burress. 1971. *Reclamation of ponds, lakes, and streams with fish toxicants: A review.* Washington, D.C.: U.S. Fish and Wildlife Service.

Leonard, J. W. 1939. Notes on the use of derris as a fish poison. *Transactions of the American Fisheries Society* 68, 269–279.

Leopold, A. 1949. *A Sand County almanac and sketches here and there.* New York: Oxford University Press.

Leopold, A. S. 1963. *Wildlife management in the national parks.* Washington, D.C.: Advisory Board for Wildlife Management, U.S. National Park Service.

Leopold, L. B., ed. 1953. *Round River: From the Journals of Aldo Leopold.* New York: Oxford University Press.

Lewis, W. M., Jr., and D. P. Morris. 1986. Toxicity of nitrite to fish: A review. *Transactions of the American Fisheries Society* 115, 183–195.

Li, H. W., and P. B. Moyle. 1981. Ecological analysis of species introductions into aquatic systems. *Transactions of the American Fisheries Society* 110, 772–782.

Liles, T. 1981. Survey of aquatic resources, Region III, Statewide fishery investigations. Federal Aid to Fisheries Restoration Project Completion Report, F-7-R-22, 24. Arizona Game and Fish Department, Phoenix.

Linnaeus, C. [Linné, Karl]. 1758. *Systema naturae*, 10th ed., vol. 1.

Litton, M. 1969. Death Valley. *Sunset* 143 (5), 78–89.

———. 1970. Saving the pupfish, a program for action. *Cry California* 5, 22–25.

Liu, R. K. 1969. The comparative behavior of allopatric species (Teleostei-Cyprinodontidae: *Cyprinodon*). Ph.D. diss., University of California, Los Angeles.

Liu, R. K., and D. L. Soltz. 1983. Letter to E. P. Pister (dated 19 February 1975). *Proceedings of the Desert Fishes Council* 7 (1975), 209–210.

Lockington, W. N. 1881. Description of a new species of *Catostomus* (*Catostomus cypho*) from the Colorado River. *Proceedings of the Philadelphia Academy of Natural Sciences* 32, 237–240.

Loiselle, P. V. 1980. Spawn recognition by male *Cyprinodon macularius californiensis*. *Proceedings of the Desert Fishes Council* 11 (1979), 46 (abstract).

———. 1982. An experimental analysis of pupfish (Teleostei: Cyprinodontidae: *Cyprinodon*) reproductive behavior. Ph.D. diss., University of California, Berkeley.

Lopinot, A. C. 1975. Summary of the use of toxicants to rehabilitate fish populations in the Midwest. In *Rehabilitation of fish populations with toxicants: A symposium*, ed. P. H. Eschmeyer, pp. 1–4. Bethesda, Md.: North Central Division, American Fisheries Society, Special Publication 4.

Lotspeich, F. B., and W. S. Platts. 1982. An integrated land-aquatic classification system. *North American Journal of Fisheries Management* 2, 138–149.

Loudenslager, E. J., and G.A.E. Gall. 1980. Geographic patterns of protein variation and subspeciation in the cutthroat trout *Salmo clarki*. *Systematic Zoology* 29, 27–42.

Loudenslager, E. J., J. N. Rinne, G.A.E. Gall, and R. E. David. 1986. Biochemical genetic studies of native Arizona and New Mexico trout. *The Southwestern Naturalist* 31, 221–234.

Loudermilk, W. E. 1985. Aspects of razorback sucker (*Xyrauchen texanus* Abbot [*sic*]) life history which help explain their decline. *Proceedings of the Desert Fishes Council* 13 (1981), 67–72.

Loudermilk, W. E., and L. C. Ulmer. 1985. A fishery inventory with emphasis on razorback sucker (*Xyrauchen texanus*) status in the lower Colorado River. Final Report for U.S. Fish and Wildlife Service Cooperative Agreement 14-16-002-80-921. California Department of Fish and Game, Blythe (also appeared as California Department of Fish and Game, Region 5 Information Bulletin 0012-9-1985, 1–27, 238 pp., appendix).

Lovejoy, T. E., R. D. Bierregaard, Jr., A. B. Reynolds, J. R. Malcolm, C. E. Quintela, L. H. Harper, K. S. Brown, Jr., A. H. Powell, G. V. N. Powell, N. O. R. Schubart, and M. B. Hays. 1986. Edge and other effects of isolation on Amazon forest fragments. In *Conservation biology: The science of scarcity and diversity*, ed. M. E. Soulé, pp. 257–285. Sunderland, Mass.: Sinauer Associates.

Lovejoy, T. E., J. M. Rankin, R. O. Bierregaard, Jr., K. S. Brown, Jr., and M. E. van der Voort. 1984. Ecosystem decay of Amazon forest remnants. In *Extinctions*, ed. M. H. Nitecki, pp. 295–325. Chicago: University of Chicago Press.

Lowe, C. H., and W. G. Heath. 1969. Behavioral and physiological responses to temperature in the desert pupfish, *Cyprinodon macularius*. *Physiological Zoology* 42, 53–59.

Lowe, C. H., D. S. Hinds, and E. A. Halpern. 1967. Experimental catastrophic selection and tolerances to low oxygen concentration in native Arizona freshwater fishes. *Ecology* 48, 1013–1017.

Lurie, E. 1960. *Louis Agassiz: A life in science*. Chicago: University of Chicago Press.

McAda, C. W. 1977. Aspects of the life history of three catostomids native to the upper Colorado River basin. Master's thesis, Utah State University, Logan.

———. 1983. Colorado squawfish, *Ptychocheilus lucius* (Cyprinidae), with a channel catfish, *Ictalurus punctatus* (Ictaluridae), lodged in its throat. *The Southwestern Naturalist* 28, 119–120.

McAda, C. W., and K. Seethaler. 1977. Investigations of the movements and ecological requirements of the Colorado squawfish and humpback sucker in the Yampa and Green rivers. Final Report for U.S. Fish and Wildlife Service Contract 14-16-008-1140. Utah Cooperative Fisheries Research Unit, Utah State University, Logan.

McAda, C. W., and R. S. Wydoski. 1980. The razorback sucker, *Xyrauchen texanus*, in the upper Colorado River basin, 1974–76. *U.S. Fish and Wildlife Service Technical Paper* 99, 1–15.

McAllister, D. E. 1970. Rare or endangered Canadian fishes. *Canadian Field-Naturalist* 84, 5–8.

McAllister, D. E., B. J. Parker, and P. M. McKee. 1985. Rare, endangered, and extinct fishes in Canada. *Syllogeus* (Canadian National Museum of Natural Science, Ottawa) 54, 1–192.

MacArthur, R. H. 1964. Environmental factors affecting bird species diversity. *American Naturalist* 98, 387–397.

MacArthur, R. H., and E. O. Wilson. 1967. *The theory of island biogeography*. Princeton: Princeton University Press.

McBroom, J. T. 1983. Pupfish taskforce report. *Proceedings of the Desert Fishes Council* 3 (1971), 21–22.

McCall, T. C. 1980. Fisheries investigation of Lake Mead, Arizona-Nevada, from Separation Rapids to Boulder Canyon, 1978–1979. Final Report for U.S. Water and Power Resources Service (U.S. Bureau of Reclamation) Contract 8-07-30-X0025. Arizona Game and Fish Department, Phoenix.

McCarthy, M. S. 1986. Age of imperiled razorback sucker (Pisces: Catostomidae) from Lake Mohave, Arizona-Nevada. Master's thesis, Arizona State University, Tempe.

McCarthy, M. S., and W. L. Minckley. 1987. Age estimation for razorback sucker (Pisces: Catostomidae) from Lake Mohave, Arizona-Nevada. *Journal of the Arizona-Nevada Academy of Sciences* 21, 87–97.

McDonald, D. B., and P. A. Dotson. 1960. Fishery investigations of the Glen Canyon and Flaming Gorge impoundment areas. *Utah Department of Fish and Game Information Bulletin* 60-3, 1–70.

McEvoy, J. III. 1973. The American public's concern with the environment. In *Environmental quality and water development*, ed. C. R. Goldman, J. McEvoy III, and P. J. Richerson, pp. 135–156. San Francisco: W. H. Freeman.

McKirdy, H. J. 1964. Evaluation of stream improvement structures. Regional Office, U.S. Forest Service, Albuquerque, N.M.

McLane, A. 1971a. The crisis of the Devils Hole pupfish, Devil's Hole, Nevada. *National Speleological Society News* 29 (4), 39–42.

———. 1971b. History of Devil's Hole, Nye County, Nevada. *Journal of Spelean History* 4, 43–49.

McMahon, T. E. and R. R. Miller. 1985. Status of the fishes of the Río Sonoyta basin, Arizona and Sonora, México. *Proceedings of the Desert Fishes Council* 14 (1982), 53–59.

McMahon, T. E., and J. C. Tash. 1988. Experimental analysis of the role of emigration in population regulation of desert pupfish. *Ecology* 69, 1871–1883.

McNatt, R. M. 1974. Re-evaluation of the native fishes of the Río Yaqui in the United States. *Proceedings of the Annual Conference of the Western Association of State Game and Fish Commissioners* 44, 273–279.

———. 1979. Discovery and loss of a new locality for the Gila topminnow, *Poeciliopsis occidentalis occidentalis* (Poeciliidae). *The Southwestern Naturalist* 24, 555–556.

McNulty, F. 1973. Kill the pupfish? Save the pupfish! *Audubon* 75 (4), 40–47.

McPhail, J. D., and C. C. Lindsey. 1986. Zoogeography of the freshwater fishes of Cascadia (the Columbia system and rivers north to the Stikine). In *The zoogeography of North American freshwater fishes*, ed. C. H. Hocutt and E. O. Wiley, pp. 615–638. New York: John Wiley and Sons.

Maddux, H. R., D. M. Kubly, J. C. deVos, Jr., W. R. Persons, R. Staedicke, and R. L. Wright. 1987. Effects of varied flow regimes on aquatic resources of Glen and Grand canyons. Final Report for U.S. Bureau of Reclamation Contract 4-AG-40-01810. Arizona Game and Fish Department, Phoenix.

Mahon, R. 1984. Divergent structure of fish taxocenes in north temperate streams. *Canadian Journal of Fisheries and Aquatic Sciences* 41, 330–350.

Maitland, P. S., and D. Evans. 1986. The role of captive breeding in the conservation of fish species. *International Zoo Yearbook* 24/25 (1986), 66–74.

Mangum, F. A. 1984. Aquatic ecosystem inventory: Macroinvertebrate analysis. Annual Progress Report to the Gila National Forest, 1984. U.S. Forest Service, Aquatic Ecosystem Analysis Laboratory, Provo, Utah.

———. 1985. Aquatic ecosystem inventory: Macroinvertebrate analysis. Annual Progress Report to the Gila National Forest, 1985. U.S. Forest Service, Aquatic Ecosystem Analysis Laboratory, Provo, Utah.

Margules, C., and M. B. Usher. 1981. Criteria used in assessing wildlife conservation potential: A review. *Biological Conservation* 21, 79–109.

Marking, L. L., and T. D. Bills. 1975. Toxicity of potassium permanganate to fish and its effectiveness for detoxifying antimycin. *Transactions of the American Fisheries Society* 104, 579–583.

Marsh, P. C., ed. 1984. Biota of Cuatro Ciénegas, Coahuila, México. *Journal of the Arizona-Nevada Academy of Sciences* 19, 1.

———. 1985. Effect of incubation temperature on survival of embryos of native Colorado River fishes. *The Southwestern Naturalist* 30, 129–140.

———. 1987a. Food of adult razorback sucker in Lake Mohave, Arizona-Nevada. *Transactions of the American Fisheries Society* 116, 117–119.

———. 1987b. Razorback sucker management in Arizona. *Proceedings of the Desert Fishes Council* 17 (1985), 181–182 (abstract).

———. 1990. Native fishes at Buenos Aires National Wildlife Refuge and Arizona State University Research Park: Opportunities for research, management, and public education on endangered species. *Proceedings of the Desert Fishes Council* 19 (1987), 67–78.

Marsh, P. C., and J. L. Brooks. 1989. Predation by ictalurid catfishes as a deterrent to re-establishment of introduced razorback suckers. *The Southwestern Naturalist* 34, 188–195.

Marsh, P. C., and D. R. Langhorst. 1988. Feeding and fate of wild larval razorback suckers. *Environmental Biology of Fishes* 21, 59–67.

Marsh, P. C., and W. L. Minckley. 1982. Fishes of the Phoenix metropolitan area in central Arizona. *North American Journal of Fisheries Management* 2, 395–404.

———. 1985. Aquatic resources of the Yuma Division, lower Colorado River. Final Report for U.S. Bureau of Reclamation Contract 2-07-30-X0214. Arizona State University, Tempe.

———. 1987. Aquatic resources of the Yuma Division, lower Colorado River. Final Report for Phase II, U.S. Bureau of Reclamation Contract 2-07-30-X0214. Arizona State University, Tempe.

———. 1989. Observations on recruitment and ecology of razorback sucker: Lower Colorado River, Arizona-California-Nevada. *Great Basin Naturalist* 49, 71–78.

———. 1990. Management of endangered Sonoran topminnow at Bylas Springs, Arizona: Description, critique, and recommendations. *Great Basin Naturalist* 50, 265–272.

———. 1991. Radiotelemetry of razorback suckers in the Gila River, eastern Arizona. *Proceedings of the Desert Fishes Council* 21 (1989), 163–171.

Marsh, P. C., and D. Papoulias. 1989. Ichthyoplankton of Lake Havasu, a Colorado River impoundment, Arizona-California. *California Fish and Game* 75, 68–73.

Martinez, P. J. 1986. White River–Taylor Draw Project: Pre- and post-impoundment fish community investigations. Final Report for Colorado River Water Conservation District Contract 5281-X. Colorado Division of Wildlife, Grand Junction.

Maughan, O. E., and K. L. Nelson. 1982. Improving stream fisheries. *Water Spectrum* 12, 11–15.

May, R. C. 1974. Larval mortality in marine fishes and the critical period concept. In *The early life history of fish*, ed. J. H. S. Blaxter, pp. 3–19. Berlin: Springer-Verlag.

Measeles, E. B. 1981. *A crossing on the Colorado, Lee's Ferry.* Boulder, Colo.: Pruett Publications.

Medel-Ulmer, L. 1983. Movement and reproduction of the razorback sucker (*Xyrauchen texanus*) inhabiting Senator Wash Reservoir, Imperial County, California. *Proceedings of the Desert Fishes Council* 12 (1980), 106 (abstract).

Meffe, G. K. 1983a. Ecology of species replacement in the Sonoran topminnow (*Poeciliopsis occidentalis*) and the mosquitofish (*Gambusia affinis*) in Arizona. Ph.D. diss., Arizona State University, Tempe.

———. 1983b. Attempted chemical renovation of an Arizona springbrook for management of the endangered Sonoran topminnow. *North American Journal of Fisheries Management* 3, 315–321.

———. 1984. Effects of abiotic disturbance on coexistence of predator-prey fish species. *Ecology* 65, 1525–1534.

———. 1985a. Life history patterns of *Gambusia marshi* (Poeciliidae) from Cuatro Cienegas, Mexico. *Copeia* 1985, 898–905.

———. 1985b. Predation and species replacement in American southwestern fishes: A case study. *The Southwestern Naturalist* 30, 173–187.

———. 1986. Conservation genetics and the management of endangered fishes. *Fisheries* 11, 14–23.

———. 1987. Conserving fish genomes: Philosophies and practices. *Environmental Biology of Fishes* 18, 3–9.

———. 1989. Fish utilization of springs and cienegas in the arid southwest. In *Freshwater wetlands and wildlife: Perspectives on natural, managed, and degraded ecosystems*, ed. R. R. Sharitz and J. W. Gibbons, pp. 475–485. Oak Ridge, Tenn.: U.S. Department of Energy, Office of Scientific and Technical Information.

Meffe, G. K., D. A. Hendrickson, W. L. Minckley, and J. N. Rinne. 1982. Factors resulting in decline of the endangered Sonoran topminnow (Atheriniformes: Poeciliidae) in the United States. *Biological Conservation* 25, 135–159.

Meffe, G. K., D. A. Hendrickson, and J. N. Rinne. 1982. Description of a new topminnow population in Arizona, with observations on topminnow/mosquitofish co-occurrence. *The Southwestern Naturalist* 27, 226–228.

Meffe, G. K., and R. C. Vrijenhoek. 1988. Conservation genetics in the management of desert fishes. *Conservation Biology* 2, 157–169.

Mehringer, P. J., Jr. 1986. Prehistoric environments. In *Handbook of North American Indians, Volume 11: Great Basin*, ed. W. L. D'Azevedo, pp. 31–50. Washington, D.C.: Smithsonian Institution.

Mello, K., and P. R. Turner. 1980. Status of Gila trout. *Endangered Species Report* 6, 1–52. U.S. Fish and Wildlife Service, Albuquerque, N.M.

Menge, B. A., and J. P. Sutherland. 1976. Species diversity gradients: Synthesis of the roles of predation, competition, and temporal heterogeneity. *American Naturalist* 110, 351–369.

Meyer, C. W., and M. Moretti. 1988. Fisheries survey of the San Juan River, Utah, 1987. Final Report for Utah Division of Wildlife Resources–U.S. Bureau of Reclamation Cooperative Agreement 7FC-40-05050. Salt Lake City.

Meyers, M. 1977. Brown trout control program, South Fork of the Kern River, 1976. California Department of Fish and Game Memorandum, Sacramento.

Millemann, R. E., W. J. Birge, J. A. Black, R. M. Cushman, K. Danilds, P. J. Franco, J. M. Giddings, J. F. McCarthy, and A. J. Stewart. 1984. Comparative acute toxicity to aquatic organisms of components of coal-derived synthetic fuels. *Transactions of the American Fisheries Society* 113, 74–85.

Miller, K. D. 1982. Colorado River fishes recovery team. In *Fishes of the upper Colorado River system: Present and future*, ed. W. H. Miller, H. M. Tyus, and C. A. Carlson, pp. 96–97. Bethesda, Md.: American Fisheries Society.

Miller, R. B. 1953. Comparative survival of wild and hatchery-reared cutthroat trout in a stream. *Transactions of the American Fisheries Society* 83, 120–130.

Miller, R. I., S. P. Bratton, and P. S. White. 1987. A regional strategy for reserve design and placement based on an analysis of rare and endangered species distribution patterns. *Biological Conservation* 39, 255–268.

Miller, R. I., and L. D. Harris. 1979. Predicting species changes in isolated wildlife reserves. In *Proceedings of the First Conference on Scientific Research in the National Parks*, ed. R. M. Linn, pp. 79–82. Washington, D.C.: U.S. National Park Service Transactions and Proceedings Series 5.

Miller, R. J., and M. E. Evans. 1965. External morphology of the brain and lips of catostomid fishes. *Copeia* 1965, 467–487.

Miller, R. R. 1943a. The status of *Cyprinodon macularius* and *Cyprinodon nevadensis*, two desert fishes of western North America. *Occasional Papers of the Museum of Zoology, University of Michigan* 473, 1–25.

———. 1943b. *Cyprinodon salinus*, a new species of fish from Death Valley, California. *Copeia* 1943, 69–78.

———. 1943c. Further data on freshwater populations of the killifish, *Fundulus parvipinnis*. *Copeia* 1943, 51–52.

———. 1946a. The need for ichthyological surveys of the major rivers of western North America. *Science* 104, 517–519.

———. 1946b. *Gila cypha,* a remarkable new species of cyprinid fish from the Colorado River in Grand Canyon, Arizona. *Journal of the Washington Academy of Sciences* 36, 409–415.

———. 1948. The cyprinodont fishes of the Death Valley system of eastern California and southeastern Nevada. *Miscellaneous Publications of the Museum of Zoology, University of Michigan* 529, 1–155.

———. 1949. Hot springs and fish life. *Aquarium Journal* 20, 286–288.

———. 1950. Notes on the cutthroat and rainbow trouts with the description of a new species from the Gila River, New Mexico. *Occasional Papers of the Museum of Zoology, University of Michigan* 68, 1–42.

———. 1952. Bait fishes of the lower Colorado River from Lake Mead, Nevada, to Yuma, Arizona, with a key for their identification. *California Fish and Game* 38, 7–42.

———. 1955. Fish remains from archaeological sites in the lower Colorado River basin, Arizona. *Papers of the Michigan Academy of Science, Arts, and Letters* 40, 125–136.

———. 1956. A new genus and species of cyprinodontid fish from San Luis Potosí, Mexico, with remarks on the subfamily Cyprinodontinae. *Occasional Papers of the Museum of Zoology, University of Michigan* 581, 1–17.

———. 1959. Origin and affinities of the freshwater fish fauna of western North America. In *Zoogeography,* ed. C. L. Hubbs, pp. 187–222. Publication 51 (1958). Washington, D.C.: American Association for the Advancement of Science.

———. 1961. Man and the changing fish fauna of the American Southwest. *Papers of the Michigan Academy of Science, Arts, and Letters* 46, 365–404.

———. 1963. Is our native underwater life worth saving? *National Parks Magazine* 37, 4–9.

———. 1964a. Fishes of Dinosaur. *Naturalist* 15, 24–29.

———. 1964b. Extinct, rare and endangered American freshwater fishes. *Proceedings of the Sixteenth International Congress of Zoology* 8, 4–16.

———. 1964c. Redescription and illustration of *Cyprinodon latifasciatus,* an extinct cyprinodontid fish from Coahuila, Mexico. *The Southwestern Naturalist* 9, 62–67.

———. 1967. Status of populations of native fishes of the Death Valley system in California and Nevada. Final Completion Report for Resource Studies, U.S. National Park Service, June–July 1967, Death Valley, California. University of Michigan, Ann Arbor.

———. 1968a. *Rare and endangered world freshwater fishes.* Morges, Switzerland: IUCN, Survival Service Commission.

———. 1968b. Records of some native freshwater fishes transplanted into various waters of California, Baja California, and Nevada. *California Fish and Game* 54, 170–179.

———. 1968c. Two new fishes of the genus *Cyprinodon* from the Cuatro Cienegas basin, Coahuila, Mexico. *Occasional Papers of the Museum of Zoology, University of Michigan* 659, 1–15.

———. 1969a. Conservation of fishes of the Death Valley system in California and Nevada. *Transactions of the California-Nevada Section, Wildlife Society* 1969, 107–122.

———. 1969b. *Red Data book. Volume 3 Pisces, freshwater fishes.* Morges, Switzerland: IUCN, Survival Service Commission.

———. 1972a. Threatened freshwater fishes of the United States. *Transactions of the American Fisheries Society* 101, 239–252.

———. 1972b. Classification of the native trouts of Arizona with the description of a new species, *Salmo apache. Copeia* 1972, 401–422.

———. 1973. Two new fishes, *Gila bicolor snyderi* and *Catostomus fumeiventris,* from the Owens River basin, California. *Occasional Papers of the Museum of Zoology, University of Michigan* 667, 1–19.

———. 1976a. Four new pupfishes of the genus *Cyprinodon* from Mexico, with a key to the *C. eximius* complex. *Bulletin of the California Academy of Sciences* 75, 68–75.

———. 1976b. An evaluation of Seth E. Meek's contributions to Mexican Ichthyology. *Fieldiana Zoology* 69, 1–31.

———. 1981. Coevolution of deserts and pupfishes (genus *Cyprinodon*) in the American Southwest. In *Fishes in North American deserts*, ed. R. J. Naiman and D. L. Soltz, pp. 39–94. New York: John Wiley and Sons.

———. 1984. *Rhinichthys deaconi*, a new species of dace (Pisces: Cyprinidae) from southern Nevada. *Occasional Papers of the Museum of Zoology, University of Michigan* 707, 1–21.

Miller, R. R., and J. R. Alcorn. 1946. The introduced fishes of Nevada, with a history of their introduction. *Transactions of the American Fisheries Society* 73 (1943), 173–193.

Miller, R. R., and B. Chernoff. 1980. Status of populations of the endangered Chihuahua chub, *Gila nigrescens*, in New Mexico and Mexico. *Proceedings of the Desert Fishes Council* 11 (1979), 74–84.

Miller, R. R., and J. E. Deacon. 1973. New localities for the rare warm spring pupfish, *Cyprinodon nevadensis pectoralis*, from Ash Meadows, Nevada. *Copeia* 1973, 137–140.

Miller, R. R., and A. A. Echelle. 1974. *Cyprinodon tularosa*, a new cyprinodontid fish from the Tularosa Basin, New Mexico. *The Southwestern Naturalist* 19, 365–377.

Miller, R. R., and L. A. Fuiman. 1987. Description and conservation status of *Cyprinodon macularius eremus*, a new subspecies of pupfish from Organ Pipe Cactus National Monument, Arizona. *Copeia* 1987, 593–609.

Miller, R. R., and C. L. Hubbs. 1960. The spiny-rayed cyprinid fishes (Plagopterini) of the Colorado River system. *Miscellaneous Publications of the Museum of Zoology, University of Michigan* 115, 1–39.

Miller, R. R., and C. H. Lowe. 1964. Part 2. Annotated check-list of the fishes of Arizona. In *The vertebrates of Arizona*, ed. C. H. Lowe, pp. 133–151. Tucson: University of Arizona Press.

———. 1967. Part 2. Fishes of Arizona. In *The vertebrates of Arizona*, 2d printing, ed. C. H. Lowe, pp. 133–151. Tucson: University of Arizona Press.

Miller, R. R., and W. L. Minckley. 1963. *Xiphophorus gordoni*, a new species of platyfish from Coahuila, México. *Copeia* 1963, 538–546.

Miller, R. R., and E. P. Pister. 1971. Management of the Owens pupfish, *Cyprinodon radiosus*, in Mono County, California. *Transactions of the American Fisheries Society* 100, 502–509.

Miller, R. R., P. G. Sanchez, and D. L. Soltz. 1985. *Fishes and aquatic resources, Death Valley system, California/Nevada. A Bibliography, 1878–1984*. San Francisco: U.S. National Park Service.

Miller, R. R., and G. R. Smith. 1981. Distribution and evolution of *Chasmistes* (Pisces: Catostomidae) in western North America. *Occasional Papers of the Museum of Zoology, University of Michigan* 696, 1–46.

Miller, R. R., and V. Walters. 1972. A new genus of cyprinodontid fish from Nuevo Leon, Mexico. *Los Angeles County Natural History Museum, Contributions to Science* 233, 1–13.

Miller, R. R., and P. W. Webb. 1986. Tactics used by *Rhinichthys* and some other freshwater fishes in response to strong currents. Abstracts, Sixty-fourth Annual Meeting, American Society of Ichthyologists and Herpetologists, Victoria, British Columbia, Canada, 1986.

Miller, R. R., J. D. Williams, and J. E. Williams. 1989. Extinction in North American fishes during the past century. *Fisheries* 14, 22–29, 31–38.

Miller, R. R., and H. E. Winn. 1951. Additions to the known fish fauna of Mexico: Three species and one subspecies from Sonora. *Journal of the Washington Academy of Sciences* 41, 83–84.

Miller, W. H., D. L. Archer, H. M. Tyus, and K. C. Harper. 1982a. White River fishes study. Final Report for Memorandum of Understanding CO-910-MU9-933, U.S. Bureau of Land Management. U.S. Fish and Wildlife Service, Salt Lake City, Utah.

Miller, W. H., D. L. Archer, H. M. Tyus, and R. M. McNatt. 1982b. Yampa River fishes study. Final Report for Memorandum of Understanding 14-16-0006-81-931(1A), U.S. National Park Service. U.S. Fish and Wildlife Service, Salt Lake City, Utah.

Miller, W. H., L. R. Kaeding, and H. M. Tyus. 1983. Windy Gap fishes study. First Annual Report for Northern Colorado Water District, Municipal Subdistrict Cooperative Agreement 14-16-0006-82-959(R). U.S. Fish and Wildlife Service, Salt Lake City, Utah.

Miller, W. H., H. M. Tyus, and C. A. Carlson, eds. 1982c. *Fishes of the upper Colorado River system: Present and future.* Bethesda, Md.: American Fisheries Society.

Miller, W. H., J. J. Valentine, D. L. Archer, H. M. Tyus, R. A. Valdez, and L. R. Kaeding. 1982d. Part 1, *Colorado River Fishery Project, final report summary.* Final Report for U.S. Bureau of Reclamation Contract 9-07-40-L-1016, Memorandum of Understanding CO-910-MU9-933, U.S. Bureau of Land Management. U.S. Fish and Wildlife Service, Salt Lake City, Utah.

Millis, R., and J. Kanaly. 1958. A study of the survival of hatchery trout in the Big Laramie River, Albany County, Wyoming. *Wyoming Game and Fish Commission Fisheries Technical Report 8,* 1–53.

Milstead, E. A. 1980. Genetic differentiation among subpopulations of three *Gambusia* species (Pisces: Poeciliidae) in the Pecos River, Texas and New Mexico. Master's thesis, Baylor University, Waco, Texas.

Minckley, C. O., and S. W. Carothers. 1980. Recent collections of the Colorado squawfish and razorback sucker from the San Juan and Colorado rivers in New Mexico and Arizona. *The Southwestern Naturalist* 24, 686–687.

Minckley, W. L. 1962. Two new species of fishes of the genus *Gambusia* (Poeciliidae) from northeastern México. *Copeia* 1962, 391–396.

———. 1963. A new poeciliid fish (genus *Gambusia*) from the Rio Grande drainage of Coahuila, México. *The Southwestern Naturalist* 8, 154–161.

———. 1965. Native fishes as natural resources. In *Native plants and animals as resources in arid lands of the southwestern United States,* ed. J. L. Gardner, pp. 48–60. Committee on Desert and Arid Zones Research, Contribution 8. Washington, D.C.: American Association for the Advancement of Science.

———., ed. 1969a. Investigations of commercial fisheries potentials in reservoirs. Final Report for Arizona Game and Fish Department, U.S. Bureau of Commercial Fisheries Contract (P.L. 88-309). Arizona State University, Tempe.

———. 1969b. Attempted re-establishment of the Gila topminnow within its former range. *Copeia* 1969, 193–194.

———. 1969c. Environments of the Bolsón of Cuatro Ciénegas, Coahuila, México, with special reference to the aquatic biota. *University of Texas, El Paso, Science Series* 2, 1–65.

———. 1973. *Fishes of Arizona.* Phoenix: Arizona Game and Fish Department.

———. 1976. Appendix VII. Fishes. In *The Hohokam, desert farmers and craftsmen: Excavations at Snaketown, 1964–74,* by E. W. Haury, p. 379. Tucson: University of Arizona Press.

———. 1978. Endemic fishes of the Cuatro Ciénegas basin, northern Coahuila, México. In *Transactions of the Symposium on the Biological Resources of the Chihuahuan Desert Region,* ed. R. H. Wauer and D. H. Riskind, pp. 383–404. Washington, D.C.: U.S. National Park Service Transactions and Proceedings Series 3.

———. 1979a. Aquatic habitats and fishes of the lower Colorado River, southwestern United States. Final Report for U.S. Bureau of Reclamation Contract 14-06-300-2529. Arizona State University, Tempe.

———. 1979b. Benthic invertebrates. In Resource inventory for the Gila River complex, eastern Arizona, ed. W. L. Minckley and M. R. Sommerfeld, pp. 502–510. Final Report for U.S. Bureau of Land Management Contract YA-512-CT6-216. Arizona State University, Tempe.

———. 1981. Ecological studies of Aravaipa Creek, central Arizona, relative to past, present, and future uses. Final Report for U.S. Bureau of Land Management Contract YA-512-CT5-98. Arizona State University, Tempe.

———. 1982. Trophic interrelations among introduced fishes in the lower Colorado River, southwestern United States. *California Fish and Game* 68, 78–89.

———. 1983. Status of the razorback sucker, *Xyrauchen texanus* (Abbott), in the lower Colorado River basin. *The Southwestern Naturalist* 28, 165–187.

———. 1984. Cuatro Ciénegas fishes: Research review and a local test of diversity versus habitat size. *Journal of the Arizona-Nevada Academy of Sciences* 19, 13–21.

———. 1985a. Memorandum (dated 16 April), Razorback sucker monitoring effort, Arizona reservoirs, March 1985. U.S. Fish and Wildlife Service, Dexter National Fish Hatchery, New Mexico.

———. 1985b. Native fishes and natural aquatic habitats of U.S. Fish and Wildlife Service Region II, west of the Continental Divide. Final Report for U.S. Fish and Wildlife Service–Arizona State University Interagency Personnel Act Agreement. Arizona State University, Tempe.

———. 1987. Fishes and aquatic habitats of the upper San Pedro River system, Arizona and Sonora. Final Report for U.S. Bureau of Land Management Purchase Order YA-558-CT7-001. Arizona State University, Tempe.

———. 1989a. Red shiner eradication from the Virgin River, Arizona-Nevada-Utah: Background, impacts, and needs for future studies. Unsolicited report for U.S. Fish and Wildlife Service, Denver, Colo. Arizona State University, Tempe.

———. 1989b. Aging of Colorado squawfish/razorback sucker by otolith examination. Final Report for U.S. Fish and Wildlife Service Purchase Order 20181-87-0026. Arizona State University, Tempe.

Minckley, W. L., and N. T. Alger. 1968. Fish remains from an archaeological site along the Verde River, Yavapai County, Arizona. *Plateau* 40, 91–97.

Minckley, W. L., and E. T. Arnold. 1969. "Pit digging," a behavorial feeding adaptation in pupfishes (genus *Cyprinodon*). *Journal of the Arizona Academy of Sciences* 4, 254–257.

Minckley, W. L., and J. E. Brooks. 1985. Transplantations of native Arizona fishes: Records through 1980. *Journal of the Arizona-Nevada Academy of Sciences* 20, 73–89.

Minckley, W. L., and D. E. Brown. 1982. Wetlands. In *Biotic communities of the American Southwest, United States and Mexico*, ed. D. E. Brown, pp. 222–277, 333–341. *Desert Plants* 4 (Special Issue).

Minckley, W. L., D. G. Buth, and R. L. Mayden. 1989. Origin of brood stock and allozyme variation in hatchery-reared bonytail, an endangered North American cyprinid fish. *Transactions of the American Fisheries Society* 118, 139–145.

Minckley, W. L., and R. Clarkson. 1979. Fishes. In Resource inventory for the Gila River complex, eastern Arizona, ed. W. L. Minckley and M. R. Sommerfeld, pp. 510–531. Final Report for U.S. Bureau of Land Management Contract YA-512-CT6-216, Arizona State University, Tempe.

Minckley, W. L., and J. E. Deacon. 1968. Southwestern fishes and the enigma of "endangered species." *Science* 159, 1424–1433.

Minckley, W. L., and E. S. Gustafson. 1982. Early development of the razorback sucker, *Xyrauchen texanus* (Abbott). *Great Basin Naturalist* 42, 533–561.

Minckley, W. L., D. A. Hendrickson, and C. E. Bond. 1986. Geography of western North American freshwater fishes: Description and relationships to intracontinental tectonism. In *Zoogeography of North American freshwater fishes*, ed. C. H. Hocutt and E. O. Wiley, pp. 519–613. New York: John Wiley and Sons.

Minckley, W. L., and B. L. Jensen. 1985. Replacement of Sonoran topminnow by Pecos gambusia under hatchery conditions. *The Southwestern Naturalist* 30, 465–466.

Minckley, W. L., and G. C. Kobetich. 1974. Recovery plan for the razorback sucker, *Xyrauchen texanus* (Abbott). Unpublished manuscript, U.S. Fish and Wildlife Service, Albuquerque, N.M.

Minckley, W. L., and G. K. Meffe. 1987. Differential selection by flooding in stream fish communities of the arid American Southwest. In *Community and evolutionary ecology of North American stream fishes*, ed. W. J. Matthews and D. C. Heins, pp. 93–104. Norman: University of Oklahoma Press.

Minckley, W. L., and P. Mihalick. 1982. Effects of chemical treatment for fish eradication on stream-dwelling invertebrates. *Journal of the Arizona-Nevada Academy of Sciences* 16, 79–82.

Minckley, W. L., and C. O. Minckley. 1986. *Cyprinodon pachycephalus*, a new species of pupfish (Cyprinodontidae) from the Chihuahuan Desert of northern México. *Copeia* 1986, 184–192.

Minckley, W. L., J. N. Rinne, and J. E. Johnson. 1977. Status of the Gila topminnow and its co-occurrence with mosquitofish. *U.S. Forest Service Research Paper* RM-198, 1–8.

Mitchill, L. C. 1981. *Witness to a vanishing America: The nineteenth century response*. Princeton: Princeton University Press.

Moffett, J. W. 1942. A fishery survey of the Colorado River below Boulder Dam. *California Fish and Game* 28, 76–86.

———. 1943. A preliminary report on the fishery of Lake Mead. *Transactions of the North American Wildlife Conference* 8, 179–186.

Molles, M. 1980. The impacts of habitat alterations and introduced species on the native fishes of the upper Colorado River basin. In *Energy development in the Southwest: Problems of water, fish and wildlife in the upper Colorado River basin*, vol. 2, ed. W. O. Spofford, Jr., A. L. Parker, and A. V. Kneese, pp. 163–191. Research Paper R-18. Washington, D.C.: Resources for the Future.

Monastersky, R. 1988. Devil's Hole fires ice age debate. *Science News* 134, 356.

Mooney, H. A., and J. A. Drake, eds. 1986. *Ecology of biological invasions of North America and Hawaii*. New York: Springer-Verlag.

Morgan, D. L., ed. 1964. *The West of William H. Ashley, 1822–1838*. Denver, Colo.: Old West Publishing.

Moring, J. R., and D. V. Buchanan. 1978. Downstream movements and catches of two strains of stocked trout. *Journal of Wildlife Management* 42, 329–333.

Moss, B. 1977. Conservation problems in the Norfolk Broads and rivers of East Anglia, England—phytoplankton, boats and the causes of turbidity. *Biological Conservation* 12, 95–112.

Motro, U., and G. Thomson. 1982. On heterozygosity and the effective size of populations subject to size changes. *Evolution* 36, 1059–1066.

Moyle, P. B. 1976. *Inland fishes of California*. Berkeley: University of California Press.

Moyle, P. B., R. A. Daniels, B. Herbold, and D. M. Baltz. 1985. Patterns in distribution and abundance of a noncoevolved assemblage of estuarine fishes in California. *U.S. Fish and Wildlife Service Fisheries Bulletin* 84, 105–117.

Moyle, P. B., and H. W. Li. 1979. Community ecology and predator-prey relationships in warmwater streams. In *Predator-prey systems in fisheries management*, ed. H. Clepper, pp. 117–180. Washington, D.C.: Sport Fisheries Institute.

Moyle, P. B., H. W. Li, and B. A. Barton. 1986. The Frankenstein effect: Impact of introduced fishes on native fishes in North America. In *Fish culture in fisheries management*, ed. R. H. Stroud, pp. 415–426. Bethesda, Md.: American Fisheries Society.

Moyle, P. B., and R. Nichols. 1973. Ecology of some native and introduced fishes of the Sierra Nevada foothills in central California. *Copeia* 1973, 478–490.

Moyle, P. B., B. Vondracek, and G. D. Grossman. 1983. Responses of fish populations in the North Fork of the Feather River, California, to treatments with fish toxicants. *North American Journal of Fisheries Management* 3, 48–60.

Moyle, P. B., E. D. Wikramanayke, and J. E. Williams. 1989. *Fish species of special concern in California*. Sacramento: California Fish and Game Department.

Mpoame, M. 1981. Parasites of some fishes native to Arizona and New Mexico, with ecological notes. Ph.D. diss., Arizona State University, Tempe.

Mpoame, M., and J. N. Rinne. 1983. Parasites of some fishes native to Arizona and New Mexico. *The Southwestern Naturalist* 28, 399–405.

Mueller, G. 1989. Observations of spawning razorback sucker (*Xyrauchen texanus*) utilizing river habitat in the lower Colorado River, Arizona-Nevada. *The Southwestern Naturalist* 34, 147–149.

Mueller, G., W. Rinne, T. Burke, and M. Delamore. 1982. Habitat selection of spawning razorback suckers observed in Arizona Bay, Lake Mohave, Arizona-Nevada. U.S. Bureau of Reclamation, Lower Colorado River Region, Boulder City, Nev.

———. 1985. Habitat selection of spawning razorback suckers observed in Arizona Bay, Lake Mohave, Arizona-Nevada. *Proceedings of the Desert Fishes Council* 14 (1982), 157 (abstract).

Mullan, J. W. 1962. Is stream improvement the answer? *Virginia Wildlife* 23, 18–21.

———. 1973. Considerations in perpetuation of greenback cutthroat trout (*Salmo c. stomias*). U.S. Bureau of Sport Fisheries and Wildlife, Vernal, Utah.

———. 1975. Annual Project Report, Fisheries Management Program in Rocky Mountain National Park. U.S. Fish and Wildlife Service, Vernal, Utah.

Murphy, M. L., J. Heifetz, S. W. Johnson, K. V. Koski, and J. F. Thedinga. 1986. Effects of clear-cut logging with and without buffer strips on juvenile salmonids in Alaskan streams. *Canadian Journal of Fisheries and Aquatic Sciences* 43, 1521–1533.

Muth, R. T. 1988. Description and identification of bonytail, humpback, and roundtail chub larvae and early juveniles. Ph.D. diss., Colorado State University, Fort Collins.

Myers, G. S. 1927. On the identity of the killifish *Fundulus meeki* Evermann with *Fundulus lima* Valliant. *Copeia* 1927, 178.

———. 1964. A brief sketch of the history of ichthyology in America to the year 1850. *Copeia* 1964, 33–41.

———. 1965. *Gambusia*, the fish destroyer. *Tropical Fish Hobbyist* 1965, 61–65.

Myers, N. 1979a. *The sinking ark: A new look at the problem of disappearing species*. Oxford: Pergamon Press.

———. 1979b. Conserving our global stock. *Environment* 21 (9), 25–33.

Naess, A. 1986. Intrinsic value: Will the defenders of nature please rise? In *Conservation biology, the science of scarcity and diversity*, ed. M. Soulé, pp. 504–515. Sunderland, Mass.: Sinauer Associates.

Naiman, R. J. 1979. Preliminary food studies of *Cyprinodon macularius* and *Cyprinodon nevadensis* (Cyprinodontidae). *The Southwestern Naturalist* 24, 538–541.

Naiman, R. J., S. D. Gerking, and T. D. Ratcliff. 1973. Thermal environment of a Death Valley pupfish. *Copeia* 1973, 366–369.

Naiman, R. J., and D. L. Soltz, eds. 1981. *Fishes in North American deserts*. New York: John Wiley and Sons.

Nankervis, J. M. 1988. Age, growth and reproduction of Gila trout in a small headwater stream in the Gila National Forest. Master's thesis, New Mexico State University, Las Cruces.

Nappe, L. 1972a. The desert pupfish: Prisoner of change. *National Parks Conservation Magazine* 46, 25–28.

———. 1972b. Desert fish in hot water. *Animals* 14, 556–559.

Nash, R. 1970. *Wilderness and the American mind*. New Haven: Yale University Press.

National Research Council. 1968. *Water and choice in the Colorado River basin*. Washington, D. C.: National Academy of Sciences, National Research Council.

National Science Foundation. 1977. The biology of aridity. *Mosaic* 8, 28–35.

Needham, P. R. 1936. Stream improvement in arid regions. *Transactions of the North American Wildlife Conference* 1, 453–460.

Neghberbon, W. O. 1959. No. 183, Toxaphene. In *Handbook of toxicology.* Philadelphia: W. B. Saunders.

Nehlsen, W., J. E. Williams, and J. A. Lichatowich. 1991. Pacific salmon at the crossroads: Stocks at risk from California, Oregon, Idaho, and Washington. *Fisheries* 16, 4–21.

Nei, M. 1975. *Molecular population genetics and evolution.* Amsterdam: North-Holland.

———. 1977. F-statistics and analysis of gene diversity in subdivided populations. *Annals of Human Genetics* 41, 225–233.

———. 1983. Genetic polymorphism and the role of mutation in evolution. In *Evolution of genes and proteins,* ed. M. Nei and R. K. Koehn, pp. 165–190. Sunderland, Mass.: Sinauer Associates.

Nei, M., T. Maruyama, and R. Chakraborty. 1975. The bottleneck effect and genetic variability in populations. *Evolution* 29, 1–10.

Nesler, T. P. 1986. Aquatic non-game research 1985–1986, Squawfish–humpback chub studies. Colorado Division of Wildlife Annual Job Progress Report, SE-3, 1–30. Fort Collins.

Nesler, T. P., R. T. Muth, and A. Wasowicz. 1988. Evidence for baseline flow spikes as spawning cues for Colorado squawfish in the Yampa River, Colorado. *American Fisheries Society Symposium* 5, 68–79.

Nevo, E., A. Beiles, and R. Ben-Shlomo. 1984. The evolutionary significance of genetic diversity: Ecological, demographic and life history correlates. *Lecture Notes in Biomathematics* 53, 12–213.

Nibley, H. 1978. *Nibley on the timely and timeless.* Religious Studies Monograph, Series 1, Religious Studies Center. Provo, Utah: Brigham Young University.

Nielson, R. S., N. Reimers, and H. D. Kennedy. 1957. A six-year study of the survival and vitality of hatchery-reared rainbow trout of catchable size in Convict Creek, California. *California Fish and Game* 43, 5–42.

Norman, J. R. 1931. *A history of fishes.* New York: F. A. Stokes.

Norman, J. R., and P. H. Greenwood. 1963. *A History of fishes.* 2d ed. New York: Hill and Wang.

———. 1975. *A History of fishes.* 3d ed. New York: John Wiley and Sons.

Obregon, H., and S. Contreras Balderas. 1988. Una nueva especie de pez del genero *Xiphophorus* del grupo *couchianus* en Coahuila, México. *Publicaciones Biologias, Faculdad Ciencias Biologias, Universidad Autónoma de Nuevo León* (México) 2, 93–124.

Odens, P. R. 1989. Southwest corner. *The Plain Speaker,* Jacumba, Calif.

Ohmart, R. D., and B. W. Anderson. 1982. North American desert riparian ecosystems. In *Reference handbook on the deserts of North America,* ed. G. L. Bender, pp. 433–479. Westport, Conn.: Greenwood Press.

Ohmart, R. D., W. O. Deason, and S. J. Freeland. 1975. Dynamics of marshland formation and succession along the lower Colorado River and their importance and management problems as related to wildlife in the arid southwest. *Transactions of the North American Natural Resources Conference* 40, 240–251.

Olson, H. F. 1962a. State-wide rough fish control: Rehabilitation of the San Juan River. Federal Aid to Fisheries Restoration Job Completion Report, F-19-D-4, 1–6. New Mexico Game and Fish Department, Santa Fe.

———. 1962b. State-wide fisheries investigations: A pre-impoundment study of Navajo Reservoir, New Mexico. Federal Aid to Fisheries Job Completion Report, F-22-R-3, 1–29. New Mexico Game and Fish Department, Santa Fe.

Ono, R. D., J. D. Williams, and A. Wagner. 1983. *Vanishing fishes of North America.* Washington, D.C.: Stone Wall Press.

Orth, D. J. 1983. Aquatic habitat measurements. In *Fisheries techniques,* ed. L. A. Nielsen and D. L. Johnson, pp. 61–84. Bethesda, Md.: American Fisheries Society.

Osmundson, D. B. 1987. Growth and survival of Colorado squawfish (*Ptychocheilus lucius*) stocked in riverside ponds, with reference to largemouth bass (*Micropterus salmoides*) predation. Master's thesis, Utah State University, Logan.

Papoulias, D. 1988. Survival and growth of larval razorback sucker, *Xyrauchen texanus*. Master's thesis, Arizona State University, Tempe.

Papoulias, D., and W. L. Minckley. 1990. Food limited survival of larval razorback suckers in the laboratory. *Environmental Biology of Fishes* 29, 73–78.

Parenti, L. 1981. A phylogenetic and biogeographic analysis of cyprinodontiform fishes (Teleostei, Atherinomorpha). *Bulletin of the American Museum of Natural History* 168, 335–557.

Parker, B. J., and C. Brousseau. 1988. Status of the aurora trout, *Salvelinus fontinalis timagamiensis*, a distinct stock endemic to Canada. *Canadian Field-Naturalist* 102, 87–91.

Patten, B. G., and D. T. Rodman. 1969. Reproductive behavior of northern squawfish (*Ptychocheilus oregonensis*). *Transactions of the American Fisheries Society* 98, 109–111.

Paulson, L. J., J. E. Deacon, and J. R. Baker. 1980a. The limnological status of Lake Mead and Lake Mohave under present and future powerplant operations of Hoover Dam. Final Report for U.S. Bureau of Reclamation Contract 14-06-300-2218. University of Nevada, Las Vegas.

Paulson, L. J., T. G. Miller, and J. R. Baker. 1980b. Influence of dredging and high discharge on the ecology of Black Canyon. Final Report for U.S. Bureau of Reclamation Contract 14-06-300-2218. University of Nevada, Las Vegas.

Pavlik, B. M. 1987. Autecological monitoring of endangered plants and damaged communities. *Proceedings of the Desert Fishes Council* 17 (1985), 91–100.

Pavlov, D. S., Yu. S. Reshetnikof, M. I. Shatunsvsky, and N. I. Shilin. 1985. Rare and disappearing fishes in the USSR and the principles of their inclusion in the *Red Book*. *Journal of Ichthyology* 25, 88–89.

Pearson, W. D. 1967. Distribution of macroinvertebrates in the Green River below Flaming Gorge Dam, 1963–1965. Master's thesis, Utah State University, Logan.

Pearson, W. D., R. H. Kramer, and D. R. Franklin. 1968. Macroinvertebrates in the Green River below Flaming Gorge Dam, 1964–65 and 1967. *Proceedings of the Utah Academy of Science, Arts, and Letters* 45, 148–167.

Peden, A. E. 1970. Courtship behavior of *Gambusia* (Poeciliidae) with emphasis on isolating mechanisms. Ph.D. diss., University of Texas, Austin.

———. 1973. Virtual extinction of *Gambusia amistadensis* n. sp., a new poeciliid fish from Texas. *Copeia* 1973, 210–221.

Persons, W. R., and R. V. Bulkley. 1980. Adult striped bass movements and feeding, Colorado River inlet, Lake Powell, 1980. Annual Report for U.S. Bureau of Reclamation Contract 9-07-40-L-1016. Utah Cooperative Fisheries Unit, Utah State University, Logan.

Persons, W. R., R. V. Bulkley, and W. R. Noonan. 1982. Movements and feeding of adult striped bass, Colorado River inlet, Lake Powell, 1980–1981. In Part 3, *Colorado River Fisheries Project, final report contracted studies*, ed. W. H. Miller, J. J. Valentine, D. L. Archer, H. M. Tyus, R. A. Valdez, and L. R. Kaeding, pp. 255–274. Final Report for U.S. Bureau of Reclamation Contract 9-07-40-L-1016, Memorandum of Understanding CO-910-MU9-933, U.S. Bureau of Land Management. U.S. Fish and Wildlife Service, Salt Lake City, Utah.

Petulla, J. M. 1980. *American Environmentalism: Values, tactics, priorities*. College Station: Texas A&M University Press.

Pickett, S. T. A., and J. N. Thompson. 1978. Patch dynamics and the design of nature reserves. *Biological Conservation* 13, 27–37.

Picton, H. D. 1979. The application of insular biogeographic theory to the conservation of large mammals in the northern Rocky Mountains. *Biological Conservation* 15, 73–79.

Pimentel, R., R. V. Bulkley, and H. M. Tyus. 1985. Choking of Colorado squawfish, *Ptychocheilus lucius* (Cyprinidae), on channel catfish, *Ictalurus punctatus* (Ictaluridae), as a cause of mortality. *The Southwestern Naturalist* 30, 154–158.

Pimm, S. L. 1980. Food web design and the effects of species deletion. *Oikos* 35, 139–149.

————. 1986. Community stability and structure. In *Conservation biology: The science of scarcity and diversity*, ed. M. E. Soulé, pp. 309–329. Sunderland, Mass.: Sinauer Associates.

Pimm, S. L., J. M. Diamond, and H. L. Jones. 1989. On the risk of extinction. *American Naturalist* 132, 757.

Pister, E. P., ed. 1970. The rare and endangered fishes of the Death Valley system, a summary of the proceedings of a symposium relating to their protection and preservation. California Department of Fish and Game, Bishop.

————., ed. 1971. The rare and endangered fishes of the Death Valley system, a summary of the proceedings of a symposium relating to their protection and preservation. California Department of Fish and Game, Bishop.

————. 1974. Desert fishes and their habitats. *Transactions of the American Fisheries Society* 103, 531–540.

————. 1976. A rationale for the management of nongame fish and wildlife. *Fisheries* 1, 11–14.

————. 1979a. The resource must come first. Keynote Address, Second Conference on Scientific Research in the National Parks. U.S. National Park Service and American Institute of Biological Sciences, San Francisco, California.

————. 1979b. Endangered species: Costs and benefits. *Environmental Ethics* 1, 341–353.

————. 1979c. Obituary, Carl Leavitt Hubbs. *Fisheries* 4, 28.

————., ed. 1981a. *Proceedings of the Desert Fishes Council* 12 (1980). Bishop, Calif.: Desert Fishes Council.

————. 1981b. The conservation of desert fishes. In *Fishes in North American deserts*, ed. R. J. Naiman and D. L. Soltz, pp. 411–445. New York: John Wiley and Sons.

————. 1983. Summary of topic assigned to Otto Aho, field solicitor, Interior Department, at "pupfish conference" at Death Valley National Monument, November 14 and 15, 1972. *Proceedings of the Desert Fishes Council* 4 (1972), 73–74.

————., ed. 1984. *Proceedings of the Desert Fishes Council* 3 (1971). Bishop, Calif.: Desert Fishes Council.

————. 1985a. Desert pupfishes: Reflections on reality, desirability, and conscience. *Environmental Biology of Fishes* 12, 3–12.

————. 1985b. Desert pupfishes: Reflections on reality, desirability, and conscience. *Fisheries* 10, 10–15.

————. 1985c. Death Valley Area Committee Report. *Proceedings of the Desert Fishes Council* 13 (1981), 15–20.

————. 1987a. A pilgrim's progress from group A to group B. In *Companion to a Sand County Almanac*, ed. J. B. Callicott, pp. 221–232. Madison: University of Wisconsin Press.

————., ed. 1987b. *Proceedings of the Desert Fishes Council* 14 (1984). Bishop, Calif.: Desert Fishes Council.

————., ed. 1990. *Proceedings of the Desert Fishes Council* 19 (1987). Bishop, Calif.: Desert Fishes Council.

Pister, E. P., and W. C. Unkel. 1989. Conservation of western wetlands and wildlife and their fishes. In *Freshwater wetlands and wildlife: Perspectives on natural, managed, and degraded ecosystems*, ed. R. R. Sharitz and J. W. Gibbons, pp. 455–473. Oak Ridge, Tenn.: U.S. Department of Energy, Office of Science and Technical Information.

Pitts, T. 1988. Conflict resolution: Endangered species protection and water development in the upper Colorado River basin. *Colorado Water Rights* (Colorado Water Congress) 7 (1), 1, 4–5, 7.

Platania, S. P., and K. R. Bestgen. 1988. Survey of the fishes of the lower San Juan River, New Mexico. Final Report for U.S. Bureau of Reclamation Contract 516.6-74-23. New Mexico Game and Fish Department, Santa Fe.

Platania, S. P., K. R. Bestgen, M. A. Moretti, D. L. Propst, and J. E. Brooks. 1990. Status of Colorado squawfish and razorback sucker in the San Juan River, Colorado, New Mexico, and Utah. *The Southwestern Naturalist*.

Platania, S. P., and D. A. Young. 1990. Summary report on the ichthyofauna of the San Juan River drainage, New Mexico and Utah. Draft Report, University of New Mexico, Albuquerque.

Platts, W. S., and J. N. Rinne. 1985. Riparian and stream enhancement, management and research in the Rocky Mountains. *North American Journal of Fisheries Management* 5, 115–125.

Pope, C. 1988. The politics of plunder. *Sierra* 73 (6), 48–55.

Post, J. R., and D. Cucin. 1984. Changes in the benthic community of a small pre-Cambrian lake following the introduction of yellow perch, *Perca flavescens*. *Canadian Journal of Fisheries and Aquatic Sciences* 41, 1496–1501.

Potter, L., and F. Cheever. 1989. Petition for a rule to list the razorback sucker as an endangered species, directed to U.S. Department of Interior, Fish and Wildlife Service, Washington, D.C. Sierra Club Legal Defense Fund, Denver, Colo.

Powell, G. C. 1958. Evaluation of the effects of a power dam release pattern upon the downstream fishery. Master's thesis, Colorado State University, Fort Collins.

Powell, J. W. 1878. *Report on the lands of the arid region of the United States*. Washington, D.C.: U.S. Government Printing Office.

Priscu, J. C., Jr., J. Verduin, and J. E. Deacon. 1981. The fate of biogenic suspensoids in a desert reservoir. In *Proceedings of the Symposium on Surface Water Impoundments*, ed. H. G. Stefa, pp. 1657–1667. Minneapolis, Minn.: American Society of Civil Engineers (1980).

———. 1982. Primary productivity and nutrient balance in a lower Colorado River reservoir. *Archive für Hydrobiologie* 94, 1–23.

Propst, D. L., K. R. Bestgen, and C. W. Painter. 1985a. Distribution, status, and biology of the spikedace (*Meda fulgida*) in New Mexico. *Endangered Species Report* 15. U.S. Fish and Wildlife Service, Albuquerque.

Propst, D. L., P. C. Marsh, and W. L. Minckley. 1985b. Arizona survey for spikedace (*Meda fulgida*) and loach minnow (*Tiaroga cobitis*): Fort Apache and San Carlos Apache Indian reservations and Eagle Creek, May 1985. U.S. Fish and Wildlife Service, Dexter National Fish Hatchery, New Mexico.

Pryde, P. R. 1972. *Conservation in the Soviet Union*. Cambridge: Cambridge University Press.

Pucherelli, M. J., R. C. Clar, and J. N. Halls. 1988. Green River backwater habitat mapping study. Draft Report, U.S. Bureau of Reclamation Engineering and Research Center, Denver, and Advanced Sciences, Inc., Lakewood, Colo.

Pumpelly, R. 1870. *Across America and Asia: Notes of a five-year journey around the world and of residence in Arizona, Japan and China*. New York: Leypoldt and Holt.

Quattro, J., and R. C. Vrijenhoek. 1989. Fitness differences among remnant populations of the endangered Sonoran topminnow. *Science* 245, 976–978.

Quinn, J. F., and A. Hastings. 1987. Extinction in subdivided habitats. *Conservation Biology* 1, 198–208.

———. 1988. Extinction in subdivided habitats: Reply to Gilpin. *Conservation Biology* 2, 293–296.

Quinn, J. F., and G. Robinson. In press. Habitat fragmentation, species diversity, extinction, and the design of nature reserves. In *Applied population biology*, ed. S. Jain and L. Botsford. The Hague: Dr. W. Junk.

Rabe, R. W., and N. L. Savage. 1979. A methodology for the selection of aquatic natural areas. *Biological Conservation* 15, 291–300.

Radant, R. D. 1982. Colorado squawfish, humpback chub, bonytail chub and razorback sucker population and habitat monitoring. Utah Division of Wildlife Resources, Federal Aid Project, SE-2, 1–3. Salt Lake City.

———. 1984. June sucker management plan. Utah Division of Wildlife Resources, Salt Lake City.

———. 1986. Colorado squawfish, humpback chub, bonytail chub and razorback sucker population and habitat monitoring. Utah Division of Wildlife Resources, Federal Aid Project, SE-2, 1–3. Salt Lake City.

Radant, R. D., and T. J. Hickman. 1985. Status of the June sucker (*Chasmistes liorus*). *Proceedings of the Desert Fishes Council* 15 (1983), 277–282.

Radant, R. D., and D. K. Sakaguchi. 1981. Utah Lake fisheries inventory. Final Report for U.S. Bureau of Reclamation Contract 8-07-40-50634. Utah Division of Wildlife Resources, Salt Lake City.

Radant, R. D., M. M. Wilson, and D. Shirley. 1987. June sucker Provo River instream flow analysis. Final Report for U.S. Bureau of Reclamation Contract 8-07-40-50634, Modification 4. Utah Division of Wildlife Resources, Salt Lake City.

Radonski, G. C., and R. G. Martin. 1986. Fish culture is a tool, not a panacea. In *Fish culture in fisheries management*, ed. R. H. Stroud, pp. 7–13. Bethesda, Md.: American Fisheries Society.

Rahel, F. J. 1986. Biogeographic influences on fish species composition of northern Wisconsin lakes with applications for lake acidification studies. *Canadian Journal of Fisheries and Aquatic Sciences* 43, 124–134.

Rapport, E. H., G. Borioli, J. A. Monjeau, J. E. Puntieri, and R. D. Oviedo. 1986. The design of natural reserves: A simulation trial for assessing specific conservation value. *Biological Conservation* 37, 269–290.

Rauchenberger, M. 1989. Appendix 1: Annotated list of species of the subfamily Poeciliinae. In *Ecology and evolution of livebearing fishes (Poeciliidae)*, ed. G. K. Meffe and F. F. Snelson, Jr., pp. 359–367. Englewood Cliffs, N.J.: Prentice-Hall.

Raup, D. M. 1986. Biological extinction in earth history. *Science* 231, 1528–1533.

———. 1988. Diversity crises in the geological past. In *Biodiversity*, ed. E. O. Wilson, pp. 51–57. Washington, D.C.: National Academy of Sciences Press.

Raup, D. M., and J. J. Sepkoski. 1982. Mass extinctions in the marine fossil record. *Science* 215, 1501–1503.

Rea, A. M. 1983. *Once a river: Bird life and habitat changes on the middle Gila*. Tucson: University of Arizona Press.

Regan, D. M. 1966. Ecology of Gila trout in Main Diamond Creek, New Mexico. *U.S. Bureau of Sport Fisheries and Wildlife Technical Paper* 5, 1–24.

Reiger, G. 1977. Native fish in troubled waters. *Audubon* 79, 18–41.

Reisner, M. 1986. *Cadillac Desert*. New York: Viking Press.

Reynolds, J. B. 1983. Electrofishing. In *Fishery techniques*, ed. L. A. Nielsen and D. L. Johnson, pp. 147–163. Bethesda, Md.: American Fisheries Society.

Reynolds, M. 1966. *The treatment of nature in English poetry between Pope and Wordsworth*. New York: Gordian Press.

Reznick, D. N., and H. Bryga. 1987. Life-history evolution in guppies (*Poecilia reticulata*). I: Phenotypic and genetic changes in an introduction experiment. *Evolution* 41, 1370–1385.

Rich, P. H., and R. G. Wetzel. 1978. Detritus in the lake ecosystem. *American Naturalist* 112, 57–69.

Richardson, J. 1836. *Fauna Boreali-Americana; or the zoology of the northern parts of British America; third part, the fish*. London: Richard Bentley.

Richardson, W. M., ed. 1976. Fishes known to be in the Colorado River drainage. In Technical committee minutes, Colorado River Wildlife Council, pp. 12–18. Las Vegas, Nev.

Richey, J. E., M. E. Perkins, and C. R. Goldman. 1975. Effects of kokanee salmon (*Oncorhynchus nerka*) decomposition on the ecology of a subalpine stream. *Journal of the Fisheries Research Board of Canada* 32, 817–820.

Richter-Dyn, N., and N. S. Goel. 1972. On the extinction of a colonizing species. *Theoretical Population Biology* 3, 406–433.

Riggs, A. C. 1984. Major carbon-14 deficiency in modern snail shells from southern Nevada springs. *Science* 224, 58–61.

Rinne, J. N. 1973. A limnological survey of central Arizona reservoirs, with reference to horizontal fish distribution. Ph.D. diss., Arizona State University, Tempe.

———. 1975. Changes in minnow populations in a small desert stream resulting from naturally and artificially induced factors. *The Southwestern Naturalist* 20, 185–195.

———. 1976. Cyprinid fishes of the genus *Gila* from the lower Colorado River basin. *Wassman Journal of Biology* 34, 65–107.

———. 1981. Stream habitat improvement and native southwestern trouts. *U.S. Forest Service Research Note* RM-409, 1–4.

———. 1982. Movement, home range, and growth of a rare southwestern trout in improved and unimproved habitat. *North American Journal of Fisheries Management* 2, 150–157.

———. 1985a. Grazing effects on southwestern streams: A complex research problem. In *Riparian ecosystems and their management: Reconciling conflicting uses*, R. R. Johnson, C. D. Ziebel, D. R. Patton, P. F. Folliott, and R. H. Hamre, technical coordinators, pp. 295–299. First North American Riparian Conference, U.S. Forest Service General Technical Report RM-120.

———. 1985b. Variation in Apache trout populations in the White Mountains, Arizona. *North American Journal of Fisheries Management* 5, 147–158.

———. 1988a. Grazing effects on stream habitat and fishes: Research design considerations. *North American Journal of Fisheries Management* 8, 240–247.

———. 1988b. Native southwestern (USA) trouts: Status, taxonomy, ecology and conservation. *Polish Archives of Hydrobiologia* 35 (3–4), 305–320.

Rinne, J. N., J. E. Johnson, B. L. Jensen, A. W. Ruger, and R. Sorenson. 1986. The role of hatcheries in the management and recovery of threatened and endangered fishes. In *Fish culture in fisheries management*, ed. R. H. Stroud, pp. 271–285. Bethesda, Md.: American Fisheries Society.

Rinne, J. N., and W. L. Minckley. 1985. Patterns of variation and distribution in Apache trout (*Salmo apache*) relative to co-occurrence with introduced salmonids. *Copeia* 1985, 285–292.

Rinne, J. N., W. L. Minckley, and J. N. Hanson. 1982. Chemical treatment of Ord Creek, Apache County, Arizona, to re-establish Arizona trout. *Journal of the Arizona-Nevada Academy of Sciences* 16 (1981), 74–79.

Rinne, J. N., B. Rickle, and D. A. Hendrickson. 1980. A new Gila topminnow locality in southern Arizona. *U.S. Forest Service, Rocky Mountain Forest and Range Experiment Station Research Note* RM-382, 1–2.

Rinne, J. N., B. Robertson, R. Major, and K. Harper. 1981. Sport fishing for the native Arizona trout, *Salmo apache* Miller, in Christmas Tree Lake: A case study. In *Wild Trout Symposium II, September 14–15, 1979*, ed. W. King, pp. 158–164. Vienna, Va.: Trout Unlimited.

Rinne, J. N., and J. Stefferud. 1982. Stream habitat improvement in the southwestern United States, Arizona and New Mexico. In *Proceedings of the Rocky Mountain Stream Habitat Management Workshop, September 7–10, 1982, Jackson, Wyoming*, unpaginated. Washington, D.C.: U.S. Forest Service, Vienna, Va.: Trout Unlimited.

Robins, C. R. 1986. Foreword to *Distribution, biology, and management of exotic fishes*, ed. W. R. Courtenay, Jr., and J. R. Stauffer, Jr., pp. vii–viii. Baltimore, Md.: Johns Hopkins University Press.

Rogers, B. D. 1975. Fish distribution in the Mimbres River, New Mexico. New Mexico Game and Fish Department, Santa Fe.

Rolston, H. 1988. *Environmental ethics*. Philadelphia: Temple University Press.

Romero, M. A., ed. 1972. *Amargosa Canyon–Dumont Dunes proposed natural area*. Montrose, Calif.: Pupfish Habitat Preservation Committee.

Ronald, A. 1989. Western literature and natural resources. *Halcyon* 11, 125–136.

Rose, K. 1984. Yampa River spawning habitat investigations. U.S. Fish and Wildlife Service, Salt Lake City, Utah.

Rose, K., and J. Hamill. 1988. Help is on the way for rare fishes of the upper Colorado River basin. *Endangered Species Technical Bulletin* 13, 1, 6–7.

Rosen, D. E. 1960. Middle-American poeciliid fishes of the genus *Xiphophorus*. *Bulletin of the Florida State Museum* 5, 57–242.

Rosenlund, B. D., and D. R. Stevens. 1988. Fisheries and aquatic management, Rocky Mountain National Park, 1987. U.S. Fish and Wildlife Service, Colorado Fish and Wildlife Assistance Office, Golden.

Ross, S. T. 1986. Resource partitioning in fish assemblages: A review of field studies. *Copeia* 1986, 352–388.

Rostlund, E. 1952. Freshwater fish and fishing in native North America. *University of California Publications in Geography* 9, 1–313.

Royce, W. F. 1972. *Introduction to fishery sciences*. New York: Academic Press.

Ruhr, E. C. 1957. Stream survey methods in the southeast. In *Symposium on the Evaluation of Fish Populations in Warm-water Streams*, ed. K. D. Carlander, pp. 4–50. Ames: Iowa State College.

Ryman, N. 1983. Patterns of distribution of biochemical genetic variation in salmonids: Differentiation between species. *Aquaculture* 33, 1–21.

Ryman, N., and F. Utter, eds. 1987. *Population genetics and fishery management*. Seattle: University of Washington Press.

Sada, D. W. 1987a. Introduction to the symposium: Rehabilitation of Ash Meadows, Nevada. *Proceedings of the Desert Fishes Council* 17 (1985), 84–87.

———. 1987b. Perspectives on an integrated approach to recover Ash Meadows, Nevada. *Proceedings of the Desert Fishes Council* 17 (1985), 88–90.

Sada, D. W., and A. Mozejko. 1984. Environmental assessment, proposed acquisition to establish Ash Meadows National Wildlife Refuge, Nye County, Nevada. U.S. Fish and Wildlife Service, Portland, Ore.

Sage, R. D., and R. K. Selander. 1975. Trophic radiation through polymorphism in cichlid fishes. *Proceedings of the National Academy of Sciences* 72, 4669–4673.

Saile, B. 1986. Squawfish not so rare in Yampa. *Denver Post*, 21 August 1986.

St. Amant, J. A., R. Hulquist, C. Marshall, and A. Pickard. 1974. Fisheries section including information on fishery resources of the Coachella Canal study area. In Inventory of the fish and wildlife resources, recreational consumptive use, and habitat in and adjacent to the upper 49 miles and ponded areas of the Coachella Canal, pp. 64–88. Final Report for U.S. Bureau of Reclamation Contract 14-06-300-2555. California Department of Fish and Game, Sacramento.

St. Amant, J. A., and S. Sasaki. 1971. Progress report on reestablishment of the Mohave chub, *Gila mohavensis* (Snyder), an endangered species. *California Fish and Game* 57, 307–308.

Sanders, J. G. 1986. Direct and indirect effects of arsenic on the survival and fecundity of estuarine zooplankton. *Canadian Journal of Fisheries and Aquatic Sciences* 43, 694–699.

Sato, G. M., and P. B. Moyle. 1987. Stream surveys of Deer Creek and Mill Creek, Tehama County, California, 1976 and 1977, with recommendations for protection of their drainages and their spring-over chinook salmon. Final Report for U.S. National Park Service Contract. University of California, Davis.

Savage, N. L., and R. W. Rabe. 1979. Stream types in Idaho: An approach to classification of streams in natural areas. *Biological Conservation* 15, 310–315.

Schartl, M., and J. H. Schroeder. 1988. A new species of the genus *Xiphophorus* Heckel 1848, endemic to northern Coahuila, Mexico. *Senckenbergiana Biologische* 68 (4/6), 311–321.

Schmidt, B. R. 1979. Life history of the Utah chub in Flaming Gorge Reservoir. Paper presented to the Bonneville Chapter, American Fisheries Society Meeting, 1–2 February, Salt Lake City, Utah (abstract).

Schoenherr, A. A. 1974. Life history of the Gila topminnow, *Poeciliopsis occidentalis* (Baird and Girard) in Arizona and an analysis of its interaction with the mosquitofish *Gambusia affinis* (Baird and Girard). Ph.D. diss., Arizona State University, Tempe.

———. 1977. Density dependent and density independent regulation of reproduction in the Gila topminnow, *Poeciliopsis occidentalis* (Baird and Girard). *Ecology* 58, 438–444.

———. 1979. Niche separation within a population of freshwater fishes in an irrigation drain near the Salton Sea, California. *Bulletin of the Southern California Academy of Sciences* 78, 46–55.

———. 1981. The role of competition in the displacement of native fishes by introduced species. In *Fishes in North American deserts*, ed. R. J. Naiman and D. L. Soltz, pp. 173–203. New York: John Wiley and Sons.

———. 1985. Replacement of *Cyprinodon macularius* by *Tilapia zilli* in an irrigation drain near the Salton Sea. *Proceedings of the Desert Fishes Council* 12 (1981), 65–66.

———. 1988. A review of the life history and status of the desert pupfish, *Cyprinodon macularius*. *Bulletin of the Southern California Academy of Sciences* 87, 104–134.

Schonewald-Cox, C. M., and J. W. Bayless. 1986. The boundary model: A geographic analysis of design and conservation of nature reserves. *Biological Conservation* 38, 305–322.

Schonewald-Cox, C. M., S. M. Chambers, B. MacBryde, and L. Thomas, eds. 1983. *Genetics and conservation: A reference for managing wild animal and plant populations*. Menlo Park, Calif.: Benjamin/Cummings.

Schuck, H. A. 1948. Survival of hatchery trout in streams and possible methods of improving the quality of hatchery trout. *Progressive Fish-Culturist* 10, 3–14.

Schultz, R. J. 1989. Origins and relationships of unisexual poeciliids. In *Ecology and evolution of livebearing fishes (Poeciliidae)*, ed. G. K. Meffe and F. F. Snelson, Jr., pp. 69–90. Englewood Cliffs, N.J.: Prentice-Hall.

Schwartz, A. 1984. Bright future for a desert refugium. *Garden* 8 (2), 26–29.

Schwartz, F. J. 1981. *World literature on fish hybrids, with an analysis by family, species, and hybrids*. Supplement 1. Washington, D.C.: National Oceanographic and Atmospheric Administration Technical Report, National Marine Fisheries Service, SSRF-750.

Schwartzschild, S. S. 1984. The unnatural Jew. *Environmental Ethics* 6, 347–362.

Scoppettone, G. G. 1988. Growth and longevity of the cui-ui (*Chasmistes cujus*) and longevity in other catostomids and cyprinids. *Transactions of the American Fisheries Society* 117, 301–307.

Scoppettone, G. G., M. Coleman, and G. A. Wedemeyer. 1986. Life history and status of the endangered cui-ui of Pyramid Lake, Nevada. *U.S. Fish and Wildlife Service Research* 1, 1–23.

Scoppettone, G. G., G. A. Wedemeyer, M. Coleman, and H. Burge. 1983. Reproduction by the endangered cui-ui in the lower Truckee River. *Transactions of the American Fisheries Society* 112, 788–793.

Scott, J. M., B. Csuti, J. J. Jacobi, and J. E. Estes. 1987. Species richness: A geographic approach to protecting future biological diversity. *BioScience* 37, 782–788.

Scott, J. M., B. Csuti, K. Smith, J. E. Estes, and S. Caicco. 1988. Beyond endangered species: An integrated conservation strategy for the preservation of biological diversity. *Endangered Species Update* 5, 43–48.

Seale, A. 1896. Note on *Deltistes*, a new genus of catostomid fishes. *Proceedings of the California Academy of Sciences* 6, 269.

Sears, P. B. 1989. *Deserts on the march.* Covelo, Calif.: Island Press.

Secretaria de Desarrollo Urbano y Ecología (SEDUE). 1988. Ley General del Equilibrio Ecológica y la Protección al Ambiente. *Diario Oficial 28 January 1988*, 23–57. México, D.F., México.

Seethaler, K. H. 1978. Life history and ecology of the Colorado squawfish (*Ptychocheilus lucius*) in the upper Colorado River basin. Master's thesis, Utah State University, Logan.

Seethaler, K. H., C. H. McAda, and R. S. Wydoski. 1979. Endangered and threatened fish in the Yampa and Green rivers of Dinosaur National Monument. In *Proceedings of the First Conference on Scientific Research in the National Parks*, ed. R. M. Linn, pp. 605–612. Washington, D.C.: U.S. National Park Service Transactions and Proceedings Series 5.

Sepkoski, J. J., and M. A. Rex. 1974. Distribution of freshwater mussels: Coastal rivers as biogeographic islands. *Systematic Zoology* 23, 165–188.

Sevon, M., and D. Delany. 1987. Development of Blue Link Spring, a refugium for Hiko White River springfish. *Proceedings of the Desert Fishes Council* 17 (1985), 168–173.

Shaffer, M. L. 1981. Minimum population sizes for species conservation. *BioScience* 31, 131–134.

Shaffer, M. L., and F. B. Samson. 1985. Population size and extinction: A note on determining critical population sizes. *American Naturalist* 125, 155–162.

Shannon, E. H. 1969. The toxicity and detoxification of the rotenone formulations used in fish management. Federal Aid to Fisheries Restoration, Project Completion Report, F-19-1, Job IV-B, 1–8. North Carolina Wildlife Research Commission, Raleigh.

Shapovalov, L., W. A. Dill, and A. J. Cordone. 1959. A check-list of the freshwater and anadromous fishes of California. *California Fish and Game* 45, 159–180.

Sharpe, F. P. 1983. Status report of the desert pupfish sanctuary constructed by the Bureau of Reclamation below Hoover Dam, Clark County, Nevada. *Proceedings of the Desert Fishes Council* 4 (1972), 78–80.

Sharpe, F. P., H. R. Guenther, and J. E. Deacon. 1973. Endangered desert pupfish at Hoover Dam. *Reclamation Era* 59 (2), 24–29.

Sheldon, A. L. 1987. Rarity: Patterns and consequences for stream fishes. In *Community and evolutionary ecology of North American stream fishes*, ed. W. J. Matthews and J. C. Heins, pp. 203–209. Norman: University of Oklahoma Press.

———. 1988. Stream fishes: Patterns of diversity, rarity and risk. *Conservation Biology* 2, 149–156.

Shields, R. H. 1982. Fishes of the upper Colorado River system: Political and resources considerations. In *Fishes of the upper Colorado River system: Present and future*, ed. W. H. Miller, H. M. Tyus, and C. A. Carlson, pp. 9–11. Bethesda, Md.: Western Division, American Fisheries Society.

Shipley, C. W. 1974. The Fish and Wildlife Coordination Act's application to wetlands. In *Environmental planning: Law of land and resources*, ed. A. W. Reitze, various pagination. Washington, D.C.: North America International.

Shirley, D. S. 1983. Spawning ecology and larval development of the June sucker. *Proceedings of the Bonneville Chapter of the American Fisheries Society* 1983, 18–36.

Sholes, W. H. 1979. An evaluation of rearing fall-run chinook salmon, *Oncorhynchus tshawytscha*, to yearlings at Feather River hatchery, with a comparison of returns from hatchery and downstream releases. *California Fish and Game* 65, 239–255.

Shumway, M. 1987. 1986 Annual Report, Owens Pupfish Recovery Project. Account 102, California Department of Fish and Game, Bishop.

Shute, J. R., and A. W. Allan. 1980. *Fundulus zebrinus* Jordan and Gilbert, Plains killifish. In *Atlas of North American freshwater fishes*, ed. D. S. Lee, C. R. Gilbert, C. H. Hocutt, R. E. Jenkins, D. E. McAllister, and J. R. Stauffer, Jr., p. 531. Raleigh: North Carolina State Museum of Natural History.

Siebert, D. J. 1980. Movements of fishes in Aravaipa Creek, Arizona. Master's thesis, Arizona State University, Tempe.

Sigler, J. W., T. C. Bjornn, and F. H. Everest. 1984. Effects of chronic turbidity on density and growth of steelheads and coho salmon. *Transactions of the American Fisheries Society* 113, 142–150.

Sigler, W. F., W. T. Helm, J. W. Angelovic, D. W. Linn, and S. S. Martin. 1966. The effects of uranium mill wastes on stream biota. *Utah Agricultural Experiment Station Bulletin* (Utah State University) 462, 1–76.

Sigler, W. F., and R. R. Miller. 1963. *Fishes of Utah*. Salt Lake City: Utah Department of Fish and Game.

Sigler, W. F., and J. W. Sigler. 1987. *Fishes of the Great Basin, a natural history*. Reno: University of Nevada Press.

Sigler, W. F., S. Vigg, and M. Bres. 1985. Life history of the cui-ui, *Chasmistes cujus* Cope, in Pyramid Lake, Nevada: A review. *Great Basin Naturalist* 45, 571–603.

Simberloff, D. S., and L. G. Abele. 1976. Island biogeography theory and conservation practice. *Science* 191, 285–286.

———. 1982. Refuge design and island biogeographic theory: Effects and fragmentation. *American Naturalist* 120, 41–50.

Simon, J. R. 1946. Wyoming fishes. *Wyoming Game and Fish Department Bulletin* 4, 1–129.

———. 1951. Wyoming fishes. 2d printing. *Wyoming Game and Fish Department Bulletin* 4, 1–129.

Simons, L. H. 1987. Status of the endangered Gila topminnow, *Poeciliopsis occidentalis*, in the United States. Project E-1, Title VI. Arizona Game and Fish Department, Phoenix.

Simons, L. H., D. A. Hendrickson, and D. Papoulias. 1989. Recovery of the Gila topminnow: A success story? *Conservation Biology* 3, 1–5.

Simpson, G. G. 1961. *Principles of animal taxonomy*. New York: Columbia University Press.

Sioli, H. 1986. Tropical continental aquatic habitats. In *Conservation biology: The biology of scarcity and diversity*, ed. M. E. Soulé, pp. 383–393. Sunderland, Mass.: Sinauer Associates.

Slatkin, M. 1985. Gene flow in natural populations. *Annual Review of Ecology and Systematics* 16, 393–430.

Slifer, G. W. 1970. Stream reclamation techniques. *Wisconsin Department of Natural Resources, Bureau of Fisheries Management Report* 33, 1–35.

Smith, G. R. 1959. Annotated checklist of fishes of Glen Canyon. In *Ecological studies of the flora and fauna in Glen Canyon*, ed. A. M. Woodbury, pp. 195–199. University of Utah Anthropological Papers 40.

———. 1960. Annotated list of fishes of the Flaming Gorge Reservoir basin, 1959. In *Ecological studies of the flora and fauna of Flaming Gorge Reservoir basin, Utah and Wyoming*, ed. A. M. Woodbury, pp. 163–168. University of Utah Anthropological Papers 48.

———. 1975. Fishes of the Pliocene Glens Ferry Formation, southwest Idaho. *University of Michigan Papers in Paleontology* 14, 1–68.

———. 1978. Biogeography of intermountain fishes. *Great Basin Naturalist Memoirs* 2, 17–46.

———. 1981a. Late Cenozoic freshwater fishes of North America. *Annual Review of Ecology and Systematics* 12, 163–193.

———. 1981b. Effects of habitat size on species richness and adult body size of desert fishes. In *Fishes in North American deserts*, ed. R. J. Naiman and D. L. Soltz, pp. 125–172. New York: John Wiley and Sons.

Smith, G. R., J. G. Hall, R. K. Koehn, and D. J. Innes. 1983. Taxonomic relationships of the Zuñi mountain sucker, *Catostomus discobolus yarrowi*. *Copeia* 1983, 37–48.

Smith, G. R., R. R. Miller, and W. D. Sable. 1979. Species relationships among fishes of the genus *Gila* in the upper Colorado River drainage. In *Proceedings of the First Conference on Scientific Research in National Parks*, ed. R. M. Linn, pp. 613–623. Washington, D.C.: U.S. National Park Service Transactions and Proceedings Series 5.

Smith, G. R., G. Musser, and D. B. McDonald. 1959. Appendix A, Aquatic survey tabulation. In *Ecological studies of the flora and fauna in Glen Canyon*, ed. A. M. Woodbury, pp. 177–194. University of Utah Anthropological Papers 40.

Smith, M. H., K. T. Scribner, J. D. Hernandez, and M. C. Wooten. 1989. Demographic, spatial, and temporal genetic variation in *Gambusia*. In *Ecology and evolution of livebearing fishes (Poeciliidae)*, ed. G. K. Meffe and F. F. Snelson, Jr., pp. 235–257. Englewood Cliffs, N.J.: Prentice-Hall.

Smith, M. L. 1980. The status of *Megupsilon aporus* and *Cyprinodon alvarezi* at El Potosí, Mexico. *Proceedings of the Desert Fishes Council* 10 (1978), 24–29.

———. 1981. Late Cenozoic fishes in the warm deserts of North America: A reinterpretation of desert adaptations. In *Fishes in North American deserts*, ed. R. J. Naiman and D. L. Soltz, pp. 11–38. New York: John Wiley and Sons.

Smith, M. L., and B. Chernoff. 1981. Breeding populations of cyprinodontoid fishes in a thermal stream. *Copeia* 1981, 701–702.

Smith, M. L., and R. R. Miller. 1980. Systematics and variation of a new cyprinodontid fish, *Cyprinodon fontinalis*, from Chihuahua, Mexico. *Proceedings of the Biological Society of Washington* 93, 405–416.

———. 1986. The evolution of the Rio Grande basin as inferred from its fish fauna. In *The zoogeography of North American freshwater fishes*, ed. C. H. Hocutt and E. O. Wiley, pp. 457–486. New York: John Wiley and Sons.

Smith, P. J., and Y. Fujio. 1982. Genetic variation in marine teleosts: High variability in habitat specialists and low variability in habitat generalists. *Marine Biology* 69, 7–20.

Snyder, D. E. 1983. Contributions to a guide to the cypriniform fish larvae of the upper Colorado River system in Colorado. Final Report for U.S. Bureau of Land Management Contract YA-512-CT8-129. Colorado State University, Fort Collins (also appeared as *U.S. Bureau of Land Management Biological Science Series* 3, 1–81, 1983).

Snyder, J. O. 1908. Relationships of the fish fauna of the lakes of southeastern Oregon. *Bulletin of the U.S. Bureau of Fisheries* 27, 69–102.

———. 1915. Notes on a collection of fishes made by Dr. Edgar A. Mearns from rivers tributary to the Gulf of California. *Proceedings of the U.S. National Museum* 49, 573–586.

———. 1917. The fishes of the Lahontan system of Nevada and northeastern California. *Bulletin of the U.S. Bureau of Fisheries* 35 (1915–16), 33–86.

———. 1919. Three new whitefishes from Bear Lake, Idaho and Utah. *Bulletin of the U.S. Bureau of Fisheries* 36 (1917–19), 1–9.

Soltz, D. L., and M. F. Hirshfield. 1981. Genetic differentiation of pupfishes (genus *Cyprinodon*) in the American Southwest. In *Fishes in North American deserts*, ed. R. J. Naiman and D. L. Soltz, pp. 291–333. New York: John Wiley and Sons.

Soltz, D. L., and R. J. Naiman. 1978. The natural history of native fishes in the Death Valley system. *Natural History Museum of Los Angeles County, California, Science Series* 30, 1–76.

Soulé, M. E., ed. 1986. *Conservation biology: The biology of scarcity and diversity*. Sunderland, Mass.: Sinauer Associates.

Soulé, M. E., and D. Simberloff. 1986. What do genetics and ecology tell us about the design of nature reserves? *Biological Conservation* 35, 19–40.

Soulé, M. E., and B. A. Wilcox. 1980. *Conservation biology: An evolutionary-ecological perspective*. Sunderland, Mass.: Sinauer Associates.

Soulé, M. E., B. A. Wilcox, and C. Holtby. 1979. Benign neglect: A model of faunal collapse in the game reserves of East Africa. *Biological Conservation* 15, 259–272.

Sousa, W. P. 1984. The role of disturbance on natural communities. *Annual Review of Ecology and Systematics* 15, 353–391.

Spofford, W. O., Jr., A. L. Parker, and A. V. Kneese, eds. 1980. *Energy development in the Southwest: Problems of water, fish and wildlife in the upper Colorado River basin.* 2 vols. Research Paper R-18. Washington, D.C.: Resources for the Future.

Stalnaker, C. B., and P. B. Holden. 1973. Changes in the native fish distribution in the Green River system, Utah-Colorado. *Proceedings of the Utah Academy of Science, Arts, and Letters* 51, 25–32.

Stegner, W. 1954. *Beyond the hundredth meridian: John Wesley Powell and the second opening of the West.* Boston: Houghton Mifflin.

Stephenson, R. L. 1985. Arizona coldwater fisheries strategic plan, 1985–1990. Federal Aid to Fisheries, Restoration Project FW-11-R. Arizona Game and Fish Department, Phoenix.

Stevenson, M. H., and T. M. Buchanan. 1973. An analysis of hybridization between the cyprinodont fishes *Cyprinodon variegatus* and *C. elegans. Copeia* 1973, 682–692.

Stewart, H., and J. E. Johnson. 1981. A refuge for southwestern fish. *New Mexico Wildlife* 26, 1–5, 29.

Stewart, K. M. 1957. Mohave fishing. *Master Key* 31, 198–203 .

Stewart, N. H. 1926. Development, growth, and food habits of the white sucker, *Catostomus commersonnii* LeSueur. *Bulletin of the U.S. Bureau of Fisheries* 42, 147–183.

Stone, C. 1974. *Should trees have standing? Toward legal rights for natural objects.* Los Altos, Calif.: William Kaufmann.

———. 1987. *Earth and other ethics.* New York: Harper and Row.

Stone, L. 1876. Report of operations during 1874 at the United States salmon-hatching establishment on the McCloud River, California. *Report of the U.S. Commissioner for Fish and Fisheries,* pt. 3, 1873–4 and 1874–5.

———. 1883. Account of operations at the McCloud River fish-breeding stations of the United States Fish Commission, from 1872 to 1882, inclusive. *Report of the U.S. Commissioner for Fish and Fisheries* 2 (1882), 217–236.

Strong, F. M. 1956. *Topics in microbial chemistry.* New York: John Wiley and Sons.

Sturgess, J. A., and P. B. Moyle. 1978. Biology of rainbow trout (*Salmo gairdneri*), brown trout (*S. trutta*), and interior dolly varden (*Salvelinus confluentus*) in the McCloud River, California, in relation to management. *Cal-Neva Wildlife* 1978, 239–250.

Sublette, J. E. 1977. A survey of the fishes of the San Juan basin with particular reference to the endangered species. Final Report for U.S. Fish and Wildlife Service Contract. Eastern New Mexico State University, Portales.

Sublette, J. E., M. D. Hatch, and M. Sublette. 1990. *The fishes of New Mexico.* Albuquerque: University of New Mexico Press.

Sumner, F. B., and V. N. Lanham. 1942. Studies of the respiratory metabolism of warm and cool spring fishes. *Biological Bulletin* 88, 313–327.

Sumner, F. B., and M. C. Sargent. 1940. Some observations on the physiology of warm spring fishes. *Ecology* 21, 45–54.

Sumner, F. H. 1940. The decline of the Pyramid Lake fishery. *Transactions of the American Fisheries Society* 69, 216–224.

Suttkus, R. D., and G. H. Clemmer. 1977. The humpback chub, *Gila cypha*, in the Grand Canyon area of the Colorado River. *Occasional Papers of the Tulane University Museum of Natural History* 1, 1–30.

———. 1979. Fishes of the Colorado River in Grand Canyon National Park. In *Proceedings of the First Conference on Scientific Research in the National Parks,* ed. R. M. Linn, pp. 599–604. Washington, D.C.: U.S. National Park Service Transactions and Proceedings Series 5.

Suttkus, R. D., G. H. Clemmer, C. Jones, and C. R. Shoop. 1976. Survey of fishes, mammals and herpetofauna of the Colorado River in Grand Canyon. *Grand Canyon National Park, Colorado River Research Series, Contribution* 34, 1–48 (also appeared as *Grand Canyon National Park, Colorado River Research Program Technical Report* 5, 1–48 [1976]).

Sweet, J. G., and O. Kinne. 1964. The effects of various temperature-salinity combinations on the body form of newly hatched *Cyprinodon macularius* (Teleostei). *Helgolaender Wissenschaften Meeresunters* 11, 49–69.

Swofford, D. L., and R. B. Selander. 1981. BIOSYS-1: A FORTRAN program for comprehensive analysis of electrophoretic data in population genetics and systematics. *Journal of Heredity* 72, 281–282.

Szaro, R. C., and J. N. Rinne. 1989. Ecosystem approach to management of southwestern riparian communities. *Transactions of the North American Wildlife and Natural Resources Conference* 53, 502–511.

Taba, S. S., J. R. Murphy, and H. H. Frost. 1965. Notes on the fishes of the Colorado River near Moab, Utah. *Proceedings of the Utah Academy of Science, Arts, and Letters* 42, 280–283.

Tanner, H. A., and M. L. Hayes. 1955. Evaluation of toxaphene as a fish poison. *Colorado Cooperative Fisheries Research Unit Report* 1, 31–39. Colorado State University, Fort Collins.

Tanner, V. M. 1936. A study of the fishes of Utah. *Utah Academy of Science, Arts, and Letters* 13, 155–183.

Tarlock, D. A. 1984. Endangered Species Act and western water rights. *Land and Water Law Review* 20, 1–30.

Tarzwell, C. M. 1938. An evaluation of the methods and results of stream improvement structures in the Southwest. *Transactions of the North American Wildlife Conference* 3, 339–364.

Taylor, D. W., and G. R. Smith. 1981. Pliocene molluscs and fishes from northeastern California and northwestern Nevada. *University of Michigan Museum of Paleontology Contributions* 25, 339–413.

Taylor, F. R., R. R. Miller, J. W. Pedretti, and J. E. Deacon. 1988. Rediscovery of the Shoshone pupfish, *Cyprinodon nevadensis shoshone* (Cyprinodontidae), at Shoshone Springs, Inyo County, California. *Bulletin of the Southern California Academy of Sciences* 87, 67–73.

Taylor, J. N., W. R. Courtenay, Jr., and J. A. McCann. 1984. Known impacts of exotic fishes in the continental United States. In *Distribution, biology, and management of exotic fishes*, ed. W. R. Courtenay, Jr., and J. R. Stauffer, Jr., pp. 322–373. Baltimore, Md.: Johns Hopkins University Press.

Taylor, T. L., and D. McGriff. 1985. Age and growth of Mohave tui chub *Gila bicolor mohavensis* from two ponds at Ft. Soda. *Proceedings of the Desert Fishes Council* 15 (1983), 299–302.

Terborgh, J. 1976. Island biogeography and conservation: Strategy and limitations. *Science* 193, 1029–1030.

Terrell, J. U. 1969. *The man who rediscovered America: A biography of John Wesley Powell.* New York: Weybright and Talley.

Theilacker, G. H. 1986. Starvation-induced mortality of young sea-caught jack mackerel, *Trachurus symmetricus*, determined with histological and morphological methods. *U.S. Fish and Wildlife Service Fisheries Bulletin* 84, 1–17.

Thibault, R. E. and R. J. Schultz. 1978. Reproductive adaptations among viviparous fishes (Cyprinodontiformes: Poeciliidae). *Evolution* 32, 320–333.

Thompson, D. H., and F. D. Hunt. 1930. The fishes of Champaign County, Illinois: A study of distribution and abundance of fishes in small streams. *Bulletin of the Illinois State Natural History Survey* 19, 5–101.

Tilman, D. S., S. Kilham, and P. Kilham. 1982. Phytoplankton community ecology: The role of limited nutrients. *Annual Review of Ecology and Systematics* 13, 349–372.

Toney, D. P. 1974. Observations on the propagation and rearing of two endangered fish species in a hatchery environment. *Proceedings of the Annual Conference of the Western Association of Fish and Game Commissioners* 54, 252–259.

Tonn, W. M., and J. J. Magnuson. 1982. Patterns in the species composition and richness of fish assemblages in northern Wisconsin lakes. *Ecology* 63, 1149–1166.

Trusso, J. J., Jr. 1972. Big troubles for a tiny fish. *Smithsonian* 3 (4), 48–51.

Tsivoglov, E. C., S. D. Shearer, R. M. Shaw, J. D. Jones, J. B. Anderson, C. E. Sponagle, and D. A. Clark. 1959. *Survey of interstate pollution of the Animas River, Colorado–New Mexico*. Cincinnati, Ohio: U.S. Public Health Service, Robert A. Taft Engineering Center.

Turner, B. J. 1974. Genetic divergence of Death Valley pupfish species: Biochemical versus morphological evidence. *Evolution* 28, 281–294.

———. 1983. Genic variation and differentiation in remnant populations of the desert pupfish, *Cyprinodon macularius*. *Evolution* 37, 690–700.

———. 1984. Evolutionary genetics of artificial refugium populations of an endangered species, the desert pupfish. *Copeia* 1984, 364–370.

Turner, B. J., and R. K. Liu. 1977. Extensive interspecific genetic compatibility in the New World genus *Cyprinodon*. *Copeia* 1977, 259–269.

Turner, P. R., and M. L. McHenry. 1985. Population ecology of Gila trout, with special emphasis on sport fish potential. Final Report for New Mexico Game and Fish Department Contract. New Mexico State University, Las Cruces.

Turner, W. R. 1959. Effectiveness of various rotenone containing preparations in eradication of farm pond fish populations. *Kentucky Department of Fish and Wildlife Resources, Fisheries Bulletin* 25, 1–22.

Tyus, H. M. 1984. Loss of stream passage as a factor in the decline of the endangered Colorado squawfish. In *Issues and technology in the management of impacted western wildlife; proceedings of a national symposium*, pp. 138–144. Boulder: Thorne Ecological Institute, Technical Publication 14.

———. 1985. Homing behavior noted for Colorado squawfish. *Copeia* 1985, 213–215.

———. 1986. Life strategies in the evolution of the Colorado squawfish (*Ptychocheilus lucius*). *Great Basin Naturalist* 46, 656–661.

———. 1987. Distribution, reproduction, and habitat use of the razorback sucker in the Green River, Utah, 1979–1986. *Transactions of the American Fisheries Society* 116, 111–116.

———. 1988. Long term retention of implanted transmitters in Colorado squawfish and razorback suckers. *North American Journal of Fisheries Management* 8, 264–267.

———. 1990. Potadromy and reproduction of Colorado squawfish *Ptychocheilus lucius*. *Transactions of the American Fisheries Society* 119, 1035–1047.

Tyus, H. M., and J. M. Beard. 1990. Non-native fish *Esox lucius* (Esocidae) and *Stizostedion vitreum* (Percidae) in the Green River basin, Colorado and Utah. *Great Basin Naturalist* 50, 33–39.

Tyus, H. M., B. D. Burdick, and C. W. McAda. 1984. Use of radiotelemetry for obtaining habitat preference data on Colorado squawfish. *North American Journal of Fisheries Management* 4, 177–180.

Tyus, H. M., B. D. Burdick, R. A. Valdez, C. M. Haynes, T. A. Lytle, and C. R. Berry. 1982a. Fishes of the upper Colorado River basin: Distribution, abundance, and status. In *Fishes of the upper Colorado River system: Present and future*, ed. W. H. Miller, H. M. Tyus, and C. A. Carlson, pp. 12–70. Bethesda, Md.: Western Division, American Fisheries Society.

Tyus, H. M., and G. B. Haines. 1991. Distribution, abundance, habitat use, and movements of young Colorado River squawfish *Ptychocheilus lucius*. *Transactions of the American Fisheries Society* 120, 79–89.

Tyus, H. M., R. L. Jones, and L. A. Trinca. 1987. Green River rare and endangered fish studies, 1982–1985. Final Report for Colorado River Fishes Monitoring Program, U.S. Bureau of Reclamation (and other agencies). U.S. Fish and Wildlife Service, Vernal, Utah.

Tyus, H. M., and C. A. Karp. 1989. Habitat use and streamflow needs of rare and endangered fishes, Yampa River, Colorado and Utah. *U.S. Fish and Wildlife Service Biological Report* 89 (14), i–vi, 1–27.

Tyus, H. M., and C. W. McAda. 1984. Migration, movements, and habitat preferences of Colorado squawfish, *Ptychocheilus lucius*, in the Green, White, and Yampa rivers, Colorado and Utah. *The Southwestern Naturalist* 29, 289–299.

Tyus, H. M., C. W. McAda, and B. D. Burdick. 1981. Radiotelemetry of Colorado squawfish and razorback sucker, Green River system of Utah, 1980. *Transactions of the Bonneville Chapter of the American Fisheries Society* 1981, 19–24.

———. 1982b. Green River fishery investigations: 1979–1981. In Part 2, *Colorado River Fisheries Project, final report field investigations*, ed. W. H. Miller, J. J. Valentine, D. L. Archer, H. M. Tyus, R. A. Valdez, and L. R. Kaeding, pp. 1–100. Final Report for U.S. Bureau of Reclamation Contract 9-07-40-L-1016, Memorandum of Understanding CO-910-MU9-933, U.S. Bureau of Land Management. U.S. Fish and Wildlife Service, Salt Lake City, Utah.

Tyus, H. M., and W. L. Minckley. 1988. Migrating Mormon crickets, *Anabrus simplex* (Orthoptera: Tettigoniidae), as food for stream fishes. *Great Basin Naturalist* 48, 25–30.

Tyus, H. M., and N. J. Nikirk. 1990. Abundance, growth, and diet of channel catfish, *Ictalurus punctatus*, in the Green and Yampa rivers, Colorado and Utah. *The Southwestern Naturalist* 35, 188–198.

Tyus, H. M., E. J. Wick, and D. L. Skates. 1985. A spawning migration of Colorado squawfish (*Ptychocheilus lucius*) in the Yampa and Green rivers, 1981. *Proceedings of the Desert Fishes Council* 13 (1981), 102–108.

Udall, S. L. 1963. Secretary of the Interior; review of Green River fish eradication program. Department of the Interior Directive (dated 25 March), Washington, D.C.

Ulmer, L. 1987. Management plan for the razorback sucker in California. *Proceedings of the Desert Fishes Council* 17 (1985), 183 (abstract).

Ulmer, L., and K. R. Anderson. 1985. Management plan for the razorback sucker (*Xyrauchen texanus*) in California. *California Department of Fish and Game, Region 5 Information Bulletin* 0013-10-1985, 1–26.

U.S. Army Corps of Engineers. 1979. Eutrophication problems of Upper Klamath Lake. U.S. Army Corps of Engineers, San Francisco.

U.S. Bureau of Land Management. 1981. San Juan River management plan: Environmental assessment. U.S. Bureau of Land Management, Moab, Utah.

———. 1987. Long term resident celebrates 20th anniversary as endangered species. News release, 87-16, 1–4. U.S. Bureau of Land Management, Ely, Nevada.

U.S. Bureau of Reclamation. 1946. *The Colorado River, A national menace becomes a national resource.* U.S. Department of the Interior. Washington, D.C.: U.S. Government Printing Office.

———. 1975. *Westwide study report on critical water problems facing the eleven western states.* Washington, D.C.: U.S. Government Printing Office.

———. 1980. Lower Colorado River maps, Colorado River frontwork and levee system, Arizona-California. Lower Colorado Region, Boulder City, Nev.

———. 1984. Endangered species biological assessment for the Dirty Devil River unit. Salt Lake City, Utah.

———. 1986. Central Arizona Project Granite Reef Aqueduct fishery investigations, progress report 1986. Arizona Project Office, Phoenix.

———. 1988. Central Arizona Project Granite Reef Aqueduct fishery investigations, progress report 1987. Arizona Project Office, Phoenix.

U.S. Bureau of Sport Fisheries and Wildlife. 1966. Rare and endangered fish and wildlife of the United States. *U.S. Bureau of Sport Fisheries Research Publication* 34, 1–180.

U.S. Congress. 1973. *Endangered Species Act.* 87 Statutes 884, Public Law 93-205. Washington, D.C.: U.S. Government Printing Office.

U.S. Department of the Interior. 1963. *Measures to improve sport fishing at Flaming Gorge Unit, Colorado River Storage Project, Wyoming and Utah*. Washington, D.C.: U.S. Government Printing Office.

———. 1970. A task force report on Let's Save the Desert Pupfish. U.S. Fish and Wildlife Service, Washington, D.C.

———. 1971. Status of the desert pupfish, a task force report. U.S. Fish and Wildlife Service, Washington, D.C.

———. 1973. Threatened wildlife of the United States. *U.S. Bureau of Sport Fisheries and Wildlife Resources Publication* 114, 1–289.

———. 1989. Box score of listings/recovery plans. *Endangered Species Technical Bulletin* 14, 16.

U.S. Fish and Wildlife Service. 1977. *Devils Hole pupfish recovery plan*. Portland: U.S. Fish and Wildlife Service.

———. 1978a. Fish transportation. In *Manual of fish culture*, pp. 1–88. Washington, D.C.: U.S. Government Printing Office.

———. 1978b. *Colorado squawfish recovery plan*. Denver: U.S. Fish and Wildlife Service.

———. 1978c. Proposed endangered status for the bonytail chub and threatened status for the razorback sucker. *Federal Register* 43, 17373–17377.

———. 1979a. *Arizona trout recovery plan*. Albuquerque: U.S. Fish and Wildlife Service.

———. 1979b. *Gila trout recovery plan*. Albuquerque: U.S. Fish and Wildlife Service.

———. 1980a. Special report on distribution and abundance of fishes of the lower Colorado River, aquatic study—Colorado River from Lees Ferry to southern International Boundary and selected tributaries, Arizona, California, and Nevada. Final Report for U.S. Water and Power Resources Service (U.S. Bureau of Reclamation) Contract 9-07-03-X0066. U.S. Fish and Wildlife Service, Phoenix.

———. 1980b. *Recovery plan: Pahrump poolfish*. Portland, Ore.: U.S. Fish and Wildlife Service.

———. 1980c. Notice of withdrawal of an expired proposal for listing of the razorback sucker. *Federal Register* 45, 35410.

———. 1981a. Aquatic study of the lower Colorado River, aquatic study—Colorado River from Lees Ferry to southern International Boundary and selected tributaries, Arizona, California, and Nevada. Final report for U.S. Water and Power Resources Service (U.S. Bureau of Reclamation) contract 9-07-03-X0066. U.S. Fish and Wildlife Service, Phoenix.

———. 1981b. *Comanche Springs pupfish recovery plan*. Albuquerque: U.S. Fish and Wildlife Service.

———. 1982a. *Mexican wolf recovery plan*. Albuquerque: U.S. Fish and Wildlife Service.

———. 1982b. *Recovery plan for the Socorro isopod* (Thermosphaeroma thermophilum). Albuquerque: U.S. Fish and Wildlife Service.

———. 1983a. *Moapa dace recovery plan*. Portland, Ore.: U.S. Fish and Wildlife Service.

———. 1983b. *Arizona trout (Apache trout) recovery plan*. Albuquerque: U.S. Fish and Wildlife Service.

———. 1983c. Endangered and threatened species; listing and recovery priority guidelines. *Federal Register* 48, 43098–43105.

———. 1983d. *Pecos gambusia recovery plan*. Albuquerque: U.S. Fish and Wildlife Service.

———. 1983e. *The Endangered Species Act, as amended by Public Law 97-304 (The Endangered Species Act amendments of 1982)*. Washington, D.C.: U.S. Government Printing Office.

———. 1984a. *Gila trout recovery plan*. Albuquerque: U.S. Fish and Wildlife Service.

———. 1984b. *Big Bend gambusia* (Gambusia gaigei) *recovery plan*. Albuquerque: U.S. Fish and Wildlife Service.

———. 1984c. *Sonoran (Gila and Yaqui) topminnow recovery plan*. Albuquerque: U.S. Fish and Wildlife Service.

———. 1984d. *Red wolf recovery plan*. Albuquerque: U.S. Fish and Wildlife Service.

———. 1984e. *American peregrine falcon recovery plan*. Denver: U.S. Fish and Wildlife Service.

———. 1984f. Endangered and threatened wildlife and plants; experimental populations, final rule. *Federal Register* 49, 33885–33894.

———. 1984g. *Recovery plan for the Mohave tui chub*, Gila bicolor mohavensis. Portland: U.S. Fish and Wildlife Service.

———. 1984h. *Recovery Plan for the Owens pupfish*, Cyprinodon radiosus. Portland: U.S. Fish and Wildlife Service.

———. 1984i. Endangered and threatened wildlife and plants; final rule to determine the Yaqui chub to be an endangered species with critical habitat, and to determine the beautiful shiner and the Yaqui catfish to be threatened species with critical habitat. *Federal Register* 49, 34490–34497.

———. 1985a. *Recovery plan for woundfin*, Plagopterus argentissimus. Albuquerque: U.S. Fish and Wildlife Service.

———. 1985b. *Leon Springs pupfish recovery plan*. Albuquerque: U.S. Fish and Wildlife Service.

———. 1985c. *Finding of no significant impact: Conservation of California condor, 12/23/85*. Washington, D.C.: U.S. Fish and Wildlife Service.

———. 1985d. *Paiute cutthroat trout recovery plan*. Portland: U.S. Fish and Wildlife Service.

———. 1985e. *Revised unarmored threespine stickleback recovery plan*. Portland: U.S. Fish and Wildlife Service.

———. 1985f. Endangered and threatened wildlife and plants; determination of experimental population status for certain introduced populations of Colorado squawfish and woundfin. *Federal Register* 50, 30188–30195.

———. 1986a. *Chihuahua chub recovery plan*. Albuquerque: U.S. Fish and Wildlife Service.

———. 1986b. *Whooping crane recovery plan*. Albuquerque: U.S. Fish and Wildlife Service.

———. 1986c. Endangered and threatened wildlife and plants; proposed listing of Virgin River chub as an endangered species with critical habitat. *Federal Register* 51, 22949–22955.

———. 1986d. Endangered and threatened wildlife and plants; determination of endangered status and critical habitat for the desert pupfish. *Federal Register* 51, 10842–10851.

———. 1986e. Endangered and threatened species listing and recovery priority guidelines. *Federal Register* 48, 16756–16759.

———. 1986f. Interagency cooperation—Endangered Species Act of 1973, as amended; final rule. *Federal Register* 51, 19925–19963.

———. 1987a. *Final recovery implementation program for endangered fish species in the upper Colorado River basin*. Denver: U.S. Fish and Wildlife Service.

———. 1987b. Endangered and threatened wildlife and plants; proposed determination of experimental population status for an introduced population of Colorado squawfish. *Federal Register* 52, 32143–32145.

———. 1987c. *Final environmental assessment: Recovery implementation program for endangered fish species in the upper Colorado River basin*. Denver: U.S. Fish and Wildlife Service.

———. 1987d. *Recovery plan for the Borax Lake chub*, Gila boraxobius. Portland: U.S. Fish and Wildlife Service.

———. 1987e. Results of the interagency standardized monitoring program, 1986. U.S. Fish and Wildlife Service, Grand Junction, Colo.

———. 1987f. San Bernardino National Wildlife Refuge management plan. U.S. Fish and Wildlife Service, Albuquerque, N. Mex.

———. 1987g. Proposed delisting for Amistad gambusia. *Federal Register* 52, 86083–86085.

———. 1988a. Endangered and threatened wildlife and plants; shortnose and Lost River suckers, Houghton's goldenrod and Pitcher's thistle; final rules. *Federal Register* 53, 27130–27134.

———. 1988b. *Black-footed ferret management plan*. Denver: U.S. Fish and Wildlife Service.

———. 1988c. Results of the interagency standardized monitoring program, 1987. U.S. Fish and Wildlife Service, Grand Junction, Colo.

———. 1989a. *Colorado squawfish* (Ptychocheilus lucius*) recovery plan* (draft). Denver: U.S. Fish and Wildlife Service.

———. 1989b. *Bonytail chub* (Gila elegans) *recovery plan* (draft). Denver: U.S. Fish and Wildlife Service.

———. 1989c. *Title 50—Wildlife and Fisheries. Part 17: Endangered and threatened wildlife and plants, Subpart B: lists, endangered and threatened wildlife (CFR 17.11 and 17.12)*. Washington, D.C.: U.S. Government Printing Office.

———. 1989d. 50 CFR part 17, Endangered and threatened wildlife and plants; animal notice of review. *Federal Register* 54, 554–579.

———. 1990a. Endangered and threatened wildlife and plants; proposal to determine the razorback sucker (Xyrauchen texanus) to be an endangered species. *Federal Register* 55, 21154–21161.

———. 1990b. Recovery plan for the endangered and threatened species of Ash Meadows, Nevada. Portland: U.S. Fish and Wildlife Service.

U.S. Water Resources Council. 1968. *The nation's water resources*. Washington, D.C.: U.S. Water Resources Council.

Utah Department of Fish and Game. 1959. The impact of Flaming Gorge Unit upon wildlife resources in northeastern Utah. Federal Aid to Wildlife Restoration Report, Salt Lake City.

Utah Water Research Laboratory. 1975. *Colorado River regional assessment study*. 4 vols. Utah Water Resources Laboratory. Logan: Utah State University.

Uyeno, T., and R. R. Miller. 1965. Middle Pliocene cyprinid fishes from the Bidahochi Formation, Arizona. *Copeia* 1965, 28–41.

Valdez, R. A. 1988. Annual summary report—1987, Fisheries biology and rafting. Annual Report for U.S. Bureau of Reclamation Contract 6-CS-40-03980. BIO/WEST, Logan, Utah.

———. 1990. The endangered fish of Cataract Canyon. Final Report for U.S. Bureau of Reclamation Contract 6-CS-40-03980. BIO/WEST, Logan, Utah.

Valdez, R. A., P. B. Holden, T. B. Hardy, and R. J. Ryel. 1987. Habitat suitability index curves for endangered fishes of the Upper Colorado River basin. Final Report for U.S. Fish and Wildlife Service HSI Curve Development Project. BIO/WEST, Logan, Utah.

Valdez, R. A., and P. G. Magnan. 1980. Fishes of the upper Colorado River, Debeque, Colorado, to Lake Powell, Utah. U.S. Fish and Wildlife Service, Salt Lake City, Utah.

Valdez, R. A., P. G. Magnan, M. McInery, and R. P. Smith. 1982a. Tributary report: Fishery investigations of the Gunnison and Dolores rivers. In Part 2, *Colorado River Fisheries Project, final report field investigations*, ed. W. H. Miller, J. J. Valentine, D. L. Archer, H. M. Tyus, R. A. Valdez, and L. R. Kaeding, pp. 321–362. Final Report for U.S. Bureau of Reclamation Contract 9-07-40-L-1016, Memorandum of Understanding CO-910-MU9-933, U.S. Bureau of Land Management. U.S. Fish and Wildlife Service, Salt Lake City, Utah.

Valdez, R. A., P. G. Magnan, R. Smith, and B. Nilson. 1982b. Upper Colorado River Investigations (Rifle, Colorado, to Lake Powell, Utah). In Part 2, *Colorado River Fisheries Project, final report field investigations*, ed. W. H. Miller, J. J. Valentine, D. L. Archer, H. M. Tyus, R. A. Valdez, and L. R. Kaeding, pp. 101–279. Final Report for U.S. Bureau of Reclamation Contract 9-07-40-L-1016, Memorandum of Understanding, CO-910-MU9-933, U.S. Bureau of Land Management. U.S. Fish and Wildlife Service, Salt Lake City, Utah.

Valdez, R. A., and W. J. Masslich. 1989. Winter habitat study of endangered fish—Green River. Final Report for U.S. Fish and Wildlife Service Contract. Report 136-2, 1–184. BIO/WEST, Logan, Utah.

Valdez, R. A., and E. J. Wick. 1981. Natural versus manmade backwaters as native fish habitat. In *Aquatic resources management of the Colorado River ecosystem*, ed. V. D. Adams and V. A. Lamarra, pp. 519–536. Ann Arbor, Mich.: Ann Arbor Scientific Publishers.

Valdez, R. A., and R. D. Williams. 1986. Cataract Canyon fish study. Final Report for U.S. Bureau of Reclamation Contract 6-CS-40-03980. U.S. Bureau of Reclamation, Salt Lake City, Utah.

———. 1987. Endangered fishes of Cataract Canyon. *Proceedings of the Desert Fishes Council* 18 (1986), 212–219.

Vanicek, C. D. 1967. Ecological studies of native Green River fishes below Flaming Gorge Dam, 1964–1966. Ph.D. diss., Utah State University, Logan.

Vanicek, C. D., and R. H. Kramer. 1969. Life history of the Colorado squawfish, *Ptychocheilus lucius*, and the Colorado chub, *Gila robusta*, in the Green River in Dinosaur National Monument, 1964–1966. *Transactions of the American Fisheries Society* 98, 193–208.

Vanicek, C. D., R. H. Kramer, and D. R. Franklin. 1970. Distribution of Green River fishes in Utah and Colorado following closure of Flaming Gorge Dam. *The Southwestern Naturalist* 14, 297–315.

Varner, G. E. 1987. Do species have standing? *Environmental Ethics* 9, 57–72.

Verdon, R., and E. Magnin. 1977. Dynamique de la population de meuniers noirs *Catostomus commersoni* (Lacepede) du Lac Croche dans les Laurentides, Quebec. *Canadian Naturalist* 104 (1972), 197–206.

Vigg, S. 1980. Seasonal benthic distribution of adult fish in Pyramid Lake, Nevada. *California Fish and Game* 66, 49–59.

Vincent, D. T. 1968. The influences of some environmental factors on the distribution of fishes in upper Klamath Lake. Master's thesis, Oregon State University, Corvallis.

Vondracek, B., D. M. Baltz, L. R. Brown, and P. B. Moyle. 1989. Spatial, seasonal, and diel distribution of fishes in a California reservoir dominated by native fishes. *U.S. Fish and Wildlife Service Fisheries Research* 7, 31–53.

Vrijenhoek, R. C., M. E. Douglas, and G. K. Meffe. 1985. Conservation genetics of endangered fish populations in Arizona. *Science* 229, 400–402.

VTN Consolidated. 1978. Fish, wildlife and habitat assessment, San Juan River, New Mexico and Utah. Final Report for U.S. Bureau of Reclamation Gallup-Navajo Indian Water Supply Project. VTN Consolidated, Salt Lake City.

Wagner, R. A. 1954. Basic survey of the Verde River and its on-stream impoundments. Federal Aid to Fisheries Restoration, Project Completion Report F-2-R-1, 9–12. Arizona Game and Fish Department, Phoenix.

Walbaum, J. J., ed. 1792. Petri Artedi renovati, i.e., bibliotheca et philosophia ichthyolegica, Ichthyologiae pars III, Petri Artedi Sueci genera piscium in quibus systema totum ichthyologiae proponitur cum observationibus plurimis, Redactis speciebus 242 ad genera 52, Grypeswaldiae. *Ant. Ferdin. Rose*, 1–723.

Walker, B. W., ed. 1961. *The ecology of the Salton Sea, California, in relation to the sportfishery.* California Department of Fish and Game, Fisheries Bulletin 113, 1–204.

Walker, B. W., R. R. Whitney, and G. W. Barlow. 1961. The fishes of the Salton Sea. In *The ecology of the Salton Sea, California, in relation to the sportfishery,* ed. B. W. Walker, pp. 77–91. California Department of Fish and Game, Fisheries Bulletin 113.

Walker, C. R., R. E. Lennon, and B. L. Berger. 1964. Investigations in Fish Control II: Preliminary observations on the toxicity of antimycin-A to fish and other aquatic animals. *U.S. Bureau of Sport Fisheries and Wildlife Circular* 186, 1–18.

Wallace, A. R. 1863. On the physical geology of the Malay Archipelago. *Journal of the Royal Geographic Society* (London) 33, 234.

Wallis, O. L. 1951. The status of the fish fauna of the Lake Mead National Recreation area, Arizona-Nevada. *Transactions of the American Fisheries Society* 80, 84–92.

Walters, L. L., and F. F. Legner. 1980. The impact of the desert pupfish, *Cyprinodon macularius*, and *Gambusia affinis affinis* on fauna in pond ecosystems. *Hilgardia* 18, 1–18.

Waring, G. A. 1920. Groundwater in Pahrump, Mesquite, and Ivanpah valleys, Nevada and California. *U.S. Geological Survey Water-supply Paper* 450, 51–86.

Weatherford, G. D., and G. C. Jacoby. 1975. Impact of energy development on the law of the Colorado River. *Natural Resources Journal* 15, 171–213.

Weber, D. T. 1959. Effects of reduced stream flows on the trout fishery below Granby Dam, Colorado. Master's thesis, Colorado State University, Fort Collins.

Welcomme, R. L. 1985. River fisheries. *FAO Technical Paper* 262, 1–330.

Wheeler, S. S. 1967. *The desert lake*. Caldwell, Ida.: Caxton Printers.

White, L., Jr. 1967. The historical roots of our ecological crises. *Science* 155, 1203–1207.

Whitmore, D. H. 1983. Introgressive hybridization of smallmouth bass and Guadalupe bass (*M. treculi*). *Copeia* 1983, 672–679.

Wick, E. J., and J. A. Hawkins. 1989. *Colorado squawfish winter habitat study, Yampa River, Colorado*. Larval Fish Laboratory, Contribution 43, Colorado State University, Fort Collins.

Wick, E. J., J. A. Hawkins, and C. A. Carlson. 1985. Colorado squawfish population and habitat monitoring, 1983 and 1984. Endangered Wildlife Investigations, Federal Aid Progress Report SE-3, 1–48. Colorado Division of Wildlife, Denver.

Wick, E. J., T. A. Lytle, and C. M. Haynes. 1981. Colorado squawfish and humpback chub population and habitat monitoring, 1979–1980. Endangered Wildlife Investigations, Federal Aid Progress Report SE-3, 1–156. Colorado Division of Wildlife, Denver.

Wick, E. J., C. W. McAda, and R. V. Bulkley. 1982. Life history and prospects for recovery of the razorback sucker. In *Fishes of the upper Colorado River system: Present and future*, ed. W. H. Miller, H. M. Tyus, and C. A. Carlson, pp. 120–126. Bethesda, Md.: Western Division, American Fisheries Society.

Wick, E. J., D. L. Stoneburner, and J. A. Hawkins. 1983. Observations on the ecology of the Colorado squawfish (*Ptychocheilus lucius*) in the Yampa River. *Colorado Water Resources Field Support Laboratory Report* 83-7, 1–55. U.S. National Park Service, Fort Collins.

Wickliff, E. L. 1938. Additional returns from tagged fish in Ohio. *Transactions of the American Fisheries Society* 67, 211.

Wilcove, D. S. 1985. Nest predation in forest tracts and the decline of migratory songbirds. *Ecology* 66, 1211–1214.

Wilcox, B. A., and D. D. Murphy. 1985. Conservation strategy: The effects of fragmentation and extinction. *American Naturalist* 125, 879–887.

Wilke, P. J. 1978. Late prehistoric human ecology at Lake Cahuilla, Coachella Valley, California. *University of California, Berkeley, Archaeological Research Facility Contribution* 38.

———. 1980. Prehistoric weir fishing on recessional shorelines of Lake Cahuilla, Salton basin, southeastern California. *Proceedings of the Desert Fishes Council* 11 (1979), 101–102.

Williams, C. D., and J. E. Williams. 1981. Distribution and status of native fishes of the Railroad Valley system, Nevada. *Transactions of the California-Nevada Section, Wildlife Society* 1981, 48–51.

———. 1989. Refuge management for the threatened Railroad Valley springfish in Nevada. *North American Journal of Fisheries Management* 9, 465–470.

Williams, J. D. 1981. Threatened desert fishes and the endangered species act. In *Fishes in North American deserts*, ed. R. J. Naiman and D. L. Soltz, pp. 447–475. New York: John Wiley and Sons.

Williams, J. E. 1977. Observations on the status of the Devils Hole pupfish in the Hoover Dam refugium. *U.S. Bureau of Reclamation, Environmental Research Center* REC-ERC-77-11, 1–15.

———. 1978. Taxonomic status of *Rhinichthys osculus* (Cyprinidae) in the Moapa River, Nevada. *The Southwestern Naturalist* 23, 511–518.

Williams, J. E., D. B. Bowman, J. E. Brooks, A. A. Echelle, R. J. Edwards, D. A. Hendrickson, and J. J. Landye. 1985. Endangered aquatic ecosystems in North American deserts, with a list of vanishing fishes of the region. *Journal of the Arizona-Nevada Academy of Sciences* 20, 1–62.

Williams, J. E., and J. E. Deacon. 1986. Subspecific identity of the Amargosa pupfish, *Cyprinodon nevadensis*, from Crystal Spring, Ash Meadows, Nevada. *Great Basin Naturalist* 46, 220–223.

Williams, J. E., J. E. Johnson, D. A. Hendrickson, S. Contreras Balderas, J. D. Williams, M. Navarro-Mendoza, D. E. McAllister, and J. E. Deacon. 1989. Fishes of North America, endangered, threatened, or of special concern: 1989. *Fisheries* 14, 2–20.

Williams, J. E., and D. W. Sada. 1985a. America's desert fishes: Increasing their protection under the Endangered Species Act. *Endangered Species Technical Bulletin* 10, 8–14.

———. 1985b. Status of two endangered fishes, *Cyprinodon nevadensis mionectes* and *Rhinichthys osculus nevadensis*, from two springs in Ash Meadows, Nevada. *The Southwestern Naturalist* 30, 475–484.

Williams, J. E., D. W. Sada, C. D. Williams, et al. 1988. American Fisheries Society guidelines for introduction of threatened and endangered fishes. *Fisheries* 13, 5–11.

Williams, J. E., and G. R. Wilde. 1981. Taxonomic status and morphology of isolated populations of the White River springfish, *Crenichthys baileyi* (Cyprinodontidae). *The Southwestern Naturalist* 25, 485–503.

Wilson, B. L., J. E. Deacon, and W. G. Bradley. 1966. Parasitism in the fishes of the Moapa River, Clark County, Nevada. *Transactions of the California-Nevada Section, Wildlife Society* 1966, 12–23.

Wilson, E. D., R. T. Moore, and R. T. O'Haire. 1960. Geologic map of Pima and Santa Cruz counties, Arizona. Arizona Bureau of Mines, University of Arizona, Tucson.

Wilson, E. O., ed. 1988. *Biodiversity.* Washington, D.C.: National Academy of Sciences Press.

Wiltzius, W. J. 1978. Some factors historically affecting the distribution and abundance of fishes in the Gunnison River. Final Report for U.S. Bureau of Reclamation Contract. Colorado Division of Wildlife, Fort Collins.

Winn, H. E., and R. R. Miller. 1954. Native postlarval fishes of the lower Colorado River basin, with a key for their identification. *California Fish and Game* 40, 273–285.

Winograd, I. J., and F. J. Pearson, Jr. 1976. Major carbon-14 anomaly in a regional carbonate aquifer: Possible evidence for megascale channels. *Water Resources Research* 12, 1125–1143.

Winograd, I. J., B. J. Szabo, T. B. Coplen, and A. C. Riggs. 1988. A 250,000-year climatic record from Great Basin calcite: Implications for Milanokovitch theory. *Science* 242, 1275–1280.

Wooding, J. 1985. *Colorado's little fishes.* Denver: Colorado Division of Wildlife.

Worts, G. F., Jr. 1963. Effect of ground-water development on the pool level in Devil's Hole, Death Valley National Monument, Nye County, Nevada. U.S. Geological Survey Open-file Report. Water Resources Division, Carson City, Nev.

Wright, D. H. 1983. Species-energy theory: An extension of species-area theory. *Oikos* 41, 496–506.

Wydoski, R. S. 1980. Potential impacts of alterations in streamflow and water quality on fish and macroinvertebrates in the upper Colorado River basin. In *Energy development in the Southwest: Problems of water, fish and wildlife in the upper Colorado River basin,* vol. 2, ed. W. O. Spofford, Jr., A. L. Parker, and A. V. Kneese, pp. 77–147. Research Paper R-18. Washington, D.C.: Resources for the Future.

Wydoski, R. S., and L. Emery. 1983. Tagging and marking. In *Fisheries techniques,* ed. A. A. Nielsen and D. L. Johnson, pp. 215–237. Bethesda, Md.: American Fisheries Society.

Wydoski, R. S., K. Gilbert, K. Seethaler, C. W. McAda, and J. A. Wydoski. 1980. Annotated bibliography for aquatic resources management of the upper Colorado River ecosystem. *U.S. Fish and Wildlife Service Research Publication* 135, 1–186.

Zallen, M. 1986. Evolution of ESA consultations on western water projects. *Natural Resources and Environment* 2, 41–42, 70–71.

Contributors

James E. Brooks
U.S. Fish and Wildlife Service
Dexter National Fish Hatchery
P.O. Box 219
Dexter, New Mexico 88230

Salvador Contreras Balderas
Universidad Autónoma de Nuevo León
Departamento de Biologia
Monterrey, Nuevo León, México

James E. Deacon
University of Nevada, Las Vegas
Department of Biological Sciences
Las Vegas, Nevada 89154

Michael E. Douglas
Arizona State University
Department of Zoology and Museum
Tempe, Arizona 85287

Anthony A. Echelle
Oklahoma State University
Department of Zoology
Stillwater, Oklahoma 74074

John Hamill
U.S. Fish and Wildlife Service
Denver Federal Center
P.O. Box 25486
Denver, Colorado 80225

Dean A. Hendrickson
Texas Memorial Museum
University of Texas
2400 Trinity
Austin, Texas 78705

Paul B. Holden
BIO/WEST, Inc.
1063 West 1400 North
Logan, Utah 84321

Clark Hubbs
University of Texas
Department of Zoology
Austin, Texas 78712

Buddy Lee Jensen
U.S. Fish and Wildlife Service
Dexter National Fish Hatchery
P.O. Box 219
Dexter, New Mexico 88230

James E. Johnson
University of Arkansas
Cooperative Fisheries Unit and Department
  of Zoology
Fayetteville, Arkansas 72701

Paul C. Marsh
Arizona State University
Center for Environmental Studies
Tempe, Arizona 85287

Gary K. Meffe
University of Georgia
Savannah River Ecology Laboratory
Drawer E
Aiken, South Carolina 29802

Frances H. Miller (deceased)
University of Michigan
Museum of Zoology
Ann Arbor, Michigan 48109

Robert Rush Miller
University of Michigan
Museum of Zoology
Ann Arbor, Michigan 48109

W. L. Minckley
Arizona State University
Department of Zoology
Tempe, Arizona 85287

Peter B. Moyle
University of California, Davis
Department of Wildlife and Fisheries Biology
Davis, California 95616

Edwin Philip Pister
Desert Fishes Council
P.O. Box 337
Bishop, California 93515

John N. Rinne
U.S. Forest Service
Rocky Mountain Forest and Range
  Experiment Station
Forestry Research Laboratory
Tempe, Arizona 85287

Holmes Rolston III
Colorado State University
Department of Philosophy
Fort Collins, Colorado 80523

Georgina M. Sato
U.S. Forest Service
Modoc National Forest
441 North Main Street
Alturas, California 96101

G. Gary Scoppettone
U.S. Fish and Wildlife Service
4600 Kietzke Lane
Reno, Nevada 89502

David L. Soltz
California State University
Department of Biology
Long Beach, California 90840

Paul R. Turner
New Mexico State University
Department of Wildlife and Fisheries
Las Cruces, New Mexico 88003

Harold M. Tyus
U.S. Fish and Wildlife Service
Colorado River Fisheries Project
1680 West Highway 40, Room 1210
Vernal, Utah 84078

Gary Vinyard
University of Nevada, Reno
Department of Biology
Reno, Nevada 89557

Cynthia Deacon Williams
U.S. Forest Service
National Fisheries Program
14th Street and Independence Avenue,
    Southwest
Washington, D.C. 20090

Jack E. Williams
U.S. Bureau of Land Management
Division of Wildlife and Fisheries
18th and C Streets, Northwest
Washington, D.C. 20240

Richard S. Wydoski
U.S. Fish and Wildlife Service
Denver Federal Center
P.O. Box 25486
Denver, Colorado 80225

Index of Common and Scientific Names

This compilation is designed to serve two purposes, a listing of the common and scientific names for plants and animals mentioned in text and an index to their occurrences. Scientific names are cross-referenced to common names (in capital letters) where pagination is provided. We and many other workers in the American West apply common names to subspecies of fishes, which can be sorted out by referring to the scientific names where intraspecific taxa are listed alphabetically under their respective species; vernacular names of species and subspecies of each taxon are provided in alphabetical order. In taxa for which both English and Spanish common names are used in text, both are provided (e.g., pupfish/cachorrito); they are indexed by the English epithet.

Index of Authors Cited

General Index[1]

---

[1]Abbreviations for Mexican states are as follows: BCN, Baja California del Norte; BCS, Baja California del Sur; CHI, Chihuahua; CHP, Chiapas; COA, Coahuila, DFE, Distrito Federal; DGO, Durango, NLE, Nuevo Leon; SLP, San Luis Potosi; SIN, Sinaloa; SON, Sonora; TAB, Tabasco; TAM, Tamaulipas; YUC, Yucatan; ZAC, Zacatecas

About the Editors

W. L. Minckley, Professor of Zoology at Arizona State University, has a long and distinguished career in systematic and ecological ichthyology and conservation biology. He was a founding member of the Desert Fishes Council and, since 1986, leader of the U.S. Fish and Wildlife Service Desert Fishes Recovery Team. Throughout his career he has emphasized the training of students, and five of the contributors to this volume received degrees in part under his direction. He has published on a variety of topics in aquatic ecology and ichthyology, most dealing with the biota of southwestern United States and northern Mexico. He received his B.S. in wildlife/fisheries management from Kansas State University, his M.A. in zoology from the University of Kansas, and his Ph.D. in geology/ecology from the University of Louisville.

James E. Deacon, Distinguished Professor of the University of Nevada, Las Vegas, has devoted a long professional career to the conservation of native fishes. He was instrumental in the establishment of the first refuge in the United States (Hot Creek Refuge in Nevada) for an endemic, non-game fish species, and he provided expert testimony in the Devils Hole Pupfish trial, which extended the doctrine of prior water rights to include groundwater and which led to the establishment of the Ash Meadows Wildlife Refuge. His publications include more than seventy scientific papers on the biology and conservation of desert fishes. He received his Ph.D. at the University of Kansas in 1960. From 1974 to 1982 he was chair of the Department of Biological Sciences at the University of Nevada, Las Vegas.